P9-ASI-931

Solutions Manual
for
Quantitative Chemical Analysis

Ninth Edition

Daniel C. Harris
Michelson Laboratory

W. H. Freeman and Company
A Macmillan Education Company
New York

ISBN-13: 9781464175633
ISBN-10: 1464175632

© 2016 W. H. Freeman and Company

All rights reserved.

Printed in the United States of America.

W. H. Freeman and Company
41 Madison Avenue
New York, NY 10010
www.whfreeman.com

Contents

CHAPTER 0
THE ANALYTICAL PROCESS

0-1. Qualitative analysis finds out what is in a sample. Quantitative analysis measures how much is in a sample.

0-2. Steps in a chemical analysis:

(1) Formulate the question: Convert a general question into a specific one that can be answered by a chemical measurement.

(2) Select the appropriate analytical procedure.

(3) Obtain a representative sample.

(4) Sample preparation: Convert the representative sample into a sample suitable for analysis. If necessary, concentrate the analyte and remove or mask interfering species.

(5) Analysis: Measure the unknown concentration in replicate analyses.

(6) Produce a clear report of results, including estimates of uncertainty.

(7) Draw conclusions: Based on the analytical results, decide what actions to take.

0-3. Masking converts an interfering species to a noninterfering species.

0-4. A calibration curve shows the response of an analytical method as a function of the known concentration of analyte in standard solutions. Once the calibration curve is known, then the concentration of an unknown can be deduced from a measured response.

0-5. (a) A homogeneous material has the same composition everywhere. In a heterogeneous material, the composition is not the same everywhere.

(b) In a segregated heterogeneous material, the composition varies on a large scale. There could be large patches with one composition and large patches with another composition. The differences are segregated into different regions. In a random heterogeneous material, the differences occur on a fine scale. If we collect a "reasonable-size" portion, we will capture each of the different compositions that are present.

(c) To sample a *segregated heterogeneous material*, we take representative amounts from each of the obviously different regions. In panel b in Box 0-1, 66% of the area has composition A, 14% is B, and 20% is C. To construct a representative bulk sample, we could take 66 randomly selected samples from region A, 14 from region B, and 20 from region C. To sample a *random heterogeneous material*, we divide the material into imaginary segments and collect random segments with the help of a table of random numbers.

0-6. We are apparently observing *interference* by Mn^{2+} in the I^- analysis by method A. The result of the I^- analysis is affected by the presence of Mn^{2+}. The greater the concentration of Mn^{2+} in the mineral water, the greater is the apparent concentration of I^- found by method A. Method B is not subject to the same interference, so the concentration of I^- is low and independent of addition of Mn^{2+}. There could be some Mn^{2+} in the original mineral water, which causes method A to give a higher result than method B even when no Mn^{2+} is deliberately added. The result from Method B (0.009 mg/L) is more reliable.

CHAPTER 1
CHEMICAL MEASUREMENTS

> *A note from Dan:* Don't worry if your numerical answers are slightly different from mine. You or I may have rounded intermediate results. In general, retain many extra digits for intermediate answers and save your roundoff until the end. We'll study this process in Chapter 3.

1-1. (a) meter (m), kilogram (kg), second (s), ampere (A), kelvin (K), mole (mol)

(b) hertz (Hz), newton (N), pascal (Pa), joule (J), watt (W)

1-2. See Table 1-3. Abbreviations above kilo are capitalized: M (mega, 10^6), G (giga, 10^9), T (tera, 10^{12}), P (peta, 10^{15}), E (exa, 10^{18}), Z (zetta, 10^{21}), and Y (yotta, 10^{24}).

1-3.

(a) mW	=	milliwatt	=	10^{-3} watt
(b) pm	=	picometer	=	10^{-12} meter
(c) kΩ	=	kiloohm	=	10^3 ohm
(d) μF	=	microfarad	=	10^{-6} farad
(e) TJ	=	terajoule	=	10^{12} joule
(f) ns	=	nanosecond	=	10^{-9} second
(g) fg	=	femtogram	=	10^{-15} gram
(h) dPa	=	decipascal	=	10^{-1} pascal

1-4. (a) 100 fJ or 0.1 pJ

(b) 43.172 8 nF

(c) 299.79 THz or 0.299 79 PHz

(d) 0.1 nm or 100 pm

(e) 21 TW

(f) 0.483 amol or 483 zmol

1-5. (a) $8 \text{ Pg} = 8 \times 10^{15}$ g. $\quad 8 \times 10^{15} \cancel{g} \times \dfrac{1 \text{ kg}}{1\,000 \cancel{g}} = 8 \times 10^{12}$ kg of C

(b) The formula mass of CO_2 is $12.010\,6 + 2(15.999\,4) = 44.009\,4$

$$8 \times 10^{12} \cancel{\text{kg C}} \times \frac{44.009\,4 \text{ kg } CO_2}{12.010\,6 \cancel{\text{ kg C}}} = 2.9 \times 10^{13} \text{ kg } CO_2$$

(c) $2.9 \times 10^{13} \cancel{\text{kg}} CO_2 \times \dfrac{1 \text{ ton}}{1\,000 \cancel{\text{kg}}} = 2.9 \times 10^{10}$ tons of CO_2

$$\frac{2.9 \times 10^{10} \text{ tons } CO_2}{7 \times 10^9 \text{ people}} = 4 \text{ tons } CO_2 \text{ per person}$$

1-6. We will convert ounces of tuna to grams of tuna and find out how many μg of mercury are in one can (6 oz) of tuna. For a body mass of 68 kg, we will compute how many days are allowed between eating tuna so that the average dose does not exceed 0.1 μg Hg/kg body weight per day.

Table 1-4 tells us that 1 lb = 0.453 6 kg

6 oz = (6/16) lb = (6/16)(0.453 6 kg) = 0.170 kg = 170 g

One part per million means 1 μg Hg per gram of tuna

There is 0.6 ppm of mercury in chunk *white* tuna = 0.6 μg Hg/g tuna.

A 6-oz can contains $(170 \text{ g tuna})\left(\dfrac{0.6 \text{ μg Hg}}{\text{g tuna}}\right) = 102$ μg Hg

A dose of 0.1 μg Hg/kg body weight per day for a 68 kg person is

$$\left(\dfrac{0.1 \text{ μg Hg}}{\text{kg·day}}\right)(68 \text{ kg}) = 6.8 \dfrac{\text{μg Hg}}{\text{day}}$$

If I eat 102 μg Hg in one day from one can of tuna, I have eaten the amount of mercury allowed in $(102 \text{ μg Hg})\left(\dfrac{1 \text{ day}}{6.8 \text{ μg Hg}}\right) = 15$ days

I should wait 15 days before consuming my next can of tuna so that my average intake does not exceed 6.8 μg Hg/day.

Chunk *light* tuna contains 0.14 ppm Hg = 0.14 μg Hg/g tuna. Substituting this number for 0.6 μg Hg/g tuna in the sequence of calculations gives a period of 3.5 days. I could eat 2 cans of chunk light tuna per week.

1-7. Table 1-4 tells us that 1 horsepower = 745.700 W = 745.700 J/s.

100.0 horsepower = $(100.0 \text{ horsepower})\left(\dfrac{745.700 \text{ J/s}}{\text{horsepower}}\right) = 7.457 \times 10^4$ J/s.

$$\dfrac{7.457 \times 10^4 \; \dfrac{\text{J}}{\text{s}}}{4.184 \; \dfrac{\text{J}}{\text{cal}}} \times 3\,600 \; \dfrac{\text{s}}{\text{h}} = 6.416 \times 10^7 \; \dfrac{\text{cal}}{\text{h}}.$$

1-8. (a) $\dfrac{\left(2.2 \times 10^6 \dfrac{\text{cal}}{\text{day}}\right)\left(4.184 \dfrac{\text{J}}{\text{cal}}\right)\left(\dfrac{1 \text{ day}}{24 \text{ h}}\right)\left(\dfrac{1 \text{ h}}{3\,600 \text{ s}}\right)}{(120 \text{ pound})\left(0.453\,6 \dfrac{\text{kg}}{\text{pound}}\right)} = 2.0$ J/(s·kg)

= 2.0 W/kg

Similarly, $3.4 \times 10^3 \dfrac{kcal}{day} \Rightarrow 3.0 \text{ J/(s·kg)} = 3.0 \text{ W/kg}$.

(b) The office worker's power output is

$$\left(2.2 \times 10^6 \frac{cal}{day}\right)\left(4.184 \frac{J}{cal}\right)\left(\frac{1\ day}{24\ h}\right)\left(\frac{h}{3\,600\ s}\right) = 1.1 \times 10^2\ \frac{J}{s} = 1.1 \times 10^2\ W$$

The person's power output is greater than that of the 100 W light bulb.

1-9. $\left(5.00 \times 10^3 \dfrac{Btu}{h}\right)\left(1055.06 \dfrac{J}{Btu}\right)\left(\dfrac{1\ h}{3\,600\ s}\right) = 1.47 \times 10^3\ \dfrac{J}{s} = 1.47 \times 10^3\ W$

1-10. (a) $\left(1\,000 \dfrac{m}{km}\right)\left(\dfrac{1\ inch}{0.025\,4\ m}\right)\left(\dfrac{1\ foot}{12\ inch}\right)\left(\dfrac{1\ mile}{5\,280\ foot}\right) = 0.621\,37\ \dfrac{mile}{km}$

(b) $\left(\dfrac{100\ km}{4.6\ L}\right)\left(\dfrac{0.621\,37\ miles}{km}\right)\left(\dfrac{3.785\,4\ L}{gallon}\right) = 51\ \dfrac{miles}{gallon}$

(c) The diesel engine produces 223 g CO_2/km, which we will convert into g/mile:

$$\left(223 \frac{g\ CO_2}{km}\right)\left(\frac{1\ km}{0.621\,37\ mile}\right) = 359\ \frac{g\ CO_2}{mile}$$

In 15 000 miles, CO_2 = (15 000 miles)(359 g/mile) = 5.38×10^6 g or 5.38×10^3 kg = 5.38 metric tons. The gasoline engine produces 266 g CO_2/km, which we convert into 428 g/mile or 6.42 metric tons in 15 000 miles.

1-11. Newton = force = mass × acceleration = $kg\left(\dfrac{m}{s^2}\right)$

Joule = energy = force × distance = $kg\left(\dfrac{m}{s^2}\right) \cdot m = kg\left(\dfrac{m^2}{s^2}\right)$

Pascal = pressure = force / area = $kg\left(\dfrac{m}{s^2}\right)/m^2 = \dfrac{kg}{m \cdot s^2}$

1-12. $\left(0.03 \dfrac{mg}{m^2 \cdot day}\right)\left(1\,000 \dfrac{m}{km}\right)^2 (535\ km^2)\left(\dfrac{1\ g}{1\,000\ mg}\right) \times$

$\left(\dfrac{1\ kg}{1\,000\ g}\right)\left(\dfrac{1\ ton}{1\,000\ kg}\right)\left(365 \dfrac{day}{year}\right) = 6\ \dfrac{ton}{year}$

1-13.   (a)   molarity = moles of solute / liter of solution

(b)   molality = moles of solute / kilogram of solvent

(c)   density = grams of substance / milliliter of substance

(d)   weight percent = $100 \times$ (mass of substance/mass of solution or mixture)

(e)   volume percent = $100 \times$ (volume of substance/volume of solution or mixture)

(f)   parts per million = $10^6 \times$ (grams of substance/grams of sample)

(g)   parts per billion = $10^9 \times$ (grams of substance/grams of sample)

(h)   formal concentration = moles of formula/liter of solution

1-14.   Acetic acid (CH_3CO_2H) is a weak electrolyte that is partially dissociated. When we dissolve 0.01 mol in a liter, the concentrations of CH_3CO_2H plus $CH_3CO_2^-$ add to 0.01 M. The concentration of CH_3CO_2H alone is less than 0.01 M.

1-15.   32.0 g / [(22.990 + 35.452) g/mol] = 0.548 mol NaCl

0.548 mol / 0.500 L = 1.10 M

1-16. $\left(1.71 \dfrac{\text{mol } CH_3OH}{\text{L solution}}\right)(0.100 \text{ L solution}) = 0.171 \text{ mol } CH_3OH$

$(0.171 \text{ mol } CH_3OH)\left(\dfrac{32.04 \text{ g}}{\text{mol } CH_3OH}\right) = 5.48 \text{ g}$

1-17.   (a)   If atmospheric pressure is 1 bar, then a concentration of 1 ppb is 10^{-9} bar. A concentration of 39 ppb is 39×10^{-9} bar. There are exactly 10^5 Pa in a bar. So, first convert bar to Pa:

$$(39 \times 10^{-9} \text{ bar})\left(\frac{10^5 \text{ Pa}}{1 \text{ bar}}\right) = 39 \times 10^{-4} \text{ Pa}$$

Then convert Pa to mPa:

$$(39 \times 10^{-4} \text{ Pa})\left(\frac{10^3 \text{ mPa}}{1 \text{ Pa}}\right) = 3.9 \text{ mPa}$$

The ground-level O_3 concentration of 3.9 mPa is about 20% of the stratospheric concentration of 19 mPa.

(b)   The pressure of the atmosphere at 16 km altitude 9.6 kPa. The pressure of O_3 is 19 mPa. From the definition of ppb, the O_3 pressure in ppb is

$$O_3 \text{ (ppb)} = \frac{O_3 \text{ pressure (Pa)}}{\text{atmospheric pressure (Pa)}} \times 10^9 = \frac{19 \times 10^{-3} \text{ Pa}}{9.6 \times 10^3 \text{ Pa}} \times 10^9$$

$$= 2.0 \times 10^3 \text{ ppb}$$

1-18. (a) $19 \text{ mPa} = 19 \times 10^{-3} \text{ Pa}.$ $19 \times 10^{-3} \text{ Pa} \times \dfrac{1 \text{ bar}}{10^5 \text{ Pa}} = 1.9 \times 10^{-7} \text{ bar}$

(b) $T \text{ (K)} = 273.15 + {}^{\circ}\text{C} = 273.15 - 70 = 203 \text{ K}$

$$\frac{n}{V} = \frac{P}{RT} = \frac{1.9 \times 10^{-7} \text{ bar}}{0.083\,14 \dfrac{\text{L} \cdot \text{bar}}{\text{mol} \cdot \text{K}} \times 203 \text{ K}} = 1.1 \times 10^{-8} \text{ M} = 11 \text{ nM}$$

1-19. (a) $PV = nRT$

$$(1.000 \text{ bar})(5.24 \times 10^{-6} \text{ L}) = n \left(0.083\,14 \frac{\text{L} \cdot \text{bar}}{\text{mol} \cdot \text{K}} \right)(298.15K)$$

$$\Rightarrow n = 2.11 \times 10^{-7} \text{ mol} \Rightarrow 2.11 \times 10^{-7} \text{ M}$$

(b) Ar: 0.934% means $0.009\,34$ L of Ar per L of air

$$(1.000 \text{ bar})(0.009\,34 \text{ L}) = n \left(0.083\,14 \frac{\text{L} \cdot \text{bar}}{\text{mol} \cdot \text{K}} \right)(298.15 \text{ K})$$

$$\Rightarrow n = 3.77 \times 10^{-4} \text{ mol} \Rightarrow 3.77 \times 10^{-4} \text{ M}$$

Kr: $1.14 \text{ ppm} \Rightarrow 1.14 \text{ μL Kr per L of air} \Rightarrow 4.60 \times 10^{-8} \text{ M}$

Xe: $87 \text{ ppb} \Rightarrow 87 \text{ nL Xe per L of air} \Rightarrow 3.5 \times 10^{-9} \text{ M}$

1-20. $1 \text{ ppm} = \dfrac{1 \text{ g solute}}{10^6 \text{ g solution}}.$ Since 1 L of dilute solution $\approx 10^3$ g,

$1 \text{ ppm} = 10^{-3} \text{ g solute/L}$ ($= 10^{-3}$ g solute / 10^3 g solution).

Since $10^{-3} \text{ g} = 10^3 \text{ μg}$, $1 \text{ ppm} = 10^3 \text{ μg/L or } 1 \text{ μg/mL}$.

Since $10^{-3} \text{ g} = 1 \text{ mg}$, $1 \text{ ppm} = 1 \text{ mg/L}$.

1-21. 0.2 ppb means 0.2×10^{-9} g of $C_{20}H_{42}$ per g of rainwater

$$= 0.2 \times 10^{-6} \frac{\text{g } C_{20}H_{42}}{1\,000 \text{ g rainwater}} \approx \frac{0.2 \times 10^{-6} \text{ g } C_{20}H_{42}}{\text{L rainwater}}.$$

$$\frac{0.2 \times 10^{-6} \text{ g/L}}{282.56 \text{ g/mol}} = 7 \times 10^{-10} \frac{\text{mol}}{\text{L}} = 7 \times 10^{-10} \text{ M}$$

1-22. $\left(0.705 \dfrac{\text{g HClO}_4}{\text{g solution}} \right)(37.6 \text{ g solution}) = 26.5 \text{ g HClO}_4$

$37.6 \text{ g solution} - 26.5 \text{ g HClO}_4 = 11.1 \text{ g H}_2\text{O}$

1-23. (a) $\left(1.67 \dfrac{\text{g solution}}{\text{mL}}\right)\left(1000 \dfrac{\text{mL}}{\text{L}}\right) = 1.67 \times 10^3$ g solution

(b) $\left(0.705 \dfrac{\text{g HClO}_4}{\text{g solution}}\right)(1.67 \times 10^3 \text{ g solution}) = 1.18 \times 10^3$ g HClO$_4$

(c) $(1.18 \times 10^3 \text{ g}) / (100.45 \text{ g /mol}) = 11.7$ mol

1-24. molality $= \dfrac{\text{mol KI}}{\text{kg solvent}}$

20.0 wt% KI $= \dfrac{200 \text{ g KI}}{1\,000 \text{ g solution}} = \dfrac{200 \text{ g KI}}{800 \text{ g H}_2\text{O}}$

To find the grams of KI in 1 kg of H$_2$O, we set up a proportion:

$\dfrac{200 \text{ g KI}}{800 \text{ g H}_2\text{O}} = \dfrac{x \text{ g KI}}{1\,000 \text{ g H}_2\text{O}} \Rightarrow x = 250$ g KI

250 g KI / 166.00 g/mol $= 1.51$ mol KI, so the molality is 1.51 m.

1-25. (a) $\dfrac{150 \times 10^{-15} \text{ mol/cell}}{2.5 \times 10^4 \text{ vesicles/cell}} = 6.0 \dfrac{\text{amol}}{\text{vessicle}}$

(b) $(6.0 \times 10^{-18} \text{ mol})\left(6.022 \times 10^{23} \dfrac{\text{molecules}}{\text{mol}}\right) = 3.6 \times 10^6$ molecules

(c) Volume $= \dfrac{4}{3}\pi (200 \times 10^{-9} \text{ m})^3 = 3.35 \times 10^{-20}$ m^3;

$\dfrac{3.35 \times 10^{-20} \text{ m}^3}{10^{-3} \text{ m}^3 / \text{L}} = 3.35 \times 10^{-17}$ L

(d) $\dfrac{10 \times 10^{-18} \text{ mol}}{3.35 \times 10^{-17} \text{ L}} = 0.30$ M

1-26. $\dfrac{80 \times 10^{-3} \text{ g}}{180.2 \text{ g / mol}} = 4.4 \times 10^{-4}$ mol; $\quad \dfrac{4.4 \times 10^{-4} \text{ mol}}{0.1 \text{ L}} = 4.4 \times 10^{-3}$ M;

Similarly, 120 mg/100 L $= 6.7 \times 10^{-3}$ M.

1-27. (a) Mass of 1.000 L $= 1.046 \dfrac{\text{g}}{\text{mL}} \times 1\,000 \dfrac{\text{mL}}{\text{L}} \times 1.000 \text{ L} = 1\,046$ g

Grams of C$_2$H$_6$O$_2$ per liter $= 6.067 \dfrac{\text{mol}}{\text{L}} \times 62.07 \dfrac{\text{g}}{\text{mol}} = 376.6 \dfrac{\text{g}}{\text{L}}$

(b) 1.000 L contains 376.6 g of C$_2$H$_6$O$_2$ and $1\,046 - 376.6 = 669$ g of H$_2$O

$$= 0.669 \text{ kg}$$

$$\text{Molality} = \frac{6.067 \text{ mol C}_2\text{H}_6\text{O}_2}{0.669 \text{ kg H}_2\text{O}} = 9.07 \frac{\text{mol C}_2\text{H}_6\text{O}_2}{\text{kg H}_2\text{O}} = 9.07 \, m$$

1-28. Shredded wheat: 1.000 g contains 0.099 g protein + 0.799 g carbohydrate

$$0.099 \, g \times 4.0 \frac{\text{Cal}}{g} + 0.799 \, g \times 4.0 \frac{\text{Cal}}{g} = 3.6 \text{ Cal}$$

Doughnut: 1.000 g contains 0.046 g protein + 0.514 g carbohydrate + 0.186 g fat

$$0.046 \, g \times 4.0 \frac{\text{Cal}}{g} + 0.514 \, g \times 4.0 \frac{\text{Cal}}{g} + 0.186 \, g \times 9.0 \frac{\text{Cal}}{g} = 3.9 \text{ Cal}$$

In a similar manner, we find $2.8 \frac{\text{Cal}}{g}$ for hamburger and $0.48 \frac{\text{Cal}}{g}$ for apple.

There are 16 ounces in 1 pound, which Table 1-4 says is equal to 453.592 37 g

$$\Rightarrow 453.592\,37 \text{ g} / 16 \text{ ounce} = 28.35 \frac{\text{g}}{\text{ounce}}.$$

To convert Cal/g to Cal/ounce, multiply (Cal/g) times (28.35 g/ounce):

	Shredded Wheat	Doughnut	Hamburger	Apple
Cal/g	3.6	3.9	2.8	0.48
Cal/ounce	102	111	79	14

1-29. Mass of water $= \pi (225 \text{ m})^2 (10.0 \text{ m}) \left(\frac{1\,000 \text{ kg}}{\text{m}^3} \right) = 1.59 \times 10^9 \text{ kg}$

$$1.6 \text{ ppm} = \frac{1.6 \times 10^{-3} \text{ g F}^-}{\text{kg H}_2\text{O}}$$

Mass of F^- required =

$$\left(1.6 \times 10^{-3} \frac{\text{g F}^-}{\text{kg H}_2\text{O}} \right) (1.59 \times 10^9 \text{ kg H}_2\text{O}) = 2.5 \times 10^6 \text{ g F}^-.$$

(If we retain three digits for the next calculation, this last number is 2.54×10^6.) The atomic mass of F is 18.998 and the formula mass of H_2SiF_6 is 144.09. One mole of H_2SiF_6 contains 6 moles of F.

$$\frac{\text{mass of F}^-}{\text{mass of H}_2\text{SiF}_6} = \frac{6 \times 18.998}{144.09} = \frac{2.54 \times 10^6 \text{ g F}}{x \text{ g H}_2\text{SiF}_6} \Rightarrow x = 3.2 \times 10^6 \text{ g H}_2\text{SiF}_6$$

1-30. $2.00 \, L \times 0.050\,0 \frac{\text{mol}}{L} \times 61.83 \frac{\text{g}}{\text{mol}} = 6.18 \text{ g in a 2 L volumetric flask}$

1-31. Weigh out $2 \times 0.050\,0 \text{ mol} = 0.100 \text{ mol} = 6.18 \text{ g B(OH)}_3$ and dissolve in 2.00 kg H_2O.

1-32. $M_{con} \cdot V_{con} = M_{dil} \cdot V_{dil}$

$$\left(0.80\,\frac{mol}{L}\right)(1.00\,L) = \left(0.25\,\frac{mol}{L}\right)\,V_{dil} \Rightarrow V_{dil} = 3.2\,L$$

1-33. We need $1.00\,L \times 0.10\,\dfrac{mol}{L} = 0.10\,mol\,NaOH = 4.0\,g\,NaOH$

$$\frac{4.0\,g\,NaOH}{0.50\,\dfrac{g\,NaOH}{g\,solution}} = 8.0\,g\,solution$$

1-34. (a) $V_{con} = V_{dil}\,\dfrac{M_{dil}}{M_{con}} = 1\,000\,mL\left(\dfrac{1.00\,M}{18.0\,M}\right) = 55.6\,mL$

(b) One liter of 98.0% H_2SO_4 contains $(18.0\,mol)(98.079\,g/mol) = 1.77 \times 10^3$ g of H_2SO_4. Since the solution contains 98.0 wt% H_2SO_4, and the mass of H_2SO_4 per mL is 1.77 g, the mass of solution per milliliter (the density) is

$$\frac{1.77\,g\,H_2SO_4\,/mL}{0.980\,g\,H_2SO_4\,/g\,solution} = 1.80\,g\,solution/mL$$

1-35. 2.00 L of 0.169 M NaOH = 0.338 mol NaOH = 13.5 g NaOH

$$density = \frac{g\,solution}{mL\,solution}$$

$$= \frac{13.5\,g\,NaOH}{(16.7\,mL\,solution)\left(0.534\,\dfrac{g\,NaOH}{g\,solution}\right)} = 1.52\,\frac{g}{mL}$$

1-36. FM of $Ba(NO_3)_2 = 261.34$ 4.35 g of solid with 23.2 wt% $Ba(NO_3)_2$ contains $(0.232)(4.35\,g) = 1.01\,g\,Ba(NO_3)_2$

$$mol\,Ba^{2+} = \frac{(1.01\,g\,Ba(NO_3)_2)}{(261.34\,g\,Ba(NO_3)_2\,/mol)} = 3.86 \times 10^{-3}\,mol$$

$mol\,H_2SO_4 = mol\,Ba^{2+} = 3.86 \times 10^{-3}\,mol$

$$volume\,of\,H_2SO_4 = \frac{(3.86 \times 10^{-3}\,mol)}{(3.00\,mol/L)} = 1.29\,mL$$

1-37. 25.0 mL of 0.023 6 M Th^{4+} contains

$(0.025\,0\,L)(0.023\,6\,M) = 5.90 \times 10^{-4}\,mol\,Th^{4+}$

mol HF required for stoichiometric reaction $= 4 \times mol\,Th^{4+} = 2.36 \times 10^{-3}\,mol$

50% excess $= 1.50(2.36 \times 10^{-3}\,mol) = 3.54 \times 10^{-3}\,mol\,HF$

Required mass of pure HF $= (3.54 \times 10^{-3}\ \text{mol})(20.01\ \text{g/mol}) = 0.070\,8\ \text{g}$

Mass of 0.491 wt% HF solution $= \dfrac{(0.070\,8\ \text{g HF})}{(0.004\,91\ \text{g HF / g solution})} = 14.4\ \text{g}$

1-38. (a) Acetic acid $C_2H_4O_2 = 60.05$ g/mol $NaHCO_3 = 84.01$ g/mol

(b) $(5\ \text{g NaHCO}_3)/(84.01\ \text{g/mol}) = 0.0595$ mol

1 mol $NaHCO_3$ reacts with 1 mol CH_3CO_2H, so we need

$(0.0595\ \text{mol CH}_3\text{CO}_2\text{H})(60.05\ \text{g/mol}) = 3.57\ \text{g CH}_3\text{CO}_2\text{H}$

(c) But vinegar is ~5 wt% acetic acid, so we need

$\dfrac{(3.57\ \text{g CH}_3\text{CO}_2\text{H})}{(0.05\ \text{g CH}_3\text{CO}_2\text{H/g vinegar})} = 71.5\ \text{g acetic acid}$

If the density is 1.0 g/mL, 71.5 g vinegar = 71.5 mL vinegar

(d) Vinegar is the limiting reagent because 71.5 mL are required, but only 50 mL are available.

(e) Vinegar is the limiting reagent. The amount of CH_3CO_2H in 50 mL (≈ 50 g) of vinegar is $(0.05\ \text{g CH}_3\text{CO}_2\text{H/g vinegar})(50\ \text{g vinegar}) = 2.5\ \text{g CH}_3\text{CO}_2\text{H}$. Mol $CH_3CO_2H = 2.5\ \text{g (CH}_3\text{CO}_2\text{H})/(60.05\ \text{g/mol}) = 0.041\,6$ mol which will create 0.041 6 mol of CO_2 whose volume is

$$V = \dfrac{nRT}{P} = \dfrac{(0.041\,6\ \text{mol CO}_2)(0.083\,14\ \text{L·bar/(mol·K)})(300\ \text{K})}{1\ \text{bar}} = 1.04\ \text{L}$$

If the air space in the bottle is 0.5 L, and it contains 0.5 L of air plus 1.04 L of CO_2, we have 1.54 L of gas in 0.5 L of volume, giving a pressure of 1.54/0.5 ≈ 3 bar. The cork pops before this much pressure builds up.

CHAPTER 2
TOOLS OF THE TRADE

2-1.   (a) The primary rule is to familiarize yourself with the hazards of what you are about to do and do not carry out a dangerous procedure without adequate precautions.

2-2.   Nonpolar organic liquids might penetrate through rubber gloves. Concentrated hydrochloric acid is a polar aqueous solution that is not likely to penetrate through rubber gloves.

2-3.   Dichromate ($Cr_2O_7^{2-}$) is soluble in water and contains carcinogenic Cr(VI). Reducing Cr(VI) to Cr(III) decreases the toxicity of the metal. Converting aqueous Cr(III) to solid $Cr(OH)_3$ decreases the solubility of the metal and therefore decreases its ability to be spread by water. Evaporation produces the minimum volume of waste.

2-4.   Green chemistry is a set of principles intended to change our behavior in a manner that will help sustain the habitability of Earth. Green chemistry seeks to design chemical products and processes to reduce the use of resources and energy and the generation of hazardous waste.

2-5.   The lab notebook must: (1) state what was done; (2) state what was observed; and (3) be understandable to a stranger.

2-6.   An object placed on the electronic balance pushes the pan down with a force $m \times g$, where m is the mass of the object and g is the acceleration of gravity. The pan pushes down on a load receptor attached to parallel guides. The force of the sample pushes one side of the force-transmitting lever down and moves the other side of the lever up. A null position sensor detects movement of the lever arm away from its equilibrium (null) position. When the null sensor detects displacement of the lever arm, a servo amplifier sends electric current through the force compensation wire coil in the field of a permanent magnet. Electric current in the coil interacts with the magnetic field to produce a downward force that exactly compensates for the upward force on the lever arm to maintain a null position. Current flowing through the coil is measured and ultimately converted to a readout in grams. The conversion between current and mass is accomplished by measuring the current required to balance an internal calibration mass.

2-7. The buoyancy correction is 1 when the substance being weighed has the same density as the weights used to calibrate the balance.

2-8.
$$m = \frac{(14.82 \text{ g})\left(1 - \dfrac{0.001\,2 \text{ g/mL}}{8.0 \text{ g/mL}}\right)}{\left(1 - \dfrac{0.001\,2 \text{ g/mL}}{0.626 \text{ g/mL}}\right)} = 14.85 \text{ g}$$

2-9. The smallest correction will be for PbO_2, whose density is closest to 8.0 g/mL. The largest correction will be for the least dense substance, lithium.

2-10.
$$m = \frac{4.236\,6 \text{ g}\left(1 - \dfrac{0.001\,2 \text{ g/mL}}{8.0 \text{ g/mL}}\right)}{\left(1 - \dfrac{0.001\,2 \text{ g/mL}}{1.636 \text{ g/mL}}\right)} = 4.239\,1 \text{ g}$$

Without correcting for buoyancy, we would think the mass of primary standard is less than the actual mass and we would think the molarity of base reacting with the standard is also less than the actual molarity. The percentage error would be

$$\frac{\text{true mass} - \text{measured mass}}{\text{true mass}} \times 100 = \frac{4.239\,1 - 4.236\,6}{4.239\,1} \times 100 = 0.06\%.$$

2-11.
$$m' = \frac{1.267 \text{ g}\left(1 - \dfrac{0.001\,2 \text{ g/mL}}{3.988 \text{ g/mL}}\right)}{\left(1 - \dfrac{0.001\,2 \text{ g/mL}}{8.0 \text{ g/mL}}\right)} = 1.266_8 \text{ g}$$

The buoyancy correction is negligible because the density of CsCl is high.

2-12. (a) One mol of He (= 4.003 g) occupies a volume of

$$V = \frac{nRT}{P} = \frac{(1 \text{ mol})\left(0.083\,14 \dfrac{\text{L} \cdot \text{bar}}{\text{mol} \cdot \text{K}}\right)(293.15 \text{ K})}{1 \text{ bar}} = 24.37 \text{ L}$$

Density = 4.003 g / 24.37 L = 0.164 g/L = 0.000 164 g/mL

(b) $$m = \frac{(0.823 \text{ g})\left(1 - \dfrac{0.000\,164 \text{ g/mL}}{8.0 \text{ g/mL}}\right)}{\left(1 - \dfrac{0.000\,164 \text{ g/mL}}{0.97 \text{ g/mL}}\right)} = 0.823 \text{ g}$$

2-13. (a) (0.42) (2 330 Pa) = 979 Pa

(b) Air density =

$$\frac{(0.003\ 485)(94\ 000) - (0.001\ 318)(979)}{293.15} = 1.11\ \text{g/L} = 0.001\ 1\ \text{g/mL}$$

(c) Mass $= 1.000\ 0\ \text{g} \left(\dfrac{1 - \dfrac{0.0011\ \cancel{\text{g/mL}}}{8.0\ \cancel{\text{g/mL}}}}{1 - \dfrac{0.0011\ \cancel{\text{g/mL}}}{1.00\ \cancel{\text{g/mL}}}} \right) = 1.001\ 0\ \text{g}$

2-14. (a) $m_b = m_a \dfrac{r_a^2}{r_b^2} = (100.000\ 0\ \text{g}) \dfrac{(6\ 370\ 000\ \text{m})^2}{(6\ 370\ 030\ \text{m})^2} = 99.999\ 1\ \text{g}$

(b) The calibrate function measures a known internal calibration mass and applies a factor to the electronic response to indicate the known mass. This calibration factor accounts for the change in gravitational force on going from one elevation to another.

2-15. (a) Area of gold electrode $= (3.3\ \text{mm}^2)(0.1\ \text{cm/mm})^2 = 0.033\ \text{cm}^2$

Moles DNA bound $= (1.2\ \text{pmol/cm}^2)(0.033\ \text{cm}^2)$
$$= (1.2 \times 10^{-12}\ \text{mol/cm}^2)(0.033\ \text{cm}^2) = 3.96 \times 10^{-14}\ \text{mol}$$

Mass of added cytosine $= (3.96 \times 10^{-14}\ \text{mol})(287.2\ \text{g/mol})$
$$= 1.14 \times 10^{-11}\ \text{g} = 11\ \text{pg}$$

(b) Binding of 1 ng/cm^2 gives a shift of -10 Hz. The observed shift of -4.4 Hz corresponds to

$$\frac{-4.4\ \text{Hz}}{-10\ \text{Hz/(ng/cm}^2)} = 0.44\ \text{ng/cm}^2 \text{ of bound cytosine}$$

$(0.44\ \text{ng/cm}^2$ of bound cytosine$)(0.033\ \text{cm}^2) = 14$ pg, which is approximately equal to the expected binding of 11 pg of cytosine. The measured frequency shift corresponds to $(14/11) = 1.3$ cytosine nucleotides.

2-16. TD means "to deliver" and TC means "to contain."

2-17. Dissolve $(0.250\ 0\ \text{L})(0.150\ 0\ \text{mol/L}) = 0.037\ 50$ mol of K_2SO_4 ($= 6.535$ g, FM 174.26 g/mol) in less than 250 mL of water in a 250-mL volumetric flask. Add more water and mix. Dilute to the 250.0 mL mark and invert the flask many times for complete mixing.

2-18. The plastic flask is needed for trace analysis of ionic analytes (especially cations) at ppb levels that might be lost by adsorption on the glass surface or contaminated by leaching of ions from the glass.

2-19. (a) With a suction device, suck liquid up past the 5.00 mL mark. Discard one or two pipet volumes of liquid to rinse the pipet. Take up a third volume past the calibration mark and quickly replace the bulb with your index finger. (Alternatively, use an automatic suction device that remains attached to the pipet.) Wipe excess liquid off the outside of the pipet with a clean tissue. Touch the tip of the pipet to the side of a beaker and drain liquid until the bottom of the meniscus reaches the center of the mark. Transfer the pipet to a receiving vessel and drain it by gravity while holding the tip against the wall. After draining stops, hold the pipet to the wall for a few more seconds to complete draining. Do not blow out the last drop. The pipet should be nearly vertical at the end of delivery.

  (b) Transfer pipet.

2-20. (a) Adjust the knob for 50.0 μL. Place a fresh tip tightly on the barrel. Depress the plunger to the first stop, corresponding to 50.0 μL. Hold the pipet vertically, dip it 3–5 mm into reagent solution, and slowly release the plunger to suck up liquid. Leave the tip in the liquid for a few more seconds. Withdraw the pipet vertically. Take up and discard three squirts of reagent to clean and wet the tip and fill it with vapor. To dispense liquid, hold the pipet nearly vertical, touch the tip to the wall of the receiver, and gently depress the plunger to the first stop. After a few seconds, depress the plunger further to squirt out the last liquid.

  (b) The procedure in (a) is called *forward mode*. For a foaming or viscous liquid, use *reverse* mode. Depress the plunger beyond the 50.0 μL stop and take in more than 50.0 μL. To deliver 50.0 μL, depress the plunger to the first stop and not beyond.

2-21. The trap prevents liquid filtrate from being sucked into the vacuum system. The watchglass keeps dust out of the sample.

2-22. Phosphorus pentoxide

2-23. (a) A 100.0 mM solution is the same as 0.100 0 M. To make 250.0 mL of 0.100 0 M solution requires $(0.100\ 0\ M)(0.250\ 0\ L) = 0.025\ 00$ mol benzoic acid. Required mass $= (0.025\ 00\ \text{mol})(122.12\ \text{g/mol}) = 3.053_0$ g.

(b) $m' = \dfrac{3.053_0\ \text{g}\left(1 - \dfrac{0.001\ 2\ \text{g/mL}}{1.27\ \text{g/mL}}\right)}{\left(1 - \dfrac{0.001\ 2\ \text{g/mL}}{8.0\ \text{g/mL}}\right)} = 3.050_6$ g

(c) We want to make a 50.0 μM solution from a 100.0 mM solution, which represents a dilution by a factor of $(100.0\ \text{mM})/(50.0\ \mu M) = (100.0 \times 10^{-3}\ M)/(50.0 \times 10^{-6}\ M) = 2\ 000$. You could make a 2 000-fold dilution with a 20-fold dilution followed by a 100-fold dilution $(20 \times 100 = 2\ 000)$. You could do this by first diluting 5.00 mL of 100.0 mM solution up to 100.0 mL to get $(5/100)(100.0\ \text{mM}) = 5.00$ mM. Then dilute 10.00 mL of the resulting solution up to 1000.0 mL to get $(10/1\ 000)(5.00\ \text{mM}) = 50.0\ \mu M$.

2-24. $20.214\ 4\ \text{g} - 10.263\ 4\ \text{g} = 9.951\ 0$ g. Column 3 of Table 2-7 tells us that the true volume is $(9.951\ 0\ \text{g})(1.002\ 9\ \text{mL/g}) = 9.979\ 9$ mL.

2-25. Expansion $= \dfrac{0.999\ 102\ 6}{0.997\ 047\ 9} = 1.002\ 060\ 8 \approx 0.2\%$. Densities were taken from Table 2-7. The 0.500 0 M solution at 25° would be $(0.500\ 0\ M)/(1.002) = 0.499\ 0$ M.

2-26. Using column 2 of Table 2-7, mass in vacuum $=$
$(50.037\ \cancel{\text{mL}})(0.998\ 207\ 1\ \text{g/}\cancel{\text{mL}}) = 49.947$ g.

Using column 3, mass in air $= \dfrac{50.037\ \cancel{\text{mL}}}{1.0029\ \cancel{\text{mL}}/\text{g}} = 49.892$ g.

2-27. When the solution is cooled to 20°C, the concentration will be higher than the concentration at 24°C by a factor of $\frac{\text{density at 20°C}}{\text{density at 24°C}}$. Therefore, the concentration needed at 24° will be lower than the concentration at 20°C.

Desired concentration at 24°C $= (1.000\ M)\left(\dfrac{0.997\ 299\ 5\ \cancel{\text{g/mL}}}{0.998\ 207\ 1\ \cancel{\text{g/mL}}}\right) = 0.999\ 1$ M

(using the quotient of densities from Table 2-7). The true mass of KNO_3 needed is $\left(0.500\ 0\ \cancel{L}\right)\left(0.999\ 1\dfrac{\cancel{\text{mol}}}{\cancel{L}}\right)\left(101.103\dfrac{\text{g}}{\cancel{\text{mol}}}\right) = 50.506$ g.

$$m' = \frac{(50.506 \text{ g}) \left(1 - \dfrac{0.001\,2 \text{ g/mL}}{2.109 \text{ g/mL}}\right)}{\left(1 - \dfrac{0.001\,2 \text{ g/mL}}{8.0 \text{ g/mL}}\right)} = 50.484 \text{ g}$$

2-28. Procedure (ii) with the 10-mL pipet and 1-L flask is more accurate than procedure (i) with the 1-mL pipet and 100-mL flask. Relative uncertainties in the larger pipet and larger flask are less than the relative uncertainties in the smaller pipet and smaller flask. You can improve the accuracy of either procedure by calibrating the specific pipet and specific volumetric flask that you use so that you know how much volume they actually contain. The calibration uncertainty should be significantly smaller than the manufacturer's tolerance for the glassware.

2-29. (a) Fraction within specifications $= e^{-t(\ln 2)/t_m}$. If $t_m = 2$ yr and $t = 2$ yr, then fraction within specifications $= e^{-2(\ln 2)/2} = e^{-\ln 2} = \frac{1}{2}$.

(b) Fraction within specifications $= 0.95 = e^{-t(\ln 2)/2 \text{ yr}}$

To solve for t, take the natural logarithm of both sides:

$\ln(0.95) = -t(\ln 2)/2 \Rightarrow t = -2 \ln(0.95)/\ln 2 = 0.148 \text{ yr} = 54 \text{ days} \approx 8 \text{ weeks}$

2-30. Al extracted from glass $= (0.200 \text{ L})(5.2 \times 10^{-6} \text{ M}) = 1.04 \times 10^{-6} \text{ mol}$

mass of Al $= (1.04 \times 10^{-6} \text{ mol})(26.98 \text{ g/mol}) = 28.1 \text{ μg}$

This much Al was extracted from 0.50 g of glass, so

$$\text{wt\% Al extracted} = 100 \times \frac{28.1 \times 10^{-6} \text{ g}}{0.50 \text{ g}} = 0.005\,6_2 \text{ wt\%}$$

$$\text{Fraction of Al extracted} = \frac{0.005\,6_2 \text{ wt\%}}{0.80 \text{ wt\%}} = 0.007\,0 \text{ (or 0.70\% of the Al)}$$

2-31.

Graph of van Deemter Equation		
	Flow rate	Plate height
Constants	(mL/min)	(mm)
A =	4	8.194
1.65	6	6.092
B =	8	5.064
25.8	10	4.466
C =	20	3.412
0.0236	30	3.218
	40	3.239
	50	3.346
	60	3.496
	70	3.671
	80	3.861
	90	4.061
	100	4.268
Formula:		
C5 = $A$6+$A$8/B5+$A$10*B5		

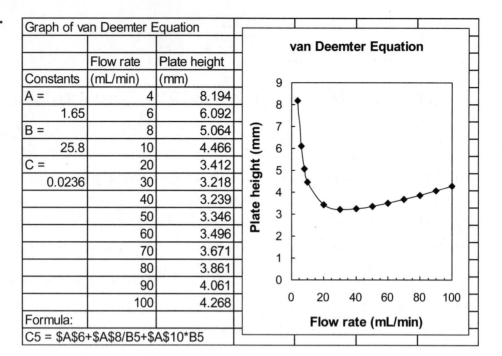

CHAPTER 3
EXPERIMENTAL ERROR

3-1. (a) 5 (b) 4 (c) 3

3-2. (a) 1.237 (b) 1.238 (c) 0.135 (d) 2.1 (e) 2.00

3-3. (a) 0.217 (b) 0.216 (c) 0.217

3-4. (b) 1.18 (3 significant figures) (c) 0.71 (2 significant figures)

3-5. (a) 3.71 (b) 10.7 (c) 4.0×10^1 (d) 2.85×10^{-6}
(e) 12.625 1 (f) 6.0×10^{-4} (g) 242

3-6. (a) $BaF_2 = 137.327 + 2(18.998\ 403\ 2) = 175.324$ because the atomic mass of Ba has only 3 decimal places.
(b) $C_6H_4O_4 = 6(12.010\ 6) + 4(1.007\ 98) + 4(15.999\ 4) = 140.093$

3-7. (a) 12.3 (b) 75.5 (c) 5.520×10^3 (d) 3.04
(e) 3.04×10^{-10} (f) 11.9 (g) 4.600 (h) 4.9×10^{-7}

3-9. All measurements have some uncertainty, so there is no way to know the true value.

3-10. Systematic error is always above or always below the "true value" if you make replicate measurements. In principle, you can find the source of this error and eliminate it in a better experiment so the measured mean equals the true mean. Random error is equally likely to be positive or negative and cannot be eliminated. Random error can be reduced in a better experiment.

3-11. The apparent mass of product is systematically low because the initial mass of the (crucible plus moisture) is higher than the true mass of the crucible. The error is systematic. There is also always some random error superimposed on the systematic error.

3-12. (a) 25.031 mL is a systematic error. The pipet always delivers more than it is rated for. The number ± 0.009 is the random error in the volume delivered. The volume fluctuates around 25.031 by ±0.009 mL.
(b) The numbers 1.98 and 2.03 mL are systematic errors. The buret delivers too little between 0 and 2 mL and too much between 2 and 4 mL. The observed

variations ±0.01 and ±0.02 are random errors.

(c) The difference between 1.9839 and 1.9900 g is random error. The mass will probably be different the next time I try the same procedure.

(d) Differences in peak area are random error based on inconsistent injection volume, inconsistent detector response, and probably other small variations in the condition of the instrument from run to run.

3-13. (a) Carmen   (b) Cynthia   (c) Chastity   (d) Cheryl

3-14. 3.124 (±0.005), 3.124 (±0.2%). It would also be reasonable to keep an additional digit: 3.123_6 (±0.005_2), 3.123_6 (±0.1_7%)

3-15. (a)   6.2 (±0.2)

$\underline{-\ 4.1\ (\pm 0.1)}$

2.1 $\pm\ e$ $e^2 = 0.2^2 + 0.1^2 \Rightarrow e = 0.2_{24}$ Answer: 2.1 ± 0.2 (or $2.1 \pm 11\%$)

(b)   9.43 (±0.05)   9.43 (±0.53%)

$\underline{\times\ 0.016\ (\pm 0.001)}$ $\Rightarrow$ $\underline{\times\ 0.016\ (\pm 6.25\%)}$ $\%e^2 = 0.53^2 + 6.25^2$

0.150 88 ($\pm\ \%e$) $\Rightarrow$ $\%e = 6.272$

Relative uncertainty = 6.27%; Absolute uncertainty = 0.150 88 × 0.062 7

= 0.009 46;   Answer: 0.151 ± 0.009 (or $0.151 \pm 6\%$)

(c) The first term in brackets is the same as part (a), so we can rewrite the problem as $2.1\ (\pm 0.2_{24}) \div 9.43\ (\pm 0.05) = 2.1\ (\pm 10.7\%) \div 9.43\ (\pm 0.53\%)$

$\%e = \sqrt{10.7^2 + 0.53^2} = 10.7\%$

Absolute uncertainty = 0.107 × 0.223 = 0.023 9

Answer: $0.22_3 \pm 0.02_4$ (±11%)

(d) The term in brackets is

6.2 (±0.2) × 10⁻³ $e = \sqrt{0.2^2 + 0.1^2} \Rightarrow e = 0.2_{24}$

$\underline{+\ 4.1\ (\pm 0.1)\ \times\ 10^{-3}}$

10.3 ($\pm 0.2_{24}$) × 10⁻³ = 10.3 × 10⁻³ (±2.17%)

9.43 (±0.53%) × 0.010 3 (±2.17%) = 0.097 13 ± 2.23% = 0.097 13 ± 0.002 17

Answer: $0.097_1 \pm 0.002_2$ (± $2._2$%)

3-16. (a) Uncertainty = $\sqrt{0.03^2 + 0.02^2 + 0.06^2} = 0.07$

Answer: 10.18 (±0.07) (±0.7%)

(b) 91.3 (±1.0) × 40.3 (±0.2)/21.1 (±0.2)

= 91.3 (±1.10%) × 40.3 (±0.50%)/21.1 (±0.95%)

$$\% \text{ uncertainty} = \sqrt{1.10^2 + 0.50^2 + 0.95^2} = 1.54\%$$

Answer: 174 (± 3) ($\pm 2\%$) or 174.4 (± 2.7) ($\pm 1.5\%$)

(c) $[4.97\,(\pm 0.05) - 1.86\,(\pm 0.01)]/21.1\,(\pm 0.2)$

$\quad = [3.11\,(\pm 0.05_{10})]/21.1\,(\pm 0.2) = [3.11\,(\pm 1.64\%)]/21.1\,(\pm 0.95\%)$

$\quad = 0.147\,(\pm 1.90\%) = 0.147\,(\pm 0.003)\,(\pm 2\%)$ or $0.147_4\,(\pm 0.002_8)\,(\pm 1.9\%)$

(d)
$$\begin{array}{l} 2.016\,4 \;\;(\pm 0.000\,8) \\ 1.233 \;\;\;\;(\pm 0.002) \\ +\;\;4.61\;\;\;\;\;(\pm 0.01) \\ \hline 7.85_{94} \;\;\sqrt{(0.000\,8)^2 + (0.002)^2 + (0.01)^2} = 0.01_{02} \end{array}$$

Answer: 7.86 (± 0.01)($\pm 0.1\%$)

(e)
$$\begin{array}{l} 2\,016.4 \;\;(\pm 0.8) \\ +\;123.3 \;\;(\pm 0.2) \\ +\;\;\;46.1 \;\;(\pm 0.1) \\ \hline 2\,185.8 \;\;\sqrt{(0.8)^2 + (0.2)^2 + (0.1)^2} = 0.8 \end{array}$$

Answer: 2 185.8 (± 0.8) ($\pm 0.04\%$)

(f) $[3.14\,(\pm 0.05)]^{1/3} = ?$

For $y = x^a$, $\%e_y = a\%e_x$

$x = 3.14 \pm 0.05 \Rightarrow \%e_x = (0.05\,/\,3.14) \times 100 = 1.592\%$

$\%e_y = \frac{1}{3}(1.592\%) = 0.531\%$

Answer: $1.464_3 \pm 0.007_8$ ($\pm 0.5_3\%$)

(g) $\log [3.14\,(\pm 0.05)] = ?$

For $y = \log x$, $e_y = 0.434\,29\,\dfrac{e_x}{x}$

$x = 3.14 \pm 0.05 \Rightarrow e_y = 0.434\,29\left(\dfrac{0.05}{3.14}\right) = 0.006\,915$

Answer: $0.496_9 \pm 0.006_9$ ($\pm 1.3_9\%$)

3-17. (a) $y = x^{1/2} \Rightarrow \%e_y = \frac{1}{2}\left(100 \times \dfrac{0.001\,1}{3.141\,5}\right) = 0.017\,5\%$

$\quad (1.75 \times 10^{-4})\sqrt{3.141\,5} = 3.1 \times 10^{-4}$ \qquad Answer: $1.772\,4_3 \pm 0.000\,3_1$

(b) $y = \log x \Rightarrow e_y = 0.434\,29\left(\dfrac{0.001\,1}{3.141\,5}\right) = 1.52 \times 10^{-4}$

Answer: $0.497\,1_4 \pm 0.000\,1_5$

(c) $y = \text{antilog } x = 10^x \Rightarrow e_y = y \times 2.302\,6\,e_x$

$$= (10^{3.141\,5})(2.302\,6)(0.001\,1) = 3.51 \qquad \text{Answer: } 1.385_2 \pm 0.003_5 \times 10^3$$

(d) $y = \ln x \Rightarrow e_y = \dfrac{0.001\,1}{3.141\,5} = 3.5 \times 10^{-4}$ \qquad Answer: $1.144\,7_0 \pm 0.000\,3_5$

(e) Numerator of log term: $y = x^{1/2} \Rightarrow e_y = \dfrac{1}{2}\left(\dfrac{0.006}{0.104} \times 100\right) = 2.88\%$

$$\dfrac{0.322\,5 \pm 2.88\%}{0.051\,1 \pm 0.0009} = \dfrac{0.322\,5 \pm 2.88\%}{0.051\,1 \pm 1.76\%}$$

$$= 6.311 \pm 3.380\% = 6.311 \pm 0.213$$

For $y = \log x$, $e_y = 0.434\,29\,\dfrac{e_x}{x} = 0.434\,29\left(\dfrac{0.213}{6.311}\right) = 0.015$

Answer: $0.80_0 \pm 0.01_5$

3-18. (a) Na = 22.989 769 28 \qquad $\pm$ 0.000 000 02

Cl = 35.452 \qquad\qquad $\pm$ 0.006

$$\underline{\hspace{8cm}}$$

58.441 770 \qquad $\sqrt{(2 \times 10^{-8})^2 + (0.006)^2} = 0.006$

58.442 $\pm$ 0.006 g/mol

(b) molarity $= \dfrac{\text{mol}}{\text{L}} = \dfrac{[2.634\,(\pm 0.002)\,\text{g}]\,/\,[58.442\,(\pm 0.006)\,\text{g/mol}]}{0.100\,00\,(\pm 0.000\,08)\,\text{L}}$

$$= \dfrac{2.634\,(\pm 0.076\%)\,/\,58.442\,(\pm 0.010\%)}{0.100\,00\,(\pm 0.08\%)}$$

relative error $= \sqrt{(0.076\%)^2 + (0.010\%)^2 + (0.08\%)^2} = 0.11\%$

molarity $= 0.450\,7\,(\pm 0.000\,5)\,\text{M}$

3-19. $m = \dfrac{m'\left(1 - \dfrac{d_a}{d_w}\right)}{1 - \dfrac{d_a}{d}}$

$$m = \dfrac{[1.034\,6\,(\pm 0.000\,2)\,\text{g}]\left(1 - \dfrac{0.001\,2(\pm 0.000\,1)\,\text{g/mL}}{8.0\,(\pm 0.5)\,\text{g/mL}}\right)}{1 - \dfrac{0.001\,2(\pm 0.000\,1)\,\text{g/mL}}{0.997\,299\,5\,\text{g/mL}}}$$

$$m = \dfrac{[1.034\,6\,(\pm 0.019\,3\%)]\left(1 - \dfrac{0.001\,2\,(\pm 8.33\%)}{8.0\,(\pm 6.25\%)}\right)}{1 - \dfrac{0.001\,2\,(\pm 8.33\%)}{0.997\,299\,5\,(\pm 0\%)}}$$

$$m = \dfrac{[1.034\,6\,(\pm 0.019\,3\%)][1 - 0.000\,150\,(\pm 10.4\%)]}{[1 - 0.001\,203\,(\pm 8.33\%)]}$$

$$m = \frac{[1.034\,6\,(\pm 0.019\,3\%)]\,[1 - 0.000\,150\,(\pm 0.000\,015\,6)]}{[1 - 0.001\,203\,(\pm 0.000\,100)]}$$

$$m = \frac{[1.034\,6\,(\pm 0.019\,3\%)]\,[0.999\,850\,0\,(\pm 0.000\,015\,6)]}{[0.998\,797\,(\pm 0.000\,100)]}$$

$$m = \frac{[1.034\,6\,(\pm 0.019\,3\%)]\,[0.999\,850\,0\,(\pm 0.001\,56\%)]}{[0.998\,797\,(\pm 0.010\,0\%)]}$$

$$m = 1.035\,7\,(\pm 0.021\,8\%) = 1.035\,7\,(\pm 0.000\,2)\ \text{g}$$

3-20.   (a)   2Na: $2(22.989\,769\,28 \pm 0.000\,000\,02) = 45.979\,538\,56 \pm 0.000\,000\,04$

1C: $1(12.010\,6 \qquad \pm 0.001\,0) \qquad = 12.010\,6 \qquad \pm 0.001\,0$

3O: $3(15.999\,4 \qquad \pm 0.000\,4) \qquad = 47.998\,2 \qquad \pm 0.001\,2$

Na_2CO_3: $105.988\,339$ $\sqrt{(4 \times 10^{-6})^2 + (0.001\,0)^2 + (0.001\,2)^2} = 0.001_{56}$

$105.9883 \pm 0.001_{56}$ g/mol

$$\text{Percent relative uncertainty} = 100 \times \frac{0.001_{56}}{105.9883} = 0.001\,5\%$$

(b)   mol $H^+ = 2 \times$ mol Na_2CO_3

$$\text{mol } Na_2CO_3 = \frac{0.967\,4\,(\pm 0.000\,9)\ \text{g}}{105.9883\,(\pm 0.001_{56})\ \text{g/mol}} = \frac{0.967\,4\,(\pm 0.093\%)\ \text{g}}{105.988\,(\pm 0.001\,5\%)\ \text{g/mol}}$$

$$= 0.009\,127\,4\,(\pm 0.093\%)\ \text{mol}$$

mol $H^+ = 2[0.009\,127\,4\,(\pm 0.093\%)] = 0.018\,255\,(\pm 0.093\%)$ mol

(Relative error is not affected by the multiplication by 2 because mol H^+ and uncertainty in mol H^+ are both multiplied by 2.)

$$\text{molarity of HCl} = \frac{0.018\,255\,(\pm 0.093\%)\ \text{mol}}{0.027\,35\,(\pm 0.000\,04)\ \text{L}} = \frac{0.018\,255\,(\pm 0.093\%)\ \text{mol}}{0.027\,35\,(\pm 0.146\%)\ \text{L}}$$

$$= 0.66746\,(\pm 0.173\%) = 0.667\,46\,(\pm 0.001\,155) = 0.667 \pm 0.001\ \text{M}$$

(c)   We can account for the uncertainty in purity of Na_2CO_3 by introducing an additional uncertainty of $\pm 0.05\%$ into the mol of Na_2CO_3:

mol $H^+ = 2 \times$ mol $Na_2CO_3\,(1.000\,0 \pm 0.05\%)$

$$\text{molarity of HCl} = \frac{0.018\,255\,(\pm 0.093\%)\ \text{mol}\,(1.000\,0 \pm 0.05\%)}{0.027\,35\,(\pm 0.146\%)\ \text{L}}$$

relative error $= \sqrt{(0.093\%)^2 + (0.05\%)^2 + (0.146\%)^2} = 0.180\%$

$$= 0.66746\,(\pm 0.180\%) = 0.667\,46\,(\pm 0.001\,20) = 0.667 \pm 0.001\ \text{M}$$

3-21.

9C:	$9(12.010\,6 \pm 0.001\,0)$	=	$108.095\,4 \pm 0.009\,0$
9H:	$9(1.007\,98 \pm 0.000\,14)$	=	$9.071\,82 \pm 0.001\,26$
6O:	$6(15.999\,4 \pm 0.000\,4)$	=	$95.996\,4 \pm 0.002\,4$
3N:	$3(14.006\,8 \pm 0.000\,4)$	=	$42.020\,4 \pm 0.001\,2$

$C_9H_9O_6N_3$: $\qquad\qquad\qquad\qquad 255.184\,02 \pm\ ?$

Uncertainty $= \sqrt{0.009\,0^2 + 0.001\,26^2 + 0.002\,4^2 + 0.001\,2^2} = 0.009\,476$

Answer: 255.184 ± 0.009

3-22. (a) Glassware tolerance:

1-mL pipet: 1.000 ± 0.006 mL $\qquad$ 10-mL pipet: 10.00 ± 0.02 mL

100-mL volumetric flask: 100.00 ± 0.08 mL

We write the dilution on one line to simplify the propagation of uncertainty:

$$\frac{\text{Final}}{\text{concentration}} = \frac{\text{initial}}{\text{concentration}} \times \frac{\text{first}}{\text{dilution}} \times \frac{\text{second}}{\text{dilution}}$$

$$= \left(150.0 \pm 0.3 \frac{\mu g}{mL}\right)\left(\frac{10.00 \pm 0.02\ mL}{100.00 \pm 0.08\ mL}\right)\left(\frac{10.00 \pm 0.02\ mL}{100.00 \pm 0.08\ mL}\right)$$

$$\frac{\text{Final}}{\text{concentration}} = \left(150.0 \frac{\mu g}{mL}\right)\left(\frac{1}{10}\right)\left(\frac{1}{10}\right) = 1.500 \frac{\mu g}{mL}$$

$$\frac{\text{Relative}}{\text{uncertainty}} = \sqrt{\left(\frac{0.3}{150}\right)^2 + \left(\frac{0.02}{10}\right)^2 + \left(\frac{0.08}{100}\right)^2 + \left(\frac{0.02}{10}\right)^2 + \left(\frac{0.08}{100}\right)^2} = 0.003\,64$$

Absolute uncertainty $= (0.003\,64)(1.500\ \mu g/mL) = 0.005\,47\ \mu g/mL$

Answer: $1.500_0 \pm 0.005_5\ \mu g/mL$

To compute relative uncertainty, you could have written each fraction in the square root as a percentage (such as $100 \times \frac{0.03}{150}$), and then convert percent uncertainty back to relative uncertainty by dividing by 100 at the end. Instead, we just used relative uncertainty without converting to and from percentage.

(b) $\frac{\text{Final}}{\text{concentration}} = \left(150.0 \pm 0.3 \frac{\mu g}{mL}\right)\left(\frac{1.000 \pm 0.006\ mL}{100.00 \pm 0.08\ mL}\right) = 1.500 \frac{\mu g}{mL}$

$$\frac{\text{Relative}}{\text{uncertainty}} = \sqrt{\left(\frac{0.3}{150}\right)^2 + \left(\frac{0.006}{1}\right)^2 + \left(\frac{0.08}{100}\right)^2} = 0.006\,37$$

Absolute uncertainty $= (0.006\,38)(1.500\ \mu g/mL) = 0.009\,56\ \mu g/mL$

Answer: $1.500_0 \pm 0.009_6\ \mu g/mL$

The second method is less precise because the uncertainty in the 1-mL pipet is 0.6%, which is 3 times larger than any other uncertainty in either step.

3-23.   To find the uncertainty in $c_0{}^3$, we use the function $y = x^a$ in Table 3-1,

where $x = c_0$ and $a = 3$. The uncertainty in $c_0{}^3$ is

$$\%e_y = a\,\%e_x = 3 \times \frac{0.000\ 000\ 33}{5.431\ 020\ 36} \times 100 = 1.823 \times 10^{-5}\%$$

So $c_0{}^3 = (5.431\ 020\ 36 \times 10^{-8}\ \text{cm})^3 = 1.601\ 932\ 796\ 0 \times 10^{-22}\ \text{cm}^3$ with a
relative uncertainty of $1.823 \times 10^{-5}\%$. We retain extra digits for now and round
off at the end of the calculations. (If your calculator cannot hold as many digits as
we need for this arithmetic, you can do the math with a spreadsheet set to display
10 decimal places.)

The value of Avogadro's number is computed as follows:

$$N_A = \frac{m_{Si}}{(\rho c_0{}^3)/8} = \frac{28.085\ 384\ 2\ \text{g/mol}}{(2.329\ 031\ 9\ \text{g/cm}^3 \times 1.601\ 932\ 79_{60} \times 10^{-22}\ \text{cm}^3)/8}$$

$$= 6.022\ 136\ 936\ 1 \times 10^{23}\ \text{mol}^{-1}$$

The relative uncertainty in Avogadro's number is found from the relative
uncertainties in m_{Si}, ρ, and $c_0{}^3$. (There is no uncertainty in the number 8
atoms/unit cell.)

percent uncertainty in $m_{Si} = 100\ (0.000\ 003\ 5/28.085\ 384\ 2) = 1.246 \times 10^{-5}\%$

percent uncertainty in $\rho\rho = 100\ (0.000\ 001\ 8/2.329\ 031\ 9) = 7.729 \times 10^{-5}\%$

percent uncertainty in $c_0{}^3 = 1.823 \times 10^{-5}\%$ (calculated before)

percent uncertainty in $N_A = \sqrt{\%e_{m_{Si}}{}^2 + \%e_r{}^2 + (\%e_{c_0{}^3})^2} =$

$$= \sqrt{(1.246 \times 10^{-5})^2 + (7.729 \times 10^{-5})^2 + (1.823 \times 10^{-5})^2} = 8.038 \times 10^{-5}\%$$

The absolute uncertainty in N_A is $(8.038 \times 10^{-5}\%)(6.022\ 136\ 936\ 1 \times 10^{23})/100$
$= 0.000\ 004\ 841 \times 10^{23}$. Now we round off N_A to the second digit of its
uncertainty to express it in a manner consistent with the other data in this
problem:

$$N_A = 6.022\ 136\ 9\ (\pm 0.000\ 004\ 8) \times 10^{23}\ \text{or}\ 6.022\ 136\ 9\ (48) \times 10^{23}$$

CHAPTER 4
STATISTICS

4-1. The smaller the standard deviation, the greater the precision. There is no necessary relationship between standard deviation and accuracy. The statistics in this chapter pertain to precision, not accuracy.

4-2. (a) $\mu \pm \sigma$ corresponds to $z = -1$ to $z = +1$. The area from $z = 0$ to $z = +1$ is 0.341 3. The area from $z = 0$ to $z = -1$ is also 0.341 3.

Total area (= fraction of population) from $z = -1$ to $z = +1 = 0.682\ 6$.

(b) $z = -2$ to $z = +2 \Rightarrow$ area $= 2 \times 0.477\ 3 = 0.954\ 6$

(c) $z = 0$ to $z = +1 \Rightarrow$ area $= 0.341\ 3$

(d) $z = 0$ to $z = 0.5 \Rightarrow$ area $= 0.191\ 5$

(e) Area from $z = -1$ to $z = 0$ is 0.341 3. Area from $z = -0.5$ to $z = 0$ is 0.191 5. Area from $z = -1$ to $z = -0.5$ is $0.341\ 3 - 0.191\ 5 = 0.149\ 8$.

4-3. (a) Mean $= \frac{1}{8}(1.526\ 60 + 1.529\ 74 + 1.525\ 92 + 1.527\ 31 + 1.528\ 94 +$

$1.528\ 04 + 1.526\ 85 + 1.527\ 93) = 1.527\ 67$

(b) Standard deviation =

$$\sqrt{\frac{(1.526\ 60 - 1.527\ 67)^2 + \cdots + (1.527\ 93 - 1.527\ 67)^2}{8 - 1}} = 0.001\ 26$$

(c) Variance $= (0.001\ 26)^2 = 1.59 \times 10^{-6}$

(d) Standard deviation of the mean $= (0.001\ 26)/\sqrt{8} = 0.000\ 45$

(e) Significant figures: $\bar{x} \pm s = 1.527_7 \pm 0.001_3$ or 1.528 ± 0.001.

4-4. (a) Mean of 16 means at left side of table $= 0.890\ 2_0$ g.
Mean of 16 means at right side of table $= 0.896\ 4_9$ g.

(b) Standard deviation of 16 means at left side of table $= 0.027\ 8_5$ g.
Standard deviation of 16 means at right side of table $= 0.011\ 9_5$ g.

(c) The standard deviation of the mean for sets of 4 candies is theoretically $\sigma_4 = \sigma/\sqrt{4}$, where σ is the population standard deviation for all candies. The standard deviation of the mean for sets of 16 candies is theoretically $\sigma_{16} = \sigma/\sqrt{16}$. The quotient σ_{16}/σ_4 is theoretically $(\sigma/\sqrt{16})/(\sigma/\sqrt{4}) = \sqrt{4}/\sqrt{16} = 0.5$. The observed quotient of standard deviations in (b) is $(0.011\ 9_5$ g$)/(0.027\ 8_5)$ $= 0.429$. If we measured many more sets of 4 and sets of 16 candies, we expect the quotient to approach 0.5.

4-5.   (a) 1005.3 hours corresponds to $z = (1005.3 - 845.2)/94.2 = 1.700$.

In Table 4-1, the area from the mean to $z = 1.700$ is 0.455 4. The area above $z = 1.700$ is therefore $0.5 - 0.455\ 4 = 0.044\ 6$.

(b) 798.1 corresponds to $z = (798.1 - 845.2)/94.2 = -0.500$.

The area from the mean to $z = -0.500$ is the same as the area from the mean to $z = +0.500$, which is 0.191 5 in Table 4-1.

901.7 corresponds to $z = (901.7 - 845.2)/94.2 = 0.600$.

The area from the mean to $z = 0.600$ is 0.225 8 in Table 4-1.

The area between 798.1 and 901.7 is the sum of the two areas:

$0.191\ 5 + 0.225\ 8 = 0.417\ 3$

(c) The following spreadsheet shows that the area from $-\infty$ to 800 h is 0.315 7 and the area from $-\infty$ to 900 h is 0.719 6. Therefore, the area from 800 to 900 h is $0.719\ 6 - 0.315\ 7 = 0.404\ 0$.

	A	B	C
1	Mean =	Std dev =	
2	845.2	94.2	
3			
4	Area from -∞ to 800 =		0.3157
5	Area from -∞ to 900 =		0.7196
6	Area from 800 to 900		0.4040
7			
8	C4 = NORMDIST(800,$A$2,$B$2,TRUE)		
9	C5 = NORMDIST(900,$A$2,$B$2,TRUE)		
10	C6 = C5-C4		

4-6.   (a) Half the people with tumors have $K < 0.92$ and would not be identified by the test. The false negative rate is 50%.

(b) The false positive rate is the fraction of healthy people with $K \geq 0.92$. To use Table 4-1, we need to convert $x = 0.92$ to a z value defined as

$$z = \frac{x - \mu}{s} = \frac{0.92 - 0.75}{0.07} = 2.43$$

In Table 4-1, area from mean ($z = 0$) to $z = 2.4$ is 0.491 8. Area from mean to $z = 2.5$ is 0.493 8. We estimate that area from mean to $z = 2.43$ is a little greater than 0.492. Area above $z = 2.43$ is therefore $0.5 - 0.492 = 0.008$. That is, 0.8% of healthy people will have a false positive indication of cancer.

In the following spreadsheet, cell E5 computes the area below $K = 0.92$ with the formula NORMDIST(0.92, $B$4,$B$5,True), where B4 contains K and

B5 contains the standard deviation. The area below 0.92 is found in cell E5 to be 0.992 4. The area above $K = 0.92$ is therefore $1 - 0.002 4 = 0.007 6$.

	A	B	C	D	E	F	G	H
1	Gaussian distribution for phase partitioning of plasma proteins							
2								
3	Healthy patients			For healthy people,			Area below cutoff	
4	Mean K =	0.75		area below 0.92 =			for people with tumors	
5	s =	0.07			0.992421		Cutoff (K)	Area
6	Malignant tumor			area above 0.92 =			0.8	0.137656
7	Mean K =	0.92			0.007579		0.81	0.158655
8	s =	0.11					0.82	0.181651
9				area below 0.845 =			0.83	0.206627
10					0.912632		0.84	0.233529
11				area above 0.845 =			0.85	0.26227
12					0.087368		0.845	0.247677
13								
14	E5 = NORMDIST(0.92,$B$4,$B$5,TRUE)				H6 = NORMDIST(G6,$B$7,$B$8,TRUE)			
15	E7 = 1 - E5							

(c) In column G, we vary the value of K and compute the area above K under the curve for people with malignant tumors in column H. We search for the value of K that gives an area of 0.25, which means that 25% of people with tumors will not be identified. The value 0.84 gives an area of 0.233 5 and the value 0.85 gives an area of 0.262 3. By trial and error, we find that $K = 0.845$ gives an area near 0.25.

In cell E10, we insert $K = 0.845$ into the NORMDIST function for healthy people and find that the area below $K = 0.845$ is 0.912 6. The area above $K = 0.845$ is $1 - 0.912 6 = 0.087 4$. That is, 8.7% of healthy people will produce a false positive result, indicating the presence of a tumor.

4-7.

	A	B	C	D	E	F	G	H
1	Gaussian curve for light bulbs							
2								
3	mean =	x (hours)	y (bulbs)	Formula for cell C2 =				
4	845.2	500	0.49	($A$8*$A$10/($A$6*$A$12))*				
5	std dev =	525	1.25	EXP(-((B4-$A$4)^2)/(2*$A$6^2))				
6	94.2	550	2.98					
7	total bulbs =	600	13.64					
8	4768	625	26.28					
9	bulbs per bar	650	47.19					
10	20	675	78.95					
11	sqrt(2 pi) =	700	123.11					
12	2.506628	725	178.92					
13		750	242.35					
14		775	305.94					
15		800	359.94					
16		825	394.68					
17		845.2	403.85					
18		850	403.33					
19		875	384.14					
20		900	340.99					
21		925	282.09					
22		950	217.50					
23		975	156.29					
24		1000	104.67					
25		1025	65.33					
26		1050	38.00					
27		1075	20.60					
28		1100	10.41					
29		1125	4.90					
30		1150	2.15					
31		1175	0.88					
32		1200	0.34					

4-8. Use the same spreadsheet as in the previous problem, but vary the standard deviation. Here are the results:

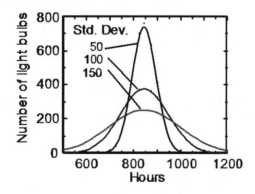

4-9. A confidence interval is a region around the measured mean in which the true mean is likely to lie: If we were to repeat a set of n measurements many times and compute the mean and standard deviation for each set, the 95% confidence interval would include the true population mean (whose value we do not know) in 95% of the sets of n measurements.

4-10. Bars are drawn at a 50% confidence level, so 50% of them ought to include the mean value if many experiments are performed. 90% of the 90% confidence bars should reach the mean value if we do enough experiments. The 90% bars must be longer than the 50% bars because more of the 90% bars must reach the mean.

4-11. Case 1: Comparing a measured result to a "known" value. See if the known value is included within the 95% confidence interval computed as in Equation 4-8.

Case 2: Comparing replicate measurements. Use the F test (Equation 4-6) to decide if the two standard deviations are significantly different.

Case 2a: If the two standard deviations are not significantly different, find the pooled standard deviation with Equation 4-10a and compute t with Equation 4-9a.

Case 2b: If the two standard deviations are significantly different, find the degrees of freedom with Equation 4-10b and compute t with Equation 4-10a.

Case 3: Comparing individual differences with Equations 4-11 and 4-12.

4-12. $\bar{x} = 0.14_8$, $s = 0.03_4$

$$90\% \text{ confidence interval} = 0.14_8 \pm \frac{(2.015)(0.03_4)}{\sqrt{6}} = 0.14_8 \pm 0.02_8$$

$$99\% \text{ confidence interval} = 0.14_8 \pm \frac{(4.032)(0.03_4)}{\sqrt{6}} = 0.14_8 \pm 0.05_6$$

4-13. $99\% \text{ confidence interval} = \bar{x} \pm \frac{(3.707)(0.000\,07)}{\sqrt{7}} = \bar{x} \pm 0.000\,10$

$(1.527\,83 \text{ to } 1.528\,03)$

4-14.   (a)   dL = deciliter = 0.1 L = 100 mL

(b) $F_{\text{calculated}} = (0.05_3/0.04_2)^2 = 1.5_9 < F_{\text{table}} = 6.26$ (for 5 degrees of freedom in the numerator and 4 degrees of freedom in the denominator). Since $F_{\text{calculated}} < F_{\text{table}}$, we can use the following equations:

$$s_{\text{pooled}} = \sqrt{\frac{0.53^2(5) + 0.42^2(4)}{6 + 5 - 2}} = 0.48_4$$

$$t = \frac{|14.5_7 - 13.9_5|}{0.48_4} \sqrt{\frac{6 \cdot 5}{6 + 5}} = 2.12 < 2.262 \text{ (listed for 95\% confidence and}$$

9 degrees of freedom). The results agree and the trainee should be released.

4-15.

	A	B	C	D	E	F
1	Comparison of two methods					
2						
3	Sample	Method 1	Method 2	d_i		
4	A	0.88	0.83	0.05	= B4-C4	
5	B	1.15	1.04	0.11		
6	C	1.22	1.39	-0.17		
7	D	0.93	0.91	0.02		
8	E	1.17	1.08	0.09		
9	F	1.51	1.31	0.20		
10			mean =	0.050	= AVERAGE(D4:D9)	
11			stdev =	0.124	= STDEV(D4:D9)	
12			$t_{\text{calculated}}$ =	0.987	= D10/D11*SQRT(6)	
13			t_{table} =	2.571	= TINV(0.05,5)	

$t_{\text{calculated}} = 0.987 < 2.571$ (Student's t for 95% confidence and 5 deg of freedom) The difference is <u>not</u> significant.

4-16.   In the following spreadsheet, we find $t_{\text{calculated}}$ (which is labeled t Stat in cell F10) is less than t_{table} (t Critical two-tail in cell F14). Therefore, the difference between the methods is <u>not</u> significant.

The probability P(T<=t) two-tail in cell F13 is 0.37. There is a 37% chance of finding the observed difference between equivalent methods by random variations in results. The probability would have to be ≤0.05 for us to conclude that the methods differ.

	A	B	C	D	E	F	G
1	Paired t test				t-Test: Paired Two Sample for Means		
2							
3	Sample	Method 1	Method 2			Variable 1	Variable 2
4	A	0.88	0.83		Mean	1.14333333	1.09333333
5	B	1.15	1.04		Variance	0.05118667	0.04818667
6	C	1.22	1.39		Observations	6	6
7	D	0.93	0.91		Pearson Correlation	0.84541418	
8	E	1.17	1.08		Hypothesized Mean Difference	0	
9	F	1.51	1.31		df	5	
10					t Stat	0.98692754	
11	Calculated t Statistic in cell F10 is				P(T<=t) one-tail	0.18449929	
12	less than critical t in cell F14.				t Critical one-tail	2.01504918	
13	Therfore, the difference between the				P(T<=t) two-tail	0.36899857	
14	methods is *not significant*.				t Critical two-tail	2.57057764	

4-17. $F_{calculated} = s_2^2 / s_1^2 = (0.039)^2/(0.025)^2 = 2.43$

$F_{table} = 9.28$ for 3 degrees of freedom in the numerator and denominator

Since $F_{calculated} < F_{table}$, the difference in standard deviation is not significant.

$$s_{pooled} = \sqrt{\frac{s_1^2\,(n_1 - 1) + s_2^2\,(n_2 - 1)}{n_1 + n_2 - 2}} = \sqrt{\frac{0.025^2\,(4-1) + 0.039^2\,(4-1)}{4 + 4 - 2}} = 0.032\,8$$

$$t_{calculated} = \frac{|\bar{x}_1 - \bar{x}_2|}{s_{pooled}}\sqrt{\frac{n_1 n_2}{n_1 + n_2}} = \frac{|1.382 - 1.346|}{0.032\,8}\sqrt{\frac{4 \cdot 4}{4 + 4}} = 1.55$$

t_{table} (4 + 4 − 2 = 6 degrees of freedom) = 2.447

Since $t_{calculated} < t_{table}$, the difference is <u>not</u> significant.

4-18. For Method 1, $\bar{x}_1 = 0.082\,605_2$, $s_1 = 0.000\,013_4$.

For Method 2, $\bar{x}_2 = 0.082\,00_5$, $s_2 = 0.000\,12_9$.

The two standard deviations differ by approximately a factor of 10. We should use the F test to compare the two standard deviations:

$F_{calculated} = s_2^2 / s_1^2 = (0.000\,12_9)^2/(0.000\,013_4)^2 = 92.7$

$F_{table} = 6.26$. Since $F_{calculated} > F_{table}$, we use Equations for Case 2b in comparison of means.

The following spreadsheet shows that $t_{calculated} = 11.3$ and $t_{table} = 2.57$.

$t_{calculated} > t_{table}$, so the difference <u>is</u> significant at the 95% confidence level.

Paired t test		t-Test: Two-Sample Assuming Unequal Variances		
Method 1	Method 2		Variable 1	Variable 2
0.082601	0.08183	Mean	0.082605	0.082005
0.082621	0.08186	Variance	1.8E-10	1.67E-08
0.082589	0.08205	Observations	5	6
0.082617	0.08206	Hypothesized Mean Difference	0	
0.082598	0.08215	df	5	
	0.08208	t Stat	11.31371	$\longleftarrow t_{calculated} = 11.3$
		P(T<=t) one-tail	4.72E-05	
		t Critical one-tail	2.015049	
		P(T<=t) two-tail	9.43E-05	
		t Critical two-tail	2.570578	$\longleftarrow t_{table} = 2.57$

4-19.   90% confidence interval $= \bar{x} \pm \dfrac{(2.353)(1\%)}{\sqrt{4}} = \bar{x} \pm 1.1_8\% < 1.2\%$.

The answer is yes.

4-20.   For indicators 1 and 2: $F_{calculated} = (0.002\ 25/0.000\ 98)^2 = 5.2_7 > F_{table} \approx 2.2$ (for 27 degrees of freedom in the numerator and 17 degrees of freedom in the denominator). Since $F_{calculated} > F_{table}$, we use the following equations:

$$\text{Degrees of freedom} = \frac{(s_1^2/n_1 + s_2^2/n_2)^2}{\dfrac{(s_1^2/n_1)^2}{n_1 - 1} + \dfrac{(s_2^2/n_2)^2}{n_2 - 1}}$$

$$= \frac{(0.002\ 25^2/28 + 0.000\ 98^2/18)^2}{\dfrac{(0.002\ 25^2/28)^2}{28 - 1} + \dfrac{(0.000\ 98^2/18)^2}{18 - 1}} = 39.8 \text{ (round to 40)}$$

$$t_{calculated} = \frac{|\bar{x}_1 - \bar{x}_2|}{\sqrt{s_1^2/n_1 + s_2^2/n_2}} = \frac{|0.095\ 65 - 0.086\ 86|}{\sqrt{0.002\ 25^2/28 + 0.000\ 98^2/18}} = 18.2$$

$t_{calculated}$ is much greater than t_{table} for ~40 degrees of freedom and 95% confidence, which is 2.02.   The difference <u>is</u> significant.

For indicators 2 and 3: $F_{calculated} = (0.001\ 13/0.000\ 98)^2 = 1.3_3 < F_{table} \approx 2.2$ (for 28 degrees of freedom in the numerator and 17 degrees of freedom in the denominator). Since $F_{calculated} < F_{table}$, we use the following equations:

$$s_{pooled} = \sqrt{\frac{s_1^2(n_1 - 1) + s_2^2(n_2 - 1)}{n_1 + n_2 - 2}} = 0.001\ 075\ 8$$

$$t_{calculated} = \frac{|\bar{x}_1 - \bar{x}_2|}{s_{pooled}} \sqrt{\frac{n_1 n_2}{n_1 + n_2}} = 1.39 < 2.02 \text{ (~40 degrees of freedom)}$$

$\Rightarrow$ difference is <u>not</u> significant.

4-21. The standard deviations are almost identical, so we use the equations

$$s_{pooled} = \sqrt{\frac{30.0^2(31) + 29.8^2(31)}{32 + 32 - 2}} = 29.9$$

$$t = \frac{52.9 - 31.4}{29.9}\sqrt{\frac{32 \cdot 32}{32 + 32}} = 2.88.$$ The table gives t for 60 degrees of freedom,

which is close to 62. The difference <u>is</u> significant at the 95 and 99% levels.

4-22. $\bar{x} = 97.0_0, \quad s = 1.6_6$

95% confidence interval $= \bar{x} \pm \dfrac{ts}{\sqrt{n}} = 97.0_0 \pm \dfrac{(2.776)(1.6_6)}{\sqrt{5}} = 97.0_0 \pm 2.0_6$

Range $= 94.9_4$ to 99.0_6

The 95% confidence interval does not include the certified value of 94.6 ppm, so the difference <u>is</u> significant at the 95% confidence level.

If we make one more measurement, the results are $\bar{x} = 96.5_8, s = 1.8_0$

95% confidence interval $= 96.5_8 \pm \dfrac{(2.571)(1.8_0)}{\sqrt{6}} = 96.5_8 \pm 1.8_9$

Range $= 94.6_9$ to 98.4_7

The 95% confidence interval still does not include the certified value of 94.6 ppm, so the difference <u>is</u> still significant at the 95% confidence level.

4-23. (a) <u>Rainwater:</u>

$F_{calculated} = (0.008/0.005)^2 = 2._{56} < F_{table} = 4.53$ (for 4 degrees of freedom in the numerator and 6 degrees of freedom in the denominator). Since $F_{calculated} < F_{table}$, we use the following equations:

$$s_{pooled} = \sqrt{\frac{0.005^2(6) + 0.008^2(4)}{7 + 5 - 2}} = 0.006_{37}$$

$$t_{calculated} = \frac{0.069 - 0.063}{0.006_{37}}\sqrt{\frac{7 \cdot 5}{7 + 5}} = 1._{61} < t_{table} = 2.228$$

The difference is <u>not</u> significant.

<u>Drinking water:</u>

$F_{calculated} = (0.008/0.007)^2 = 1._{31} < F_{table} = 6.39$ (for 4 degrees of freedom in the numerator and 4 degrees of freedom in the denominator).

Since $F_{calculated} < F_{table}$, we use the following equations:

$$s_{pooled} = \sqrt{\frac{0.007^2(4) + 0.008^2(4)}{5 + 5 - 2}} = 0.007_{52}$$

$$t = \frac{0.087 - 0.078}{0.007_{52}} \sqrt{\frac{5 \cdot 5}{5 + 5}} = 1._{89} < 2.306. \text{ The difference is } \underline{\text{not}} \text{ significant.}$$

(b) <u>Gas chromatography</u>:

$F_{\text{calculated}} = (0.007/0.005)^2 = 1._{96} < F_{\text{table}} = 4.53$ (for 4 degrees of freedom in the numerator and 6 degrees of freedom in the denominator). Since $F_{\text{calculated}} < F_{\text{table}}$, we use the following equations:

$$s_{\text{pooled}} = \sqrt{\frac{0.005^2(6) + 0.007^2(4)}{7 + 5 - 2}} = 0.005_{88}$$

$$t = \frac{0.078 - 0.069}{0.005_{88}} \sqrt{\frac{7 \cdot 5}{7 + 5}} = 2._{61} > 2.228. \text{ The difference } \underline{\text{is}} \text{ significant.}$$

<u>Spectrophotometry</u>: (the two standard deviations are equal)

$$s_{\text{pooled}} = \sqrt{\frac{0.008^2(4) + 0.008^2(4)}{5 + 5 - 2}} = 0.008_{00}$$

$$t = \frac{0.087 - 0.063}{0.008_{00}} \sqrt{\frac{5 \cdot 5}{5 + 5}} = 4._{74} > 2.306. \text{ The difference } \underline{\text{is}} \text{ significant.}$$

4-24. $\bar{x} = 201.8;\ s = 9.34$

$G_{\text{calculated}} = |216 - 201.8| / 9.34 = 1.52$

$G_{\text{table}} = 1.672$ for five measurements

Because $G_{\text{calculated}} < G_{\text{table}}$, we should retain 216.

4-25. Statement (i) is true. The null hypothesis for the F test is that the two sets of data are drawn from populations with the same population standard deviation ($\sigma_1 = \sigma_2$). There must be strong evidence to *reject* the null hypothesis, or we do not reject it. We retain the null hypothesis if there is more than a 5% chance that $\sigma_1 = \sigma_2$. The statement that "there is at least a 95% probability that the two sets of data are drawn from populations with $\sigma_1 = \sigma_2$" is not being tested and is a much stronger statement than the one that is being tested.

If the null hypothesis were that "the data comes from populations with different population standard deviations ($\sigma_1 > \sigma_2$)," then we would reject this null hypothesis if there were less than a 5% chance that $\sigma_1 > \sigma_2$.

4-26. Slope $= -1.298\,72 \times 10^4\ (\pm 0.001\,319\,0 \times 10^4)$

$\qquad\qquad = -1.299\ (\pm 0.001) \times 10^4 \text{ or } -1.298_7\ (\pm 0.001_3) \times 10^4$

Intercept $= 256.695\ (\pm 323.57) = 3\ (\pm 3) \times 10^2$

4-27.

x_i	y_i	x_iy_i	x_i^2	d_i	d_i^2
0	1	0	0	0.071 43	0.005 10
2	2	4	4	−0.214 29	0.045 92
3	3	9	9	0.142 86	0.020 41
sums: 5	6	13	13	0	0.071 43

$$m = \frac{n\,\Sigma\,(x_iy_i) - \Sigma\,x_i\,\Sigma\,y_i}{n\,\Sigma\,(x_i^2) - (\Sigma\,x_i)^2} = \frac{3 \times 13 - 5 \times 6}{3 \times 13 - 5^2} = \frac{9}{14} = 0.642\,86$$

$$b = \frac{\Sigma\,(x_i^2)\,\Sigma\,y_i - \Sigma\,(x_iy_i)\,\Sigma\,x_i}{n\,\Sigma\,(x_i^2) - (\Sigma\,x_i)^2} = \frac{13 \times 6 - 13 \times 5}{3 \times 13 - 5^2} = \frac{13}{14} = 0.928\,57$$

$$s_y = \sqrt{\frac{\Sigma\,(d_i^2)}{n-2}} = \sqrt{\frac{0.071\,43}{3-2}} = 0.267\,26$$

$$u_m = s_y\sqrt{\frac{n}{D}} = (0.267\,26)\sqrt{\frac{3}{14}} = 0.123\,71$$

$$u_b = s_y\sqrt{\frac{\Sigma\,(x_i^2)}{D}} = (0.267\,26)\sqrt{\frac{13}{14}}$$

$$= 0.257\,54$$

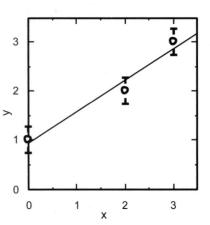

slope $= 0.6_4 \pm 0.1_2$ intercept $= 0.9_3 \pm 0.2_6$

4-28.

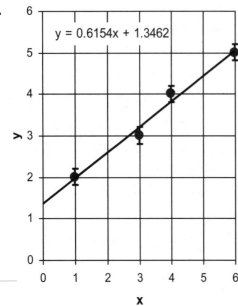

$y = 0.6154x + 1.3462$

4-29.

	A	B	C	D	E	F	G	H	I
1	x	y							
2	3.0	-0.074							
3	10.0	-1.411							
4	20.0	-2.584							
5	30.0	-3.750							
6	40.0	-5.407							
7									
8		LINEST output							
9	m	-0.13789	0.195343	b					
10	u_m	0.006635	0.162763	u_b					
11	R^2	0.993102	0.197625	s_y					
12									
13	Highlight cells B9:C11								
14	Type								
15	"=LINEST(B2:B6,A2:A6,TRUE,TRUE)"								
16	Press CTRL+SHIFT+ENTER (on PC)								
17	Press APPLE(⌘)+ENTER (on Mac)								

4-30. We must measure how an analytical procedure responds to a known quantity of analyte (or a known quantity of a related compound) before the procedure can be used for an unknown. Therefore, we must be able to measure out the analyte (or a related compound) in pure form to use as a calibration standard.

4-31. Hopefully, the negative value is within experimental error of 0. If so, no detectable analyte is present. If the negative concentration is beyond experimental error, there is something wrong with your analysis. The same is true for a value above 100% of the theoretical maximum concentration of an analyte. Another possible way to get values below 0 or above 100% is if you extrapolated the calibration curve past the range covered by standards, and the curve is not linear.

4-32. For 8 degrees of freedom, $t_{90\%} = 1.860$ and $t_{99\%} = 3.355$.

90% confidence interval: $15.2_2 (\pm 1.860 \times 0.4_6) = 15.2_2 \pm 0.8_6$ μg

99% confidence interval: $15.2_2 (\pm 3.355 \times 0.4_6) = 15._2 \pm 1._5$ μg

4-33. (a) $x = \dfrac{y-b}{m} = \dfrac{2.58 - 1.3_5}{0.61_5} = 2.00$

$\bar{y} = (2+3+4+5)/4 = 3.5 \qquad \bar{x} = (1+3+4+6)/4 = 3.5$

$\Sigma(x_i - \bar{x})^2 = (1-3.5)^2 + (3-3.5)^2 + (4-3.5)^2 + (6-3.5)^2 = 13.0$

$$u_x = \frac{s_y}{|m|} \sqrt{\frac{1}{k} + \frac{1}{n} + \frac{(y-\bar{y})^2}{m^2 \, \Sigma(x_i-\bar{x})^2}}$$

$$= \frac{0.196\,12}{|0.615\,38|} \sqrt{\frac{1}{1} + \frac{1}{4} + \frac{(2.58-3.5)^2}{(0.615\,38)^2\,(13.0)}} = 0.38$$

Answer: $2.0_0 \pm 0.3_8$

(b) For $k = 4$ replicate measurements,

$$u_x = \frac{s_y}{|m|} \sqrt{\frac{1}{k} + \frac{1}{n} + \frac{(y-\bar{y})^2}{m^2 \, \Sigma(x_i-\bar{x})^2}}$$

$$= \frac{0.196\,12}{|0.615\,38|} \sqrt{\frac{1}{4} + \frac{1}{4} + \frac{(2.58-3.5)^2}{(0.615\,38)^2\,(13.0)}} = 0.26$$

Answer: $2.0_0 \pm 0.2_6$

(c) There are only 4 calibration points, so there are $4 - 2 = 2$ degrees of freedom. Student's t for 95% confidence and 2 degrees of freedom is 4.303. For (a), 95% confidence interval $= \pm(4.303)(0.38) = \pm 1.6$. For (b), 95% confidence interval $= \pm(4.303)(0.26) = \pm 1.1$. You really need more than 4 points in a calibration curve to cut down the breadth of the confidence interval.

4-34. (a) Corrected absorbance $= 0.264 - 0.095 = 0.169$

Equation of line: $0.169 = 0.016\,30\,x + 0.004\,7 \Rightarrow x = 10.1\ \mu g$

(b) In the spreadsheet below, $x = 10.082\ \mu g$ in cell B30 and $u_x = 0.204\,5\ \mu g$ in cell B31. For $14 - 2 = 12$ degrees of freedom and 95% confidence, Student's $t = 2.179$ in cell B32. The 95% confidence interval is $\pm 0.4455\ \mu g$ in cell B33. A reasonable answer is $10.08 \pm 0.45\ \mu g$.

	A	B	C	D	E	F	G	H	I
1	Least-Squares Spreadsheet								
2									
3	Highlight cells B10:C12	x	y						
4	Type "= LINEST(C4:C7,	0	-0.0003						
5	B4:B7,TRUE,TRUE)	0	-0.0003						
6	For PC, press	0	0.0007						
7	CTRL+SHIFT+ENTER	5	0.0857						
8	For Mac, press	5	0.0877						
9	APPLE(⌘) + ENTER	5	0.0887						
10		10	0.1827						
11		10	0.1727						
12		10	0.1727						
13		15	0.2457						
14		15	0.2477						
15		20	0.3257						
16		20	0.3257						
17		20	0.3307						
18									
19		LINEST output:							
20	m	0.0163	0.0047	b					
21	u_m	0.0002	0.0026	u_b					
22	R^2	0.9978	0.0059	s_y					
23									
24	n =	14	B24 = COUNT(B4:B17)						
25	Mean y =	0.16184	B25 = AVERAGE(C4:C17)						
26	$\Sigma(x_i$ - mean x$)^2$ =	723.214	B26 = DEVSQ(B4:B17)						
27									
28	Measured y =	0.169	Input						
29	k = Number of replicate measurements of y =	4	Input						
30	Derived x =	10.082	B30 = (B28-C20)/B20						
31	u_x =	0.2045	B31 = (C22/ABS(B20))*SQRT((1/B29)+(1/B24)+((B28-B25)^2)/(B20^2*B26))						
32	t =	2.17881	B32 = TINV(0.05,12) (Student's t for 12 degrees freedom, 95% confidence)						
33	Confidence interval = ±	0.44553	B33 = B32*B31 = t*u_x						

The chart (columns E–I, rows 3–22) shows the least-squares line with equation $y = 0.0163x + 0.0047$ and $R^2 = 0.9978$, plotting y versus x over the range 0 to 20.

4-35. (a)

	A	B	C	D	E	F	G	H	I
1	Least-Squares Spreadsheet								
2									
3		x	$y_{corrected}$	y					
4		0	0	9.1					
5		0.062	38.4	47.5					
6		0.122	86.5	95.6					
7		0.245	184.7	193.8					
8		0.486	378.4	387.5					
9		0.971	803.4	812.5					
10		1.921	1662.8	1671.9					
11									
12		LINEST output:							
13	m	869.13	-22.0852	b					
14	u_m	10.6422	8.9474	u_b					
15	R^2	0.9993	18.0527	s_y					
16									
17	n =	7	B17 = COUNT(B4:B10)						
18	Mean y =	450.6	B18 = AVERAGE(C4:C10)						
19	$\Sigma(x_i - \text{mean } x)^2$ =	2.87757	B19 = DEVSQ(B4:B10)						
20									
21	Measured y =	145.0	Input						
22	Number of replicate measurements of y (k) =	4	Input						
23	Derived x	0.19224	B23 = (B21-C13)/B13						
24	u_x =	0.0137	B24 = (C15/B13)*SQRT((1/B22)+(1/B17)+((B21-B18)^2)/(B13^2*B19))						
25	t =	2.57058	B25 = TINV(0.05,5) (Student's t for 5 degrees of freedom, 95% confidence)						
26	Conf. interval =	0.03525	B26 = B25*B24 = tu_x						

Chart: $y = 869.13x - 22.085$; Signal (mV) vs Methane (vol%)

(b)   Corrected signal = $154.0 - 9.0 = 145.0$ mV

(c)   Cell B23 gives $[CH_4] = 0.0192_2$ vol%

Cell B24 gives $u_x = 0.013_7$ vol%

Cell B26 gives confidence interval = $t^* \, u_x = 0.035_2$ vol%

Rounded answers: $x \pm u_x = 0.19_2 \ (\pm 0.01_4)$ vol%

95% confidence interval: $0.19_2 \ (\pm 0.03_5)$ vol%

4-36. $0.350 = -1.17 \times 10^{-4} x^2 + 0.018\,58\,x - 0.000\,7$

$1.17 \times 10^{-4} x^2 - 0.018\,58\,x + 0.350\,7 = 0$

$$x = \frac{+0.018\,58 \pm \sqrt{0.018\,58^2 - 4\,(1.17 \times 10^{-4})\,(0.350\,7)}}{2\,(1.17 \times 10^{-4})} = 137 \text{ or } 21.9 \text{ μg}$$

Correct answer is 21.9 μg.

4-37. (a) The logarithmic graph spreads out the data and is linear over the entire range.

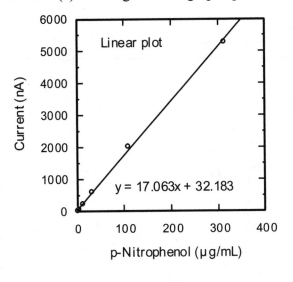

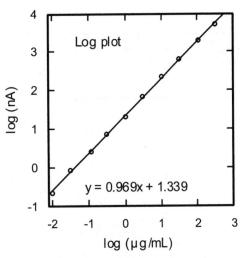

(b) log (current, nA) = 0.969 2 log (concentration, µg/mL) + 1.338 9

(c) log (99.9) = 0.969 2 log [X] + 1.338 9

$\Rightarrow$ log [X] = 0.681 6 $\Rightarrow$ [X] = 4.80 µg/mL

(d) concentration = $y = 10^{0.681\ 65\ \pm\ 0.044\ 04}$

The function is of the form $y = 10^x$, for which Table 3-1 states

$e_y/y = (\ln 10)e_x = (2.302\ 6)(0.044\ 04) = 0.101\ 41$

Substituting $y = 10^{0.681\ 65} = 4.805$ for y gives

$e_y/4.805 = 0.101\ 41 \Rightarrow e_y = 0.487$

Answer: concentration = 4.80 (±0.49) µg/mL

The uncertainty is an estimate of the standard deviation of the mean. If you need to find a confidence interval, multiply (±0.49) µg/mL by Student's t.

CHAPTER 5
QUALITY ASSURANCE AND CALIBRATION METHODS

5-1. Get the right data: Measure what is relevant to the question at hand.

Get the data right: Sampling and analytical procedures must be satisfactory to measure what we intend to measure.

Keep the data right: Record keeping should document that samples were collected properly and data has demonstrated reliability.

5-2. The three parts of quality assurance are defining use objectives, setting specifications, and assessing results.

Use objectives:

Question: Why do I want the data and results and how will I use them?

Actions: Write use objectives.

Specifications:

Question: How good do the numbers have to be?

Actions: Write specifications and pick an analytical method to meet the specifications. Consider requirements for sampling, precision, accuracy, selectivity, sensitivity, detection limit, robustness, and allowed rate of false results. Plan to employ blanks, fortification, calibration checks, quality control samples, and control charts. Write and follow standard operating procedures.

Assessment:

Question: Did I meet the specifications?

Actions: Compare data and results with specifications. Document procedures and keep records suitable for meeting use objectives. Verify that the use objectives were met.

5-3. *Precision* is demonstrated by the repeatability of analyses of replicate samples and replicate portions of the same sample. Accuracy is demonstrated by analyzing certified reference materials, by comparing results from different analytical methods, by fortification (spike) recovery, by standard additions, by calibration checks, blanks, and quality control samples (blind samples).

5-4. *Raw data* are directly measured quantities, such as peak area in a chromatogram or volume from a buret. *Treated data* are concentrations or amounts found by applying a calibration method to the raw data. *Results*, such as the mean and standard deviation, are what we ultimately report after applying statistics to treated data.

5-5. A calibration check is an analysis of a solution *formulated by the analyst* to contain a known concentration of analyte. It is the analyst's own check that procedures and instruments are functioning correctly. A performance test sample is an analysis of a solution *formulated by someone other than the analyst* to contain a known concentration of analyte. It is a test to see if the analyst gets the right answer when he or she does not know the right answer.

5-6. A blank is a sample intended to contain no analyte. Its purpose is to find the response of a method when no analyte is deliberately present. Positive response to the blank arises from analyte impurities in reagents and equipment and from interference by other species. A method blank is taken through all steps in a chemical analysis. A reagent blank is the same as a method blank, but it has not been subjected to all sample preparation procedures. A field blank is similar to a method blank, but it has been taken into the field and exposed to the same environment as samples collected in the field and transported to the lab.

5-7. Linear range is the analyte concentration interval over which the analytical signal is proportional to analyte concentration. Dynamic range is the concentration range over which there is a useable response to analyte, even if it is not linear. Range is the analyte concentration interval over which an analytical method has specified linearity, accuracy, and precision.

5-8. A false positive is a conclusion that analyte exceeds a certain limit when, in fact, it is below the limit. A false negative is a conclusion that analyte is below a certain limit when, in fact, it is above the limit.

5-9. ~1% of the area under the curve for blanks lies to the right of the detection limit. Therefore, ~1% of samples containing no analyte will give a signal above the detection limit. 50% of the area under the curve for samples containing analyte at the detection limit lies below the detection limit. Therefore, 50% of samples containing analyte at the detection limit will be reported as not containing analyte at a level above the detection limit.

5-10. A control chart tracks the performance of a process to see if it remains within expected bounds. Six indications that a process might be out of control are (1) a reading outside the action lines, (2) 2 out of 3 consecutive readings between the warning and action lines, (3) 7 consecutive measurements all above or all below the center line, (4) six consecutive measurements, all steadily increasing or all steadily decreasing, wherever they are located, (5) 14 consecutive points alternating up and down, regardless of where they are located, and (6) an obvious nonrandom pattern.

5-11. Question (iii) is correct. The purpose of the analysis is to see if concentrations of haloacetates are in compliance with (i.e., do not exceed) levels set by a certain rule. The purpose is not just to achieve a specified precision and accuracy (question i) or to just see if detectable levels of haloacetates are present (question ii).

5-12. The *instrument detection limit* is obtained by replicate measurements of aliquots from one sample. The *method detection limit* is obtained by preparing and analyzing many independent samples. There is more variability in the latter procedure, so the method detection limit should be higher than the instrument detection limit.

Robustness is the ability of an analytical method to be unaffected by small, deliberate changes in operating parameters. *Intermediate precision* is the variation observed when an assay is performed by different people on different instruments on different days in the same lab. Each analysis might incorporate independently prepared reagents and different lots of the same chromatography column from one manufacturer. When demonstrating intermediate precision, experimental conditions are intended to be the same in each analysis. When measuring robustness, conditions are intentionally varied by small amounts.

5-13. *Repeatability* describes the spread in results when one person uses one procedure to analyze the same sample by the same method multiple times. *Reproducibility* describes the spread in results when different people in different labs using different instruments each try to follow the same procedure.

Instrument precision is the reproducibility observed when the same quantity of one sample is repeatedly introduced into an instrument.

Intra-assay precision is evaluated by analyzing aliquots of a homogeneous material several times by one person on one day with the same equipment.

Intermediate precision is the variation observed when an assay is performed by different people on different instruments on different days in the same lab.

Interlaboratory precision is the reproducibility observed when aliquots of the same sample are analyzed by different people in different laboratories at different times using equipment and reagents belonging to each lab.

Interlaboratory precision is the same as reproducibility.

5-14. Criteria:

- Observations outside action lines — no
- 2 out of 3 consecutive measurements between warning and action lines — no
- 7 consecutive measurements all above or all below the center line — YES: observations 2–10 (starting from the left side) are all below the center line and observations 27-33 are all above the center line
- 6 consecutive measurements steadily increasing or steadily decreasing — no
- 14 consecutive points alternating up and down — no
- Obvious nonrandom pattern — no

5-15. LINEST gives m, b, u_m, u_b, R^2, and s_y in cells B19:C21. TRENDLINE produces the same value of m, b, and R^2, which are printed inside the graph. The 95% confidence interval for y is computed in cell C24.

	A	B	C	D	E	F	G	H	I
1	Graphing data with random Gaussian noise								
2	Generating equation: y = 26.4x+1.37								
3	Gaussian std dev =		80						
4									
5	x	y							
6	0	14							
7	10	350							
8	20	566							
9	30	957							
10	40	1067							
11	50	1354							
12	60	1573							
13	70	1732							
14	80	2180							
15	90	2330							
16	100	2508							
17									
18		LINEST output							
19	m	24.80727	89.72727	b					
20	u_m	0.683532	40.43827	u_b					
21	R^2	0.993214	71.6894	s_y					
22									
23	Student's t =		2.262157	=TINV(0.05,9)					
24	t^*s_y		162.1727	=C23*C21					

Graph (cells E3:I20):
y = 24.807x + 89.727
R^2 = 0.9932
Error bars are ± t*s_y in cell C24
(y-axis labeled y, ranging 0 to 2500; x-axis labeled x, ranging 0 to 100)

5-16. (a) For the fortification level of 22.2 ng/mL, the mean of the 5 values is 23.6_6 ng/mL and the standard deviation is 5.6_3 ng/mL.

$$\text{Precision} = 100 \times \frac{5.63}{23.66} = 23.8\%.$$

$$\text{Accuracy} = 100 \times \frac{23.66 - 22.2}{22.2} = 6.6\%$$

For the fortification level of 88.2 ng/mL, the mean of the 5 values is 82.4_8 ng/mL and the standard deviation is 11.4_9 ng/mL.

$$\text{Precision} = 100 \times \frac{11.49}{82.48} = 13.9\%.$$

$$\text{Accuracy} = 100 \times \frac{82.48 - 88.2}{88.2} = -6.5\%$$

For the fortification level of 314 ng/mL, the mean of the 5 values is $302._8$ ng/mL and the standard deviation is 23.5_1 ng/mL.

$$\text{Precision} = 100 \times \frac{23.51}{302.8} = 7.8\%.$$

$$\text{Accuracy} = 100 \times \frac{302.8 - 314}{314} = -3.6\%$$

(b) Standard deviation of 10 samples: $s = 28._2$; mean blank: $y_{\text{blank}} = 45._0$
Signal detection limit $= y_{\text{blank}} + 3s = 45._0 + (3)(28._2) = 129.6$

$$\text{Concentration detection limit} = \frac{3s}{m} = \frac{(3)(28._2)}{1.75 \times 10^9 \text{ M}^{-1}} = 4.8 \times 10^{-8} \text{ M}$$

$$\text{Lower limit of quantitation} = \frac{10s}{m} = \frac{(10)(28._2)}{1.75 \times 10^9 \text{ M}^{-1}} = 1.6 \times 10^{-7} \text{ M}$$

5-17. (a) 1 wt% $\Rightarrow$ $C = 0.01$: $\text{CV}(\%) \approx 2^{(1 - 0.5 \log 0.01)} = 2^2 = 4\%$
If $C = 10^{-12}$, $\text{CV}(\%) \approx 2^7 = 128\%$.

(b) If class CV is 50% of the value given by the Horwitz curve, it would be $0.5 \times 2^{(1 - 0.5 \log 0.1)} = 1.4\%$.

5-18. Mean $= 0.383$ µg/L and standard deviation $= 0.021_4$ µg/L

$$\% \text{ recovery} = \frac{0.383 \text{ µg/L}}{0.40 \text{ µg/L}} \times 100 = 96\%$$

The measurements are already expressed in concentration units. The concentration detection limit is 3 times the standard deviation $= 3(0.021_4 \text{ µg/L}) = 0.064$ µg/L.

5-19. For a concentration of 0.2 µg/L, the relative standard deviation of 14.4% corresponds to $(0.144)(0.2 \text{ µg/L}) = 0.028\,8 \text{ µg/L}$. The detection limit is $3(0.028\,8 \text{ µg/L}) = 0.086 \text{ µg/L}$. Here are the results for the other concentrations:

Concentration (µg/L)	Relative standard deviation (%)	Concentration standard deviation (µg/L)	Detection limit (µg/L)
0.2	14.4	0.028 8	0.086
0.5	6.8	0.034 0	0.102
1.0	3.2	0.032 0	0.096
2.0	1.9	0.038 0	0.114
		mean detection limit:	0.10

5-20. If an athlete tests positive for drugs, the test should be repeated with a second sample that was drawn at the same time as the first sample and preserved in an appropriate manner. If there is a 1% chance of a false positive in each test, the chances of observing a false positive twice in a row are 1% of 1% or 0.01%. Instead of falsely accusing 1% of innocent athletes, we would be falsely accusing 0.01% of innocent athletes.

5-21. *Comparison of Lab C with Lab A:*

First, use the F test to see if the standard deviations are significantly different:
$F_{\text{calculated}} = s_C^2/s_A^2 = 0.78^2/0.14^2 = 31._0 > F_{\text{table}} = 3.88$ (with 2 degrees of freedom for s_C and 12 degrees of freedom for s_A)

Standard deviations are not equivalent, so use the following t test:

$$\text{Degrees of freedom} = \frac{(s_1^2/n_1 + s_2^2/n_2)^2}{\dfrac{(s_1^2/n_1)^2}{n_1 - 1} + \dfrac{(s_2^2/n_2)^2}{n_2 - 1}} = \frac{(0.14^2/13 + 0.78^2/3)^2}{\dfrac{(0.14^2/13)^2}{13 - 1} + \dfrac{(0.78^2/3)^2}{3 - 1}} = 2.03 \approx 2$$

$$t_{\text{calculated}} = \frac{|\bar{x}_1 - \bar{x}_2|}{\sqrt{s_1^2/n_1 + s_2^2/n_2}} = \frac{|1.59 - 2.68|}{\sqrt{0.14^2/13 + 0.78^2/3}} = 2.4_1$$

For 2 degrees of freedom, $t_{\text{table}} = 4.303$ for 95% confidence. Since $t_{\text{calculated}} < t_{\text{table}}$, we conclude that the difference between Lab C and Lab A is <u>not</u> significant.

Comparison of Lab C with Lab B:

$F_{calculated} = s_C^2/s_B^2 = 0.78^2/0.56^2 = 1.9_4 < F_{table} = 4.74$ (with 2 degrees of freedom for s_C and 7 degrees of freedom for s_A). The standard deviations are not significantly different, so we use the following t test:

$$s_{pooled} = \sqrt{\frac{0.56^2\,(8-1) + 0.78^2\,(3-1)}{8+3-2}} = 0.61_6$$

$$t_{calculated} = \frac{|1.65 - 2.68|}{0.61_6}\sqrt{\frac{8 \cdot 3}{8+3}} = 2.4_7$$

$t_{table} = 2.262$ for 95% confidence and $8 + 3 - 2 = 9$ degrees of freedom. $t_{calculated} > t_{table}$, so the difference <u>is</u> significant at the 95% confidence level.

It makes no sense to conclude that Lab C [2.68 ± 0.78 (3)] > Lab B [1.65 ± 0.56 (8)], but Lab C = Lab A [1.59 ± 0.14 (13)]. The problem with the comparison of Labs C and A is that the standard deviation of C is much greater than the standard deviation of A and the number of replicates for C is much less than the number of replicates for A. The result is that we used a large composite standard deviation and a small composite number of degrees of freedom. The conclusion is biased by a large standard deviation and a small number of degrees of freedom. I would tentatively conclude that results from Lab C are greater than results from Labs B and A. I would also ask for more replicate results from Lab C. With just 3 replications, it is hard to reach any statistically significant conclusions.

5-22. A small volume of standard will not change the sample matrix very much, so matrix effects remain nearly constant. If large, variable volumes of standard are used, the matrix is different in every mixture and the matrix effects will be different in every sample.

5-23. (a) $[Cu^{2+}]_f = [Cu^{2+}]_i \dfrac{V_i}{V_f} = 0.950\,[Cu^{2+}]_i$

(b) $[S]_f = [S]_i \dfrac{V_i}{V_f} = (100.0 \text{ ppm})\left(\dfrac{1.00 \text{ mL}}{100.0 \text{ mL}}\right) = 1.00 \text{ ppm}$

(c) $\dfrac{[Cu^{2+}]_i}{1.00 \text{ ppm} + 0.950[Cu^{2+}]_i} = \dfrac{0.262}{0.500} \Rightarrow [Cu^{2+}]_i = 1.04 \text{ ppm}$

5-24. (a) All solutions were made up to the same final volume. Therefore, we prepare a graph of signal versus concentration of added standard. The line in the graph was drawn by the method of least squares with the spreadsheet on the next page. The x-intercept, 8.72 ppb, is the concentration of unknown in the 10-mL solution. In cell B27 of the spreadsheet, the standard uncertainty of the x-intercept is 0.427 ppm. A reasonable answer is $8.7_2 \pm 0.4_3$ ppb.

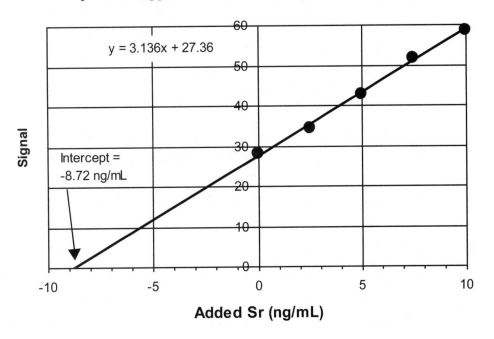

(b) Unknown solution volume = 10.0 mL with Sr = 8.72 ppb = 8.72 ng/mL. In 10.0 mL, there are (10 mL)(8.72 ng/mL) = 87.2 ng. Solution was made from 0.750 mg of tooth enamel. Sr (ppm) in tooth enamel is

$$\text{Concentration (ppm)} = \frac{\text{mass of Sr}}{\text{mass of enamel}} \times 10^6$$

$$= \frac{87.2 \times 10^{-9} \text{ g}}{0.750 \times 10^{-3} \text{ g}} \times 10^6 = 116 \text{ ppm}$$

(c) In cell B27 we calculate the standard uncertainty in the intercept $u_x = 0.43$. Relative uncertainty = $100 \times 0.43/8.72 = 4.9\%$, which leads to a 4.9% uncertainty in the concentration of Sr in the tooth enamel. 0.049×116 ppm = 5.7 ppm. Final answer: 116 ± 6 ppm.

(d) Student's t for $n - 2 = 5 - 2 = 3$ degrees of freedom and 95% confidence is 3.182. We found standard deviation = 5.7 ppm. 95% confidence interval is $\pm tu_x = (3.182)(5.7 \text{ ppm}) = 18.1$ ppm. Answer: 116 ± 18 ppm.

Spreadsheet for 5-25 (a). To execute LINEST, highlight cells B16-C18, enter "=LINEST(B7:B11,A7:A11,TRUE,TRUE", and press CONTROL + SHIFT + ENTER on a PC or COMMAND(⌘) + RETURN on a Mac.

	A	B	C	D	E
1	Standard Addition Constant Volume Least-Squares Spreadsheet				
2					
3					
4	x	y			
5	Added Sr				
6	(ng/mL)	Signal			
7	0.00	28.000			
8	2.50	34.300			
9	5.00	42.800			
10	7.50	51.500			
11	10.00	58.600			
12					
13	B16:C18 = LINEST(B7:B11,A7:A11,TRUE,TRUE)				
14					
15		LINEST output:			
16	m	3.1360	27.3600	b	
17	u_m	0.0945	0.5790	u_b	
18	R^2	0.9973	0.7474	s_y	
19					
20	x-intercept = -b/m =	-8.724			
21					
22	n =		5	B22 = COUNT(A7:A11)	
23	Mean y =		43.040	B23 = AVERAGE(B7:B11)	
24	$\Sigma(x_i - \text{mean } x)^2$ =		62.5	B24 = DEVSQ(A7:A11)	
25					
26	Std uncertainty of				
27	x-intercept = u_x =	0.427			
28	B27 =(C18/ABS(B16))*SQRT((1/B22) + B23^2/(B16^2*B24))				

5-25. (a) The intercept for tap water is –6.0 mL, corresponding to an addition of (6.0 mL)(0.152 ng/mL) = 0.91_2 ng Eu(III). This much Eu(III) is in 10.00 mL of tap water, so the concentration is 0.91_2 ng/10.00 mL = 0.091 ng/mL. For pond water, the intercept of –14.6 mL corresponds to an addition of (14.6 mL)(15.2 ng/mL) = 2.22×10^2 ng/10.00 mL pond water = 22.2 ng/mL.

(b) Added standard Eu(III) gives a response of 3.03 units/ng for tap water and 0.0822 units/ng for pond water. The relative response is 3.03/0.0822 = 36.9 times greater in tap water than in pond water. There is probably a *matrix effect* in which something in pond water decreases the Eu(III) emission. By using standard addition, we measure the response in the actual sample matrix. Even though Eu(III) in pond water and tap water do not give equal signals, we measure the actual signal in each matrix and can therefore carry out accurate analyses.

5-26.

	A	B	C	D
2				
3	V_{total} =	V_s (mL) =	x	y
4	50	NaCl	Concentration of	
5	$[S]_i$ (M) =	standard	added NaCl	I(s+x) =
6	2.64	added	$[S]_f$	signal
7	Vo =	0.000	0	3.13
8	25.00	1.000	0.0528	5.40
9		2.000	0.1056	7.89
10		3.000	0.1584	10.30
11		4.000	0.2112	12.48
12	C7 = $A$6*B7/$A$4			
13	B16:D18 = LINEST(D7:D11,C7:C11,TRUE,TRUE)			
14				
15		LINEST output:		
16	m	44.6970	3.1200	b
17	u_m	0.5511	0.0713	u_b
18	R^2	0.9995	0.0920	s_y
19				
20	x-intercept = -b/m =	-0.06980		
21				
22	n =		5	B22 = COUNT(B7:B11)
23	Mean y =		7.84	B23 = AVERAGE(D7:D11)
24	$\Sigma(x_i$ - mean $x)^2$ =		0.0278784	B24 = DEVSQ(C7:C11)
25				
26	Std uncertainty of			
27	x-intercept = u_x =	0.00235		
28	B27 =(C18/ABS(B16))*SQRT((1/B22) + B23^2/(B16^2*B24))			

Spreadsheet for 5-27. To execute LINEST, highlight cells B16-C18, enter "=LINEST(D7:D11,C7:C11,TRUE,TRUE", and press CONTROL + SHIFT + ENTER on a PC or COMMAND(⌘) + RETURN on a Mac.

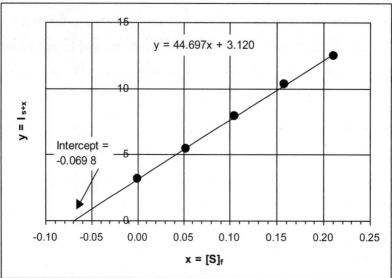

(a)   All solutions are made to constant volume, so we plot I_{S+X} vs. $[S]_f$. The negative intercept is $[X]_f = 0.069\ 8$ M. The initial concentration of NaCl is larger by the dilution factor of 2 (50.00 mL/25.00 mL). The initial concentration of NaCl in serum was $2.000 \times 0.069\ 8$ M $= 0.013\ 96$ M.

(b)   The x-intercept is computed in cell B20 and its standard uncertainty is in cell B27. The relative uncertainty is $100 \times (0.002\ 35)/(0.069\ 8) = 3.37\%$. This uncertainty is much larger than the relative uncertainties in volume measurement, so the uncertainty in the original concentration of Na^+ should be 3.37%. A reasonable expression of $[Na^+]$ in the original serum is 0.140 $(\pm 3.37\%)$ M $= 0.140\ (\pm 0.004_7)$ M.

95% confidence interval $= \pm tu_x = \pm (3.182)(0.004_7\ \text{M}) = \pm 0.015$ M, where t is taken for $5 - 2 = 3$ degrees of freedom.

5-27.

	A	B	C	D
1	Standard Addition Constant Volume Least-Squares Spreadsheet			
2				
3		x	y	
4		Spike (mg/g)	I(s+x) =	
5		$[S]_f$	signal	
6		0.00	15.6	
7		3.12	21.1	
8		7.18	25.5	
9		8.48	30.0	
10		20.0	48.8	
11		38.2	83.4	
12				
13	B16:D18 = LINEST(C6:C11,B6:B11,TRUE,TRUE)			
14				
15		LINEST output:		
16	m	1.7776	14.5928	b
17	u_m	0.0449	0.8190	u_b
18	R^2	0.9974	1.4246	s_y
19				
20	x-intercept = -b/m =	-8.20906		
21				
22	n =	6	B22 = COUNT(B6:B11)	
23	Mean y =	37.40	B23 = AVERAGE(C6:C11)	
24	$\Sigma(x_i - \text{mean } x)^2$ =	1004.7838	B24 = DEVSQ(B6:B11)	
25				
26	Std uncertainty of			
27	x-intercept = u_x =	0.62445		
28	B27 =(C18/ABS(B16))*SQRT((1/B22) + B23^2/(B16^2*B24))			

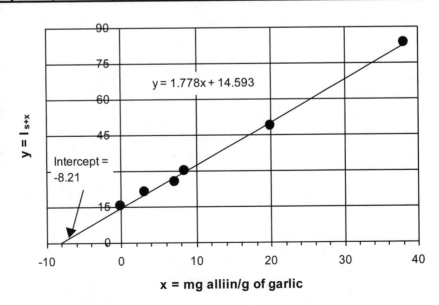

y = 1.778x + 14.593

Intercept = -8.21

y = I$_{s+x}$

x = mg alliin/g of garlic

(a) In cells B20 and B27 of the spreadsheet, the negative x-intercept of the standard addition graph is 8.21 mg alliin/g garlic with a standard uncertainty of 0.62 mg alliin/g garlic. The 95% confidence interval is $tu_x =$ (2.776)(0.624) = 1.73 mg alliin/g garlic, where Student's t is taken for 95% confidence and $6 - 2 = 4$ degrees of freedom. A reasonable answer is 8.2 ± 1.7 mg alliin/g garlic.

(b) Two moles of alliin (FM 177.2) produce one mole of allicin (FM 162.3) in the assay. Therefore, the quantity of allicin in garlic is ½(162.3/177.2)(8.21 ± 1.73 mg/g) = 3.76 ± 0.79 mg allicin/g garlic or 3.8 ± 0.8 mg allicin/g garlic.

5-28. (a) In the spreadsheet, data appear in columns B and D. The abscissa and ordinate functions are in columns C and E.

	A	B	C	D	E
1	Standard Addition Experiment				
2	Add 2.50 ppm standard Pb to river sediment extract				
3					
4		$V_s =$			y-axis function
5	V_0 (mL) =	mL Pb std	x-axis function	$I_{s+x} =$	$I_{s+x}*V/V_0$
6	4.60	added	$[S]_i*V_s/V_0$ (ppm)	signal	$(V = V_0 + V_s)$
7	$[S]_i$ (ppm) =	0.000	0	1.10	1.100
8	2.50	0.025	0.0136	1.66	1.669
9		0.050	0.0272	2.20	2.224
10		0.075	0.0408	2.81	2.856
11	V = total volume = $V_0 + V_s$				
12					
13			C16:D18 = LINEST(E7:E10,C7:C10,TRUE,TRUE)		
14					
15			LINEST output:		
16		m	42.8524	1.0888	b
17		u_m	0.8720	0.0222	u_b
18		R^2	0.9992	0.0265	s_y
19					
20	x-intercept = -b/m =		-0.02541	C20 = -D16/C16	
21					
22		n =	4	C22 = COUNT(C7:C10)	
23		Mean y =	1.962	C23 = AVERAGE(E7:E10)	
24	$\Sigma(x_i$ - mean x$)^2$ =		0.000923027	C24 = DEVSQ(C7:C10)	
25					
26	Std uncertainty of				
27	x-intercept = u_x =		0.00098		
28	C27 = (D18/ABS(C16))*SQRT((1/C22) + C23^2/(C16^2*C24))				

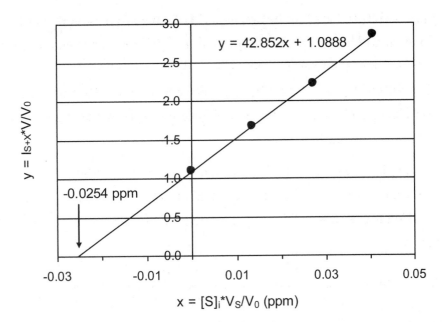

The negative x-intercept (0.025 4 ppm) is the concentration of Pb(II) in the initial sample volume $V_0 = 4.60$ mL. The concentration of Pb(II) in the 1.00-mL extract is found from the dilution formula:

$(x$ ppm$)(1.00$ mL$) = (0.025$ 4 ppm$)(4.60$ mL$) \Rightarrow x = 0.116$ 9 ppm

(b) The standard uncertainty in the intercept in cell C27 is 0.000 98 ppm. With only 4 points and 2 degrees of freedom, Student's t for 95% confidence is 4.303, so the 95% confidence interval is (4.303)(0.000 98 ppm) = 0.004 2 ppm. The relative uncertainty in the intercept is $100 \times (0.004$ 2 ppm/0.025 4 ppm) = 16.6%. If other sources of uncertainty are negligible, the relative uncertainty in concentration in the 1.00-mL extract is also 16.6%, giving an absolute uncertainty of (0.166)(0.116 9 ppm) = 0.019 ppm. A reasonable expression of the answer is 0.117 ± 0.019 ppm or 0.12 ± 0.02 ppm.

5-29. Standard addition is appropriate when the sample matrix is unknown or complex and hard to duplicate, and unknown matrix effects are anticipated. An internal standard can be added to an unknown at the start of a procedure in which uncontrolled losses of sample will occur. The relative amounts of unknown and standard remain constant. The internal standard is excellent if instrument conditions vary from run to run. Variations affect the analyte and standard equally, so the relative signal remains constant. In chromatography the amount of sample injected into the instrument is very small and not very reproducible. However, the relative quantities of standard and analyte remain constant regardless of the sample size.

5-30. (a) $\dfrac{A_X}{[X]} = F\left(\dfrac{A_S}{[S]}\right) \Rightarrow \dfrac{3\,473}{[3.47\text{ mM}]} = F\left(\dfrac{10\,222}{[1.72\text{ mM}]}\right) \Rightarrow F = 0.168_4$

(b) $[S] = (8.47\text{ mM})\left(\dfrac{1.00\text{ mL}}{10.0\text{ mL}}\right) = 0.847\text{ mM}$

(c) $\dfrac{A_X}{[X]} = F\left(\dfrac{A_S}{[S]}\right) \Rightarrow \dfrac{5\,428}{[X]} = 0.168_4\left(\dfrac{4\,431}{[0.847\text{ mM}]}\right) \Rightarrow [X] = 6.16\text{ mM}$

(d) The original concentration of [X] was twice as great as the diluted concentration, so $[X] = 12.3$ mM.

5-31. For the standard mixture:

$\dfrac{A_X}{[X]} = F\left(\dfrac{A_S}{[S]}\right) \Rightarrow \dfrac{10.1\ \mu A}{[0.800\text{ mM}]} = F\left(\dfrac{15.3\ \mu A}{[0.500\text{ mM}]}\right) \Rightarrow F = 0.412_6$

Chloroform added to unknown $= (10.2 \times 10^{-6}\text{ L})(1\,484\text{ g/L}) = 0.015\,1_4\text{ g} = 0.126_8$ mmol in 0.100 L $= 1.26_8$ mM

For the unknown mixture:

$\dfrac{A_X}{[X]} = F\left(\dfrac{A_S}{[S]}\right) \Rightarrow \dfrac{8.7\ \mu A}{[X]} = 0.412_6\left(\dfrac{29.4\ \mu A}{[1.26_8\text{ mM}]}\right) \Rightarrow [X] = 0.909\text{ mM}$

$[DDT]$ in unknown $= (0.909\text{ mM})\left(\dfrac{100\text{ mL}}{10.0\text{ mL}}\right) = 9.09\text{ mM}$

5-32. (a) and (b) In the spreadsheet at the end of this problem, the internal standard graph is a plot of A_X/A_S versus $[X]/[S]$ using data in cells A5:B10. We use LINEST in cells B12:C14 to compute slope, intercept, and uncertainties.

(c) For $A_X/A_S = 1.98$, cell A22 computes $[X]/[S] = (1.98 - b)/m = 0.560$.

Standard uncertainty in the quotient $[X]/[S]$ is computed in cell B23:

$$u_x = \frac{s_y}{|m|}\sqrt{\frac{1}{k} + \frac{1}{n} + \frac{(y - \bar{y})^2}{m^2\,\Sigma(x_i - \bar{x})^2}} = 0.024\,6$$

where $s_y = 0.072$, $m = 3.47$, k = number of replicate measurements of unknown = 1, n = number of points in calibration line = 6, $y = 1.98$, $\bar{y}$ is the mean value of y for the points on the calibration line = 1.179 in cell B17, and $\Sigma(x_i - \bar{x})^2 = 0.233$ in cell B18. For 4 degrees of freedom and 95% confidence, Student's t is 2.776, which you can find in the table of Student t values or you can calculate in Excel with the formula "=TINV(0.05,4)". The 95% confidence interval for $[X]/[S]$ is $\pm t u_x = \pm(2.776)(0.024\,6) = \pm0.068$. A reasonable expression of the mole ratio is $[X]/[S] = 0.56 \pm 0.07$.

(d) The intercept with its 95% confidence interval is $b \pm tu_b = 0.038 \pm (2.776)(0.057\ 3) = 0.038 \pm 0.159$. The 95% interval for the intercept is -0.12 to $+0.20$, which includes 0.

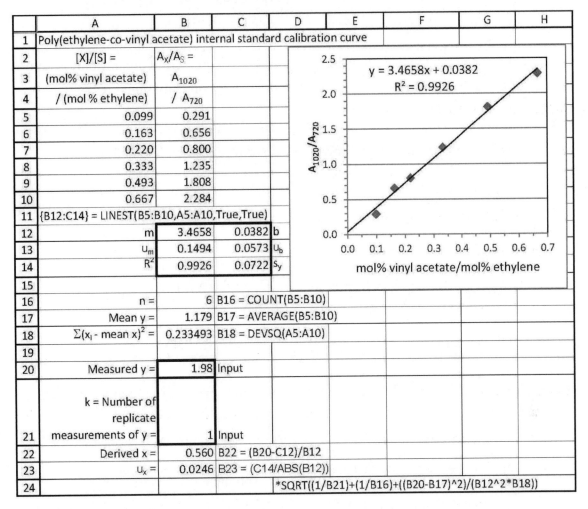

	A	B	C	D	E	F	G	H
1	Poly(ethylene-co-vinyl acetate) internal standard calibration curve							
2	[X]/[S] =	A_x/A_S =						
3	(mol% vinyl acetate)	A_{1020}						
4	/ (mol % ethylene)	/ A_{720}						
5	0.099	0.291						
6	0.163	0.656						
7	0.220	0.800						
8	0.333	1.235						
9	0.493	1.808						
10	0.667	2.284						
11	{B12:C14} = LINEST(B5:B10,A5:A10,True,True)							
12	m	3.4658	0.0382	b				
13	u_m	0.1494	0.0573	u_b				
14	R^2	0.9926	0.0722	s_y				
15								
16	n =	6	B16 = COUNT(B5:B10)					
17	Mean y =	1.179	B17 = AVERAGE(B5:B10)					
18	$\Sigma(x_i - \text{mean } x)^2$ =	0.233493	B18 = DEVSQ(A5:A10)					
19								
20	Measured y =	1.98	Input					
21	k = Number of replicate measurements of y =	1	Input					
22	Derived x =	0.560	B22 = (B20-C12)/B12					
23	u_x =	0.0246	B23 = (C14/ABS(B12))					
24			*SQRT((1/B21)+(1/B16)+((B20-B17)^2)/(B12^2*B18))					

5-33. To use the internal standard, a known quantity of internal standard is added to a sample of sewage. Analyte signal is measured relative to the signal from internal standard. Since the internal standard is so similar to analyte (differing only by the substitution of D for H), matrix effects are likely to be essentially identical for both compounds. If matrix enhances the response to analyte, it enhances the response to internal standard to the same extent. By measuring unknown relative to the known injection of standard, an accurate measurement of unknown can be made. The key is to use a standard that is almost identical to the analyte in physical properties.

CHAPTER 6
CHEMICAL EQUILIBRIUM

6-1. Concentrations in an equilibrium constant are really dimensionless <u>ratios</u> of actual concentrations divided by standard state concentrations. Since standard states are 1 M for solutes, 1 bar for gases, and pure substances for solids and liquids, these are the units we must use. A solvent is approximated as a pure liquid.

6-2. All concentrations in equilibrium constants are expressed as dimensionless ratios of actual concentrations divided by standard-state concentrations.

6-3. Predictions based on free energy or Le Châtelier's principle tell us which way a reaction will go (thermodynamics), but not how long it will take (kinetics). A reaction could be over instantly or it could take forever.

6-4. (a) $K = 1/[Ag^+]^3 [PO_4^{3-}]$ (b) $K = P_{CO_2}^6 / P_{O_2}^{15/2}$

6-5. $K = \dfrac{P_E^3}{P_A^2 [B]} = \dfrac{\left(\dfrac{3.6 \times 10^4 \text{ Torr}}{760 \text{ Torr/atm}} \times 1.013 \dfrac{\text{bar}}{\text{atm}}\right)^3}{\left(\dfrac{2.8 \times 10^3 \text{ Pa}}{10^5 \text{ Pa/bar}}\right)^2 (1.2 \times 10^{-2} \text{ M})} = 1.2 \times 10^{10}$

6-6.

HOBr + OCl$^-$	$\rightleftharpoons$ HOCl + OBr$^-$	$K_1 = 1/15$
HOCl	$\rightleftharpoons$ H$^+$ + OCl$^-$	$K_2 = 3.0 \times 10^{-8}$
HOBr	$\rightleftharpoons$ H$^+$ + OBr$^-$	$K = K_1 K_2 = 2.0 \times 10^{-9}$

6-7. (a) Decrease (b) give off

  (c) $\Delta G = \Delta H - T\Delta S$; ΔG is negative for a spontaneous change

6-8. $K = e^{-(59.0 \times 10^3 \text{ J/mol})/(8.314472 \text{ J/(K·mol)})(298.15 \text{ K})} = 5 \times 10^{-11}$

6-9. (a) Right because reactant is added

  (b) Right because product is removed

  (c) Neither because solid graphite does not appear in the reaction quotient

  (d) Right. The pressure of reactant and product both increase by a factor of 8. However, reactant appears to the second power in the reaction quotient and product only appears to the first power. Increasing each pressure by the same factor decreases the reaction quotient.

(e) Smaller. An exothermic reaction liberates heat. Adding heat is like adding a product.

6-10. (a) $K = P_{H_2O} = e^{-\Delta G°/RT} = e^{-(\Delta H° - T\Delta S°)/RT}$

$= e^{-\{[(63.11 \times 10^3 \text{ J/mol}) - (298.15K)(148 \text{ J K}^{-1} \text{ mol}^{-1})]/(8.314 \text{ J K}^{-1} \text{ mol}^{-1})(298.15 \text{ K})\}}$

$= 4.7 \times 10^{-4} \text{ bar}$

(b) $P_{H_2O} = 1 = e^{-(\Delta H° - T\Delta S°)/RT} \Rightarrow \Delta H° - T\Delta S°$ must be zero.

$\Delta H° - T\Delta S° = 0 \Rightarrow T = \dfrac{\Delta H°}{\Delta S°} = 426 \text{ K} = 153°C$

6-11. (a) Let's designate the equilibrium constant at temperature T_1 as K_1 and the equilibrium constant at temperature T_2 as K_2.

$K_1 = e^{-\Delta G°/RT_1} = e^{-(\Delta H° - T_1\Delta S°)/RT_1} = e^{-\Delta H°/RT_1} \cdot e^{\Delta S°/R}$

Similarly, $K_2 = e^{-\Delta H°/RT_2} \cdot e^{\Delta S°/R}$

Dividing K_1 by K_2 gives $\dfrac{K_1}{K_2} = e^{-(\Delta H°/R)(1/T_1 - 1/T_2)}$

$\Rightarrow \Delta H° = \left(\dfrac{1}{T_2} - \dfrac{1}{T_1}\right)^{-1} R \ln \dfrac{K_1}{K_2}$

Putting in $K_1 = 1.479 \times 10^{-5}$ at $T_1 = 278.15$ K and
$K_2 = 1.570 \times 10^{-5}$ at $T_2 = 283.15$ K gives $\Delta H° = +7.82$ kJ / mol.

(b) $K = e^{-\Delta H°/RT} \cdot e^{\Delta S°/R}$

$\ln K = -\dfrac{\Delta H°}{R}\left(\dfrac{1}{T}\right) + \dfrac{\Delta S°}{R}$

$\quad y \qquad m \quad x \qquad b$

A graph of $\ln K$ vs. $1/T$ will have a slope of $-\Delta H°/R$

6-12. (a) $Q = \left(\dfrac{48.0 \text{ Pa}}{10^5 \text{ Pa/bar}}\right)^2 \bigg/ \left(\dfrac{1\,370 \text{ Pa}}{10^5 \text{ Pa/bar}}\right)\left(\dfrac{3\,310 \text{ Pa}}{10^5 \text{ Pa/bar}}\right)$

$= 5.08 \times 10^{-4} < K \qquad$ The reaction will go to the right.

It was not really necessary to convert Pa to bar, since the units cancel.

(b)
	H_2	+	Br_2	$\rightleftharpoons$	$2HBr$
Initial pressure:	1 370		3 310		48.0
Final pressure:	$1\,370 - x$		$3\,310 - x$		$48.0 + 2x$

Note that $2x$ Pa of HBr are formed when x Pa of H_2 are consumed.

$$\frac{(48.0 + 2x)^2}{(1\,370 - x)(3\,310 - x)} = 7.2 \times 10^{-4} \Rightarrow x = 4.50 \text{ Pa}$$

$$P_{H_2} = 1\,366 \text{ Pa}, \quad P_{Br_2} = 3\,306 \text{ Pa}, \quad P_{HBr} = 57.0 \text{ Pa}$$

(c) Neither, since Q is unchanged.

(d) HBr will be formed, since $\Delta H°$ is positive.

6-13. The concentration of MTBE in solution is 100 μg/mL = 100 mg/L. The molarity is $[MTBE] = \dfrac{0.100 \text{ g/L}}{88.15 \text{ g/mol}} = 1.13_4 \times 10^{-3}$ M. The pressure in the gas phase is $P =$ $[MTBE]/K_h = (1.13_4 \times 10^{-3} \text{ M})/(1.71 \text{ M/bar}) = 0.663$ mbar.

6-14. $[Cu^+][Br^-] = K_{sp}$

$[Cu^+][0.10] = 5 \times 10^{-9} \Rightarrow [Cu^+] = 5 \times 10^{-8}$ M

6-15. $[Ag^+]^4[Fe(CN)_6^{4-}] = K_{sp}$

$[1.0 \times 10^{-6}]^4[Fe(CN)_6^{4-}] = 8.5 \times 10^{-45} \Rightarrow [Fe(CN)_6^{4-}] = 8.5 \times 10^{-21}$ M = 8.5 zM

6-16. If we let $x = [Cu^{2+}]$, then $[SO_4^{2-}] = \frac{1}{4}x$.

$$K = [Cu^{2+}]^4 [OH^-]^6 [SO_4^{2-}] = (x)^4 (1.0 \times 10^{-6})^6 (\tfrac{1}{4}x) = 2.3 \times 10^{-69}$$

$$\Rightarrow x = [Cu^{2+}] = \left(\frac{(4)(2.3 \times 10^{-69})}{(1.0 \times 10^{-6})^6}\right)^{1/5} = 3.9 \times 10^{-7} \text{ M}$$

6-17. (a) $[Zn^{2+}]^2[Fe(CN)_6^{4-}] = (0.000\,10)^2[Fe(CN)_6^{4-}] = 2.1 \times 10^{-16}$

$\Rightarrow [Fe(CN)_6^{4-}] = 2.1 \times 10^{-8}$ M

(b) $[Zn^{2+}]^2[Fe(CN)_6^{4-}] = (5.0 \times 10^{-7})^2[Fe(CN)_6^{4-}] = 2.1 \times 10^{-16}$

$\Rightarrow [Fe(CN)_6^{4-}] = 8.4 \times 10^{-4}$ M

6-18. BX_2 coprecipitates with AX_3. This means that some BX_2 is trapped in the AX_3 during precipitation of AX_3.

6-19. For $CaSO_4$, $K_{sp} = 2.4 \times 10^{-5}$. For Ag_2SO_4, $K_{sp} = 1.5 \times 10^{-5}$.

Removing 99% of the Ca^{2+} reduces $[Ca^{2+}]$ to 0.000 500 M. The concentration of SO_4^{2-} in equilibrium with 0.000 500 M Ca^{2+} is $2.4 \times 10^{-5}/0.000\,500 = 0.048$ M. This much SO_4^{2-} __will__ precipitate Ag_2SO_4, because $Q = [Ag^+]^2 [SO_4^{2-}] =$ $(0.030\,0)^2 (0.048) = 4.3 \times 10^{-5} > K_{sp}$. The separation is not feasible.

When Ag^+ first precipitates, $[SO_4^{2-}] = 1.5 \times 10^{-5}/(0.030\ 0)^2 = 1.67 \times 10^{-2}$ M.

$[Ca^{2+}] = 2.4 \times 10^{-5}/1.67 \times 10^{-2} = 0.001\ 4$ M.   97% of the Ca^{2+} has precipitated.

6-20. $BaCrO_4(s) \rightleftharpoons Ba^{2+} + CrO_4^{2-}$ $K_{sp} = 2.1 \times 10^{-10}$

$Ag_2CrO_4(s) \rightleftharpoons 2Ag^+ + CrO_4^{2-}$ $K_{sp} = 1.2 \times 10^{-12}$

The stoichiometries are not identical, so it is not clear that the salt with lower K_{sp} will precipitate first. Let's try each possibility. Suppose that $BaCrO_4$ precipitates first. The concentration of CrO_4^{2-} that will reduce Ba^{2+} to 0.1% of its initial concentration is

$[Ba^{2+}][CrO_4^{2-}] = [1.0 \times 10^{-5}][CrO_4^{2-}] = 2.1 \times 10^{-10} \Rightarrow [CrO_4^{2-}] = 2.1 \times 10^{-5}$ M.

Will this much chromate precipitate 0.010 M Ag^+? We test by evaluating the reaction quotient for Ag_2CrO_4:

$Q = [Ag^+]^2[CrO_4^{2-}] = (0.010)^2(2.1 \times 10^{-5}) = 2.1 \times 10^{-9} > K_{sp}$ for Ag_2CrO_4

Since $Q > K_{sp}$ for Ag_2CrO_4, Ag^+ will precipitate.

Let's try the reverse calculation. If Ag_2CrO_4 precipitates first, the concentration of CrO_4^{2-} that will reduce Ag^+ to 1.0×10^{-5} M is

$[Ag^+]^2[CrO_4^{2-}] = [1.0 \times 10^{-5}]^2[CrO_4^{2-}] = 1.2 \times 10^{-12} \Rightarrow [CrO_4^{2-}] = 0.012$ M.

This concentration of CrO_4^{2-} exceeds the concentration required to precipitate 99.90% of Ba^{2+}. Neither Ag^+ nor Ba^{2+} can be 99.90% precipitated without precipitating the other ion.

6-21.

Salt	K_{sp}		$[Ag^+]$	
			(M, in equilibrium with 0.1 M anion)	
AgCl	1.8×10^{-10}	$K_{sp}/0.10$	$=$	1.8×10^{-9}
AgBr	5.0×10^{-13}	$K_{sp}/0.10$	$=$	5.0×10^{-12}
AgI	8.3×10^{-17}	$K_{sp}/0.10$	$=$	8.3×10^{-16}
Ag_2CrO_4	1.2×10^{-12}	$\sqrt{K_{sp}/0.10}$	$=$	3.5×10^{-6}

I^- requires the lowest concentration of Ag^+ to begin precipitating, so I^- precipitates first. The order of precipitation is: I^- before Br^- before Cl^- before CrO_4^{2-}.

6-22. At low I^- concentration, $[Pb^{2+}]$ decreases with increasing $[I^-]$ because of the reaction $Pb^{2+} + 2I^{2-} \rightarrow PbI_2(s)$. Concentrations of other $Pb^{2+}- I^-$ species are negligible. At high I^- concentration, complex ions form by reactions such as $PbI_2(s) + I^- \rightarrow PbI_3^-$.

6-23. (a) BF_3 (b) AsF_5

6-24. $\dfrac{[SnCl_2(aq)]}{[Sn^{2+}][Cl^-]^2} = \beta_2 \Rightarrow [SnCl_2(aq)] = \beta_2[Sn^{2+}][Cl^-]^2 = (12)(0.20)(0.20)^2 = 0.096 \text{ M}$

6-25. $[Zn^{2+}] = K_{sp}/[OH^-]^2 = 2._{93} \times 10^{-3} \text{ M}$

$[ZnOH^+] = \beta_1[Zn^{2+}][OH^-] = \beta_1 K_{sp}/[OH^-] = 9 \times 10^{-6} \text{ M}$

$[Zn(OH_2(aq)] = \beta_2[Zn^{2+}][OH^-]^2 = \beta_3 K_{sp}[OH^-] = 6 \times 10^{-6} \text{ M}$

$[Zn(OH)_3^-] = \beta_3[Zn^{2+}][OH^-]^3 = \beta_3 K_{sp}[OH^-] = 8 \times 10^{-9} \text{ M}$

$[Zn(OH)_4^{2-}] = \beta_4[Zn^{2+}][OH^-]^4 = \beta_4 K_{sp}[OH^-]^2 = 9 \times 10^{-14} \text{ M}$

6-26.

	Na^+	$+$	OH^-	$\rightleftharpoons$	$NaOH(aq)$
Initial concentration:	1		1		0
Final concentration:	$1-x$		$1-x$		x

$\dfrac{x}{(1-x)^2} = 0.2 \Rightarrow x = 0.15 \text{ M}$

$x = 0.2 - 0.4x + 0.2x^2$

$\underset{a}{0.2x^2} - \underset{b}{1.4x} + \underset{c}{0.2} = 0$

$x = \dfrac{-b \pm \sqrt{b^2 - 4ac}}{2a} = \dfrac{1.4 \pm \sqrt{1.4^2 - 4(0.2)(0.2)}}{2(0.2)} = 6.85 \text{ or } 0.15$

x cannot be greater than 1 (the initial or formal concentration of NaOH), so the correct answer must be 0.15. That is, 15% of the sodium is in the form $NaOH(aq)$.

6-27. $PbI_2(s) \overset{K_{sp}}{\rightleftharpoons} Pb^{2+} + 2I^-$ $K_{sp} = [Pb^{2+}][I^-]^2 = 7.9 \times 10^{-9}$

$\quad + $

$\quad Pb^{2+} + 2I^- \overset{\beta_2}{\rightleftharpoons} PbI_2(aq)$ $\beta_2 = [PbI_2(aq)] / [Pb^{2+}][I^-]^2 = 1.4 \times 10^3$

$\quad PbI_2(s) \rightleftharpoons PbI_2(aq)$ $K = K_{sp}\beta_2 = 1.1 \times 10^{-5} = [PbI_2(aq)]$

6-28. Lewis acids and bases are electron pair acceptors and donors, respectively:

$$F_3B \ + \ :\ddot{O}(CH_3)_2 \ \rightarrow \ F_3\bar{B} - \overset{+}{O}(CH_3)_2$$

Lewis   Lewis   Adduct
acid   base

Brønsted-Lowry acids and bases are proton donors and acceptors, respectively:

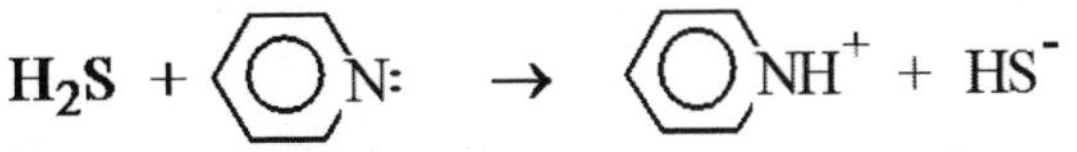

$$H_2S \ + \ \bigcirc N: \ \rightarrow \ \bigcirc NH^+ \ + \ HS^-$$

Brønsted   Brønsted
acid   base

6-29. (a) An adduct   (b) dative or coordinate covalent

(c) conjugate   (d) $[H^+] > [OH^-]$; $[OH^-] > [H^+]$

6-30. Dissolved CO_2 from the atmosphere lowers the pH by reacting with water to form carbonic acid. Water can be distilled under an inert atmosphere to exclude CO_2, or most CO_2 can be removed by boiling the distilled water.

6-31. SO_2 in the atmosphere reacts with moisture to make H_2SO_3, which is a weak acid. H_2SO_3 can be oxidized to H_2SO_4, which is a strong acid.

6-32.

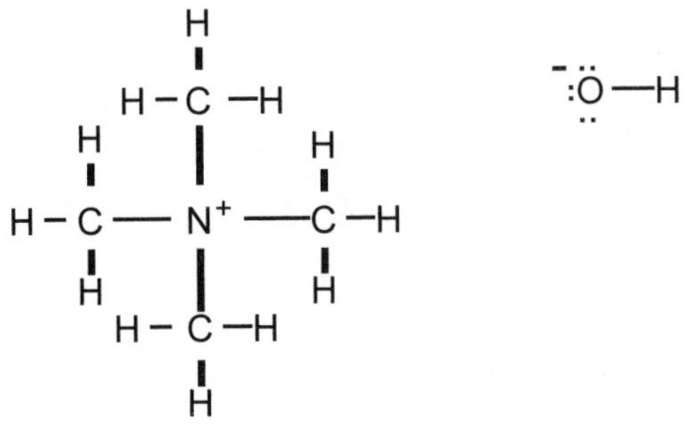

There is no place for OH^- to bond to $(CH_3)_4N^+$.

6-33. (a) HI   (b) H_2O

6-34. $2H_2SO_4 \ \rightleftharpoons \ HSO_4^- \ + \ H_3SO_4^+$

6-35.

acid	base
(a) H_3O^+	H_2O
(a) $H_3\overset{+}{N}CH_2CH_2\overset{+}{N}H_3$	$H_3\overset{+}{N}CH_2CH_2NH_2$
(b) $C_6H_5CO_2H$	$C_6H_5CO_2^-$
(b) $C_5H_5NH^+$	C_5H_5N

6-36. (a) $[H^+] = 0.010$ M $\Rightarrow$ pH $= -\log[H^+] = 2.00$

(b) $[OH^-] = 0.035\,M \Rightarrow [H^+] = K_w / [OH^-] = 2.8_6 \times 10^{-13}\,M \Rightarrow pH = 12.54$

(c) $[H^+] = 0.030\,M \Rightarrow pH = 1.52$

(d) $[H^+] = 3.0\,M \Rightarrow pH = -0.48$

(e) $[OH^-] = 0.010\,M \Rightarrow [H^+] = 1.0 \times 10^{-12}\,M \Rightarrow pH = 12.00$

6-37. (a) $K_w = [H^+][OH^-] = 1.01 \times 10^{-14}$ at 25°C

$\qquad\quad x \qquad x$

$\qquad x^2 = 1.01 \times 10^{-14} \Rightarrow x = [H^+] = 1.00_5 \times 10^{-7}\,M \Rightarrow pH = -\log[H^+] = 6.998$

(b) At 100°C, pH = 6.132

6-38. Since $[H^+][OH^-] = 1.0 \times 10^{-14}$, $K = [H^+]^4 [OH^-]^4 = 1.0 \times 10^{-56}$

6-39. $[La^{3+}][OH^-]^3 = K_{sp} = 2 \times 10^{-21}$

$[OH^-]^3 = K_{sp} / (0.010) \Rightarrow [OH^-] = 5.8 \times 10^{-7}\,M \Rightarrow pH = 7.8$

6-40. (a) At 25°C, K_w increases as temperature increases $\Rightarrow$ endothermic

(b) At 100°C, K_w increases as temperature increases $\Rightarrow$ endothermic

(c) At 300°C, K_w decreases as temperature increases $\Rightarrow$ exothermic

6-41. See Table 6-2.

6-42.

	RCO_2H	$R_3NH^+X^-$	M^{n+}
Weak acids:	Carboxylic acids	Ammonium ions	Metal ions

	R_3N:	$RCO_2^-M^+$	
Weak bases:	Amines	Carboxylate ions	

6-43. In the reaction $(H_2O)_6Fe^{3+} \rightleftharpoons (OH)(H_2O)_5Fe^{2+} + H^+$, positive charge concentrated in $(H_2O)_6Fe^{3+}$ is reduced when H^+ dissociates. This process is favored by Coulombic repulsion of like charges. In the reaction $(H_2O)_6Cl^- \rightleftharpoons (OH)(H_2O)_5Cl^{2-} + H^+$, negative charge is concentrated in the product when H^+ dissociates. This process is disfavored by Coulombic repulsion of like charges.

6-44. $Cl_3CCO_2H \rightleftharpoons Cl_3CCO_2^- + H^+$

$La^{3+} + H_2O \rightleftharpoons LaOH^{2+} + H^+$

6-45.

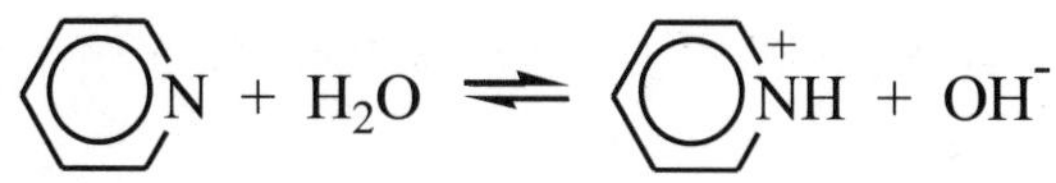

$$\text{C}_5\text{H}_5\text{N} + \text{H}_2\text{O} \rightleftharpoons \text{C}_5\text{H}_5\text{NH}^+ + \text{OH}^-$$

$$\text{HOCH}_2\text{CH}_2\text{S}^- + \text{H}_2\text{O} \rightleftharpoons \text{HOCH}_2\text{CH}_2\text{SH} + \text{OH}^-$$

6-46. K_a: $\text{HCO}_3^- \rightleftharpoons \text{H}^+ + \text{CO}_3^{2-}$ K_b: $\text{HCO}_3^- + \text{H}_2\text{O} \rightleftharpoons \text{H}_2\text{CO}_3 + \text{OH}^-$

6-47. (a) $\text{H}_3\overset{+}{\text{N}}\text{CH}_2\text{CH}_2\overset{+}{\text{N}}\text{H}_3 \overset{K_{a1}}{\rightleftharpoons} \text{H}_2\text{NCH}_2\text{CH}_2\overset{+}{\text{N}}\text{H}_3 + \text{H}^+$

  $\text{H}_2\text{NCH}_2\text{CH}_2\overset{+}{\text{N}}\text{H}_3 \overset{K_{a2}}{\rightleftharpoons} \text{H}_2\text{NCH}_2\text{CH}_2\text{NH}_2 + \text{H}^+$

  (b) $^-\text{O}_2\text{CCH}_2\text{CO}_2^- + \text{H}_2\text{O} \overset{K_{b1}}{\rightleftharpoons} \text{HO}_2\text{CCH}_2\text{CO}_2^- + \text{OH}^-$

  $\text{HO}_2\text{CCH}_2\text{CO}_2^- + \text{H}_2\text{O} \overset{K_{b2}}{\rightleftharpoons} \text{HO}_2\text{CCH}_2\text{CO}_2\text{H} + \text{OH}^-$

6-48. (a), (c)

6-49. $\text{CN}^- + \text{H}_2\text{O} \rightleftharpoons \text{HCN} + \text{OH}^-$ $K_b = \dfrac{K_w}{K_a} = 1.6 \times 10^{-5}$

6-50. $\text{H}_2\text{PO}_4^- \overset{K_{a2}}{\rightleftharpoons} \text{HPO}_4^{2-} + \text{H}^+$ $\text{HC}_2\text{O}_4^- + \text{H}_2\text{O} \overset{K_{b2}}{\rightleftharpoons} \text{H}_2\text{C}_2\text{O}_4 + \text{OH}^-$

6-51. $K_{a1} = \dfrac{K_w}{K_{b3}} = 7.04 \times 10^{-3}$ $K_{a2} = \dfrac{K_w}{K_{b2}} = 6.25 \times 10^{-8}$

  $K_{a3} = \dfrac{K_w}{K_{b1}} = 4.3 \times 10^{-13}$

6-52. Add the two reactions and multiply their equilibrium constants to get $K = 3.0 \times 10^{-6}$.

6-53. (a) $\text{Ca(OH)}_2\,(s) \rightleftharpoons \underset{x}{\text{Ca}^{2+}} + \underset{2x}{2\text{OH}^-}$

  $x(2x)^2 = K_{sp} = 10^{-5.19} \Rightarrow x = 1.2 \times 10^{-2} \text{ M}$

  (b) Since some Ca^{2+} reacts with OH^- to form CaOH^+, the K_{sp} reaction will be drawn to the right, and the solubility of Ca(OH)_2 will be greater than we would expect just on the basis of K_{sp}.

6-54. Reversing the first reaction and then adding the four reactions gives

  $\text{Ca}^{2+} + \text{CO}_2(g) + \text{H}_2\text{O}(l) \rightleftharpoons \text{CaCO}_3(s) + 2\text{H}^+$ $K = K_{\text{CO}_2}K_1K_2/K_{sp}$

  $K = (3.4 \times 10^{-2})(4.5 \times 10^{-7})(4.7 \times 10^{-11})/(6.0 \times 10^{-9}) = 1.2 \times 10^{-10}$

  $\dfrac{[\text{H}^+]^2}{[\text{Ca}^{2+}]P_{\text{CO}_2}} = \dfrac{(1.8 \times 10^{-7})^2}{[\text{Ca}^{2+}][0.10]} = K = 1.2 \times 10^{-10}$

  $[\text{Ca}^{2+}] = 2.7 \times 10^{-3} \text{ M} = 0.22 \text{ g}/2.00 \text{ L}$

CHAPTER 7
LET THE TITRATIONS BEGIN

7-1. Concentrations of reagents used in an analysis are determined either by weighing out supposedly pure primary standards or by reaction with such standards. If the standards are not pure, none of the concentrations will be correct.

7-2. The equivalence point occurs when the exact stoichiometric quantities of reagents have been mixed. The end point, which comes near the equivalence point, is marked by a sudden change in a physical property brought about by the disappearance of a reactant or appearance of a product.

7-3. In a blank titration, the quantity of titrant required to reach the end point in the absence of analyte is measured. By subtracting this quantity from the amount of titrant needed in the presence of analyte, we reduce the systematic error.

7-4. In a direct titration, titrant reacts directly with analyte. In a back titration, a known excess of reagent that reacts with analyte is used. The excess is then measured with a second titrant.

7-5. Primary standards are purer than reagent-grade chemicals. The assay of a primary standard must be very close to the nominal value (such as 99.95–100.05%), whereas the assay on a reagent chemical might be only 99%. Primary standards must have very long shelf lives.

7-6. Since a relatively large amount of acid might be required to dissolve a small amount of sample, we cannot tolerate even modest amounts of impurities in the acid for trace analysis. Otherwise, the quantity of impurity could be greater than quantity of analyte in the sample.

7-7. 40.0 mL of 0.040 0 M $Hg_2(NO_3)_2$ = 1.60 mmol of Hg_2^{2+}, which will require 3.20 mmol of KI. This is contained in volume $= \dfrac{3.20 \text{ mmol}}{0.100 \text{ mmol/mL}} = 32.0$ mL.

7-8. 108.0 mL of 0.165 0 M oxalic acid = 17.82 mmol, which requires

$$\left(\frac{2 \text{ mol MnO}_4^-}{5 \text{ mol H}_2\text{C}_2\text{O}_4} \right) (17.82 \text{ mol H}_2\text{C}_2\text{O}_4) = 7.128 \text{ mmol of MnO}_4^-.$$

7.128 mmol / (0.165 0 mmol/mL) = 43.20 mL of $KMnO_4$.

Another way to see this is to note that the reagents are both 0.165 0 M.

Therefore, volume of $MnO_4^- = \frac{2}{5}$(volume of oxalic acid).

For the second part of the question,

$$\text{volume of oxalic acid} = \frac{5}{2}(\text{volume of MnO}_4^-) = 270.0 \text{ mL}.$$

7-9. 1.69 mg of NH_3 = 0.0992 mmol of NH_3. This will react with $\frac{3}{2}(0.0992) = 0.149$ mmol of OBr^-. The molarity of OBr^- is 0.149 mmol/1.00 mL = 0.149 M.

7-10. $\text{mol sulfamic acid} = \dfrac{0.3337 \text{ g}}{97.094 \text{ g/mol}} = 3.436_9 \text{ mmol}$

$\text{molarity of NaOH} = \dfrac{3.436_9 \text{ mmol}}{34.26 \text{ mL}} = 0.1003 \text{ M}$

7-11. HCl added to powder = (10.00 mL)(1.396 M) = 13.96 mmol

NaOH required = (39.96 mL)(0.1004 M) = 4.012 mmol

HCl consumed by carbonate = 13.96 − 4.012 = 9.94$_8$ mmol

$\text{mol CaCO}_3 = \frac{1}{2} \text{ mol HCl consumed} = 4.974_4 \text{ mmol} = 0.497_8 \text{ g CaCO}_3$

$\text{wt\% CaCO}_3 = \dfrac{0.497_8 \text{ g CaCO}_3}{0.5413 \text{ g limestone}} \times 100 = 92.0 \text{ wt\%}$

7-12. (a) Theoretical molarity = 3.214 g/158.034 g/mol = 0.02034 M.

(b) 25.00 mL of 0.02034 M $KMnO_4$ = 0.508 5 mmol. But two moles of MnO_4^- react with five moles of H_3AsO_3, which comes from $\frac{5}{2}$ moles of As_2O_3.

The moles of As_2O_3 needed to react with 0.508 5 mmol of MnO_4^- =

$(^1/_2)(^5/_2)(0.508 5 \text{ mmol}) = 0.635 6 \text{ mmol} = 0.125 7 \text{ g of As}_2O_3$.

(c) $\dfrac{0.508 5 \text{ mmol KMnO}_4}{0.125 7 \text{ g As}_2O_3} = \dfrac{x \text{ mmol KMnO}_4}{0.146 8 \text{ g As}_2O_3} \Rightarrow x = 0.593 9 \text{ mmol}$

$KMnO_4$ in (29.98 − 0.03) = 29.95 mL $\Rightarrow$ $[KMnO_4]$ = 0.019 83 M.

7-13. FM of NaCl = 58.443. FM of KBr = 119.002. 48.40 mL of 0.048 37 M Ag^+ = 2.341 1 mmol. This must equal the mmol of (Cl^- + Br^-). Let x = mass of NaCl and y = mass of KBr. $x + y$ = 0.238 6 g.

$$\underbrace{\frac{x}{58.443}}_{\text{moles of Cl}^-} + \underbrace{\frac{y}{119.002}}_{\text{moles of Br}^-} = 2.341 1 \times 10^{-3} \text{ mol}$$

Substituting x = 0.238 6 − y gives y = 0.2000 g of KBr = 1.681 mmol of KBr = 1.681 mmol of Br = 0.1343 g of Br = 56.28% of the sample.

7-14. Let x = mg of $FeSO_4 \cdot (NH_4)_2SO_4 \cdot 6H_2O$ and (54.85 − x) = mg of $FeCl_2 \cdot 6H_2O$.

mmol of Ce^{4+} = mmol $FeSO_4 \cdot (NH_4)_2SO_4 \cdot 6H_2O$ + mmol $FeCl_2 \cdot 6H_2O$.

$$(13.39 \text{ mL})(0.01234 \text{ M}) = \frac{x \text{ mg}}{392.13 \text{ mg/mmol}} + \frac{(54.85 - x)}{234.84 \text{ mg/mmol}}$$

$$\Rightarrow x = 40.01 \text{ mg FeSO}_4 \cdot (\text{NH}_4)_2\text{SO}_4 \cdot 6\text{H}_2\text{O}.$$

mass of $\text{FeCl}_2 \cdot 6\text{H}_2\text{O} = 14.84 \text{ mg} = 0.06319 \text{ mmol} = 4.48 \text{ mg Cl}.$

$$\text{wt\% Cl} = \frac{4.48 \text{ mg}}{54.85 \text{ mg}} \times 100 = 8.17\%.$$

7-15. 30.10 mL of Ni^{2+} reacted with 39.35 mL of 0.01307 M EDTA. Therefore, the Ni^{2+} molarity is

$$[\text{Ni}^{2+}] = \frac{(39.35 \text{ mL})(0.01307 \text{ mol/L})}{30.10 \text{ mL}} = 0.01709 \text{ M}.$$

25.00 mL of Ni^{2+} contains 0.4272 mmol of Ni^{2+}. 10.15 mL of EDTA = 0.1327 mmol of EDTA. The amount of Ni^{2+} which must have reacted with CN^- was $0.4272 - 0.1327 = 0.2945$ mmol. The cyanide which reacted with Ni^{2+} must have been $(4)(0.2945) = 1.178$ mmol. $[\text{CN}^-] = 1.178 \text{ mmol}/12.73 \text{ mL} = 0.09254 \text{ M}.$

7-16. (a) mol $\text{O}_2 = (2.9 \times 10^6 \text{ L})(2.2 \times 10^{-4} \text{ M}) = 638$ mol

Reaction 1 requires 2 mol CH_3OH for 3 mol O_2, so the required CH_3OH is $(2/3)(638 \text{ mol}) = 425 \text{ mol CH}_3\text{OH} = 13.6 \text{ kg CH}_3\text{OH}.$

$$\text{volume of CH}_3\text{OH} = \frac{13.6 \text{ kg}}{0.791 \text{ kg/L}} = 17._2 \text{ L}$$

(b) $\qquad 6\text{NO}_3^- + 2\text{CH}_3\text{OH} \rightarrow 6\text{NO}_2^- + 2\text{CO}_2 + 4\text{H}_2\text{O}$

$\qquad\qquad 6\text{NO}_2^- + 3\text{CH}_3\text{OH} \rightarrow 3\text{N}_2 + 3\text{CO}_2 + 3\text{H}_2\text{O} + 6\text{OH}^-$

net: $6\text{NO}_3^- + 5\text{CH}_3\text{OH} \rightarrow 3\text{N}_2 + 5\text{CO}_2 + 7\text{H}_2\text{O} + 6\text{OH}^-$

mol $\text{NO}_3^- = (2.9 \times 10^6 \text{ L})(8.1 \times 10^{-3} \text{ M}) = 2.35 \times 10^4$ mol

Net reaction requires 5 mol CH_3OH for 6 mol NO_3^-, so the required CH_3OH is $(5/6)(2.35 \times 10^4 \text{ mol}) = 1.96 \times 10^4 \text{ mol CH}_3\text{OH} = 627 \text{ kg CH}_3\text{OH}.$

$$\text{volume of CH}_3\text{OH} = \frac{627 \text{ kg}}{0.791 \text{ kg/L}} = 7.9_3 \times 10^2 \text{ L}$$

(c) total volume required $= 1.30 (17._2 \text{ L} + 7.9_3 \times 10^2 \text{ L}) = 1.05 \times 10^3 \text{ L}$

7-17. A known solution of I^- titrated with known volumes of a known concentration of Ag^+. Before the equivalence point, the reaction is $\text{I}^- + \text{Ag}^+ \rightarrow \text{AgI}(s)$. There is a known excess of I^- left after each addition of Ag^+, so $[\text{Ag}^+] = K_{\text{sp}}/[\text{I}^-]$. At the equivalence point, just enough Ag^+ has been added to react with all of I^- to precipitate $\text{AgI}(s)$. The concentrations of dissolved Ag^+ and I^- are equal and are the amounts in equilibrium with $AgI(s)$: $[Ag^+][I^-] = [Ag^+]2 = K_{sp}$. After the equivalence point, there is a known excess of Ag^+, so $[I^-] = K_{sp}/[Ag^+]$.

7-18. At V_e, moles of Ag^+ = moles of I^-

$$(V_e \text{ mL})(0.0511 \text{ M}) = (25.0 \text{ mL})(0.0823 \text{ M}) \Rightarrow V_e = 40.26 \text{ mL}$$

(a) When $V_{Ag^+} = 39.00 \text{ mL}$, $[I^-] = \dfrac{40.26 - 39.00}{40.26}(0.08230)\left(\dfrac{25.00}{25.00 + 39.00}\right)$

$= 1.006 \times 10^{-3}$ M. $[Ag^+] = K_{sp}/[I^-] = 8.3 \times 10^{-14}$ M $\Rightarrow pAg^+ = 13.08$.

(b) When $V_{Ag^+} = V_e$, $[Ag^+][I^-] = x^2 = K_{sp} \Rightarrow x = [Ag^+] = 9.1 \times 10^{-9}$ M

$\Rightarrow pAg^+ = 8.04$.

(c) When $V_{Ag^+} = 44.30 \text{ mL}$, there is an excess of $(44.30 - 40.26) = 4.04 \text{ mL}$ of

Ag^+. $[Ag^+] = \left(\dfrac{4.04}{25.00 + 44.30}\right)(0.05110) = 2.98 \times 10^{-3}$ M $\Rightarrow pAg^+ = 2.53$.

7-19. Moles of Ca^{2+} = moles of $C_2O_4^{2-}$

$$(V_e)(0.0257 \text{ M}) = (25.00 \text{ mL})(0.0311 \text{ M}) \Rightarrow V_e = 30.25 \text{ mL}$$

(a) The fraction of $C_2O_4^{2-}$ remaining when 10.00 mL of Ca^{2+} have been added is

$(30.25 - 10.00)/(30.25) = 0.6694$.

$[C_2O_4^{2-}] = (0.6694)(0.03110 \text{ M})\left(\dfrac{25.00}{35.00}\right) = 0.01487$ M

$[Ca^{2+}] = K_{sp}/[C_2O_4^{2-}] = (1.3 \times 10^{-8})/(0.01487) = 8.7 \times 10^{-7}$

$\Rightarrow pCa^{2+} = -\log(8.7 \times 10^{-7}) = 6.06$

(b) At the equivalence point, there are equal numbers of moles of Ca^{2+} and $C_2O_4^{2-}$

dissolved. Call each concentration x:

$[Ca^{2+}][C_2O_4^{2-}] = (x)(x) = K_{sp} \Rightarrow x = \sqrt{K_{sp}} = 1.1_4 \times 10^{-4}$ M

$pCa^{2+} = -\log(1.1_4 \times 10^{-4}) = 3.94$

(c) $[Ca^{2+}] = (0.02570 \text{ M})\left(\dfrac{35.00 - 30.25}{60.00}\right) = 0.00203$ M. $pCa^{2+} = 2.69$

7-20. Equilibrium constants for ion pair formation:

$$\frac{[AgX(aq)]}{[Ag^+][X^-]} = \begin{cases} 10^{3.31} & (X = Cl) \\ 10^{4.6} & (X = Br) \\ 10^{6.6} & (X = I) \end{cases}$$

Calling the ion pair formation constant K_f, we can write

$[AgX(aq)] = K_f[Ag^+][X^-]$. But the product $[Ag^+][X^-]$ is just K_{sp}. So,

$[AgX(aq)] = K_fK_{sp}$. Putting in the values $K_{sp} = 10^{-9.74}$ for AgCl, $10^{-12.30}$ for AgBr, and $10^{-16.08}$ for AgI gives

$$[AgCl(aq)] = 10^{3.31}10^{-9.74} = 10^{-6.43} \text{ M} = 370 \text{ nM}$$

$$[AgBr(aq)] = 10^{4.6}10^{-12.30} = 10^{-7.7} \text{ M} = 20 \text{ nM}$$

$$[AgI(aq)] = 10^{6.6}10^{-16.08} = 10^{-9.5} \text{ M} = 0.32 \text{ nM}$$

7-21. (a) Titration reaction: SO_4^{2-} [from soil] + Ba^{2+} [from $BaCl_2$ (s)] $\rightarrow BaSO_4(s)$

(b) Prior to adding $BaCl_2$, the 25-mL solution contained 0.000 19 M Cl^-. At the end point, the solution contained 0.009 6 M Cl^- from addition of $BaCl_2$.

$[Cl^-]$ added from $BaCl_2 = 0.009 6 - 0.000 19$ M $= 0.009 4_1$ M

mmol $Cl^- = (0.009 4_1$ M$)(25$ mL$) = 0.23_5$ mmol

$BaCl_2$ contains 1 mol Ba^{2+} for every 2 mol Cl^-, so

mmol Ba^{2+} added $= \frac{1}{2}(0.23_5$ mmol$) = 0.11_8$ mmol

(c) One mole of Ba^{2+} consumes one mole of SO_4^{2-} in the titration. Therefore, 0.11_8 mmol SO_4^{2-} must have been present in the 25-mL aqueous extract.

(d) Mass of SO_4^{2-} in extract $= (0.11_8$ mmol $SO_4^{2-})(96.06$ mg/mmol$) = 11._3$ mg

wt% SO_4^{2-} in soil $= 100 \times (11._3$ mg $SO_4^{2-})/(1\,000$ mg soil$) = 1.1$ wt%

7-22. (i) I^-(excess) + Ag^+ $\rightarrow$ $AgI(s)$ $[Ag^+] = K_{sp}$ (for AgI) / $[I^-]$

(ii) A stoichiometric quantity of Ag^+ has been added that would be just equivalent to I^-, if no Cl^- were present. Instead, a tiny amount of AgCl precipitates and a slight amount of I^- remains in solution.

(iii) Cl^-(excess) + Ag^+ $\rightarrow$ $AgCl(s)$ $[Ag^+] = K_{sp}$ (for AgCl) / $[Cl^-]$

(iv) Virtually all I^- and Cl^- have precipitated.

$$[Ag^+] \approx [Cl^-] \Rightarrow [Ag^+] = \sqrt{K_{sp} \text{ (for AgCl)}}$$

(v) There is excess Ag^+ delivered from the buret.

$$[Ag^+] = [Ag^+]_{titrant} \cdot \frac{\text{volume added past 2nd equivalence point}}{\text{total volume}}$$

7-23. At the equivalence point, $[Ag^+][I^-] = K_{sp} \Rightarrow (x)(x) = 8.3 \times 10^{-17}$

$\Rightarrow [Ag^+] = 9.1 \times 10^{-9}$ M. The concentration of Cl^- in the titration solution is the initial concentration (0.0500 M) corrected for dilution from an initial volume of 40.00 mL up to ~63.85 mL at the equivalence point:

$$[Cl^-] = (0.0500 \text{ M})\left(\frac{40.00}{63.85}\right) = 0.0313 \text{ M}$$

Is the solubility of AgCl exceeded? The reaction quotient is $Q = [Ag^+][Cl^-] = (9.1 \times 10^{-9})(0.0313) = 2.8 \times 10^{-10}$, which is greater than K_{sp} for AgCl $(= 1.8 \times 10^{-10})$. Therefore, AgCl begins to precipitate before AgI finishes precipitating. If the concentration of Cl⁻ were about two times lower, AgCl would not precipitate prematurely.

7-24. $\text{mmol of BrCH}_2\text{CH}_2\text{CH}_2\text{CH}_2\text{Cl} = \dfrac{82.67 \text{ mg}}{171.46 \text{ mg/mmol}} = 0.4822 \text{ mmol}$

There will be 0.4822 mmol of Cl⁻ and 0.4822 mmol of Br⁻ liberated by reaction with $CH_3O^-Na^+$.

$$\text{Ag}^+ \text{ required for Br}^- = \frac{0.4822 \text{ mmol}}{0.02570 \text{ mmol/mL}} = 18.76 \text{ mL}$$

The same amount of Ag⁺ is required to react with Cl⁻, so the second equivalence point is at $18.76 + 18.76 = 37.52 \text{ mL}$.

7-25. Titration of 40.00 mL of 0.0502 M KI + 0.0500 M KCl with 0.0845 M $AgNO_3$

$$I^- + Ag^+ \rightarrow AgI(s) \qquad V_{e1} = (40.00 \text{ mL})\left(\frac{0.0502 \text{ M}}{0.0845 \text{ M}}\right) = 23.76 \text{ mL}$$

$$Cl^- + Ag^+ \rightarrow AgCl(s) \qquad V_{e2} = (40.00 \text{ mL})\left(\frac{0.0502 + 0.0500 \text{ M}}{0.0845 \text{ M}}\right) = 47.43 \text{ mL}$$

The figure gives $V_{e2} = 47.41$ mL, which we will use as a more accurate value.

(a) 10.00 mL: A fraction of the I⁻ has reacted.

$$[I^-] = \underbrace{\left(\frac{23.76 - 10.00}{23.76}\right)}_{\substack{\text{Fraction} \\ \text{remaining}}} \underbrace{(0.0502 \text{ M})}_{\substack{\text{Initial} \\ \text{concentration}}} \underbrace{\left(\frac{40.00}{50.00}\right)}_{\substack{\text{Dilution} \\ \text{factor}}} = 0.0233 \text{ M}$$

$$[Ag^+] = \frac{K_{sp}(AgI)}{[I^-]} = \frac{8.3 \times 10^{-17}}{0.0233} = 3.57 \times 10^{-15} \text{ M} \Rightarrow$$

$$pAg^+ = -\log[Ag^+] = 14.45$$

(b) 20.00 mL: A fraction of the I⁻ has reacted.

$$[I^-] = \left(\frac{23.76 - 20.00}{23.76}\right)(0.0502 \text{ M})\left(\frac{40.00}{60.00}\right) = 0.00530 \text{ M}$$

$$[Ag^+] = \frac{8.3 \times 10^{-17}}{0.00530} = 1.57 \times 10^{-14} \text{ M} \Rightarrow pAg^+ = 13.80$$

(c) 30.00 mL: I^- has been consumed and a fraction of Cl^- has reacted.

$$[Cl^-] = \left(\frac{47.41 - 30.00}{47.41 - 23.76}\right) (0.0500\ M) \left(\frac{40.00}{70.00}\right) = 0.0210\ M$$

$$\underbrace{\phantom{\frac{47.41 - 30.00}{47.41 - 23.76}}}_{\substack{\text{Fraction}\\ \text{remaining}}} \underbrace{}_{\substack{\text{Initial}\\ \text{concentration}}} \underbrace{\phantom{\frac{40.00}{70.00}}}_{\substack{\text{Dilution}\\ \text{factor}}}$$

$$[Ag^+] = \frac{K_{sp}(AgCl)}{[Cl^-]} = \frac{1.8 \times 10^{-10}}{0.0210} = 8.56 \times 10^{-9}\ M \Rightarrow pAg^+ = 8.07$$

(d) $[Ag^+][Cl^-] = x^2 = 1.8 \times 10^{-10} \Rightarrow [Ag^+] = 1.3 \times 10^{-5} \Rightarrow pAg^+ = 4.87$

(e) 50.00 mL: There is excess Ag^+.

$$[Ag^+] = (0.0845\ M) \left(\frac{50.00 - 47.41}{90.00}\right) = 0.00243\ M \Rightarrow pAg^+ = 2.61$$

$$\underbrace{}_{\substack{\text{Initial}\\ \text{concentration}}} \underbrace{\phantom{\frac{50.00 - 47.41}{90.00}}}_{\substack{\text{Dilution}\\ \text{factor}}}$$

7-26. (a) $Hg_2^{2+} + 2CN^- \rightarrow Hg_2(CN)_2(s)$ $K_{sp} = 5 \times 10^{-40}$

$Ag^+ + CN^- \rightarrow AgCN(s)$ $K_{sp} = 2.2 \times 10^{-16}$

K_{sp} for $Hg_2(CN)_2$ is much smaller than K_{sp} for AgCN, so it looks like $Hg_2(CN)_2$ will precipitate first. We will check that assumption soon. The equivalence point occurs at 20.00 mL. The second equivalence point is at 30.00 mL. At 5.00, 10.00, 15.00, and 19.90 mL, there is excess, unreacted Hg_2^{2+}.

At 5.00 mL, $[Hg_2^{2+}] = \left(\frac{20.00 - 5.00}{20.00}\right) (0.1000\ M)\left(\frac{10.00}{10.00 + 5.00}\right) = 0.05000\ M$

$[CN^-] = \sqrt{K_{sp}\ (\text{for } Hg_2(CN)_2)/([Hg_2^{2+}])} = 1.0 \times 10^{-19} \Rightarrow pCN^- = 19.00$

Now let's check to be sure that this much cyanide does not precipitate AgCN:

Q for $AgCN(s) = [Ag^+][CN^-] = \frac{10.00}{15.00} (0.100\ M)\ (1.0 \times 10^{-19}) = 7 \times 10^{-21} <$

K_{sp}. Since $Q < K_{sp}$, AgCN does not precipitate.

By similar calculations we find   10.00 mL: $pCN^- = 18.85$

15.00 mL: $pCN^- = 18.65$

19.90 mL: $pCN^- = 17.76$

At 20.10 mL, AgCN has begun to precipitate. The Ag^+ remaining is

$[Ag^+] = \left(\frac{30.00 - 20.10}{10.00}\right) (0.1000\ M) \left(\frac{10.00}{10.00 + 20.10}\right) = 0.0329\ M$

$[CN^-] = K_{sp}\ (\text{for AgCN})/[Ag^+] = 6.7 \times 10^{-15}\ M \Rightarrow pCN^- = 14.17$

By similar reasoning we find $pCN^- = 13.81$ at 25.00 mL.

At the second equivalence point (30.00 mL), $[Ag^+] = [CN^-] = x$
$\Rightarrow x^2 = K_{sp}$ (for AgCN) $\Rightarrow [CN^-] = 1.5 \times 10^{-8}$ M $\Rightarrow pCN^- = 7.83$.

At 35.00 mL, there are 5.00 mL of excess CN^-.

$$[CN^-] = \left(\frac{5.00}{10.00 + 35.00}\right)(0.1000 \text{ M}) = 0.0111 \text{ M} \Rightarrow pCN^- = 1.95$$

(b) The last question is "Will Ag^+ precipitate when 19.90 mL of CN^- have been added?" We calculated above that $pCN^- = 17.76$ ($\Rightarrow [CN^-] = 1.7 \times 10^{-18}$ M) at 19.90 mL if no Ag^+ had precipitated. We can check to see whether the solubility product of AgCN is exceeded if $[CN^-] = 1.7 \times 10^{-18}$ M and

$$[Ag^+] = \left(\frac{10.00}{10.00 + 19.90}\right) \times (0.1000 \text{ M}) = 0.0334 \text{ M}.$$

$[Ag^+][CN^-] = 5.7 \times 10^{-20} < K_{sp}$ (for AgCN).

The Ag^+ will not precipitate at 19.90 mL.

7-27.
$$\underset{\substack{\text{Analyte} \\ C_M^o, V_M^o}}{M^+} + \underset{\substack{\text{Titrant} \\ C_X^o, V_X^o}}{X^-} \rightleftharpoons MX(s)$$

Mass balance for M: $\qquad C_M^o\, V_M^o = [M^+](V_M^o + V_X) + \text{mol } MX(s)$

Mass balance for X: $\qquad C_X^o V_X = [X^-](V_M^o + V_X) + \text{mol } MX(s)$

Equating mol $MX(s)$ from both mass balances gives

$$C_M^o\, V_M^o - [M^+](V_M^o + V_X) = C_X^o\, V_X - [X^-](V_M^o + V_X)$$

which can be rearranged to $V_X = V_M^o \left(\dfrac{C_M^o - [M^+] + [X^-]}{C_X^o + [M^+] - [X^-]}\right)$

7-28. Your graph should look like the figure in the text.

7-29. Mass balance for M: $C_M^o\, V_M = [M^{m+}](V_M + V_X^o) + x\{\text{mol } M_xX_m(s)\}$

Mass balance for X: $C_X^o V_X^o = [X^{x-}](V_M + V_X^o) + m\{\text{mol } M_xX_m(s)\}$

Equating mol M_xX_m from the two equations gives

$$\frac{1}{x}\{C_M^o\, V_M - [M^{m+}](V_M + V_X^o)\} = \frac{1}{m}\{C_X^o V_X^o - [X^{x-}](V_M + V_X^o)\}$$

which can be rearranged to the required form.

7-30.   Titration of chromate with Ag$^+$:

	A	B	C	D	E	F	G	H	I
1	K$_{sp}$ =	pM	[M]	[X]	V$_M$				
2	1.2E-12	5.46	3.47E-06	9.98E-02	0.013		C2 = 10^-B2		
3	C$_M$ =	5.4	3.98E-06	7.57E-02	1.932		D2 = $A$2/C2^2		
4	0.1	5.3	5.01E-06	4.78E-02	5.342		E2 = $A$8*(2*$A$6+C2-2*D2)		
5	C$_X$ =	5.2	6.31E-06	3.01E-02	8.717			/($A$4-C2+2*D2)	
6	0.1	5.1	7.94E-06	1.90E-02	11.734				
7	V$_X$ =	5	1.00E-05	1.20E-02	14.195				
8	10	4.9	1.26E-05	7.57E-03	16.057				
9		4.8	1.58E-05	4.78E-03	17.388				
10		4.6	2.51E-05	1.90E-03	18.908				
11		4.4	3.98E-05	7.57E-04	19.564				
12		4	1.00E-04	1.20E-04	19.958				
13		3.6	2.51E-04	1.90E-05	20.064				
14		3.2	6.31E-04	3.01E-06	20.189				
15		2.8	1.58E-03	4.78E-07	20.483				
16		2.4	3.98E-03	7.57E-08	21.244				
17		2.3	5.01E-03	4.78E-08	21.583				
18		2.2	6.31E-03	3.01E-08	22.020				
19		2.1	7.94E-03	1.90E-08	22.589				
20		2	1.00E-02	1.20E-08	23.333				
21		1.9	1.26E-02	7.57E-09	24.321				
22		1.8	1.58E-02	4.78E-09	25.650				

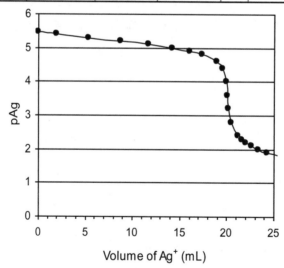

7-31.   Consider the titration of C$^+$ (in a flask) by A$^-$ (from a buret). Before the equivalence point, there is excess C$^+$ in solution. Selective adsorption of C$^+$ on the CA crystal surface gives the crystal a positive charge. After the equivalence point, there is excess A$^-$ in solution. Selective adsorption of A$^-$ on the CA crystal surface gives it a negative charge.

7-32. Beyond the equivalence point, there is excess $Fe(CN)_6^{4-}$ in solution. Selective adsorption of this ion by the precipitate will give the particles a negative charge.

7-33. Add a known excess of standard $AgNO_3$ solution to vigorously stirred unknown I^- in ~0.5 M HNO_3 to precipitate AgI (s). Add some Fe^{3+} and titrate the excess Ag^+ with standard KSCN to precipitate $AgSCN(s)$. When Ag^+ is consumed, the next drop of SCN^- reacts with Fe^{3+} to form the red complex, $FeSCN^{2+}$.

7-34. 50.00 mL of 0.3650 M $AgNO_3$ = 18.25 mmol of Ag^+

3.60 mL of 0.2870 M KSCN = 1.03 mmol of SCN^-

Difference = 18.25 – 1.03 = 17.22 mmol of Br^-

HBr concentration = 17.22 mmol/30.00 mL = 0.5740 M

mg Br^- = (17.22 mmol)(79.904 mg/mmol) = 1 376 mg

7-35. Carbonate is a weak base that reacts with 2 mol HNO_3 to form carbonic acid (H_2CO_3), which dissociates to CO_2 and H_2O. $CO_2(g)$ bubbles out of the solution when HNO_3 is added. Even if all H_2CO_3 remained in solution, the concentration of free CO_3^{2-} would be so low that the solubility product for Ag_2CO_3 would not be exceeded. It would be necessary to precipitate Ag_2CO_3 in a neutral solution, separate and wash the precipitate, and then acidify the filtrate prior to titration with KSCN.

8-1. As ionic strength increases, the charges of the ionic atmospheres increase and the net ionic attractions decrease. There is less tendency for ions to bind to each other.

8-2. (a) true   (b) true   (c) true

8-3. $HBG^- \rightleftharpoons BG^{2-} + H^+$
yellow   blue

The reaction $HBG^- \rightleftharpoons BG^{2-} + H^+$ is an ion dissociation. Increasing ionic strength promotes ion dissociation by forming an ionic atmosphere around each ion, thus decreasing the attraction of H^+ for BG^{2-}. That is, adding NaCl drives the reaction to the right. The reactant is yellow and the product is blue. The initial pale green solution is an equilibrium mixture of yellow HBG^- and blue BG^{2-} (yellow + blue makes green). As NaCl is added, the equilibrium is displaced to the right, causing blue color to increase and yellow color to decrease.

8-4. (a) $\frac{1}{2}[0.008\,7 \cdot 1^2 + 0.008\,7 \cdot (-1)^2] = 0.008\,7$ M

(b) $\frac{1}{2}[0.000\,2 \cdot 3^2 + 0.000\,6 \cdot (-1)^2] = 0.001\,2$ M

8-5. (a) 0.660   (b) 0.54   (c) 0.18 (Eu^{3+} is a lanthanide ion) (d) 0.83

8-6. The ionic strength 0.030 M is halfway between the values 0.01 and 0.05 M. Therefore, the activity coefficient will be halfway between the tabulated values: $\gamma = \frac{1}{2}(0.914 + 0.86) = 0.88_7$.

8-7. (a) $\log \gamma = \dfrac{-0.51 \cdot 2^2 \cdot \sqrt{0.083}}{1 + (600\sqrt{0.083}\,/\,305)} = -0.375 \Rightarrow \gamma = 10^{-0.375} = 0.42_2$

(b) $\gamma = \left(\dfrac{0.083 - 0.05}{0.1 - 0.05}\right)(0.405 - 0.485) + 0.485 = 0.43_2$

8-8. $\gamma = \left(\dfrac{0.083 - 0.05}{0.1 - 0.05}\right)(0.18 - 0.245) + 0.245 = 0.20_2$

8-9. If [ether(aq)] becomes smaller, γ_{ether} must become larger, since $K\,(= [ether(aq)]\,\gamma_{ether})$ is a constant.

8-10. The solubility of Hg_2Br_2 is small, so we assume that Hg_2Br_2 contributes negligible Br^- to 0.001 00 M KBr.

$$\mu = 0.001\ 00\ M,\ [Br^-] = 0.001\ 00\ M,\ \gamma_{Hg_2^{2+}} = 0.867,\ \gamma_{Br^-} = 0.964$$

$$K_{sp} = 5.6 \times 10^{-23} = [Hg_2^{2+}]\gamma_{Hg_2^{2+}}[Br^-]^2\gamma_{Br^-}^2$$

$$= [Hg_2^{2+}](0.867)(0.001\ 00)^2(0.964)^2 \quad \Rightarrow \quad [Hg_2^{2+}] = 7.0 \times 10^{-17}\ M$$

Check our assumption: Yes, Br^- from Hg_2Br_2 is negligible.

8-11. The solubility of $Ba(IO_3)_2$ is small, so we assume that $Ba(IO_3)_2$ contributes negligible IO_3^- to $0.100\ M\ (CH_3)_4NIO_3$.

$$\mu = 0.100\ M,\ [IO_3^-] = 0.100\ M,\ \gamma_{Ba^{2+}} = 0.38,\ \gamma_{IO_3^-} = 0.775$$

$$K_{sp} = 1.5 \times 10^{-9} = [Ba^{2+}]\ \gamma_{Ba^{2+}}[IO_3^-]^2\ \gamma_{IO_3^-}^2$$

$$= [Ba^{2+}](0.38)\ (0.100)^2\ (0.775)^2 \quad \Rightarrow \quad [Ba^{2+}] = 6.6 \times 10^{-7}\ M$$

8-12. Ionic strength $= 0.010\ M$ (from HCl) $+ 0.040\ M$ (from $KClO_4$ that gives $K^+ + ClO_4^-$) $= 0.050\ M$. Using Table 8-1, $\gamma_{H^+} = 0.86$.

$$pH = -\log([H^+]\gamma_{H^+}) = -\log[(0.010)(0.86)] = 2.07.$$

8-13. Ionic strength $= 0.010\ M$ from NaOH $+ 0.012\ M$ from $LiNO_3 = 0.022\ M$. Interpolating in Table 8-1 gives $\gamma_{OH^-} = 0.873$.

$$[H^+]\gamma_{H^+} = \frac{K_w}{[OH^-]\gamma_{OH^-}} = \frac{1.0 \times 10^{-14}}{(0.010)(0.873)} = 1.15 \times 10^{-12}$$

$$pH = -\log(1.15 \times 10^{-12}) = 11.94$$

If we had neglected activities, $pH \approx -\log[H^+] = -\log\dfrac{K_w}{[OH^-]} = 12.00$

8-14. $\varepsilon = 79.755\ e^{-4.6 \times 10^{-3}(323.15 - 293.15)} = 69.474$

$$\log \gamma = \frac{(-1.825 \times 10^6)[(69.474)\ (323.15)]^{-3/2}(-2)^2\ \sqrt{0.100}}{1 + \dfrac{400\ \sqrt{0.100}}{2.00\ \sqrt{(69.474)(323.15)}}}$$

$$= -0.4826 \quad \Rightarrow \quad \gamma = 0.329 \quad \text{(In the table, } \gamma = 0.355 \text{ at } 25°C.)$$

8-15. The equilibrium constant is

$$HA \rightleftharpoons H^+ + A^- \qquad K_a = \frac{[H^+]\gamma_{H^+}\ [A^-]\gamma_{A^-}}{[HA]\gamma_{HA}}.$$

From the table of activity coefficients, $\gamma_{H^+} = 0.83$ and $\gamma_{A^-} = 0.80$ at $\mu = 0.1$ M. For HA, we estimate

$$\log \gamma_{HA} = k\mu = (0.2)(0.1) = 0.02, \text{ or } \gamma_{HA} = 10^{0.02} = 1.05.$$

Putting activity coefficients for $\mu = 0.1$ M into the equilibrium expression gives

$$K_a = \frac{[H^+]\gamma_{H^+}[A^-]\gamma_{A^-}}{[HA]\gamma_{HA}} = \frac{[H^+](0.83)[A^-](0.80)}{[HA](1.05)} = 0.63\frac{[H^+][A^-]}{[HA]}$$

or $\quad \dfrac{[H^+][A^-]}{[HA]} (\mu = 0.1 \text{ M}) = K_a/0.63.$

The activity coefficients at $\mu = 0$ are all 1, so the equilibrium expression is

$$K_a = \frac{[H^+]\gamma_{H^+}[A^-]\gamma_{A^-}}{[HA]\gamma_{HA}} = \frac{[H^+](1)[A^-](1)}{[HA](1)} = \frac{[H^+][A^-]}{[HA]} (\mu = 0).$$

The concentration quotient at the two different ionic strengths is

$$\text{Concentration quotient} = \frac{\dfrac{[H^+][A^-]}{[HA]} (\mu = 0)}{\dfrac{[H^+][A^-]}{[HA]} (\mu = 0.1 \text{ M})} = \frac{K_a}{K_a/0.63} = 0.63$$

in agreement with the observed value of 0.63 ± 0.03.

8-16. The charge balance states the magnitude of positive charge equals the magnitude of negative charge in a solution. That is, the solution must be neutral. The mass balance states that atoms are conserved. If we deliver a certain number of atom A into a solution, then the sum of atom A in all species must equal the atoms of A delivered to the solution. If we deliver a certain ratio of atoms A and B into the solution, then the sum of atoms of A and B in all species must be in that same ratio.

8-17. Charge and mass are proportional to molarity, not to activity.

8-18. $[H^+] + 2[Ca^{2+}] + [Ca(HCO_3)^+] + [Ca(OH)^+] + [K^+] =$
$[OH^-] + [HCO_3^-] + 2[CO_3^{2-}] + [ClO_4^-]$

8-19. Charge: $[H^+] = [OH^-] + [HSO_4^-] + 2[SO_4^{2-}]$

8-20. $[H^+] = [OH^-] + [H_2AsO_4^-] + 2[HAsO_4^{2-}] + 3[AsO_4^{3-}]$

$$\begin{array}{c} \quad\quad O \\ \quad\quad \| \\ H - O - As - O^- \\ \quad\quad | \\ \quad\quad O^- \end{array}$$

8-21. (a) Charge balance: $2[Mg^{2+}] + [H^+] + [MgBr^+] + [MgOH^+] = [Br^-] + [OH^-]$

Mass balance: total Br = 2(total Mg)

$$[MgBr^+] + [Br^-] = 2\{[Mg^{2+}] + [MgBr^+] + [MgOH^+]\}$$

(b) $[Mg^{2+}] + [MgBr^+] + [MgOH^+] = 0.2$ M

$[MgBr^+] + [Br^-] = 0.4$ M

8-22. 250 mL of 1.0×10^{-6} M charge $= 0.25 \times 10^{-6}$ moles of charge.

$$(0.25 \times 10^{-6} \text{ moles of charge})\left(9.648 \times 10^4 \frac{\text{coulombs}}{\text{mole of charge}}\right) = 0.024\,12 \text{ C.}$$

The dielectric constant of air is $\varepsilon = 1$ and the separation is 1.5 m.

$$\text{Force} = -(8.988 \times 10^9)\frac{(0.024\,12)(-0.024\,12)}{(1)(1.5^2)} = 2.3 \times 10^6 \text{ N.}$$

$(2.3 \times 10^6 \text{ N})(0.224\,8 \text{ pounds/N}) = 5.2 \times 10^5$ pounds.

Two elephants do not weigh enough to keep the beakers apart.

8-23. $[CH_3CO_2^-] + [CH_3CO_2H] = 0.1$ M

8-24. $Y_{total} = \frac{3}{2}X_{total}$

$$2[X_2Y_2^{2+}] + [X_2Y^{4+}] + 3[X_2Y_3] + [Y^{2-}] = \frac{3}{2}\{2[X_2Y_2^{2+}] + 2[X_2Y^{4+}] + 2[X_2Y_3]\}$$

$[Y^{2-}] = [X_2Y_2^{2+}] + 2[X_2Y^{4+}]$

8-25. 3 (total Fe) = 2 (total sulfur)

$$3\{[Fe^{3+}] + [Fe(OH)^{2+}] + [Fe(OH)_2^+] + 2[Fe_2(OH)_2^{4+}] + [FeSO_4^+]\}$$
$$= 2\{[FeSO_4^+] + [SO_4^{2-}] + [HSO_4^-]\}$$

We write 2 in front of $[Fe_2(OH)_2^{4+}]$ because $Fe_2(OH)_2^{4+}$ contains 2 Fe.

8-26. Here is the spreadsheet after executing Goal Seek:

	A	B	C	D	E
1	Using Goal Seek for Ammonia Equilibrium				
2					
3	$pK_b =$	4.755	$K_b =$	1.76E-05	= 10^-B3
4	$K_w =$	14.00	$K_w =$	1.00E-14	= 10^-B4
5	$F =$	0.05			
6					
7	$pH =$	10.96791342	Initial value is estimate		
8	$[H^+] =$	1.08E-11	= 10^-B7		
9	$[NH_4^+] = K_w/[H^+] - [H^+] =$	9.29E-04	= D4/B8-B8		
10	$[OH^-] = K_w/[H^+] =$	9.29E-04	= D4/B8		
11	$[NH_3] = F - K_w/[H^+] + [H^+] =$	4.91E-02	= B5-D4/B8+B8		
12	$K_b = [NH_4^+][OH^-]/[NH_3] =$	1.76E-05	= B9*B10/B11		
13	$K_b - [NH_4^+][OH^-]/[NH_3] =$	3.82E-14	= D3-B12		

Concentrations in cells B8:B11 are $[H^+] = 1.08 \times 10^{-11}$ M, $[NH_4^+] = 9.29 \times 10^{-4}$ M, $[OH^-] = 9.29 \times 10^{-4}$ M, and $[NH_3] = 4.91 \times 10^{-2}$ M. The pH is 10.97 and the fraction of hydrolysis is $[NH_4^+]/([NH_4^+] + [NH_3]) = 9.29 \times 10^{-4}$ M $/0.05$ M $= 1.86\%$. Increasing the formal concentration of ammonia increased the pH and decreases the fraction of hydrolysis.

	0.01	0.05
$F =$	0.01	0.05
$pH =$	10.6134	10.96791
$[H^+] =$	2.44E-11	1.08E-11
$[NH_4^+] = K_w/[H^+] - [H^+] =$	0.000411	0.000929
$[OH^-] = K_w/[H^+] =$	0.000411	0.000929
$[NH_3] = F - K_w/[H^+] + [H^+] =$	0.009589	0.049071
$K_b = [NH_4^+][OH^-]/[NH_3] =$	1.76E-05	1.76E-05
% of hydrolysis = $100*[NH_4^+]/F$	4.11	1.86

8-27. Here is the spreadsheet after executing Solver:

	A	B	C	D	E	F
1	Ammonia equilibrium					F =
2	1. *Estimate* values of pC = -log[C] for NH_4^+ and OH^- in cells B6 and B7					0.01
3	2. Use Solver to adjust the values of pC to minimize the sum in cell F8					
4						
5	Species	pC			Mass and charge balances	b_i
6	NH_4^+	3.38660381	4.11E-04	C6 = 10^-B6	$b_1 = 0 = F - [NH_4^+] - [NH_3] =$	9.90E-14
7	OH^-	3.38660378	0.000410579	C7 = 10^-B7	$b_2 = 0 = [NH_4^+] + [H^+] - [OH^-] =$	4.99E-15
8	NH_3		0.009589422	C8 = C6*C7/D12	$\Sigma b_i^2 =$	9.83E-27
9	H^+		2.43559E-11	C9 = D13/C7		F6 = F2-C6-C8
10						F7 = C6+C9-C7
11						F8 = F6^2+F7^2
12	$pK_b =$	4.755	$K_b =$	1.76E-05	=10^-B12	
13	$pK_w =$	14.00	$K_w =$	1.00E-14	=10^-B13	

Concentrations in cells C6:C9 are $[NH_4^+] = 4.11 \times 10^{-4}$ M, $[OH^-] = 4.11 \times 10^{-4}$ M, $[NH_3] = 9.59 \times 10^{-3}$ M and $[H^+] = 2.44 \times 10^{-11}$ M. The fraction of hydrolysis is $[NH_4^+]/([NH_4^+] + [NH_3]) = 4.11 \times 10^{-4}/0.01 = 4.11\%$.

8-28. (a) Pertinent reactions:

$$A^- + H_2O \underset{}{\overset{K_b}{\rightleftharpoons}} HA + OH^- \qquad K_b = \frac{[HA]\gamma_{HA}[OH^-]\gamma_{OH^-}}{[A^-]\gamma_{A^-}} = 10^{-9.244} \qquad (A)$$

$$H_2O \overset{K_w}{\rightleftharpoons} H^+ + OH^- \qquad K_w = [H^+]\gamma_{H^+}[OH^-]\gamma_{OH^-} = 10^{-14.00} \qquad (B)$$

Charge balance: $[H^+] + [Na^+] = [OH^-] + [A^-]$ (C)

Mass balance: $[Na^+] = 0.01$ M $\equiv$ F (D)

Mass balance: $[HA] + [A^-] = 0.01$ M $\equiv$ F (E)

(b) For initial estimates, I chose pA = 2 and pOH = 5. It took several cycles of solving for pA and pOH one-at-a-time to find the values in the spreadsheet below. Concentrations appear in cells C8:C11. The pH is in cell B16 and μ is in cell B5. The fraction of hydrolysis is $[HA]/F = 2.4 \times 10^{-6}/0.010 = 0.024\%$.

	A	B	C	D	E	F	G	H
1	Sodium acetate hydrolysis						F = [Na$^+$] =	
2	1. *Estimate* values of pC = -log[C] for A$^-$ and OH$^-$ in cells B8 and B9						0.01	
3	2. Use Solver to adjust the values of pC to minimize the sum in cell H16							
4	Ionic strength							
5	μ	1.000E-02				Extended	Activity	
6				Size		Debye-Hückel	coefficient	
7	Species	pC	C (M)	α (pm)	Charge	log γ	γ	
8	A$^-$	2.00010375	9.998E-03	450	-1	-4.444E-02	9.027E-01	G8 = 10^F8
9	OH$^-$	5.62093675	2.394E-06	350	-1	-4.575E-02	9.000E-01	
10	HA		2.389E-06		0	0.000E+00	1.000E+00	
11	H$^+$		5.082E-09	900	1	-3.938E-02	9.133E-01	
12								
13	pK$_b$ =	9.244	K$_b$ =	5.70E-10		Mass and charge balances:		b$_i$
14	pK$_w$ =	14.00	K$_w$ =	1.00E-14		b$_1$ = 0 = F - [HA] - [A$^-$] =		-9.18E-16
15						b$_2$ = 0 = [Na$^+$] + [H$^+$] - [A$^-$] - [OH$^-$]		-2.441E-15
16	pH = -log([H+]γ$_{H+}$)=	8.33					Σb$_i^2$ =	6.80E-30
17							H14 = G2-C10-C8	
18							H15 = G2+C11-C8-C9	
19	Initial values:						H16 = H14^2 + H15^2	
20	pA$^-$ =	2					C8 = 10^-B8	
21	pOH =	5					C9 = 10^-B9	
22						C10 = D13*C8*G8/(C9*G9*G10)		
23						C11 = D14/(C9*G9*G11)		
24					F8 = -0.51*E8^2*SQRT($B$5)/(1+D8*SQRT($B$5)/305)			
25					B5 = 0.5*(G2+E8^2*C8+E9^2*C9+E10^2*C10+E11^2*C11)			

8-29. (a) Pertinent reactions:

$$Ca(OH)_2(s) \overset{K_{sp}}{\rightleftharpoons} Ca^{2+} + 2OH^- \quad K_{sp} = [Ca^{2+}]\gamma_{Ca2+}[OH^-]^2\gamma_{OH^-}^2 = 10^{-5.19}$$

$$Ca^{2+} + OH^- \overset{K_1}{\rightleftharpoons} CaOH^+ \quad K_1 = \frac{[CaOH^+]\gamma_{CaOH^+}}{[Ca^{2+}]\gamma_{Ca2+}[OH^-]\gamma_{OH^-}} = 10^{1.30}$$

$$H_2O \overset{K_w}{\rightleftharpoons} H^+ + OH^- \quad K_w = [H^+]\gamma_{H^+}[OH^-]\gamma_{OH^-} = 10^{-14.00}$$

Charge balance: $2[Ca^{2+}] + [CaOH^+] + [H^+] = [OH^-]$

Mass balance: $[OH^-] + [CaOH^+] = 2\{[Ca^{2+}] + [CaOH^+]\} + [H^+]$

$\underbrace{\qquad\qquad\qquad\qquad}$ $\underbrace{\qquad\qquad\qquad\qquad}$

species containing OH$^-$ species containing Ca^{2+}

(Mass balance gives the same result as charge balance.)

There are 4 equations (3 equilibria and charge balance) and 4 unknowns: $[Ca^{2+}]$, $[CaOH^+]$, $[H^+]$, and $[OH^-]$.

(b) There are 4 unknowns and 3 equilibria, so we need to estimate $4 - 3 = 1$ concentration. I choose $pCa^{2+} \approx 3$ as an initial estimate. The spreadsheet after executing Solver twice in a row is shown below.

	A	B	C	D	E	F	G	H
1	Calcium hydroxide equilibria							
2	1. *Estimate* pCa^{2+} in cell B8							
3	2. Use Solver to adjust B8 to minimize sum in cell H15							
4	Ionic strength							
5	μ	5.201E-02				Extended	Activity	
6				Size		Debye-Hückel	coefficient	
7	Species	pC	C (M)	α (pm)	Charge	log γ	γ	
8	Ca^{2+}	1.8077757	1.557E-02	600	2	-3.212E-01	4.773E-01	G8 = 10^F8
9	OH^-		3.645E-02	350	-1	-9.219E-02	8.087E-01	
10	$CaOH^+$		5.311E-03	500	1	-8.466E-02	8.229E-01	
11	H^+		3.982E-13	900	1	-6.952E-02	8.521E-01	
12								
13	pK_{sp} =	5.19	K_{sp} =	6.46E-06		Mass and charge balances:		b_i
14	pK_1 =	-1.30	K_1 =	2.00E+01		$b_1 = 0 = 2[Ca^{2+}]+[CaOH^+]+[H^+]-[OH^-]$ =		6.25E-17
15	pK_w =	14.00	K_w =	1.00E-14			Σb_i^2 =	3.90E-33
16							H14 = 2*C8+C10+C11-C9	
17	Ion size estimate:						H15 = H14^2	
18	$CaOH^+$ size ≈ 500 pm						C8 = 10^-B8	
19							C9 = SQRT(D13/(C8*G8))/G9	
20	Initial value:						C10 = D14*C8*G8*C9*G9/G10	
21	pCa^{2+} =	3					C11 = D15/(C9*G9*G11)	
22						F8 = -0.51*E8^2*SQRT($B$5)/(1+D8*SQRT($B$5)/305)		
23						B5 = 0.5*(E8^2*C8+E9^2*C9+E10^2*C10+E11^2*C11)		

Results: $[Ca^{2+}]$ = 0.015 6 M $[CaOH^+]$ = 0.005 3 M

$[OH^-]$ = 0.036 4 M $[H^+]$ = $K_w/[OH^-]$ = 3.98 × 10^{-13} M

Total dissolved Ca = 0.015 6 + 0.005 3 = 0.020 9 M

Fraction of hydrolysis = $[CaOH^+]/\{[Ca^{2+}] + [CaOH^+]\}$ = 25%

The formula mass of $Ca(OH)_2$ is 74.09 g/mol, so 0.020 9 M is 1.5$_5$ g/L. The calculated solubility makes sense because it is between the reported values of 1.85 g/L at 0°C and 0.77 g/L at 100°C.

8-30. The spreadsheet shows $[Na^+] = [Cl^-] = 0.024\ 86$ M and $[NaCl(aq)] = 0.000\ 143$ M in cells C8:C10. Ionic strength $= 0.024\ 86$ M in cell B5. Fraction of ion pairing $= [NaCl(aq)]/F = 0.57\%$ in cell D15.

	A	B	C	D	E	F	G	H
1	Sodium Chloride ion pairing with activities							
2	1. *Estimate* pNa$^+$ and PCl$^-$ in cells B8 and B9					Formal conc = F =	0.025	M
3	2. Use Solver to adjust B8 and B9 to minimize sum in cell H16							
4	Ionic strength							
5	μ	2.486E-02				Extended	Activity	
6				Size		Debye-Hückel	coefficient	
7	Species	pC	C (M)	α (pm)	Charge	log γ	γ	
8	Na$^+$	1.604555369	2.486E-02	450	1	-6.523E-02	0.861	G8 = 10^F8
9	Cl$^-$	1.604555369	2.486E-02	300	-1	-6.961E-02	0.852	
10	NaCl(aq)		1.432E-04		0	0.000E+00	1.000	
11								
12								
13	pK$_{ip}$ =	0.50		K$_{ip}$ =	3.16E-01	Mass and charge balances:		b$_i$
14						b$_1$ = 0 = F - [Na$^+$] - [NaCl] =		1.44E-13
15	Ion pair fraction = [NaCl(aq)]/F =			0.0057		b$_2$ = 0 = [Na$^+$] - [Cl$^-$] =		1.888E-13
16				D15 = C10/G2			Σb$_i^2$ =	5.65E-26
17	Ion size estimate:						H14 = G2-C8-C10	
18							H15 = C8-C9	
19	Initial value:						H16 = H14^2+H15^2	
20	pNa$^+$ =	2					C8 = 10^-B8	
21	pCl$^-$ =	2					C9 = 10^-B9	
22	check:						C10 = D13*C8*G8*C9*G9/G10	
23	[Na$^+$] + [NaCl(aq)] =		0.025000			F8 = -0.51*E8^2*SQRT($B$5)/(1+D8*SQRT($B$5)/305)		
24	[Cl$^-$] + [NaCl(aq)] =		0.025000			B5 = 0.5*(E8^2*C8+E9^2*C9+E10^2*C10)		

8-31. The spreadsheet shows $[Na^+] = 0.047\ 75$ M, $[SO_4^{2-}] = 0.022\ 75$M, and $[NaSO_4^-(aq)]$ $= 0.002\ 246$ M in cells C8:C10. Ionic strength $= 0.070\ 51$ M in cell B5. Ion pair fraction $= [NaSO_4^-(aq)]/F = 8.98\%$ in cell D15.

	A	B	C	D	E	F	G	H
1	Sodium Sulfate ion pairing with activities							
2	1. *Estimate* pNa and PSO4 in cells B8 and B9					Formal conc = F =	0.025	M
3	2. Use Solver to adjust B8 and B9 to minimize cell H16					$[Na+] + [NaSO_4^-(aq)] = 2F$		
4	Ionic strength					$[SO4^{2-}] + [NaSO_4^-(aq)] = F$		
5	μ	7.051E-02				Extended	Activity	
6				Size		Debye-Hückel	coefficient	
7	Species	pC	C (M)	α (pm)	Charge	log γ	γ	
8	Na^+	1.320987425	4.775E-02	450	1	-9.730E-02	0.799	G8 = 10^F8
9	SO_4^{2-}	1.642936328	2.275E-02	400	-2	-4.018E-01	0.396	
10	$NaSO_4^-$		2.246E-03	500	-1	-9.435E-02	0.805	
11								
12								
13	pK_{ip} =	-0.72		K_{ip} =	5.25E+00	Mass and charge balances:		b_i
14						$b_1 = 0 = 2F - [Na^+] - [NaSO_4-] =$		5.36E-13
15	Ion pair fraction = $[NaSO_4^-(aq)]/F$ =			0.0898		$b_2 = 0 = F - [SO_4^{2-}] - [NaSO_4-] =$		-2.33E-13
16				D15 = C10/G2			Σb_i^2 =	3.42E-25
17	Ion size estimate:					H14 = 2*G2-C8-C10		
18	$NaSO_4^-$ =	500				H15 = G2-C9-C10		
19	Initial value:					H16 = H14^2 + H15^2		
20	pNa^+ =	1.4				C8 = 10^-B8		
21	pSO_4^{2-} =	1.5				C9 = 10^-B9		
22	Check:					C10 = D13*C8*G8*C9*G9/G10		
23	$[Na^+] + [NaSO_4^-]$ =		0.050000			F8 = -0.51*E8^2*SQRT($B$5)/(1+D8*SQRT($B$5)/305)		
24	$[SO_4^{2-}] + [NaSO_4^-]$ =		0.025000			B5 = 0.5*(E8^2*C8+E9^2*C9 +E10^2*C10)		

8-32. (a) The spreadsheet shows $[Mg^{2+}] = [SO_4^{2-}] = 0.016\ 16$ M and $[MgSO_4(aq)] = 0.008\ 844$ M in cells C8:C10. Ionic strength $= 0.064\ 63$ M in cell B5. Fraction of ion pairing $= 35.4\%$ in cell D15.

(b) $Mg^{2+} + OH^- \rightleftharpoons MgOH^+$ $\qquad K_1 = 10^{2.6}$ in Appendix I

$\qquad SO_4^{2-} + H_2O \rightleftharpoons HSO_4^- + OH^-$ $\qquad pK_b = 12.01$

For hydrolysis of Mg^{2+} with $[Mg^{2+}] = 0.016$ M and $[OH^-] = 10^{-7}$ M, $[MgOH^+] \approx K_1[Mg^{2+}][OH^-] = 10^{2.6}[0.016]10^{-7} = 6 \times 10^{-7}$ M, which is negligible in comparison with $[Mg^{2+}] = 0.016$ M.

For hydrolysis of SO_4^{2-}, the equilibrium constant is $K_b = K_w/K_a$, with $pK_a = 1.99$ in Appendix G. $K_b = K_w/K_a = 10^{-14.00}/10^{-1.99} = 10^{-12.01}$. Putting in the concentrations $[SO_4^{2-}] = 0.016$ M and $[OH^-] = 10^{-7}$ M, we estimate $[HSO_4^-] \approx K_b[SO_4^{2-}]/[OH^-] = 10^{-12.01}[0.016]/10^{-7} = 2 \times 10^{-7}$ M, which is negligible in comparison with $[SO_4^{2-}] = 0.016$ M.

	A	B	C	D	E	F	G	H
1	Magnesium Sulfate ion pairing with activities							
2	1. *Estimate* pMg^{2+} and pSO_4^{2-} in cells B8 and B9					Formal conc = F =	0.025	M
3	2. Use Solver to adjust B8 and B9 to minimize sum in cell H16							
4	Ionic strength							
5	μ	6.463E-02				Extended	Activity	
6				Size		Debye-Hückel	coefficient	
7	Species	pC	C (M)	α (pm)	Charge	log γ	γ	
8	Mg^{2+}	1.79165324	1.616E-02	800	2	-3.111E-01	4.885E-01	G8 = 10^F8
9	SO_4^{2-}	1.79165324	1.616E-02	400	-2	-3.889E-01	4.084E-01	
10	$MgSO_4(aq)$		8.844E-03			0.000E+00	1.000E+00	
11								
12								
13	pK_{ip} =	-2.23		K_{ip} =	1.70E+02	Mass and charge balances:		b_i
14						$b_1 = 0 = F - [Mg^{2+}] - [MgSO_4] =$		9.48E-13
15	Ion pair fraction = $[MgSO_4(aq)]/F$ =			0.3537		$b_2 = 0 = [Mg^{2+}] - [SO4^{2-}] =$		-3.06E-13
16				D15 = C10/G2		Σb_i^2 =		9.92E-25
17	Ion size estimate:					H14 = G2-C8-C10		
18						H15 = C8-C9		
19	Initial value:					H16 = H14^2+H15^2		
20	pMg^{2+} =	1.9				C8 = 10^-B8		
21	pSO_4^{2-} =	1.9				C9 = 10^-B9		
22	Check:					C10 = D13*C8*G8*C9*G9/G10		
23	$[Mg^{2+}] + [MgSO_4(aq)]$ =	0.025000		F8 = -0.51*E8^2*SQRT($B$5)/(1+D8*SQRT($B$5)/305)				
24						B5 = 0.5*(E8^2*C8+E9^2*C9)		

8-33. (a) Equilibria:

$$LiF(s) \rightleftharpoons Li^+ + F^- \qquad K_{sp} = [Li^+]\gamma_{Li^+}[F^-]\gamma_{F^-} \qquad pK_{sp} = 2.77$$

$$LiF(s) \rightleftharpoons LiF(aq) \qquad K_{ion\ pair} = [LiF(aq)]\gamma_{LiF(aq)} \qquad pK_{ion\ pair} = 2.54$$

$$F^- + H_2O \rightleftharpoons HF + OH^- \quad K_b = \frac{K_w}{K_a} = \frac{[HF]\,\gamma_{HF}\,[OH^-]\gamma_{OH^-}}{[F^-]\gamma_{F^-}} \quad pK_b = 10.83$$

$$H_2O \overset{K_w}{\rightleftharpoons} H^+ + OH^- \qquad K_w = [H^+]\gamma_{H^+}[OH^-]\gamma_{OH^-} \qquad pK_w = 14.00$$

Derivation of $pK_{ion\ pair}$:

$$LiF(s) \rightleftharpoons Li^+ + F^- \qquad pK_{sp} = 2.77 \text{ from Appendix F}$$

$$Li+ + F^- \rightleftharpoons LiF(aq) \qquad pK_{formation} = -0.23 \text{ from Appendix J}$$

$$LiF(s) \rightleftharpoons LiF(aq) \qquad K_{ion\ pair} = K_{sp}K_{formation}$$

$$pK_{ion\ pair} = pK_{sp} + pK_{formation} = 2.54$$

Charge balance: $[Li^+] + [H^+] = [F^-] + [OH^-]$

Mass balance: $[Li^+] + [LiF(aq)] = [F^-] + [LiF(aq)] + [HF]$

(b) There are six equations and six unknowns: $[Li^+]$, $[F^-]$, $[LiF(aq)]$, $[HF]$, $[H^+]$, and $[OH^-]$. For the spreadsheet, we need to estimate pC values for (6 unknowns) – (4 equilibria) = 2 unknowns. I choose to estimate pF and pOH because F^- and OH^- are involved in multiple equilibria. It does not work to chose F^- and Li^+ because either concentration fixes that of the other through the relation $K_{sp} = [Li^+]\gamma_{Li^+}[F^-]\gamma_{F^-}$. The equations in the spreadsheet are:

$[Li^+] = K_{sp}/([F^-]\gamma_{F^-}\,\gamma_{Li^+})$ $[HF] = K_b[F^-]\gamma_{F^-}/([OH^-]\gamma_{OH^-}\gamma_{F^-})$

$[LiF(aq)] = K_{ion\ pair}/\gamma_{LiF(aq)}$ $[H^+] = K_w/[OH^-]\gamma_{OH^-}\,\gamma_{H^+}$

	A	B	C	D	E	F	G	H	I
1	Lithium fluoride equilibria								
2	1. *Estimate* values in cells B8 and B9								
3	2. Use Solver to adjust B8 and B9 to minimize sum in cell I19								
4	Ionic strength					Extended			
5	μ =	0.050126				Debye-			
6			Size			Hückel	Activity		
7	Species	pC	C (M)	α (pm)	Charge	log γ	coefficient, γ		
8	F⁻	1.299947	5.012E-02	350	-1	-9.084E-02	0.811		
9	OH⁻	6.060783	8.694E-07	350	-1	-9.084E-02	0.811		
10	Li⁺		5.013E-02	600	1	-7.927E-02	0.833		
11	LiF(aq)		2.884E-03		0	0.000E+00	1.000		
12	HF		8.528E-07		0	0.000E+00	1.000		
13	H⁺		1.661E-08	900	1	-6.876E-02	0.854		
14									
15									
16							Mass and charge balances:		b_i
17	pK_{sp} =	2.77	K_{sp} =	1.70E-03			$b_1 = 0 = [Li^+] - [F^-] - [HF]$ =		-6.81E-11
18	$pK_{ion\ pair}$ =	2.54	$K_{ion\ pair}$ =	2.88E-03			$b_2 = 0 = [Li^+] + [H^+] - [F^-] - [OH^-]$ =		-7.66E-11
19	pK_{base} =	10.83	K_{base} =	1.48E-11			Σb_i^2 =		1.05E-20
20	pK_w =	14.00	K_w =	1.00E-14				I17 = C10-C8-C12	
21								I18 = C10+C13-C8-C9	
22	Initial values:							I19 = I17^2 + I18^2	
23	pF = 1.5	pOH = 6						C8 = 10^-B8	
24								C9 = 10^-B9	
25	Optimize both pF and pOH together for a few cycles						C10 = D17/(C8*G8*G10)		
26	Then optimize just pF and just pOH alternately							C11 = D18	
27	Continue optimization as long as Σb_i^2 keeps getting smaller						C12 = D19*C8*G8/(G9*C9*G12)		
28							C13 = D20/(C9*G9*G13)		
29			B5 = 0.5*(E8^2*C8+E9^2*C9+E10^2*C10+E11^2*C11+E12^2*C12+E13^2*C13)							

8-34. (a)

$$CaCO_3(s) \rightleftharpoons Ca^{2+} + CO_3^{2-} \qquad K_{sp} = 4.5 \times 10^{-9}$$

$$CO_2(aq) + H_2O \rightleftharpoons HCO_3^- + H^+ \qquad K_1 = 4.46 \times 10^{-7}$$

$$CO_3^{2-} + H^+ \rightleftharpoons HCO_3^- \qquad 1/K_2 = 1/(4.69 \times 10^{-11})$$

$$\overline{CaCO_3(s) + CO_2(aq) + H_2O \rightleftharpoons Ca^{2+} + 2HCO_3^-} \qquad \begin{array}{l} K = K_{sp} K_1 / K_2 \\ = 4.2_8 \times 10^{-5} \end{array}$$

(b) The equilibrium constant for the net reaction is

$$\frac{[Ca^{2+}][HCO_3^-]^2}{[CO_2(aq)]} = K = 4.2_8 \times 10^{-5}$$

We can substitute into this equation $[HCO_3^-] = 2[Ca^{2+}]$ and $[CO_2(aq)] = K_{CO_2}P_{CO_2}$ (where $K_{CO_2} = 0.032$ and $P_{CO_2} = 4.0 \times 10^{-4}$ bar) to get

$$\frac{[Ca^{2+}](2[Ca^{2+}])^2}{K_{CO_2}P_{CO_2}} = K \Rightarrow [Ca^{2+}] = 5.1_5 \times 10^{-4} \text{ M} = 21 \text{ mg/L}$$

(c) If $[Ca^{2+}] = 80$ mg/L $= 2.0 \times 10^{-3}$ M, then

$$P_{CO_2} = \frac{[Ca^{2+}](2[Ca^{2+}])^2}{K_{CO_2}K} = 0.023 \text{ bar}$$

The partial pressure of CO_2 in the river is about (0.023 bar)/(4.0 × 10^{-4} bar) = 58 times higher than the atmospheric pressure of CO_2. There must be a source of extra CO_2 such as respiration in the river or inflow of ground water that is very rich in CO_2 and not in equilibrium with the atmosphere.

CHAPTER 9
MONOPROTIC ACID-BASE EQUILIBRIA

9-1.   HBr (or any other acid or base) drives the reaction $H_2O \rightleftharpoons H^+ + OH^-$ to the left, according to Le Châtelier's principle. If, for example, the solution contains 10^{-4} M HBr, the concentration of OH^- from H_2O is $K_w/[H^+] = 10^{-10}$ M. The concentration of H^+ from H_2O must also be 10^{-10} M, since H^+ and OH^- are created in equimolar quantities.

9-2.   (a) $pH = -\log [H^+] = -\log (1.0 \times 10^{-3}) = 3.00$

   (b) $[H^+] = K_w /[OH^-] = (1.0 \times 10^{-14})/(1.0 \times 10^{-2}) = 1.0 \times 10^{-12}$ M

   $pH = -\log [H^+] = 12.00$

9-3.   Charge balance: $[H^+] = [OH^-] + [ClO_4^-] \Rightarrow [OH^-] = [H^+] - 5.0 \times 10^{-8}$

   Mass balance is the same as charge balance.

   Equilibrium: $[H^+] [OH^-] = K_w$

   $[H^+] ([H^+] - 5.0 \times 10^{-8}) = 1.0 \times 10^{-14} \Rightarrow [H^+] = 1.28 \times 10^{-7}$ M

   $pH = -\log [H^+] = 6.89$

   $[OH^-] = K_w/[H^+] = 7.8 \times 10^{-8}$ M $\Rightarrow [H^+]$ from $H_2O = 7.8 \times 10^{-8}$ M

   Fraction of $[H^+]$ from $H_2O = \dfrac{7.8 \times 10^{-8} \text{ M}}{1.28 \times 10^{-7} \text{ M}} = 0.61$

9-4.   (a) $pH = -\log [H^+]\gamma_{H^+}$

   $1.092 = -\log (0.100) \gamma_{H^+} \Rightarrow \gamma_{H^+} = 0.809$

   The tabulated activity coefficient is 0.83.

   (b) $2.102 = -\log (0.0100)\gamma_{H^+} \Rightarrow \gamma_{H^+} = 0.791$

   (c) The activity coefficient depends somewhat on the identity of the counterions.

9-5.   (a) 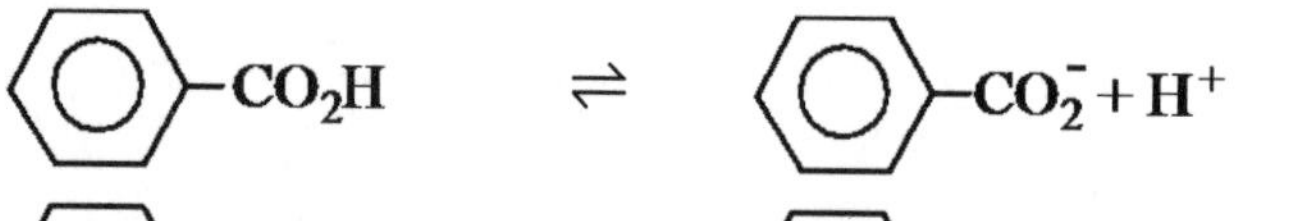 K_a

   (b) 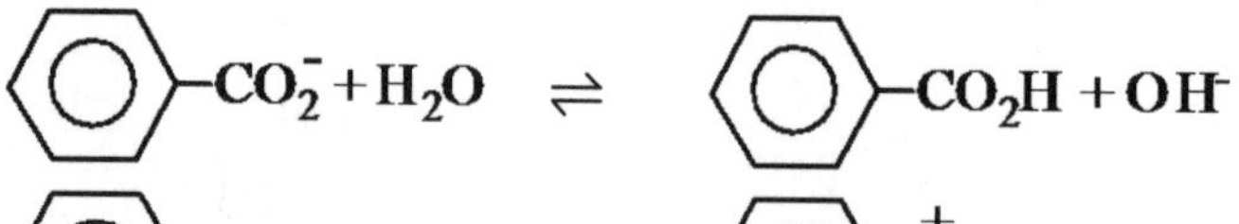 K_b

   (c) 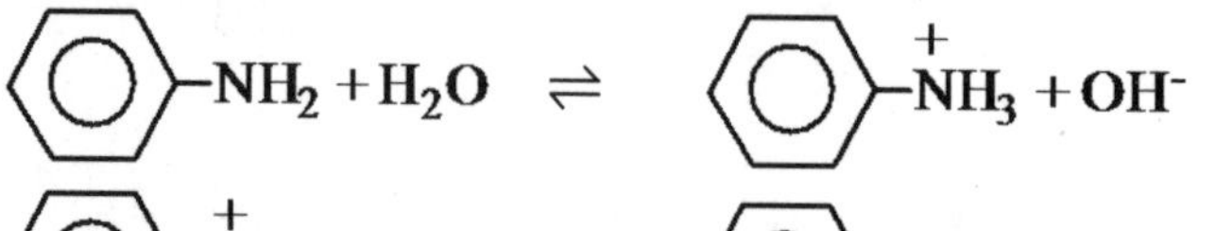 K_b

   (d) 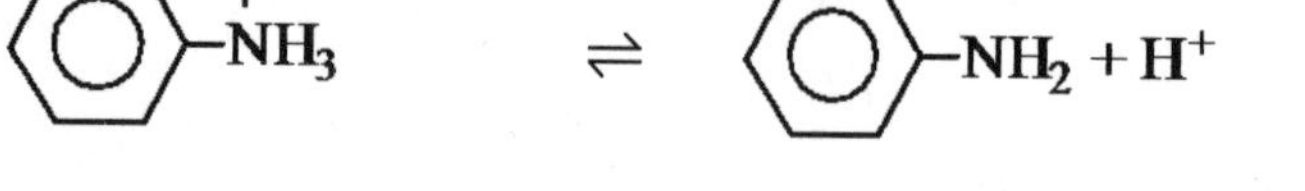 K_a

9-6. Let $x = [H^+] = [A^-]$ and $0.100 - x = [HA]$.

$$\frac{x^2}{0.100 - x} = 1.00 \times 10^{-5} \Rightarrow x = 9.95 \times 10^{-4} \text{ M} \Rightarrow \text{pH} = -\log x = 3.00$$

$$\alpha = \frac{[A^-]}{[A^-] + [HA]} = \frac{9.95 \times 10^{-4}}{0.100} = 9.95 \times 10^{-3}$$

9-7.
$$BH^+ \overset{K_a}{\rightleftharpoons} B + H^+ \qquad K_a = K_w/K_b = 1.00 \times 10^{-10}$$
$$\quad 0.100 - x \qquad\quad x \quad\; x$$

$$\frac{x^2}{0.100 - x} = 1.00 \times 10^{-10} \Rightarrow x = [B] = [H^+] = 3.16 \times 10^{-6} \text{ M} \Rightarrow \text{pH} = 5.50$$

9-8.
$$(CH_3)_3NH^+ \rightleftharpoons (CH_3)_3N + H^+ \qquad K_a = 1.59 \times 10^{-10}$$
$$\quad F - x \qquad\qquad x \qquad x$$

$$\frac{x^2}{0.060 - x} = K_a \Rightarrow x = 3.0_9 \times 10^{-6} \Rightarrow \text{pH} = 5.51$$

$$[(CH_3)_3N] = x = 3.1 \times 10^{-6} \text{ M}, \quad [(CH_3)_3NH^+] = F - x = 0.060 \text{ M}$$

9-9. $HA \overset{K_a}{\rightleftharpoons} H^+ + A^-$. $Q = \frac{[A^-][H^+]}{[HA]}$. When the system is at equilibrium, $Q = K_a$.

Let's call the concentrations at equilibrium $[A^-]_e$, $[H^+]_e$, and $[HA]_e$. If the solution is diluted by a factor of 2, the concentrations become $\frac{1}{2}[A^-]_e$, $\frac{1}{2}[H^+]_e$, and $\frac{1}{2}[HA]_e$.

The reaction quotient becomes $Q = \frac{\frac{1}{2}[A^-]_e \frac{1}{2}[H^+]_e}{\frac{1}{2}[HA]_e} = \frac{1}{2}\frac{[A^-]_e[H^+]_e}{[HA]_e} = \frac{1}{2}K_a$.

Since $Q < K_a$, the concentrations of products must increase and the concentration of reactant must decrease to attain equilibrium. That means that the weak acid dissociates further as it is diluted in order to stay in equilibrium.

9-10.
$$HA \overset{K_a}{\rightleftharpoons} H^+ + A^- \qquad K_a = \frac{[H^+][A^-]}{[HA]} = \frac{x^2}{F - x}$$
$$F - x \qquad\; x \quad\; x$$

For $F = \dfrac{K_a}{10}$, $\dfrac{x^2}{\frac{K_a}{10} - x} = K_a \Rightarrow x = 0.092\,K_a$; $\alpha = \dfrac{x}{F} = \dfrac{0.092\,K_a}{0.100\,K_a} = 92\%$

For $F = 10\,K_a$, $\dfrac{x^2}{10K_a - x} = K_a \Rightarrow x = 2.7\,K_a$; $\alpha = \dfrac{x}{F} = \dfrac{2.7\,K_a}{10\,K_a} = 27\%$

For 99% dissociation, $x = 0.99\,F \Rightarrow K_a = \dfrac{(0.99\,F)^2}{F - 0.99\,F} \Rightarrow F = (0.010\,2)K_a$

9-11.

$$\text{(benzene ring)}-CO_2H \;\rightleftharpoons\; \text{(benzene ring)}-CO_2^- + H^+$$

$$F - 10^{-2.78} \qquad\qquad 10^{-2.78} \qquad 10^{-2.78}$$

$$K_a = \frac{(10^{-2.78})^2}{0.045\,0 - 10^{-2.78}} = 6.35 \times 10^{-5} = pK_a = 4.20$$

9-12.

$$HA \;\rightleftharpoons\; H^+ + A^- \qquad \alpha = 0.006\,0 = \frac{x}{F}$$

$$F - x \qquad\quad x \qquad x$$

$$F = 0.045\,0\ M \text{ and } x = (0.006\,0)(0.045\,0\ M) = 2.7 \times 10^{-4}\ M$$

$$\Rightarrow K_a = \frac{x^2}{F - x} = 1.6 \times 10^{-6} \Rightarrow pK_a = 5.79$$

9-13. (a)

$$HA \;\rightleftharpoons\; H^+ + A^-$$

$$F - x \qquad x \qquad x$$

$$\frac{x^2}{F - x} = K_a \qquad \frac{x^2}{0.010 - x} = 9.8 \times 10^{-5} \Rightarrow x = 9.4 \times 10^{-4}$$

$$\Rightarrow pH = 3.03$$

$$\alpha = \frac{[A^-]}{[HA] + [A^-]} = \frac{x}{F} = 9.4\%$$

(b) pH = 7.00 because the acid is so dilute. From the K_a equilibrium we write

$$[A^-] = \frac{K_a}{[H^+]}[HA] = \frac{9.8 \times 10^{-5}}{1.0 \times 10^{-7}}[HA] = 980\,[HA]$$

$$\alpha = \frac{[A^-]}{[HA] + [A^-]} = \frac{980\,[HA]}{[HA] + 980\,[HA]} = \frac{980}{981} = 99.9\%$$

9-14. Phenol is a weak acid, so it will contribute negligible ionic strength. The ionic strength of the solution is 0.050 M.

$$HA \;\rightleftharpoons\; H^+ + A^- \qquad\qquad K_a = 1.01 \times 10^{-10}$$

$$F - x \qquad x \qquad x$$

$$\frac{[H^+]\,\gamma_{H^+}\,[A^-]\gamma_{A^-}}{[HA]\,\gamma_{HA}} = K_a \Rightarrow \frac{(x)(0.86)(x)(0.835)}{(0.050\,0 - x)(1.00)} = 1.01 \times 10^{-10} \Rightarrow x = 2.65 \times 10^{-6}$$

$$pH = -\log[H^+]\gamma_{H^+} = -\log[2.65 \times 10^{-6}](0.86) = 5.64$$

$$\alpha = \frac{[A^-]}{[HA] + [A^-]} = \frac{2.65 \times 10^{-6}}{0.050\,0} = 5.30 \times 10^{-5} = 0.005\,3\%$$

9-15.

$$\text{Cr}^{3+} + \text{H}_2\text{O} \overset{K_{a1}}{\rightleftharpoons} \text{Cr(OH)}^{2+} + \text{H}^+$$
$$\quad 0.010 - x \qquad\qquad\qquad x \qquad\quad x$$

$$\frac{x^2}{0.010 - x} = 10^{-3.66} \Rightarrow x = 1.3_7 \times 10^{-3} \text{ M}$$

$$\text{pH} = -\log x = 2.86 \qquad\qquad \alpha = \frac{x}{0.010} = 0.14$$

9-16.

$$\text{HNO}_3 \rightleftharpoons \text{H}^+ + \text{NO}_3^-$$
$$\quad \text{F} - x \qquad x \qquad x$$

$$\frac{x^2}{\text{F} - x} = 26.8 \Rightarrow x = 0.099\,6 \text{ M when F} = 0.100 \text{ M} \Rightarrow \alpha = \frac{x}{\text{F}} = 99.6\%$$

$$\Rightarrow x = 0.965 \text{ M when F} = 1.00 \text{ M} \Rightarrow \alpha = \frac{x}{\text{F}} = 96.5\%$$

9-17.   The initial spreadsheet follows (on the left). Guess a value for x in cell A4. The formula in cell B4 is "=A4^2/(A6-A4)". Before using Goal Seek in Excel 2010, click the File menu and select Options. In the Options window, select Formulas. Set Maximum Change to 1e-15 to find an answer with high precision. In Excel 2007, click the Microsoft Office button at the top left of the spreadsheet, click on Excel Options, and then on Formulas. Set Maximum Change to 1e-15. Highlight cell B4 and select Goal Seek. Set cell <u>B4</u> To value <u>1e-5</u> By changing cell <u>A4</u>. Click OK and Goal Seek finds the solution in the second spreadsheet (on the right). The value $x = 9.95 \times 10^{-5}$ makes the quotient $x^2/(\text{F} - x)$ equal to 1.00×10^{-5}.

	A	B
1	Using Excel GOAL SEEK	
2		
3	x =	x^2/(F-x) =
4	0.01	1.1111E-03
5	F =	
6	0.1	

	A	B
1	Using Excel GOAL SEEK	
2		
3	x =	x^2/(F-x) =
4	0.00099501	1.0000E-05
5	F =	
6	0.1	

Before executing Goal Seek   After executing Goal Seek

9-18.   The "fishy" smell comes from volatile amines (RNH_2). Lemon juice protonates the amines, giving less volatile ammonium ions (RNH_3^+).

9-19.   Let $x = [\text{OH}^-] = [\text{BH}^+]$ and $0.100 - x = [\text{B}]$. $\dfrac{x^2}{0.100 - x} = 1.00 \times 10^{-5}$

$$\Rightarrow x = 9.95 \times 10^{-4} \text{ M} \Rightarrow [\text{H}^+] = \frac{K_w}{x} = 1.005 \times 10^{-11} \Rightarrow \text{pH} = 11.00$$

$$\alpha = \frac{[\text{BH}^+]}{[\text{B}] + [\text{BH}^+]} = \frac{9.95 \times 10^{-4}}{0.100} = 9.95 \times 10^{-3}$$

9-20. $(CH_3)_3N + H_2O \rightleftharpoons (CH_3)_3NH^+ + OH^-$ $K_b = K_w/K_a = 6.3 \times 10^{-5}$

$\quad\quad$ F – x $\quad\quad\quad\quad\quad\quad\quad$ x $\quad\quad$ x

$$\frac{x^2}{0.060 - x} = K_b \Rightarrow x = 1.9_1 \times 10^{-3} \Rightarrow pH = -\log\frac{K_w}{x} = 11.28$$

$$[(CH_3)_3NH^+] = x = 1.9_1 \times 10^{-3}\ M, \quad [(CH_3)_3N] = F - x = 0.058\ M$$

9-21. $CN^- + H_2O \rightleftharpoons HCN + OH^-$ $K_b = K_w/K_a = 1.6 \times 10^{-5}$

$\quad\quad$ F – x $\quad\quad\quad\quad\quad$ x $\quad\quad$ x

$$\frac{x^2}{0.050 - x} = K_b \Rightarrow x = 8.9 \times 10^{-4} \Rightarrow pH = -\log\frac{K_w}{x} = 10.95$$

9-22. $CH_3CO_2^- + H_2O \rightleftharpoons CH_3CO_2H + OH^-$ $K_b = K_w/K_a = 5.7 \times 10^{-10}$

$\quad\quad$ F – x $\quad\quad\quad\quad\quad\quad\quad$ x $\quad\quad$ x

$$\frac{x^2}{(1.00 \times 10^{-1}) - x} = K_b \Rightarrow x = 7.6 \times 10^{-6} \Rightarrow \alpha = \frac{x}{F} = 0.0076\%$$

$$\frac{x^2}{(1.00 \times 10^{-2}) - x} = K_b \Rightarrow x = 2.4 \times 10^{-6} \Rightarrow \alpha = \frac{x}{F} = 0.024\%$$

For 1.00×10^{-12} M sodium acetate, pH = 7.00 and we can say

$$[HA] = \frac{K_b[A^-]}{[OH^-]} = (5.7 \times 10^{-3})[A^-]$$

$$\alpha = \frac{[HA]}{[HA] + [A^-]} = \frac{(5.7 \times 10^{-3})[A^-]}{(5.7 \times 10^{-3})[A^-] + [A^-]} = 0.57\%$$

The more dilute the solution, the greater is α.

9-23. $\quad\quad$ B $\quad$ + $\quad$ H_2O $\rightleftharpoons$ $\quad$ BH^+ $\quad$ + $\quad$ OH^-

$\quad\quad$ F – $(K_w/10^{-9.28})$ $\quad\quad\quad\quad\quad$ $K_w/10^{-9.28}$ $\quad$ $K_w/10^{-9.28}$

$$K_b = \frac{(K_w/10^{-9.28})^2}{F - (K_w/10^{-9.28})} = \frac{(K_w/10^{-9.28})^2}{0.10 - (K_w/10^{-9.28})} = 3.6 \times 10^{-9}$$

9-24. $\quad\quad$ B + H_2O $\rightleftharpoons$ BH^+ + OH^- $\quad\quad$ $\alpha = 0.020 = \dfrac{x}{F} \Rightarrow x = 2.0 \times 10^{-3}\ M$

$\quad\quad$ 0.10 – x $\quad\quad\quad$ x $\quad\quad$ x

$$K_b = \frac{x^2}{0.10 - x} = \frac{(2.0 \times 10^{-3})^2}{0.10 - (2.0 \times 10^{-3})} = 4.1 \times 10^{-5}$$

9-25. As $[B] \rightarrow 0$, pH $\rightarrow 7$ and $[OH^-] \rightarrow 10^{-7}$ M.

$$K_b = \frac{[BH^+][OH^-]}{[B]} = 10^{-7}\frac{[BH^+]}{[B]} \Rightarrow [BH^+] = 10^7 K_b [B]$$

$$\alpha = \frac{[BH^+]}{[B] + [BH^+]} = \frac{10^7 K_b [B]}{[B] + 10^7 K_b [B]} = \frac{10^7 K_b}{1 + 10^7 K_b}$$

$$\text{For } K_b = 10^{-4}, \text{ we have } \alpha = \frac{10^7 K_b}{1 + 10^7 K_b} = \frac{10^7 \, 10^{-4}}{1 + 10^7 \, 10^{-4}} = 0.999$$

$$\text{For } K_b = 10^{-10}, \text{ we have } \alpha = \frac{10^7 K_b}{1 + 10^7 K_b} = \frac{10^7 \, 10^{-10}}{1 + 10^7 \, 10^{-10}} = 0.000\,999$$

9-26. I would weigh out 0.020 0 mol of acetic acid (= 1.201 g) and place it in a beaker with ~75 mL of water. While monitoring the pH with a pH electrode, I would add 3 M NaOH (~4 mL is required) until the pH is exactly 5.00. I would then pour the solution into a 100-mL volumetric flask and wash the beaker several times with a few milliliters of distilled water. Each washing would be added to the volumetric flask, to ensure quantitative transfer from the beaker to the flask. After swirling the volumetric flask to mix the solution, I would carefully add water up to the 100 mL mark, insert the cap, and invert 20 times to ensure complete mixing.

9-27. The inside cover of the book states that 67.6 mL of 28 wt% ammonia are required to make 1 L of ~1.0 M solution. To make 250 mL of 1 M NH_3 will require 1/4 as much = 16.9 mL of 28 wt% ammonia. (Note that the composition of 28 wt% ammonia is only approximate and the resulting buffer molarity will only be approximate.) Add 16.9 mL of 28 wt% ammonia to ~160 mL of H_2O in a beaker containing a pH electrode and stir bar. We need somewhat more than half the moles of HCl to lower the pH of ammonia to 9, which is below $pK_a = 9.24$. 250 mL of 1 M NH_3 = 0.25 mol NH_3, which will require a little more than 0.125 mol HCl. The back of the book tells us that 82.4 mL of 37.2% HCl contains 1 mol HCl. We will need somewhat more than 1/8 of this volume of HCl ≈ 12 mL of HCl. While measuring the pH of the solution, add about 9 mL of HCl. Then add HCl dropwise until the pH is 9.00. Transfer the liquid to a 250-mL volumetric flask and wash the beaker and stir bar many times with small quantities of H_2O. Pour all of the washings into the volumetric flask to make a quantitative transfer. Then dilute to exactly 250 mL. The buffer will only be ~1.0 M because the concentration of 28 wt% NH_3 is not precisely known.

9-28. The reaction solution contains 100 mL of ~0.050 0 M $[H_3BO_3]$ and $[H_2BO_3^-]$, which amounts to ~5 mmol of $[H_3BO_3]$ and ~5 mmol of $[H_2BO_3^-]$. We do not want the acid generated by the chemical reaction to consume more than half of the $[H_2BO_3^-]$, which would be about 2.5 mmol of acid. The pH would be reduced from pH = pK_a = 9.24 down to pH = pK_a + log $[H_2BO_3^-]/[H_3BO_3]$ = 9.24 + log (2.5 mmol/7.5 mmol) = 8.76.

9-29.   The pH of a buffer depends on the ratio of the concentrations of HA and A^- (pH = pK_a + log $[A^-]/[HA]$). When the volume of solution is changed, both concentrations are affected equally and their ratio does not change.

9-30.   Buffer capacity measures the ability to maintain the original $[A^-]/[HA]$ ratio when acid or base is added. A more concentrated buffer has more A^- and HA, so a smaller fraction of A^- or HA is consumed by added acid or base. Therefore, there is a smaller change in the ratio $[A^-]/[HA]$.

9-31.   At very low or very high pH, there is so much acid or base in the solution already that small additions of acid or base will hardly have any effect. At low pH, the buffer is H_3O^+/H_2O; and at high pH, the buffer is H_2O/OH^-.

9-32.   When pH = pK_a, the ratio of concentrations $[A^-]/[HA]$ is unity. A given increment of added acid or base has the least effect on the ratio $[A^-]/[HA]$ when the concentrations of A^- and HA are initially equal.

9-33.   The Henderson-Hasselbalch is just a rearranged form of the K_a equilibrium expression, which is always true. When we make the approximation that [HA] and $[A^-]$ are unchanged from what we added, we are neglecting acid dissociation and base hydrolysis, which can change the concentrations in dilute solutions of moderately strong acids or bases.

9-34.

acid	pK_a	
hydrogen peroxide	11.65	
propanoic acid	4.87	
cyanoacetic acid	2.47	
4-aminobenzenesulfonic acid	3.23	← most suitable because pK_a is closest to desired pH

H_2O_2 would never be a good choice for a buffer because it is unstable. Propanoic acid is flammable and harmful in case of contact. Cyanoacetic acid is hygroscopic, so it would not be easy to weigh. It also burns skin and eyes on contact. 4-Amino-benzenesulfonic acid is hazardous in case of skin contact or inhalation.

9-35. $\text{pH} = pK_a + \log \dfrac{[A^-]}{[HA]} = 5.00 + \log \dfrac{0.050}{0.100} = 4.70$

9-36. $\quad pH = 3.744 + \log \dfrac{[HCO_2^-]}{[HCO_2H]}$

pH:	3.000	3.744	4.000
$[HCO_2^-]/[HCO_2H]$:	0.180	1.00	1.80

9-37. $\quad pH = 3.57 + \log \dfrac{[HCO_2^-]}{[HCO_2H]}$, where 3.57 is pK_a at $\mu = 0.1$ M

Substituting pH = 3.744 gives $[HCO_2^-]/[HCO_2H] = 1.5$

9-38. $\quad pH = pK_a + \log \dfrac{[NO_2^-]}{[HNO_2]}$, where $pK_a = 14.00 - pK_b = 3.15$

(a) If pH = 2.00, $[HNO_2]/[NO_2^-] = 14$

(b) If pH = 10.00, $[HNO_2]/[NO_2^-] = 1.4 \times 10^{-7}$

9-39. (a) HEPES is an acid with $pK_a = 7.56$. When it is dissolved in water, the solution will be acidic and will require NaOH to raise the pH to 7.45.

(b) 1. Weigh out (0.250 L)(0.0500 M) = 0.0125 mol of HEPES and dissolve in ~200 mL.

2. Adjust the pH to 7.45 with NaOH.

3. Dilute to 250 mL.

9-40. 213 mL of 0.00666 M 2,2′-bipyridine = 1.41_9 mmol base. We will add x mol H^+ to get a pH of 4.19.

	2,2′-bipyridine	+	H^+	→	2,2′-bipyridineH$^+$
Initial mmol:	1.41_9		x		—
Final mmol:	$1.41_9 - x$		—		x

$pH = pK_a + \log \dfrac{[\text{bipyridine}]}{[\text{bipyridineH}^+]}$

$4.19 = 4.34 + \log \dfrac{1.41_9 - x}{x} \Rightarrow x = 0.831$ mmol

$\text{volume} = \dfrac{0.831 \text{ mmol}}{0.246 \text{ mmol/mL}} = 3.38$ mL

9-41. (a)

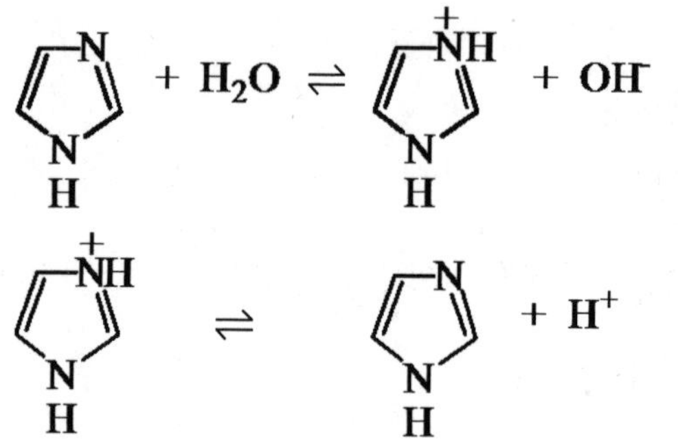

(b) FM of imidazole = 68.08.   FM of imidazole hydrochloride = 104.54.

$$pH = 6.993 + \log \frac{1.00/68.08}{1.00/104.54} = 7.18$$

(c)

	B	+	H$^+$	→	BH$^+$
Initial mmol:	14.6$_9$		2.46		9.57
Final mmol:	12.2$_3$		—		12.0$_3$

$$pH = 6.993 + \log \frac{12.2_3}{12.0_3} = 7.00$$

(d) The imidazole must be half neutralized to obtain pH = pK_a. Since there are 14.6$_9$ mmol of imidazole, this will require $\frac{1}{2}$(14.6$_9$) = 7.34 mmol of HClO$_4$ = 6.86 mL.

9-42. (a) $pH = 2.865 + \log \dfrac{0.0400}{0.0800} = 2.56$

(b) Using Equations (9-20) and (9-21), and neglecting [OH$^-$], we can write

$$K_a = 1.36 \times 10^{-3} = \frac{[H^+](0.0400 + [H^+])}{0.0800 - [H^+]} \Rightarrow [H^+] = 2.48 \times 10^{-3}\ M$$

$$\Rightarrow pH = 2.61$$

(c) 0.080 mol of HNO$_3$ + 0.080 mol of Ca(OH)$_2$ react completely, leaving an excess of 0.080 mol of OH$^-$. This much OH$^-$ converts 0.080 mol of ClCH$_2$CO$_2$H into 0.080 mol of ClCH$_2$CO$_2^-$. The final concentrations are [ClCH$_2$CO$_2^-$] = 0.020 + 0.080 = 0.100 M and [ClCH$_2$CO$_2$H] = 0.180 – 0.080 = 0.100 M. So pH = pK_a = 2.86.

9-43.

$$HA \quad + \quad OH^- \quad \rightarrow \quad A^- \quad + \quad H_2O$$

Initial moles: 0.022 4 x —

Final moles: 0.022 4 – x — x

$$pH = 7.40 = pK_a + \log \frac{[A^-]}{[HA]} = 7.48 + \log \frac{x}{0.022\,4 - x} \Rightarrow x = 0.010\,17 \text{ mol.}$$

$$\text{volume} = \frac{0.010\,17 \text{ mol}}{0.626 \text{ M}} = 16.2 \text{ mL}$$

9-44. (a) Since pK_a for acetic acid is 4.756, we expect the solution to be acidic and will ignore $[OH^-]$ in comparison to $[H^+]$.

$$[HA] = 0.002\,0 - [H^+] \qquad\qquad [A^-] = 0.004\,00 + [H^+]$$

$$K_a = 1.75 \times 10^{-5} = \frac{[H^+](0.004\,00 + [H^+])}{0.002\,00 - [H^+]} \Rightarrow [H^+] = 8.69 \times 10^{-6} \text{ M}$$

$$\Rightarrow pH = 5.06 \qquad [HA] = 0.001\,99 \text{ M} \qquad [A^-] = 0.004\,01 \text{ M}$$

If you used $K_a = 10^{-4.756}$ instead of rounding to 1.75×10^{-5}, then $[H^+] = 8.71 \times 10^{-6}$ M.

(b) Use Goal Seek to vary cell B5 until cell D4 is equal to K_a.

	A	B	C	D	E
1	Ka = 10^-pKa =	1.75E-05		Reaction quotient	
2	Kw =	1.00E-14		for Ka =	
3	FHA =	0.002000		[H+][A-]/[HA] =	
4	FA =	0.004000		1.75E-05	
5	H =	8.693E-06		<-Goal Seek solution	
6	OH = Kw/H =	1.15E-09		D4 = H*(FA+H-OH)/(FHA-H+OH)	
7	pH = -logH =	5.0608262			
8	[HA] =	0.0019913		B8 = FHA-H+OH	
9	[A-] =	0.0040087		B9 = FA+H-OH	

9-45. (a) If we dissolve B and $BH^+ Br^-$ (where Br^- is an inert anion), the mass balance is $F_{BH^+} + F_B = [BH^+] + [B]$ and the charge balance is $[Br^-] + [OH^-] = [BH^+] + [H^+]$. Noting that $[Br^-] = F_{BH^+}$, the charge balance can be rewritten as

$$[BH^+] = F_{BH^+} + [OH^-] - [H^+] \qquad\qquad \text{(A)}$$

Substituting this expression into the mass balance gives

$$[B] = F_B - [OH^-] + [H^+] \qquad\qquad \text{(B)}$$

If we assume that $[B] = 0.010\,0$ M and $[BH^+] = 0.020\,0$ M, we calculate

$$pH = pK_a + \log \frac{[B]}{[BH^+]} = 12.00 + \log \frac{0.010\,0}{0.020\,0} = 11.70$$

If we do not assume that $[B] = 0.010\,0$ M and $[BH^+] = 0.020\,0$ M, we use Equations A and B. Since the solution is basic, we neglect $[H^+]$ relative to $[OH^-]$ and write $[B] = 0.010\,0 - x$ and $[BH^+] = 0.020\,0 + x$, where $x = [OH^-]$.

Then we can say $K_b = 10^{-2.00} = \dfrac{[BH^+][OH^-]}{[B]} = \dfrac{(0.020\,0 + x)\,(x)}{(0.010\,0 - x)} \Rightarrow$

$x = 0.003\,0_3$ M $pH = -\log \dfrac{K_w}{x} = 11.48.$

(b) We use Goal Seek to vary cell B5 until cell D4 is equal to K_b.

	A	B	C	D	E
1	Kb =	1.00E-02		Reaction quotient	
2	Kw =	1.00E-14		for Kb =	
3	FBH =	0.02		[OH-][BH+]/[B] =	
4	FB =	0.01		0.01	
5	OH =	3.028E-03		<-Goal Seek solution	
6	H = Kw/OH =	3.303E-12		D4 = OH*(FBH-H+OH)/(FB+H-OH)	
7	pH = -logH =	11.481121			
8	BH =	0.0230278		C8 = FBH-H+OH	
9	B =	0.0069722		C9 = FB+H-OH	

9-46. $K_a = \dfrac{[HPO_4^{2-}][H^+]\gamma_{HPO_4^{2-}}\,\gamma_{H^+}}{[H_2PO_4^-]\,\gamma_{H_2PO_4^-}} = 10^{-7.20}$

To find pH, rearrange the K_a expression to solve for the activity of H^+, which is $[H^+]\gamma_{H^+}$:

$$[H^+]\gamma_{H^+} = \dfrac{K_a[H_2PO_4^-]\gamma_{H_2PO_4^-}}{[HPO_4^{2-}]\gamma_{HPO_4^{2-}}}$$

At $\mu = 0.1$ M, $\gamma_{H^+} = 0.83$, $\gamma_{H_2PO_4^-} = 0.775$, and $\gamma_{HPO_4^{2-}} = 0.355$, so

$$[H^+]\gamma_{H^+} = \dfrac{10^{-7.20}[H_2PO_4^-](0.775)}{[HPO_4^{2-}](0.355)} = 1.38 \times 10^{-7} \text{ when } [H_2PO_4^-] = [HPO_4^{2-}]$$

$pH = -\log \mathcal{A}_{H^+} = -\log[H^+]\gamma_{H^+} = -\log(1.38 \times 10^{-7}) = 6.86$

9-47. As the pH of the solution is adjusted by addition of acid or base, the reaction is

Pyridine   pyridinium ion
at high pH   at low pH

The chemical shift of H_4 at low pH levels off near 8.67 ppm in the graph. This value would be the chemical shift of H_4 of the pyridinium ion. The chemical shift of H_4 at high pH levels off near 7.89, which would be the chemical shift of H_4 of pyridine.

The fraction of pyridine in each solution would be the fraction of the chemical shift change from 8.67 to 7.89. When pH = pK_a, there will be equal amounts of pyridine and pyridinium ion. The chemical shift will be half way between 8.67 and 7.89, which is 8.28 ppm. Draw a horizontal line on the graph at a chemical shift of 8.28. This horizontal line intersects a smooth curve drawn between the data points at approximately pH = 5.2. We estimate $pK_a \approx 5.2$ for the pyridinium ion, which agrees with the value in the appendix.

9-48. (a) Equilibria: $\beta_1 = \dfrac{[AlOH^{2+}][H^+]}{[Al^{3+}]}$ (a)

$\beta_2 = \dfrac{[Al(OH)_2^+][H^+]^2}{[Al^{3+}]}$ (b)

$\beta_3 = \dfrac{[Al(OH)_3(aq)][H^+]^3}{[Al^{3+}]}$ (c)

$\beta_4 = \dfrac{[Al(OH)_4^-][H^+]^4}{[Al^{3+}]}$ (d)

$K_{22} = \dfrac{[Al_2(OH)_2^{4+}][H^+]^2}{[Al^{3+}]^2}$ (e)

$K_{43} = \dfrac{[Al_3(OH)_4^{5+}][H^+]^4}{[Al^{3+}]^3}$ (f)

$K_w = [H^+][OH^-]$ (g)

Charge balance: $3[Al^{3+}] + 2[AlOH^{2+}] + [Al(OH)_2^+] + 4[Al_2(OH)_2^{4+}] +$
$5[Al_3(OH)_4^{5+}] + [H^+] = [Al(OH)_4^-] + [OH^-] + [ClO_4^-]$ (h)

Mass balances: $3F = [ClO_4^-]$ (i)

$F = [Al^{3+}] + [AlOH^{2+}] + [Al(OH)_2^+] + 2[Al_2(OH)_2^{4+}] +$
$[Al(OH)_3(aq)] + [Al(OH)_4^-] + 3[Al_3(OH)_4^{5+}]$ (j)

We have 10 equations and 10 unknowns, so the problem can, in principle, be solved.

(b) We know that $[ClO_4^-] = 3F$, so there are just 9 unknown concentrations. With seven equilibrium expressions, we would use the spreadsheet to find $9 - 7 = 2$ concentrations, for which I choose $[H^+]$ and $[Al^{3+}]$. From guessed values for $[H^+]$ and $[Al^{3+}]$, we can write expressions for the concentrations of all other species. Then use Solver to vary $[H^+]$ and $[Al^{3+}]$ until the combined sum of squares of the mass and charge balances (h) and (j) are satisfied. Now all concentrations of all species must be correct.

CHAPTER 10
POLYPROTIC ACID-BASE EQUILIBRIA

10-1. The K_a reaction, with a much greater equilibrium constant than K_b, releases H^+:

$$HA^- \rightleftharpoons H^+ + A^{2-} \qquad K_a$$

Each mole of H^+ reacts with one mole of OH^- from the K_b reaction:

$$HA^- + H_2O \rightleftharpoons H_2A + OH^-.$$

The net result is that the K_b reaction is driven almost as far toward completion as the K_a reaction.

10-2.

pK values apply to $-NH_3^+$, $-CO_2H$, and, in some cases, R.

10-3.

$$K_{b1} = \frac{K_w}{K_2} = 4.37 \times 10^{-4}$$

$$K_{b2} = \frac{K_w}{K_1} = 8.93 \times 10^{-13}$$

10-4. (a) $\dfrac{x^2}{0.100-x} = K_1 \Rightarrow x = 3.11 \times 10^{-3} = [H^+] = [HA^-] \Rightarrow pH = 2.51$

$[H_2A] = 0.100 - x = 0.0969\,M \qquad [A^{2-}] = \dfrac{K_2[HA^-]}{[H^+]} = 1.00 \times 10^{-8}\,M$

(b) $[H^+] \approx \sqrt{\dfrac{K_1K_2F + K_1K_w}{K_1 + F}} = 1.00 \times 10^{-6} \Rightarrow pH = 6.00$

$[HA^-] \approx 0.100\,M$

$[H_2A] = \dfrac{[H^+][HA^-]}{K_1} = 1.00 \times 10^{-3}\,M \qquad [A^{2-}] = \dfrac{K_2[HA^-]}{[H^+]} = 1.00 \times 10^{-3}\,M$

(c) $\dfrac{x^2}{0.100-x} = \dfrac{K_w}{K_2} \Rightarrow x = [OH^-] = [HA^-] = 3.16 \times 10^{-4}\,M \Rightarrow pH = 10.50$

$[A^{2-}] = 0.100 - x = 9.97 \times 10^{-2}\,M \qquad [H_2A] = \dfrac{[H^+][HA^-]}{K_1} = 1.00 \times 10^{-10}\,M$

	pH	$[H_2A]$	$[HA^-]$	$[A^{2-}]$
0.100 M H_2A	2.51	9.69×10^{-2}	3.11×10^{-3}	1.00×10^{-8}
0.100 M NaHA	6.00	1.00×10^{-3}	1.00×10^{-1}	1.00×10^{-3}
0.100 M Na_2A	10.50	1.00×10^{-10}	3.16×10^{-4}	9.97×10^{-2}

10-5. (a) $H_2M = H^+ + HM^-$ $\quad K_1 = 1.42 \times 10^{-3}$

$ F - x \quad\; x \quad\;\; x$

$$\frac{x^2}{0.100 - x} = K_1 \Rightarrow x = 1.12 \times 10^{-2} \Rightarrow pH = -\log x = 1.95$$

$[H_2M] = 0.100 - x = 0.089\ M$

$[HM^-] = x = 1.12 \times 10^{-2}\ M[M^{2-}] = \dfrac{[HM^-]\,K_2}{[H^+]} = 2.01 \times 10^{-6}\ M$

(b) $[H^+] = \sqrt{\dfrac{K_1 K_2 (0.100) + K_1 K_w}{K_1 + 0.100}} = 5.30 \times 10^{-5} \Rightarrow pH = 4.28$

$[HM^-] \approx 0.100\ M \quad [H_2M] = \dfrac{[HM^-][H^+]}{K_1} = 3.7 \times 10^{-3}\ M$

$[M^{2-}] = \dfrac{K_2[HM^-]}{[H^+]} = 3.8 \times 10^{-3}\ M$

The method of Box 10-2 gives more accurate answers, since $[HM^-]$ is not that much greater than $[H_2M]$ or $[M^{2-}]$ in this case. Successive approximations give pH = 4.28, $[H_2M] = 0.003\,44\ M$, $[HM^-] = 0.093\,1\ M$, $[M^{2-}] = 0.003\,53\ M$.

(c) $M^{2-} + H_2O \rightleftharpoons HM^- + OH^-$ $\qquad K_{b1} = K_w/K_{a2} = 4.98 \times 10^{-9}$

$ F - x \qquad\qquad x \qquad\; x$

$$\frac{x^2}{0.100 - x} = K_{b1} \Rightarrow x = 2.23 \times 10^{-5} \Rightarrow pH = -\log\frac{K_w}{x} = 9.35$$

$[M^{2-}] = 0.100 - x = 0.100\ M \qquad [HM^-] = x = 2.23 \times 10^{-5}\ M$

$[H_2M] = \dfrac{[H^+][HM^-]}{K_1} = 7.04 \times 10^{-12}\ M$

10-6. $HN\!\!\bigcirc\!\!NH + H_2O \rightleftharpoons HN\!\!\bigcirc\!\!\overset{+}{N}H_2 + OH^- \quad K_{b1} = \dfrac{K_w}{K_2} = 5.38 \times 10^{-5}$

$ F - x \qquad\qquad\qquad x \qquad\; x$

$$\frac{x^2}{0.300 - x} = K_{b1} \Rightarrow x = 3.99 \times 10^{-3}\ M \Rightarrow pH = -\log K_w/x = 11.60$$

$[B] = 0.300 - x = 0.296\ M \qquad\qquad [BH^+] = x = 3.99 \times 10^{-3}\ M$

$[BH_2^{2+}] = \dfrac{[BH^+][H^+]}{K_1} = 2.15 \times 10^{-9}\ M$

10-7. For H_2A, $K_1 = 5.62 \times 10^{-2}$ and $K_2 = 5.42 \times 10^{-5}$

First approximation $([HA^-]_1 \approx 0.001\,00\ M)$:

$$[H^+]_1 = \sqrt{\frac{K_1 K_2 (0.001\,00) + K_1 K_w}{K_1 + 0.001\,00}} = 2.31 \times 10^{-4}\ M \Rightarrow pH_1 = 3.64$$

$$[H_2A]_1 = \frac{[H^+]_1\,[HA^-]_1}{K_1} = 4.10 \times 10^{-6}\ M$$

$$[A^{2-}]_1 = \frac{K_2\,[HA^-]_1}{[H^+]} = 2.35 \times 10^{-4}\ M$$

Second approximation:

$$[HA^-]_2 \approx 0.00100 - [H_2A]_1 - [A^{2-}]_1 = 0.000761\ M$$

$$[H^+]_2 = \sqrt{\frac{K_1K_2(0.000761) + K_1K_w}{K_1 + 0.000761}} = 2.02 \times 10^{-4}\ M \Rightarrow pH_2 = 3.70$$

$$[H_2A]_2 = \frac{[H^+]_2\,[HA^-]_2}{K_1} = 2.73 \times 10^{-6}\ M$$

$$[A^{2-}]_2 = \frac{K_2\,[HA^-]_2}{[H^+]_2} = 2.04 \times 10^{-4}\ M$$

Third approximation:

$$[HA^-]_3 \approx 0.00100 - [H_2A]_2 - [A^{2-}]_2 = 0.000793\ M$$

$$[H^+]_3 = \sqrt{\frac{K_1K_2(0.000793) + K_1K_w}{K_1 + 0.000793}} = 2.06 \times 10^{-4}\ M \Rightarrow pH_3 = 3.69$$

$$[H_2A]_3 = \frac{[H^+]_3\,[HA^-]_3}{K_1} = 2.90 \times 10^{-6}\ M$$

$$[A^{2-}]_3 = \frac{K_2\,[HA^-]_3}{[H^+]_3} = 2.09 \times 10^{-4}\ M$$

10-8. (a) Charge balance: $[K^+] + [H^+] = [OH^-] + [HP^-] + 2[P^{2-}]$ (1)

Mass balance: $[K^+] = [H_2P] + [HP^-] + [P^{2-}]$ (2)

Equilibria: $$K_1 = \frac{[H^+]\gamma_{H^+}\,[HP^-]\,\gamma_{HP^-}}{[H_2P]\,\gamma_{H_2P}} \qquad (3)$$

$$K_2 = \frac{[H^+]\gamma_{H^+}\,[P^{2-}]\,\gamma_{P^{2-}}}{[HP^-]\,\gamma_{HP^-}} \qquad (4)$$

$$K_w = [H^+]\,\gamma_{H^+}\,[OH^-]\,\gamma_{OH^-} \qquad (5)$$

Solving for $[K^+]$ in Equations (1) and (2) and equating the results gives

$$[H_2P] + [H^+] - [P^{2-}] - [OH^-] = 0$$

Making substitutions from Equations (3), (4), and (5), we can write

$$\frac{[H^+]\gamma_{H^+}\,[HP^-]\,\gamma_{HP^-}}{K_1\,\gamma_{H_2P}} + [H^+] - \frac{K_2[HP^-]\,\gamma_{HP^-}}{[H^+]\,\gamma_{H^+}\,\gamma_{P^{2-}}} - \frac{K_w}{[H^+]\,\gamma_{H^+}\,\gamma_{OH^-}} = 0$$

which can be rearranged to

$$[H^+] = \sqrt{\frac{\dfrac{K_1 K_2 [HP^-]\gamma_{HP^-}\gamma_{H_2P}}{\gamma_{H^+}\gamma_{P^{2-}}} + \dfrac{K_1 K_w \gamma_{H_2P}}{\gamma_{H^+}\gamma_{OH^-}}}{K_1\gamma_{H_2P} + [HP^-]\gamma_{H^+}\gamma_{HP^-}}} \qquad (6)$$

(b)   The ionic strength of 0.050 M KHP is 0.050 M because the only major ions are K^+ and HP^-.

$[HP^-] \approx 0.050$ M, $\gamma_{HP^-} = 0.835$, $\gamma_{P^{2-}} = 0.485$, $\gamma_{H_2P} \approx 1.00$,

$\gamma_{H^+} = 0.86$, $\gamma_{OH^-} = 0.81$. Using these values in Eq. (6) gives $[H^+] = 1.09 \times 10^{-4} \Rightarrow pH = -\log[H^+]\gamma_{H^+} = 4.03$.

10-9.   Case (a) is shown in column G and case (b) is in column H.

	A	B	C	D	E	F	G	H
1	Successive approximations by circular reference							
2	for intermediate form of diprotic acid							
3								
4	H_2A malic acid							
5	$K_1 =$	3.50E-04					1.00E-04	1.00E-04
6	$K_2 =$	7.90E-06					1.00E-08	1.00E-05
7	$K_w =$	1.00E-14					1.00E-14	1.00E-14
8	$F =$	1.000E-03					1.000E-02	1.000E-02
9	$[H^+] =$	4.356E-05	=SQRT((K_1*K_2*[HA$^-$]+K_1*K_w)/(K_1+[HA$^-$]))				9.950E-07	3.137E-05
10	$[H_2A] =$	9.532E-05	= $[H^+]$[HA$^-$]/K_1				9.755E-05	1.921E-03
11	$[HA^-] =$	7.658E-04	=F-[H_2A]-[A^{2-}]				9.804E-03	6.126E-03
12	$[A^{2-}] =$	1.389E-04	= K_2[HA$^-$]/$[H^+]$				9.853E-05	1.953E-03
13	pH =	4.360887	=-log$[H^+]$				6.002182	4.503516
14								
15	1. Set Excel Options Formulas to Enable Iterative Calculation							
16	with Maximum Change = 1E-12							
17	2. Begin by setting [HA$^-$] = F							
18	3. Write correct formulas in for other concentrations							
19	4. Then change [HA$^-$] to =F-[H_2A]-[A^{2-}]							
20	5. Correct answers now appear in all cells							

10-10. $[``H_2CO_3"] = [CO_2(aq)] = KP_{CO_2} = 10^{-1.5} \cdot 10^{-3.4} = 10^{-4.9}$ M

$$H_2CO_3 \rightleftharpoons HCO_3^- + H^+ \qquad K_{a1} = 4.46 \times 10^{-7}$$

$$10^{-4.9} - x \qquad\qquad x \qquad x$$

$$\frac{x^2}{10^{-4.9} - x} = K_{a1} \Rightarrow x = 2.1_6 \times 10^{-6} \text{ M} \Rightarrow \text{pH} = 5.67$$

10-11. (a) From Equation C, $[CO_3^{2-}] = \dfrac{K_{a2} [HCO_3^-]}{[H^+]}$ \hfill (F)

From Equation B, $[HCO_3^-] = \dfrac{K_{a1} [CO_2(aq)]}{[H^+]}$ \hfill (G)

Substituting $[HCO_3^-]$ from Equation G into Equation F gives

$$[CO_3^{2-}] = \frac{K_{a2} K_{a1} [CO_2(aq)]}{[H^+]^2} \tag{H}$$

Substituting for $[CO_2(aq)]$ from Equation A into Equation H gives

$$[CO_3^{2-}] = \frac{K_{a2} K_{a1} K_H P_{CO_2}}{[H^+]^2} \tag{I}$$

(b) For $P_{CO_2} = 800$ μbar, pH = 7.8, 0°C, we find

$$[CO_3^{2-}] = \frac{K_{a2} K_{a1} K_H P_{CO_2}}{[H^+]^2} =$$

$$\frac{10^{-9.3762} \text{ mol kg}^{-1} \; 10^{-6.1004} \text{ mol kg}^{-1} \; 10^{-1.2073} \text{ mol kg}^{-1} \text{ bar}^{-1} \; (800 \times 10^{-6} \text{ bar})}{[10^{-7.8} \text{ mol kg}^{-1}]^2}$$

$$= 6.6 \times 10^{-5} \text{ mol kg}^{-1}$$

For $P_{CO_2} = 800$ μbar, pH = 7.8, 30°C, we find

$$[CO_3^{2-}] = \frac{K_{a2} K_{a1} K_H P_{CO_2}}{[H^+]^2} =$$

$$\frac{10^{-8.8324} \text{ mol kg}^{-1} \; 10^{-5.8008} \text{ mol kg}^{-1} \; 10^{-1.6048} \text{ mol kg}^{-1} \text{ bar}^{-1} \; (800 \times 10^{-6} \text{ bar})}{[10^{-7.8} \text{ mol kg}^{-1}]^2}$$

$$= 1.8 \times 10^{-4} \text{ mol kg}^{-1}$$

(c) The equilibrium expressions for aragonite and calcite are

$$CaCO_3(s, \textit{aragonite}) \rightleftharpoons Ca^{2+} + CO_3^{2-} \tag{D}$$

$$K_{sp}^{arg} = [Ca^{2+}][CO_3^{2-}] \; = \; 10^{-6.1113} \text{ mol}^2 \text{ kg}^{-2} \text{ at } 0°C$$

$$= \; 10^{-6.1391} \text{ mol}^2 \text{ kg}^{-2} \text{ at } 30°C$$

$$CaCO_3(s, \textit{calcite}) \rightleftharpoons Ca^{2+} + CO_3^{2-} \tag{E}$$

$$K_{sp}^{cal} = [Ca^{2+}][CO_3^{2-}] \; = \; 10^{-6.3652} \text{ mol}^2 \text{ kg}^{-2} \text{ at } 0°C$$

$$= \; 10^{-6.3713} \text{ mol}^2 \text{ kg}^{-2} \text{ at } 30°C$$

The reaction quotient at 0°C is

$$[Ca^{2+}][CO_3^{2-}] = [0.010 \text{ mol kg}^{-1}][6.6 \times 10^{-5} \text{ mol kg}^{-1}]$$

$$= 6.6 \times 10^{-7} \text{ mol}^2 \text{ kg}^{-2} = 10^{-6.18} \text{ mol}^2 \text{ kg}^{-2} < 10^{-6.1113} \text{ mol}^2 \text{ kg}^{-2},$$

so aragonite will dissolve.

But $10^{-6.18}$ mol^2 kg$^{-2} > 10^{-6.3652}$ mol^2 kg^{-2}, so calcite will not dissolve.

The reaction quotient at 30°C is

$$[Ca^{2+}][CO_3^{2-}] = [0.010 \text{ mol kg}^{-1}][1.84 \times 10^{-4} \text{ mol kg}^{-1}]$$

$$= 1.84 \times 10^{-6} \text{ mol}^2 \text{ kg}^{-2} = 10^{-5.74} \text{ mol}^2 \text{ kg}^{-2} > 10^{-6.1113} \text{ mol}^2 \text{ kg}^{-2},$$

so neither aragonite nor calcite dissolve.

10-12. $pH = pK_a + \log \dfrac{[CO_3^{2-}]}{[HCO_3^-]}$

$$10.00 = 10.329 + \log \frac{(x \text{ g})/(105.99 \text{ g/mol})}{(5.00 \text{ g})/(84.01 \text{ g/mol})} \Rightarrow x = 2.96 \text{ g}$$

10-13. We begin with $(25.0 \text{ mL})(0.023\,3 \text{ M}) = 0.582_5$ mmol salicylic acid (H_2A, $pK_1 = 2.972$, $pK_2 = 13.7$). At pH 3.50, there will be a mixture of H_2A and HA^-.

	H_2A	+	OH^-	→	HA^-	+	H_2O
Initial mmol:	0.582_5		x		—		
Final mmol:	$0.582_5 - x$		—		x		

$$3.50 = 2.972 + \log \frac{x}{0.582_5 - x} \Rightarrow x = 0.449_3 \text{ mmol}$$

$$(0.449_3 \text{ mmol})/(0.202 \text{ M}) = 2.223 \text{ mL NaOH}$$

10-14. Picolinic acid is HA, the intermediate form of a diprotic system with $pK_1 = 1.01$ and $pK_2 = 5.39$. To achieve pH 5.50, we need a mixture of $HA + A^-$.

	HA	+	OH^-	→	A^-
Initial mmol:	10.0		x		—
Final mmol:	$10.0 - x$		—		x

$$5.50 = 5.39 + \log \frac{x}{10.0 - x} \Rightarrow x = 5.63 \text{ mmol} \approx 5.63 \text{ mL NaOH}$$

Procedure: Dissolve 10.0 mmol (1.23 g) picolinic acid in ~75 mL H_2O in a beaker. Add NaOH (~5.63 mL) until the measured pH is 5.50. Transfer to a 100 mL volumetric flask and use small portions of H_2O to rinse the contents of the beaker into the flask. Dilute to 100.0 mL and mix well.

10-15. At pH 2.80, we have a mixture of SO_4^{2-} and HSO_4^-, since pK_a for HSO_4^- is 1.99.

$$2.80 = 1.987 + \log \frac{[SO_4^{2-}]}{[HSO_4^-]} \Rightarrow HSO_4^- = 0.153_8 \, [SO_4^{2-}]$$

The reaction between H_2SO_4 and SO_4^{2-} produces 2 moles of HSO_4^-:

$$H_2SO_4 + SO_4^{2-} \rightarrow 2HSO_4^-$$

Initial mmol: x y —
Final mmol: — $y - x$ $2x$

The Henderson-Hasselbalch equation told us that $[HSO_4^-] = 0.153_8 \, [SO_4^{2-}]$
$\Rightarrow 2x = 0.153_8 \, (y - x)$. Since the total sulfur is 0.200 M, $x + y = 0.200$ mol.
Substituting $x = 0.200 - y$ into the equation $2x = 0.153_8 \, (y - x)$ gives
$Na_2SO_4 = y = 0.186_7$ mol $= 26.5_2$ g and $H_2SO_4 = x = 0.013_3$ mol $= 1.31$ g.

10-16. pK_2 for phosphoric acid is 7.20, so it has a high buffer capacity at pH 7.45 (from the buffer pair $H_2PO_4^-/HPO_4^{2-}$). At pH 8.5, the buffer capacity of phosphate would be low and it would not be very useful.

10-17.

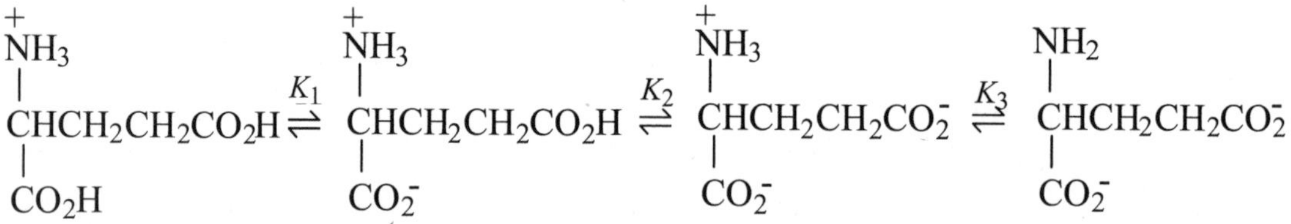

glutamic acid

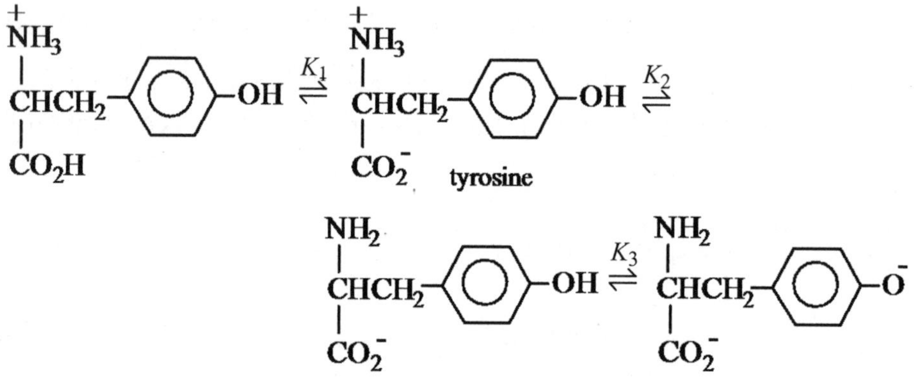

tyrosine

10-18. (a) For 0.0500 M KH_2PO_4, $[H^+] = \sqrt{\dfrac{K_1K_2(0.0500) + K_1K_w}{K_1 + 0.0500}} = 1.99 \times 10^{-5}$

$\Rightarrow$ pH $= 4.70$

$$4.70 = 2.148 + \log \frac{[H_2PO_4^-]}{[H_3PO_4]} \Rightarrow \frac{[H_3PO_4]}{[H_2PO_4^-]} = 2.8 \times 10^{-3}$$

(b)   For 0.0500 M K_2HPO_4, $[H^+] = \sqrt{\dfrac{K_2K_3(0.0500) + K_2K_w}{K_2 + 0.0500}} = 1.99 \times 10^{-10}$

$\Rightarrow$ pH = 9.70

$9.70 = 2.148 + \log \dfrac{[H_2PO_4^-]}{[H_3PO_4]} \Rightarrow \dfrac{[H_3PO_4]}{[H_2PO_4^-]} = 2.8 \times 10^{-8}$

10-19. (a) $H_3PO_4 \overset{pK_1 = 2.148}{\longrightarrow} H_2PO_4^- \overset{pK_2 = 7.198}{\longrightarrow} HPO_4^{2-} \overset{pK_3 = 12.375}{\longrightarrow} PO_4^{3-}$

$$pH \approx (2.148+7.198)/2 \qquad pH \approx (7.198+12.375)/2$$
$$= 4.67 \qquad\qquad\qquad = 9.79$$

pH 7.45 corresponds to a mixture of NaH_2PO_4 and Na_2HPO_4. (You could get the same result by mixing other combinations such as H_3PO_4 and Na_3PO_4 or H_3PO_4 and Na_2HPO_4.)

(b) pH $= pK_2 + \log \dfrac{[HPO_4^{2-}]}{[H_2PO_4^-]}$

$7.45 = 7.198 + \log \dfrac{[HPO_4^{2-}]}{[H_2PO_4^-]} \Rightarrow \dfrac{[HPO_4^{2-}]}{[H_2PO_4^-]} = 1.78_6$

Combining this last result with $[HPO_4^{2-}] + [H_2PO_4^-] = 0.0500$ M gives $[HPO_4^{2-}] = 0.0320_5$ M and $[H_2PO_4^-] = 0.0179_5$ M. Use 4.55 g of Na_2HPO_4 and 2.15 g of NaH_2PO_4.

(c) Here is one of several ways: Weigh out 0.0500 mol Na_2HPO_4 and dissolve it in 900 mL of water. Add HCl while monitoring the pH with a pH electrode. When the pH is 7.45, stop adding HCl and dilute up to exactly 1 L with H_2O.

10-20. Lysine hydrochloride (H_2L^+) is

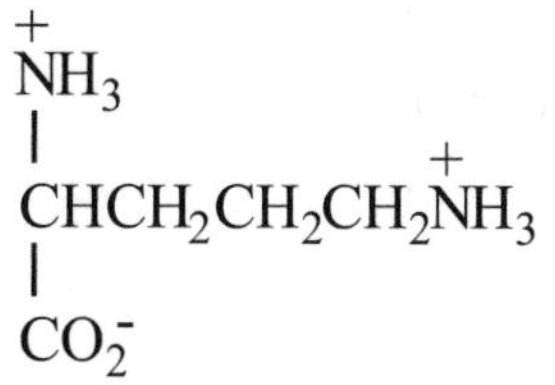

$$\begin{array}{l} \overset{+}{N}H_3 \\ | \\ CHCH_2CH_2CH_2\overset{+}{N}H_3 \\ | \\ CO_2^- \end{array}$$

for which $[H^+] = \sqrt{\dfrac{K_1K_2(0.0100)+K_1K_w}{K_1+0.0100}} = 2.32 \times 10^{-6}$ M $\Rightarrow$ pH = 5.64

$[H_2L^+] = 0.0100$ M

$$[H_3L^{2+}] = \frac{[H^+][H_2L^+]}{K_1} = 1.36 \times 10^{-6} \text{ M}$$

$$[HL] = \frac{K_2[H_2L^+]}{[H^+]} = 3.68 \times 10^{-6} \text{ M} \qquad [L^-] = \frac{K_3[HL]}{[H^+]} = 2.40 \times 10^{-11} \text{ M}$$

10-21.

Histidine hydrochloride (FM 191.62) is $H_2His^+Cl^-$, the intermediate form between pK_1 and pK_2. A pH of 9.30 requires neutralizing all of the H_2His^+ to HHis and then adding more KOH to create a mixture of HHis and His^-. Therefore, we must add 1 mol KOH for each mol of His·HCl to get to HHis and then add some more KOH to obtain the mixture of HHis and His^-. Initial mol of His·HCl = 10.0 g/(191.62 g/mol) = 0.052 1$_9$ mol. We require 0.052 1$_9$ mol of KOH plus the amount x in the following table to obtain the correct mixture:

	HHis	+	OH^-	→	His^-
Initial mol:	0.052 1$_9$		x		—
Final mol:	0.052 1$_9$ − x		—		x

$$pH = pK_3 + \log\frac{[His^{2-}]}{[HHis^-]} \Rightarrow 9.30 = 9.28 + \log\frac{x}{0.052\ 1_9 - x}$$

$$\Rightarrow x = 0.026\ 7_0$$

Total mol KOH required = 0.052 1$_9$ + 0.026 7$_0$ = 0.078 9 mol

$\qquad$ = 78.9 mL of 1.00 M KOH

10-22. (a) $\quad pH = pK_3$ (citric acid) $+ \log\dfrac{[C^{3-}]\gamma_{C^{3-}}}{[HC^{2-}]\gamma_{HC^{2-}}}$

$$pH = 6.396 + \log\frac{(1.00)(0.405)}{(2.00)(0.665)} = 5.88$$

(b) If the ionic strength is raised to 0.10 M,

$$pH = 6.396 + \log \frac{(1.00)(0.115)}{(2.00)(0.37)} = 5.59$$

10-23. (a) HA (b) A^-

(c) $pH = pK_a + \dfrac{[A^-]}{[HA]}$

$7.00 = 7.00 + \log \dfrac{[A^-]}{[HA]} \Rightarrow [A^-]/[HA] = 1.0$

$6.00 = 7.00 + \log \dfrac{[A^-]}{[HA]} \Rightarrow [A^-]/[HA] = 0.10$

10-24. (a) 4.00 (b) 8.00 (c) H_2A (d) HA^- (e) A^{2-}

10-25. (a) 9.00 (b) 9.00 (c) BH^+

(d) $12.00 = 9.00 + \log \dfrac{[B]}{[BH^+]} \Rightarrow [B]/[BH^+] = 1.0 \times 10^3$

10-26.

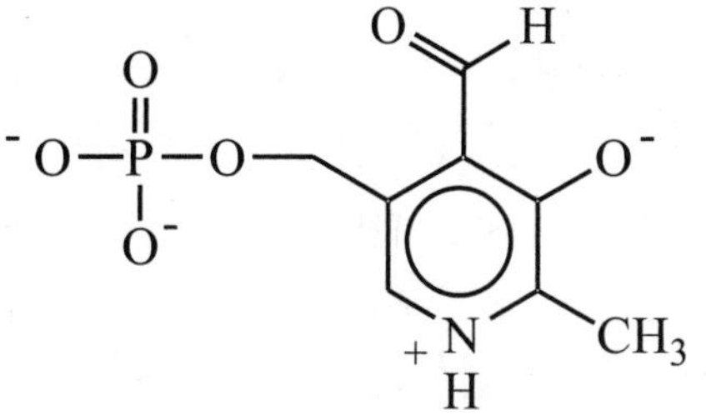

10-27. Fraction in form HA $= \alpha_{HA} = \dfrac{[H^+]}{[H^+] + K_a} = \dfrac{10^{-5}}{10^{-5} + 10^{-4}} = 0.091.$

Fraction in form $A^- = \alpha_{A^-} = \dfrac{K_a}{[H^+] + K_a} = 0.909.$

$\dfrac{[A^-]}{[HA]} = \dfrac{\alpha_{A^-}}{\alpha_{HA}} = 10$, which makes sense.

10-28. $\alpha_{H_2A} = \dfrac{[H^+]^2}{[H^+]^2 + [H^+]K_1 + K_1K_2}$, where $[H^+] = 10^{-7.00}$, $K_1 = 10^{-8.00}$, and

$K_2 = 10^{-10.00} \Rightarrow \alpha_{H_2A} = 0.91$

10-29. Use Equations for a diprotic acid with $pK_1 = 8.85$ and $pK_2 = 10.43$.

$$\alpha_{H_2A} = \frac{[H_2A]}{F} = \frac{[H^+]^2}{[H^+]^2 + [H^+]K_1 + K_1K_2}$$

$$\alpha_{HA^-} = \frac{[HA^-]}{F} = \frac{K_1[H^+]}{[H^+]^2 + [H^+]K_1 + K_1K_2}$$

$$\alpha_{A^{2-}} = \frac{[A^{2-}]}{F} = \frac{K_1K_2}{[H^+]^2 + [H^+]K_1 + K_1K_2}$$

	pH 8.00	pH 10.00
α_{H_2A}	0.876	0.0491
α_{HA^-}	0.124	0.693
$\alpha_{A^{2-}}$	4.60×10^{-4}	0.258

10-30.

pH :	1.00	1.92	6.00	6.27	10.00
α_{H_2A}	0.893	0.500	5.41×10^{-5}	2.23×10^{-5}	1.55×10^{-12}
α_{HA^-}	0.107	0.500	0.651	0.500	1.86×10^{-4}
$\alpha_{A^{2-}}$	5.76×10^{-7}	2.23×10^{-5}	0.349	0.500	0.9998

10-31. (a)   The derivation follows the outline of Equations 10-19 through 10-21. The results are

$$\alpha_{H_3A} = \frac{[H_3A]}{F} = \frac{[H^+]^3}{[H^+]^3 + [H^+]^2K_1 + [H^+]K_1K_2 + K_1K_2K_3}$$

$$\alpha_{H_2A^-} = \frac{[H_2A^-]}{F} = \frac{[H^+]^2 K_1}{[H^+]^3 + [H^+]^2K_1 + [H^+]K_1K_2 + K_1K_2K_3}$$

$$\alpha_{HA^{2-}} = \frac{[HA^{2-}]}{F} = \frac{[H^+] K_1K_2}{[H^+]^3 + [H^+]^2K_1 + [H^+]K_1K_2 + K_1K_2K_3}$$

$$\alpha_{A^{3-}} = \frac{[A^{3-}]}{F} = \frac{K_1K_2K_3}{[H^+]^3 + [H^+]^2K_1 + [H^+]K_1K_2 + K_1K_2K_3}$$

(b)   For phosphoric acid, $pK_1 = 2.148$, $pK_2 = 7.198$, and $pK_3 = 12.375$. At pH = 7.00, the previous expressions give $\alpha_{H_3A} = 8.6 \times 10^{-6}$, $\alpha_{H_2A^-} = 0.61$, $\alpha_{HA^{2-}} = 0.39$, and $\alpha_{A^{3-}} = 1.6 \times 10^{-6}$.

10-32. $pH = pK_{NH_4^+} + \log\dfrac{[NH_3]}{[NH_4^+]} \Rightarrow 9.00 = 9.245 + \log\dfrac{[NH_3]}{[NH_4^+]} \Rightarrow \dfrac{[NH_3]}{[NH_4^+]} = 0.56_9$

Fraction unprotonated $= \dfrac{[NH_3]}{[NH_3] + [NH_4^+]} = \dfrac{0.56_9}{0.56_9 + 1} = 0.36$

The presence of other acids and bases has no bearing on the fraction of ammonia that is unprotonated.

10-33. The quantity of morphine in the solution is negligible compared to the quantity of cacodylic acid. The pH is determined by the reaction of cacodylic acid (HA) with NaOH:

$$HA \quad + \quad OH^- \quad \rightarrow \quad A^- \quad + \quad H_2O$$

	HA	OH⁻	A⁻	H₂O
Initial mmol:	1.000	0.800	—	—
Final mmol:	0.200	—	0.800	—

$$pH = pK_a + \log \frac{[A^-]}{[HA]} = 6.19 + \log \frac{0.800}{0.200} = 6.79$$

For morphine (B), $K_a = K_w/K_b = 1.0 \times 10^{-14}/1.6 \times 10^{-6} = 6.2_5 \times 10^{-9}$

$$\Rightarrow pK_a = 8.20$$

At pH 6.79, we can write $pH = pK_{BH^+} + \log \frac{[B]}{[BH^+]} \Rightarrow$

$$6.79 = 8.20 + \log \frac{[B]}{[BH^+]} \Rightarrow \frac{[B]}{[BH^+]} = 0.039 \Rightarrow [B] = 0.039\,[BH^+]$$

$$\text{Fraction in form } BH^+ = \frac{[BH^+]}{[B] + [BH^+]} = \frac{[BH^+]}{0.039\,[BH^+] + [BH^+]} = 96\%$$

10-34.

	A	B	C	D	E	F
1	Fractional composition for diprotic acid					
2						
3	K1 =	pH	[H+]	α(H2A)	α(HA⁻)	α(A²⁻)
4	9.55E-04	1	0.1	9.91E-01	0.00946	3.13E-06
5	K2 =	2	0.01	0.912562	0.087149	0.000289
6	3.31E-05	3	0.001	0.503369	0.480713	0.015918
7	pK1 =	4	0.0001	0.072928	0.696455	0.230618
8	3.02	5	0.00001	0.002423	0.231386	0.766191
9	pK2 =	6	0.000001	3.07E-05	0.029313	0.970656
10	4.48	7	1E-07	3.15E-07	0.003011	0.996989
11						
12	A4 = 10^-A8	D4 = \$C4^2/(\$C4^2+\$C4*\$A\$4+\$A\$4*\$A\$6)				
13	A6 = 10^-A10	E4 = \$C4*\$A\$4/(\$C4^2+\$C4*\$A\$4+\$A\$4*\$A\$6)				
14	C4 = 10^-B4	F4 = \$A\$4*\$A\$6/(\$C4^2+\$C4*\$A\$4+\$A\$4*\$A\$6)				

10-35.

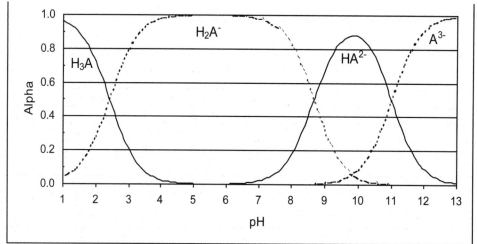

	A	B	C	D	E	F	G
1	Fractional composition for triprotic acid						
2							
3	K1 =	pH	[H+]	α(H3A)	α(H2A$^-$)	α(HA^{2-})	α(A^{3-})
4	3.89E-03	1	1.00E-01	9.63E-01	3.74E-02	8.01E-10	7.82E-20
5	K2 =	2	1.00E-02	7.20E-01	2.80E-01	5.99E-08	5.85E-17
6	2.14E-09	3	1.00E-03	2.04E-01	7.96E-01	1.70E-06	1.66E-14
7	K3 =	4	1.00E-04	2.51E-02	9.75E-01	2.08E-05	2.04E-12
8	9.77E-12	5	1.00E-05	2.56E-03	9.97E-01	2.13E-04	2.08E-10
9	pK1 =	6	1.00E-06	2.56E-04	9.98E-01	2.13E-03	2.08E-08
10	2.41	7	1.00E-07	2.52E-05	9.79E-01	2.09E-02	2.05E-06
11	pK2 =	8	1.00E-08	2.12E-06	8.24E-01	1.76E-01	1.72E-04
12	8.67	9	1.00E-09	8.14E-08	3.17E-01	6.77E-01	6.61E-03
13	pK3 =	10	1.00E-10	1.05E-09	4.09E-02	8.74E-01	8.54E-02
14	11.01	11	1.00E-11	6.07E-12	2.36E-03	5.05E-01	4.93E-01
15		12	1.00E-12	1.12E-14	4.34E-05	9.28E-02	9.07E-01
16		13	1.00E-13	1.22E-17	4.74E-07	1.01E-02	9.90E-01
17	A4 = 10^-A10						
18	C4 = 10^-B4						
19	D4 = \$C4^3/(\$C4^3+\$C4^2*\$A\$4+\$C4*\$A\$4*\$A\$6+\$A\$4*\$A\$6*\$A\$8)						
20	E4 = \$C4^2*\$A\$4/(\$C4^3+\$C4^2*\$A\$4+\$C4*\$A\$4*\$A\$6+\$A\$4*\$A\$6*\$A\$8)						
21	F4 = \$C4*\$A\$4*\$A\$6/(\$C4^3+\$C4^2*\$A\$4+\$C4*\$A\$4*\$A\$6+\$A\$4*\$A\$6*\$A\$8)						
22	G4 = \$A\$4*\$A\$6*\$A\$8/(\$C4^3+\$C4^2*\$A\$4+\$C4*\$A\$4*\$A\$6+\$A\$4*\$A\$6*\$A\$8)						

10-36.   (a) Fractional composition in tetraprotic system

	A	B	C	D	E	F	G	H	I
1	Fractional composition in tetraprotic system								
2									
3	Ka1 =	pH	[H+]	Denom.	Alph(H4A)	Alph(H3A)	Alph(H2A)	Alph(HA)	Alph(A)
4	1.58E-04	1	1E-01	1.0E-04	1.0E+00	1.6E-03	6.3E-09	2.5E-14	9.9E-25
5	Ka2 =	2	1E-02	1.0E-08	9.8E-01	1.6E-02	6.2E-07	2.5E-11	9.8E-21
6	3.98E-07	3	1E-03	1.2E-12	8.6E-01	1.4E-01	5.4E-05	2.2E-08	8.6E-17
7	Ka3 =	4	1E-04	2.6E-16	3.9E-01	6.1E-01	2.4E-03	9.7E-06	3.9E-13
8	3.98E-07	5	1E-05	1.7E-19	5.7E-02	9.1E-01	3.6E-02	1.4E-03	5.7E-10
9	Ka4 =	6	1E-06	2.5E-22	4.1E-03	6.4E-01	2.5E-01	1.0E-01	4.0E-07
10	3.98E-12	7	1E-07	3.3E-24	3.0E-05	4.8E-02	1.9E-01	7.6E-01	3.0E-05
11		8	1E-08	2.6E-25	3.9E-08	6.2E-04	2.4E-02	9.7E-01	3.9E-04
12		9	1E-09	2.5E-26	4.0E-11	6.3E-06	2.5E-03	9.9E-01	4.0E-03
13		10	1E-10	2.6E-27	3.8E-14	6.1E-08	2.4E-04	9.6E-01	3.8E-02
14		11	1E-11	3.5E-28	2.9E-17	4.5E-10	1.8E-05	7.2E-01	2.8E-01
15		12	1E-12	1.2E-28	8.0E-21	1.3E-12	5.0E-07	2.0E-01	8.0E-01
16		13	1E-13	1.0E-28	9.8E-25	1.5E-15	6.2E-09	2.5E-02	9.8E-01
17									
18	C4 = 10^-B4								
19	D4 = C4^4+$A$4*C4^3+$A$4*$A$6*C4^2+$A$4*$A$6*$A$8*C4								
20			+$A$4*$A$6*$A$8*$A$10						
21	E4 = C4^4/D4								
22	F4 = $A$4*C4^3/D4				H4 = $A$4*$A$6*$A$8*C4/D4				
23	G4 = $A$4*$A$6*C4^2/D4				I4 = $A$4*$A$6*$A$8*$A$10/D4				

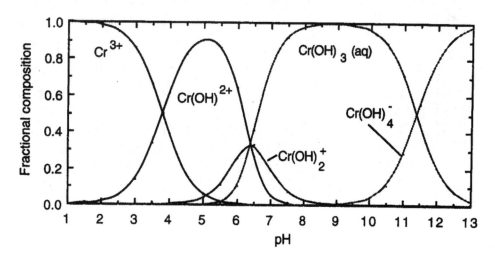

(b) $K = 10^{-6.84} = [Cr(OH)_3(aq)]$, so $[Cr(OH)_3(aq)] = 10^{-6.84}$ M $= 1.4_5 \times 10^{-7}$ M

(c) $K_{a3} = 10^{-6.40} = \dfrac{[Cr(OH)_3(aq)][H^+]}{[Cr(OH)_2^+]} = \dfrac{[10^{-6.84}][10^{-4.00}]}{[Cr(OH)_2^+]}$

$\Rightarrow [Cr(OH)_2^+] = \dfrac{[10^{-6.84}][10^{-4.00}]}{10^{-6.40}} = 10^{-4.44}$ M

$$K_{a2} = 10^{-6.40} = \frac{[Cr(OH)_2^+][H^+]}{[Cr(OH)^{2+}]} = \frac{[10^{-4.44}][10^{-4.00}]}{[Cr(OH)^{2+}]}$$

$$\Rightarrow [Cr(OH)^{2+}] = \frac{[10^{-4.44}][10^{-4.00}]}{10^{-6.40}} = 10^{-2.04} \text{ M}$$

10-37. Acidic substituents: aspartic acid, cysteine, glutamic acid, tyrosine

Basic substituents: arginine, histidine, lysine

10-38. The isoelectric pH is the pH at which the protein has no net charge, even though it has many positive and negative sites. The isoionic pH is the pH of a solution containing only protein, H^+, and OH^-.

10-39. The <u>average</u> charge is zero. There is no pH at which <u>all</u> molecules have zero charge.

10-40.

Isoionic $[H^+] = \sqrt{\dfrac{K_1 K_2(0.010) + K_1 K_w}{K_1 + (0.010)}} \Rightarrow$ pH $= 5.72$

Isoelectric pH $= \dfrac{pK_1 + pK_2}{2} = 5.59$

10-41. A mixture of proteins is exposed to a strong electric field in a medium with a pH gradient. Positively charged molecules move toward the negative pole and negatively charged molecules move toward the positive pole. Each protein migrates until it reaches the point where the pH is the same as its isoelectric pH. At this point, the protein has no net charge and no longer moves. Each protein is therefore focused in one region at its isoelectric pH. If a protein diffuses out of its isoelectric zone, it becomes charged and migrates back into the zone.

CHAPTER 11
ACID-BASE TITRATIONS

11-1. The equivalence point occurs when the quantity of titrant is exactly the stoichiometric amount needed for complete reaction with analyte. The end point occurs when there is an abrupt change in a physical property, such as pH or indicator color. Ideally, the end point is chosen to occur at the equivalence point.

11-2.

V_a	0	1	5	9	9.9	10	10.1	12
pH	13.00	12.95	12.68	11.96	10.96	7.00	3.04	1.75

Representative calculations:

0 mL:
$$pH = -\log \frac{K_w}{[OH^-]} = -\log \frac{10^{-14}}{0.100} = 13.00$$

1 mL:
$$[OH^-] = \frac{9}{10}(0.100)\frac{100}{101} = 0.0891 \text{ M} \Rightarrow pH = 12.95$$

10 mL:
$$[OH^-] = [H^+] = 10^{-7} \text{ M}$$

10.1 mL:
$$[H^+] = \left(\frac{0.1}{110.1}\right)(1.00) = 9.08 \times 10^{-4} \text{ M} \Rightarrow pH = 3.04$$

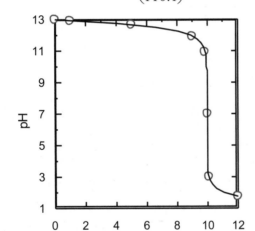

11-3. Consider the titration curve near the equivalence point. If we titrate strong acid with strong base, the concentration of H^+ is close to 1% of its initial value when we are 99% of the way to the equivalence point (i.e., when $V_b = 0.99V_e$). (This statement would be exactly true if there were no dilution occurring. We will neglect dilution.) If the initial acid concentration were, say, 0.1 M, then $[H^+] = 1\%$ of 0.1 M = 0.001 M at $V_b = 0.99 V_e$. The pH is $-\log(0.001) = 3$. At 99.9% completion, $[H^+] = 0.1\%$ of 0.1 M = 0.0001 M and the pH is 4. When the titration is 0.1% past the equivalence point, $[OH^-] = 0.0001$ M and the pH is $-\log(K_w/0.0001) = 10$. The pH jumps from 4 to 10 in the interval from $V_b =$

$0.999V_e$ to $1.001V_e$. Even though the concentration of H^+ hardly changes, its logarithm changes rapidly around the equivalence point because $[H^+]$ decreases by orders of magnitude with tiny additions of OH^- when there is hardly any H^+ present.

11-4. The sketch should look like Figure 11-2. Before base is added, the pH is determined by the acid dissociation reaction of HA. Between the initial point and the equivalence point, each mole of OH^- converts an equivalent quantity of HA into A^-. The resulting buffer containing HA and A^- determines the pH. At the equivalence point, all HA has been converted to A^-. The pH is controlled by the base hydrolysis reaction of A^- with H_2O. After the equivalence point, excess OH^- is being added to the solution. To a good approximation, the pH is determined just by the concentration of excess OH^-.

11-5. If the analyte is too weak or too dilute, there is very little change in pH at the equivalence point.

11-6.

V_b	0	1	5	9	9.9	10	10.1	12
pH	3.00	4.05	5.00	5.95	7.00	8.98	10.96	12.25

Representative calculations:

<u>0 mL:</u> $HA \ = \ H^+ + \ A^-$ $\dfrac{x^2}{0.100 - x} = 10^{-5.00} \Rightarrow x = 9.95 \times 10^{-4}$ M
                 $0.100 - x$ x x

$\Rightarrow$ pH = 3.00

<u>1 mL:</u> $pH \ = \ pK_a \ + \ \log \dfrac{[A^-]}{[HA]} = 5.00 + \log \dfrac{1}{9} = 4.05$

<u>10 mL:</u> $A^- \ + \ H_2O = HA + OH^-$ $\dfrac{x^2}{0.090\,9 - x} = \dfrac{K_w}{K_a}$
        $\left(\dfrac{100}{110}\right)(0.100) - x$ x x

$\Rightarrow x = 9.53 \times 10^{-6}$

$\Rightarrow [H^+] = \dfrac{K_w}{x} \Rightarrow$ pH = 8.98

<u>10.1 mL:</u> $[OH^-] \ = \ \left(\dfrac{0.1}{110.1}\right)(1.00) = 9.08 \times 10^{-4}$ M $\Rightarrow$ pH = 10.96

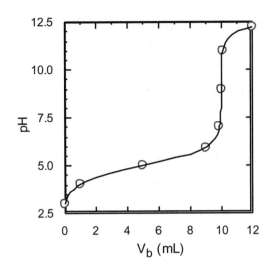

11-7. $pH = pK_a + \log\dfrac{[A^-]}{[HA]}$ $pK_a - 1 = pK_a + \log\dfrac{[A^-]}{[HA]} \Rightarrow \dfrac{[A^-]}{[HA]} = \dfrac{1}{10}$

If the ratio $\dfrac{[A^-]}{[HA]}$ is to be $\dfrac{1}{10}$, then $\dfrac{1}{11}$ of the initial HA must remain as HA.

At this point, $[A^-]/[HA] = (1/11)/(10/11) = 1/10$. So $pH = pK_a - 1$ when $V_b = V_e/11$.

In a similar manner, $pH = pK_a + 1$ when $V_b = 10V_e/11$.

For anilinium ion, $pK_a = 4.601$. For the titration of 100 mL of 0.100 M

anilinium ion with 0.100 M OH⁻, the reaction is

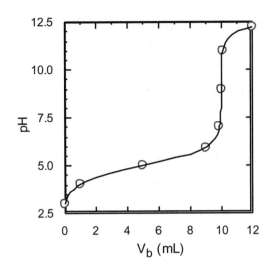

$\langle\bigcirc\rangle\!-\!\overset{+}{N}H_3 + O\bar{H} \rightarrow \langle\bigcirc\rangle\!-\!NH_2 + H_2O$ and $V_e = 100$ mL.

$\underline{0\ mL}:$ $\langle\bigcirc\rangle\!-\!\overset{+}{NH_3} \rightleftharpoons \langle\bigcirc\rangle\!-\!NH_2 + H^+$

                 $0.100 - x$ x x

$$\dfrac{x^2}{0.100 - x} = K_a = 10^{-4.60} \Rightarrow x = 1.57 \times 10^{-3} \Rightarrow pH = 2.80$$

$\underline{V_e/11 = 9.09\ mL}$: $pH = pK_a - 1 = 3.60$

$\underline{V_e/2 = 50.0\ mL}$: $pH = pK_a = 4.60$

$\underline{10V_e/11 = 90.91\ mL}$: $pH = pK_a + 1 = 5.60$

$\underline{V_e = 100.0\ mL}$: BH⁺ has been converted to B.

$\qquad B + H_2O \rightleftharpoons BH^+ + OH^- \qquad K_b = \dfrac{K_w}{K_a} = \dfrac{x^2}{\left(\dfrac{100}{200}\right)(0.100) - x}$

$\qquad F - x \qquad\qquad x \qquad x$

$\qquad\qquad \Rightarrow x = 4.46 \times 10^{-6}\ M \qquad\qquad pH = -\log\dfrac{K_w}{x} = 8.65$

$\underline{1.2V_e = 120.0\ mL}$: There are 20.0 mL of excess NaOH.

$$[OH^-] = \left(\frac{20}{220}\right)(0.100) = 9.09 \times 10^{-3} \text{ M} \Rightarrow \text{pH} = 11.96$$

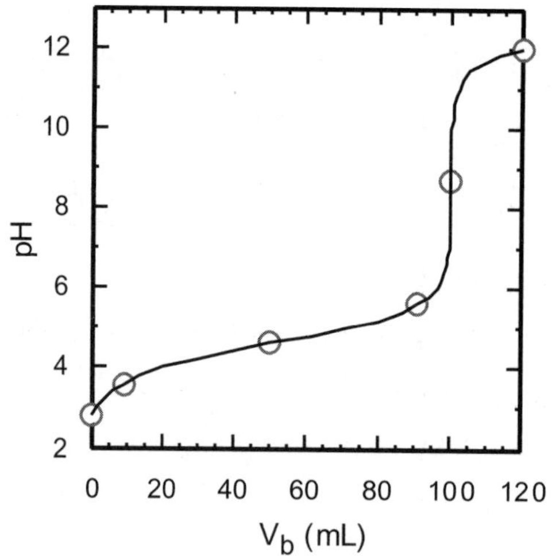

11-8. The titration reaction is $HA + OH^- \rightarrow A^- + H_2O$. A volume of V mL of HA will require $2V$ mL of KOH to reach the equivalence point, because $[HA] = 0.100$ M and $[KOH] = 0.050\ 0$ M. The formal concentration of A^- at the equivalence point will be $\left(\frac{V}{V+2V}\right)(0.100) = 0.033\ 3$ M. The pH is found by writing

$$A^- + H_2O \rightleftharpoons HA + OH^-$$

$$0.033\ 3 - x \qquad x \qquad x$$

$$\frac{x^2}{0.033\ 3 - x} = K_b = \frac{K_w}{K_a} = \frac{1.0 \times 10^{-14}}{1.48 \times 10^{-4}}$$

$$\Rightarrow x = 1.50 \times 10^{-6}\text{ M} \Rightarrow \text{pH} = 8.18$$

11-9.

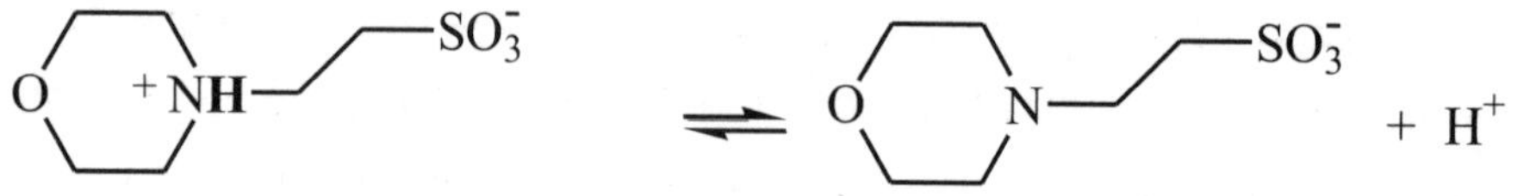

$$K_a = 10^{-6.27}$$

$$1/K_w = 10^{14}$$

$$H^+ + OH^- \rightleftharpoons H_2O$$

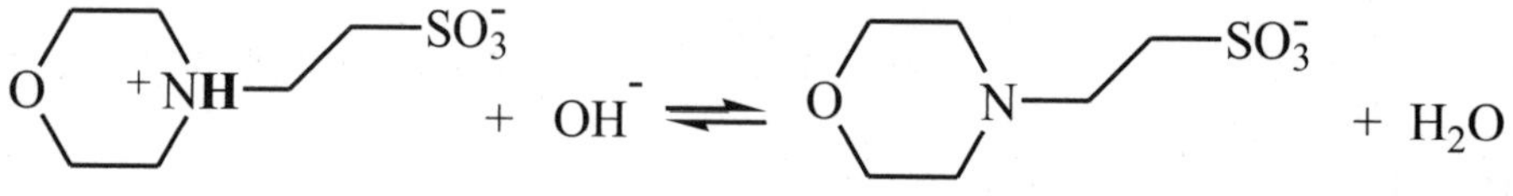

$$K = \frac{K_a}{K_w} = 5.4 \times 10^7$$

11-10.

	HA	+	OH^-	$\rightarrow$	A^-	+	H_2O
Initial mmol:	5.857		x		—		
Final mmol:	$5.857 - x$		—		x		

$$\text{pH} = 9.24 = pK_a + \log\frac{[A^-]}{[HA]} = 9.39 + \log\frac{x}{5.857 - x} \Rightarrow x = 2.4_{28} \text{ mmol}$$

$$[OH^-] = \frac{2.4_{28} \text{ mmol}}{22.63 \text{ mL}} = 0.107 \text{ M}$$

11-11.
$$(CH_3)_3NH^+ + OH^- \rightarrow (CH_3)_3N + H_2O$$

Initial mmol:	1.00	0.40	—	
Final mmol:	0.60	—	0.40	

First, find the ionic strength:

$$[(CH_3)_3NH^+] = 0.60 \text{ mmol}/14.0 \text{ mL} = 0.042\,86 \text{ M}$$

$$[Br^-] = 1.00 \text{ mmol}/14.0 \text{ mL} = 0.071\,43 \text{ M}$$

$$[Na^+] = 0.40 \text{ mmol}/14.0 \text{ mL} = 0.028\,57 \text{ M}$$

$$\mu = \tfrac{1}{2}\Sigma c_i z_i^2 = 0.071 \text{ M}$$

$$pH = pK_a + \log\frac{[B]\,\gamma_B}{[BH^+]\,\gamma_{BH^+}}$$

$$pH = 9.799 + \log\frac{(0.028\,6)(1.00)}{(0.042\,9)(0.80)} = 9.72$$

In the previous calculation, we used the size of $(CH_3)_3NH^+$ (400 pm) and the activity coefficient interpolated from Table 8-1.

11-12. The sketch should look like Figure 11-9. Before base is added, the pH is determined by the base hydrolysis reaction of B with H_2O. Between the initial point and the equivalence point, each mole of H^+ converts an equivalent quantity of B into BH^+. The resulting buffer containing B and BH^+ determines the pH. At the equivalence point, all B has been converted to BH^+. The pH is controlled by the acid dissociation reaction of BH^+. After the equivalence point, excess H^+ is being added to the solution. To a good approximation, the pH is determined just by the concentration of excess H^+.

11-13. At the equivalence point, the weak base, B, is converted completely to the conjugate acid, BH^+, which is necessarily acidic.

11-14.

V_a	0	1	5	9	9.9	10	10.1	12
pH	11.00	9.95	9.00	8.05	7.00	5.02	3.04	1.75

Representative calculations:

<u>0 mL</u>:
$$\begin{array}{ccccc} B & + & H_2O & \rightleftharpoons & BH + OH^- \\ 0.100 - x & & & & x \quad\quad x \end{array}$$
$$\frac{x^2}{0.100 - x} = 10^{-5.00} \text{ M} \Rightarrow x = 9.95 \times 10^{-4} \text{ M}$$

$$[H^+] = \frac{K_w}{x} \Rightarrow pH = 11.00$$

1 mL: $pH = pK_{BH^+} + \log \dfrac{[B]}{[BH^+]} = 9.00 + \log \dfrac{9}{1} = 9.95$

10 mL: $BH^+ \rightleftharpoons B + H^+$ $\dfrac{x^2}{0.090\,9 - x} = 10^{-9.00}$ M $\Rightarrow x = 9.53 \times 10^{-6}$ M

$\left(\dfrac{100}{110}\right)(1.00) - x$ x x $[H^+] = x \Rightarrow pH = 5.02$

10.1 mL: $[H^+] = \left(\dfrac{0.1}{110.1}\right)(1.00) = 9.08 \times 10^{-4}$ M $\Rightarrow pH = 3.04$

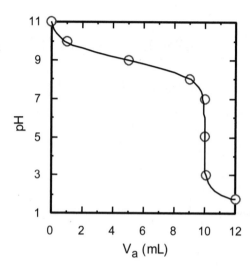

11-15. The maximum buffer capacity is reached when

$V = \frac{1}{2}V_e$, at which time $\dfrac{[B]}{[BH^+]} = 1$ and $pH = pK_a$ (for BH^+).

11-16.

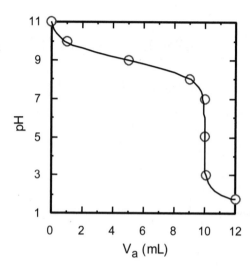

$-CH_2NH_2 + H^+ \rightleftharpoons$ $-CH_2\overset{+}{N}H_3$ Reverse of K_a reaction

$K = 1/K_a$ (for $C_6H_5CH_2NH_3^+$) $= 2.2 \times 10^9$

11-17. Titration reaction: $B + H^+ \rightarrow BH^+$. To find the equivalence point, we write
$(50.0)(0.031\,9) = (V_e)(0.050\,0) \Rightarrow V_e = 31.9$ mL.

0 mL: $B + H_2O \rightleftharpoons BH^+ + OH^-$ $\dfrac{x^2}{0.031\,9 - x} = K_b = \dfrac{K_w}{K_a} = 2.22 \times 10^{-5}$

  $0.031\,9 - x$ x x

  $\Rightarrow x = 8.31 \times 10^{-4}$ M $\Rightarrow pH = 10.92$

12.0 mL:

	B	+	H$^+$	$\rightarrow$	BH$^+$
Initial :	31.9		12.0		—
Final:	19.9		—		12.0

$pH = pK_a + \log \dfrac{[B]}{[BH^+]} = 9.35 + \log \dfrac{19.9}{12.0} = 9.57$

$1/2V_e$: $pH = pK_a = 9.35$

$\underline{30\ mL}$: $pH = pK_a + \log\dfrac{1.9}{30.0} = 8.15$

$\underline{V_e}$: B has been converted to BH^+ at a concentration of $\left(\dfrac{50.0}{81.9}\right)(0.031\,9)$

$\qquad = 0.019\,5\ M$

$$BH^+ \rightleftharpoons B + H^+ \qquad\qquad \frac{x^2}{0.019\,5 - x} = K_a \Rightarrow x = 2.96 \times 10^{-6}\ M$$

$0.019\,5 - x \quad\ x \quad\ x \qquad\qquad\qquad\qquad\qquad \Rightarrow pH = 5.53$

$\underline{35.0\ mL}$: $[H^+] = \left(\dfrac{3.1}{85.0}\right)(0.050\,0) = 1.82 \times 10^{-3}\ M \Rightarrow pH = 2.74$

11-18. Titration reaction: $CN^- + H^+ \rightarrow HCN$

At the equivalence point, moles of $CN^- =$ moles of H^+

$(0.100\ M)(50.00\ mL) = (0.438\ M)(V_e) \Rightarrow V_e = 11.42\ mL$

(a)

	CN^-	$+$	H^+	$\rightarrow$	HCN
Initial:	11.42		4.20		—
Final:	7.22		—		4.20

$$pH = pK_a + \log\frac{7.22}{4.20} = 9.45$$

(b) 11.82 mL is 0.40 mL past the equivalence point.

$$[H^+] = \left(\frac{0.40}{61.82}\right)(0.438\ M) = 2.83 \times 10^{-3}\ M \Rightarrow pH = 2.55$$

(c) At the equivalence point, we have made HCN at a formal concentration of $\left(\dfrac{50.00}{61.42}\right)(0.100) = 0.081\,4\ M$.

$$HCN \rightleftharpoons H^+ + CN^- \qquad\qquad \frac{x^2}{0.081\,4 - x} = K_a \Rightarrow x = 7.1 \times 10^{-6}$$

$0.0814 - x \qquad\ x \qquad\ x \qquad\qquad\qquad\qquad\qquad \Rightarrow pH = 5.15$

11-19.

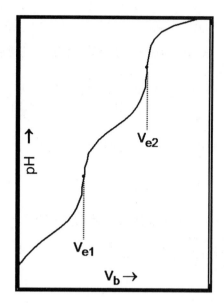

The pH of the initial solution before base is added is determined by the first acid dissociation reaction of H_2A. As base is added, it converts H_2A into an equivalent amount of HA^-. The buffer consisting of H_2A and HA^- governs the pH. At the first equivalence point, we have a solution of "pure" HA^-, the intermediate form of a diprotic acid. The pH is determined by the competitive acid and base reactions of HA^-. Between the two equivalence points there is a mixture of HA^- and A^{2-}, which is another buffer. At the second

equivalence point, we have converted all HA^- into A^{2-}, whose base hydrolysis reaction determines the pH. After the second equivalence point, the excess OH^- added from the buret is mainly responsible for determining the pH, with negligible contribution from A^{2-}.

11-20. The protein has an average charge of 0 at the isoelectric point. The isoionic pH is lower, so the average charge of the protein must be positive at the isoionic point because some basic groups react with H^+ as acid is added to lower the pH.

11-21. The equivalence point could be attained by mixing pure HA plus NaCl. Neglecting effects of ionic strength, the pH is equivalent to that of a solution of pure HA. This is the isoionic pH.

11-22. (a)

$$HA \rightleftharpoons A^- + H^+ \qquad\qquad K_a$$

$$\underline{B + H^+ \rightleftharpoons BH^+ \qquad\qquad K = K_b/K_w}$$

$$B + HA \rightleftharpoons BH^+ + A^- \qquad K = K_aK_b/K_w$$

$$= 10^{-2.86}\,10^{-3.36} / 10^{-14.00} = 10^{7.78}$$

(b) In the upper curve, $\frac{3}{2}V_e$ is halfway between the first and second equivalence points. The pH is simply pK_2, since there is a 1:1 mixture of HA^- and A^{2-}. In the lower curve, pK_2 ($= pK_{BH^+}$) occurs when there is a 1:1 mixture of B and BH^+. To achieve this condition, all of B is first transformed into BH^+ by reaction with HA until V_e is reached. Then, at $2V_e$ one more equivalent of B has been added, giving a 1:1 mole ratio $B:BH^+$, so $pH = pK_{BH^+}$.

11-23.

V_a	0	1	5	9	10	11	15	19	20	22
pH	11.49	10.95	10.00	9.05	8.00	6.95	6.00	5.05	3.54	1.79

Representative calculations:

0 mL: $B + H_2O \overset{K_{b1}}{\rightleftharpoons} BH^+ + OH^-$ $\dfrac{x^2}{0.100 - x} = 10^{-4.00} \Rightarrow x = 3.11 \times 10^{-3}$ M

 $0.100 - x$ x x

 pH $= -\log \dfrac{K_w}{x} = 11.49$

1 mL: pH $= pK_{BH^+} + \log \dfrac{[B]}{[BH^+]} = 10.00 + \log \dfrac{9}{1} = 10.95$

10 mL: Predominant form is BH^+ with formal concentration $\dfrac{100}{110}(0.100) = 0.090\,9$ M

$[H^+] \approx \sqrt{\dfrac{10^{-6.00}\,10^{-10.00}\,(0.090\,9) + 10^{-6.00}\,10^{-14.00}}{10^{-6.00} + 0.090\,9}}$

$ = 1.00 \times 10^{-8} \Rightarrow pH = 8.00$

11 mL: pH $= pK_{BH_2^{2+}} + \log \dfrac{[BH^+]}{[BH_2^{2+}]} = 6.00 + \log \dfrac{9}{1} = 6.95$

20 mL: $BH_2^{2+} \rightleftharpoons BH^+ + H^+$ $\dfrac{x^2}{0.083\,3 - x} = 10^{-6.00} \Rightarrow x = 2.88 \times 10^{-4}$

 $\dfrac{100}{120}(0.100) - x$ x x

$ \Rightarrow pH = 3.54$

22 mL: $[H^+] = \left(\dfrac{2}{122}\right)(1.00) = 1.64 \times 10^{-2}$ M $\Rightarrow pH = 1.79$

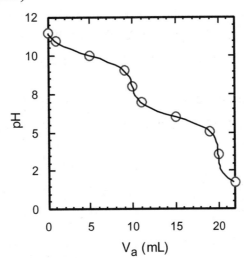

11-24.

V_b	0	1	5	9	10	11	15	19	20	22
pH	2.51	3.05	4.00	4.95	6.00	7.05	8.00	8.95	10.46	12.21

Representative calculations:

<u>0 mL</u>: $H_2A \rightleftharpoons HA^- + H^+$ $\dfrac{x^2}{0.100 - x} = 10^{-4.00} \Rightarrow x = 3.11 \times 10^{-3}$ M

  $0.100 - x \qquad x \qquad x$

$\Rightarrow$ pH = 2.51

<u>1 mL</u>:   pH $= pK_1 + \log \dfrac{[HA^-]}{[H_2A]} = 4.00 + \log \dfrac{1}{9} = 3.05$

<u>10 mL</u>: Predominant form is HA^- with formal concentration $\left(\dfrac{100}{110}\right)(0.100)$

= 0.090 9 M.

$$[H^+] \approx \sqrt{\dfrac{10^{-4.00}\ 10^{-8.00}\ (0.090\ 9) + 10^{-4.00}\ 10^{-14.00}}{10^{-4.00} + 0.090\ 9}}$$

$= 9.99 \times 10^{-7} \Rightarrow$ pH = 6.00

<u>11 mL</u>:   pH $= pK_2 + \log \dfrac{[A^{2-}]}{[HA^-]} = 8.00 + \log \dfrac{1}{9} = 7.05$

<u>20 mL</u>: $A^{2-} + H_2O \rightleftharpoons HA^- + OH^-$ $\dfrac{x^2}{0.083\ 3 - x} = \dfrac{K_w}{K_2} \Rightarrow x = 2.88 \times 10^{-4}$ M

  $\left(\dfrac{100}{120}\right)(0.100) - x \qquad\qquad x \qquad x$

$pH = -\log \dfrac{K_w}{x} = 10.46$

<u>22 mL</u>: $[OH^-] = \left(\dfrac{2}{122}\right)(1.00) = 1.64 \times 10^{-2}$ M $\Rightarrow$ pH = 12.21

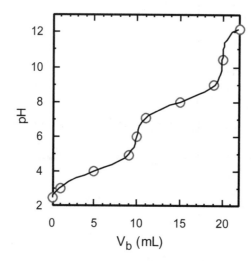

11-25. Titration reactions:

$$\text{HN}\bigcirc\text{NH} + \text{H}^+ \longrightarrow \text{H}_2\overset{+}{\text{N}}\bigcirc\text{NH} \qquad V_e = 40.0 \text{ mL}$$

$$\text{H}_2\overset{+}{\text{N}}\bigcirc\text{NH} + \text{H}^+ \longrightarrow \text{H}_2\overset{+}{\text{N}}\bigcirc\overset{+}{\text{N}}\text{H}_2 \qquad V_e = 80.0 \text{ mL}$$

<u>0 mL</u>: $\quad$ B + H$_2$O $\rightleftharpoons$ BH$^+$ + OH$^-$ $\qquad \dfrac{x^2}{0.100-x} = K_{b1} = \dfrac{K_w}{K_{a2}} = \dfrac{1.0 \times 10^{-14}}{1.86 \times 10^{-10}}$

$\quad\quad\quad 0.100-x \qquad\qquad x \qquad x$

$$\Rightarrow x = 2.29 \times 10^{-3} \text{ M} \Rightarrow \text{pH} = 11.36$$

<u>10.0 mL</u>: pH $= \text{p}K_2 + \log\dfrac{[\text{B}]}{[\text{BH}^+]} = 9.731 + \log\dfrac{3}{1} = 10.21$

<u>20.0 mL</u>: pH $= \text{p}K_2 = 9.73$

<u>30.0 mL</u>: $\,$ pH $= \text{p}K_2 + \log\dfrac{1}{3} = 9.25$

<u>40.0 mL</u>: B has been converted to BH$^+$ at a formal concentration of

$$F = \left(\dfrac{40.0}{80.0}\right)(0.100) = 0.0500 \text{ M}$$

$$[\text{H}^+] = \sqrt{\dfrac{K_1 K_2 F + K_1 K_w}{K_1 + F}}$$

$$= \sqrt{\dfrac{(4.65 \times 10^{-6})(1.86 \times 10^{-10})(0.0500) + (4.65 \times 10^{-6})(1.0 \times 10^{-14})}{4.65 \times 10^{-6} + 0.0500}}$$

$$\Rightarrow \text{pH} = 7.53$$

<u>50.0 mL</u>: pH $= \text{p}K_1 + \log\dfrac{[\text{BH}^+]}{[\text{BH}_2^{2+}]} = 5.333 + \log\dfrac{3}{1} = 5.81$

<u>60.0 mL</u>: pH $= \text{p}K_1 = 5.33$

<u>70.0 mL</u>: pH $= \text{p}K_1 + \log\dfrac{1}{3} = 4.86$

<u>80.0 mL</u>: B has been converted to BH$_2^{2+}$ at a formal concentration of

$$\left(\dfrac{40.0}{120.0}\right)(0.100) = 0.0333 \text{ M}$$

$\quad\quad$ BH$_2^{2+}$ $\rightleftharpoons$ $\;$ BH$^+$ $\;+\;$ H$^+$ $\qquad\qquad \dfrac{x^2}{0.0333-x} = K_1 = 4.65 \times 10^{-6} \Rightarrow$

$\quad\quad 0.0333-x \qquad x \qquad\quad x$

$$x = 3.91 \times 10^{-4} \text{ M} \Rightarrow \text{pH} = 3.41$$

<u>90.0 mL</u>: $\;$ [H$^+$] $= \left(\dfrac{10.0}{130.0}\right)(0.100) \Rightarrow$ pH $= 2.11$

<u>100.0 mL</u>: [H$^+$] $= \left(\dfrac{20.0}{140.0}\right)(0.100) \Rightarrow$ pH $= 1.85$

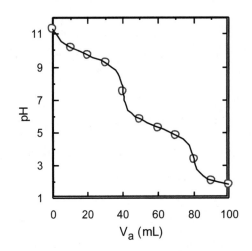

11-26.

Initial mmol: 0.500   0.164   —
Final mmol: 0.336   —   0.164

$$\text{pH} = pK_1 + \log\frac{[B]}{[BH^+]} = 4.70 + \log\frac{0.336}{0.164} = 5.01$$

11-27. (a) Titration reactions:

$$H_2NCH_2CO_2^- + H^+ \rightarrow H_3^+NCH_2CO_2^- \qquad V_e = 50.0 \text{ mL}$$
$$H_3^+NCH_2CO_2^- + H^+ \rightarrow H_3^+NCH_2CO_2H \qquad V_e = 100.0 \text{ mL}$$

At the second equivalence point, the formal concentration of

$$H_3^+NCH_2CO_2H \text{ is } \left(\frac{50.0}{150.0}\right)(0.100) = 0.033\ 3 \text{ M}$$

$$H_3^+NCH_2CO_2H \rightleftharpoons H_3^+NCH_2CO_2^- + H^+ \qquad \frac{x^2}{0.033\ 3 - x} = K_1 = 0.004\ 47 \Rightarrow$$

0.0333 − x   x   x

$$x = 1.02 \times 10^{-2} \text{ M} \Rightarrow \text{pH} = 1.99$$

(b) At V_a = 90.0 mL, the approximation gives pH = $pK_1 + \log\frac{[HG]}{[H_2G^+]}$ = 2.35 +

$\log\frac{1}{4}$ = 1.75, which is <u>lower</u> than the correct value at 100.0 mL. The pH

calculated with the equation in Table 11-5 is 2.16, which is higher than the

equivalence point pH of 1.99.

At V_a = 101.0 mL, the approximation gives $[H^+] = \left(\frac{1.0}{151.0}\right)(0.100) = 6.62 \times$

10^{-4} M $\Rightarrow$ pH = 3.18, which is <u>higher</u> than the correct value at 100.0 mL.

The pH calculated with the equation in Table 11-5 is 1.98, which is lower

than the equivalence point pH of 1.99.

11-28. (a)

$$\overset{+}{N}H_3 \\ | \\ CHCH_2CH_2CO_2H + OH^- \rightarrow \\ | \\ CO_2^-$$

$$\overset{+}{N}H_3 \\ | \\ CHCH_2CH_2CO_2^- + H_2O \\ | \\ CO_2^-$$

(b) V mL of glutamic acid will require $\dfrac{0.100}{0.025}V = 4.00\,V$ mL of RbOH to reach the equivalence point. The formal concentration of product will be

$$\left(\frac{V}{V + 4.00\,V}\right)(0.100) = 0.020\,0 \text{ M.}$$

$$[H^+] = \sqrt{\frac{K_2K_3F + K_2K_w}{K_2 + F}}$$

$$= \sqrt{\frac{(5.0 \times 10^{-5})(1.1 \times 10^{-10})(0.020\,0) + (5.0 \times 10^{-5})(1.0 \times 10^{-14})}{(5.0 \times 10^{-5}) + 0.020\,0}}$$

$$= 7.42 \times 10^{-8} \text{ M} \Rightarrow pH = 7.13$$

11-29.

$$\overset{+}{N}H_3 \\ | \\ CHCH_2-\!\!\bigcirc\!\!-OH + H^+ \rightarrow \\ | \\ CO_2^- \\ H_2T$$

$$\overset{+}{N}H_3 \\ | \\ CHCH_2-\!\!\bigcirc\!\!-OH \\ | \\ CO_2H \\ H_3T^+$$

One volume of tyrosine (0.010 0 M) requires 2.5 volumes of $HClO_4$ (0.004 00 M), so the formal concentration of tyrosine at the equivalence point is

$\left(\dfrac{1}{1 + 2.5}\right)(0.010\,0 \text{ M}) = 0.002\,86$ M. The pH is calculated from the acid dissociation of H_3T^+.

$$H_3T^+ \rightleftharpoons H_2T + H^+ \qquad\qquad \frac{x^2}{0.002\,86 - x} = K_1 = 3.9 \times 10^{-3} \Rightarrow$$
$$\quad 0.002\,86 - x \quad\quad x \quad\quad x \qquad\qquad\qquad\qquad x = 0.001\,92 \text{ M} \Rightarrow pH = 2.72$$

11-30. (a) $C^{2-} + H^+ \rightarrow HC^-$. $V_e = 20.0$ mL. At the equivalence point, the formal concentration of HC^- is $\left(\dfrac{40.0}{60.0}\right)(0.030\,0) = 0.020\,0$ M.

$$[H^+] = \sqrt{\frac{K_2K_3F + K_2K_w}{K_2 + F}}$$

$$= \sqrt{\frac{(4.4 \times 10^{-9})(1.82 \times 10^{-11})(0.020\,0) + (4.4 \times 10^{-9})(1.0 \times 10^{-14})}{(4.4 \times 10^{-9}) + 0.020\,0}}$$

$$= 2.86 \times 10^{-10} \text{ M} \Rightarrow pH = 9.54$$

(b) $H_3C^+ \rightleftharpoons H_2C + H^+$ $\dfrac{x^2}{0.0500 - x} = K_1 = 0.02$

   0.0500 − x   x   x $\Rightarrow x = 0.023\ M \Rightarrow pH = 1.64$

$pH = pK_3 + \log \dfrac{[C^{2-}]}{[HC^-]}$

$1.64 = 10.74 + \log \dfrac{[C^{2-}]}{[HC^-]} \Rightarrow \dfrac{[C^{2-}]}{[HC^-]} = 7.9 \times 10^{-10}$

11-31. The two values of pK_a for oxalic acid are 1.250 and 4.266. At a pH of 4.40, the $C_2O_4^{2-}$ has not yet been half-neutralized.

	$C_2O_4^{2-}$ +	H^+	→	$HC_2O_4^-$
Initial mmol:	x	16.0		—
Final mmol:	x − 16.0	—		16.0

$pH = 4.40 = pK_2 + \log \dfrac{[C_2O_4^{2-}]}{[HC_2O_4^-]} = 4.266 + \log \dfrac{x - 16.0}{16.0}$

$\Rightarrow x = 37.8$ mmol of $K_2C_2O_4 = 6.28$ g

11-32. Neutral alanine is designated HA.

	HA +	OH^-	→	A^-	+	H_2O
Initial mmol:	1.260 5	0.516		—		
Final mmol:	0.744 5	—		0.516		

$pH = pK_2 + \log \dfrac{[A^-]\gamma_{A^-}}{[HA]\gamma_{HA}}$ $9.57 = pK_2 + \log \dfrac{(0.516)(0.77)}{(0.744\ 5)(1)} \Rightarrow pK_2 = 9.84$

11-33. A Gran plot allows us to find the equivalence point by extrapolating from points measured prior to the equivalence point.

11-34. It is evident from the following table of data that the end point is near 23.4 mL, where the derivative dpH/dV_b is greatest. A graph of $V_b 10^{-pH}$ versus V_b follows.

The points from 21.01 to 23.30 mL were fit by the method of least squares to give the equation shown in the graph. The intercept is found by setting $y = 0$ in the equation, giving $x = V_e = 23.39$ mL.

V_b (mL)	$V_b \, 10^{-pH}$	V_b (mL)	$V_b \, 10^{-pH}$	V_b (mL)	$V_b \, 10^{-pH}$
21.01	15.22×10^{-6}	22.10	8.40×10^{-6}	22.97	2.76×10^{-6}
21.10	14.94	22.27	7.37	23.01	2.41
21.13	14.62	22.37	6.60	23.11	1.79
21.20	14.33	22.48	5.91	23.17	1.46
21.30	13.75	22.57	5.29	23.21	1.16
21.41	12.90	22.70	4.53	23.30	0.75
21.51	12.10	22.76	4.14	23.32	0.42
21.61	11.61	22.80	3.78	23.40	0.12
21.77	10.42	22.85	3.46	23.46	0.01
21.93	9.35	22.91	3.16	23.55	0.003

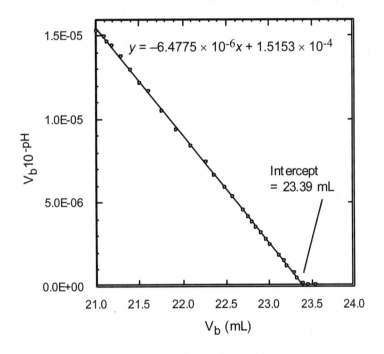

$$y = -6.4775 \times 10^{-6}x + 1.5153 \times 10^{-4}$$

Intercept
= 23.39 mL

 11-35.

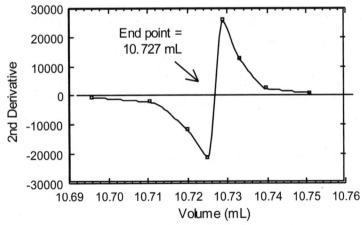

End point =
10.727 mL

Calculations are shown in the spreadsheet:

	A	B	C	D	E	F
1	Derivatives in a titration curve					
2	Data		1st derivative		2nd derivative	
3	mL NaOH	pH	mL	$\Delta pH/\Delta mL$	mL	$\Delta(\Delta pH/\Delta mL)$
4	10.679	7.643				ΔmL
5			10.6875	-11.5		
6	10.696	7.447			10.6960	-553.6
7			10.7045	-20.9		
8	10.713	7.091			10.7108	-2234.7
9			10.7170	-48.9		
10	10.721	6.700			10.7200	-11770.8
11			10.7230	-119.5		
12	10.725	6.222			10.7250	-21375.0
13			10.7270	-205.0		
14	10.729	5.402			10.7290	25687.5
15			10.7310	-102.2		
16	10.733	4.993			10.7333	12411.1
17			10.7355	-46.4		
18	10.738	4.761			10.7398	2351.0
19			10.7440	-26.4		
20	10.750	4.444			10.7508	885.2
21			10.7575	-14.5		
22	10.765	4.227				
23	Representative formulas:					
24	C5 = (A6+A4)/2		E6 = (C7+C5)/2			
25	D5 = (B6−B4)/(A6−A4)		F6 = (D7−D5)/(C7−C5)			

11-36. The quotient [HIn]/[In⁻] changes from 10:1 when pH = pK_{HIn} − 1 to 1:10 when pH = pK_{HIn} + 1. This change is generally sufficient to cause a complete color change.

11-37. The indicator has its acidic color when pH = pK_{HIn} − 1 because HIn is the dominant species. The indicator has its basic color when pH = pK_{HIn} + 1 because In⁻ is the dominant species. The color changes from the acidic color to the intermediate color to the basic color as the pH rises through the range pK_{HIn} − 1 to pK_{HIn} + 1. If the indicator is chosen correctly for the titration, this indicator pH transition range coincides with the steep part of the titration curve. The color change occurs near the equivalence point, which is the center of the steep portion of the titration curve.

11-38. The Henderson-Hasselbalch equation for the indicator, HIn, is

$$pH = pK_{HIn} + \log \frac{[In^-]}{[HIn]}.$$ If we know pK_{HIn} and we measure $\frac{[In^-]}{[HIn]}$ spectroscopically, then we can calculate the pH.

11-39. Strong acids, such as H_2SO_4, HCl, HNO_3, and $HClO_4$ have $pK_a < 0$.

11-40. yellow, green, blue

11-41. (a) red   (b) orange   (c) yellow

11-42. (a) red   (b) orange   (c) yellow   (d) red

11-43. No. When a weak acid is titrated with a strong base, the solution contains A^- at the equivalence point. A solution of A^- must have a pH above 7.

11-44. (a) The titration reaction is $F^- + H^+ \rightarrow HF$.

If V mL of NaF are used, $V_e = \frac{1}{2}V$, since the concentration of $HClO_4$ is twice as great as the concentration of NaF. The formal concentration of HF at the equivalence point is $\left(\dfrac{V}{V + \frac{1}{2}V}\right)(0.030\,0) = 0.020\,0$ M.

The pH is determined by the acid dissociation of HF.

$$HF \rightleftharpoons H^+ + F^-$$
$$0.0200 - x \qquad x \qquad x$$

$\dfrac{x^2}{0.020\,0 - x} = K_a \Rightarrow x = 3.36 \times 11\text{-}3$

$\Rightarrow pH = 2.47$

(b) The pH is so low that there would not be much (if any) break (inflection) in the titration curve at the equivalence point. A sharp change in indicator color will not be seen.

11-45. (a) violet (red + blue)   (b) blue   (c) yellow

11-46. (a) $NH_4^+ \rightleftharpoons NH_3 + H^+$
$$0.010 - x \qquad x \qquad x$$

$\dfrac{x^2}{0.010 - x} = K_a \Rightarrow x = 2.38 \times 10^{-6}$ M

$\Rightarrow pH = 5.62$

(b) One possible indicator is methyl red, using the yellow end point.

For a more complete analysis of this problem, we could compute the titration curve for a mixture of HCl and NH_4^+. For simplicity, we consider the mixture to be a "diprotic" acid with $K_1 = 100$ (i.e., a "strong" acid) and $K_2 = 5.7 \times 10^{-10}$ for the ammonium ion. We use the spreadsheet equation in Table 11-5 for titrating H_2A with strong base, taking $C_a = 0.01$ M and $C_b = 0.1$ M. An indicator with a color change in the range ~4.5–7.0 would find the HCl end point without titrating a significant amount of NH_4^+.

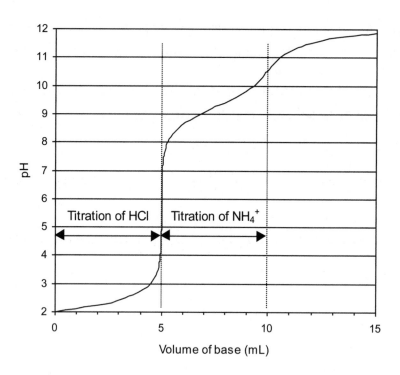

11-47. Grams of cleaner titrated $= \left(\dfrac{4.373}{10.231 + 39.466}\right)(10.231 \text{ g}) = 0.900\,3 \text{ g}$

mol HCl used = mol NH_3 present $= (0.014\,22 \text{ L})(0.106\,3 \text{ M}) = 1.512 \text{ mmol}$

$1.512 \text{ mmol } NH_3 = 25.74 \text{ mg } NH_3$

wt% $NH_3 = \dfrac{2.574 \times 10^{-2} \text{ g}}{0.900\,3 \text{ g}} \times 100 = 2.859\%$

11-48. *Alkalinity* is the capacity of natural water to react with H^+ to reach pH 4.5, which is the second equivalence point in the titration of carbonate (CO_3^{2-}) with H^+. Alkalinity measures $[OH^-] + [CO_3^{2-}] + [HCO_3^-]$ plus any other bases that are present. Bromocresol green is blue above pH 5.4 and yellow below pH 3.8. It will be at its green end-point color between these two pH values, which approximates pH 4.5.

11-49. Tris(hydroxymethyl) aminomethane ($H_2NC(CH_2OH)_3$), mercuric oxide (HgO), sodium carbonate (Na_2CO_3), and borax ($NaB_4O_7 \cdot 10H_2O$) can be used to standardize HCl. Potassium acid phthalate ($HO_2C\text{-}C_6H_4\text{-}CO_2^-K^+$), HCl azeotrope, potassium hydrogen iodate ($KH(IO_3)_2$), sulfosalicylic acid double salt ($C_7H_5SO_6K\cdot C_7H_4SO_6K_2$), and sulfamic acid ($^+H_3NSO_3^-$) can be used to standardize NaOH.

11-50. The greater the equivalent mass, the more primary standard is required. There is less relative error in weighing a large mass of reagent than a small mass.

11-51. Potassium acid phthalate is dried at 105° and weighed accurately into a flask. It is titrated with NaOH, using a pH electrode or phenolphthalein to observe the end point.

11-52. Grams of Tris titrated $= \dfrac{4.963}{(1.023 + 99.367)}(1.023) = 0.050\,57 = 0.417\,5$ mmol

Concentration of $HNO_3 = \dfrac{0.417\,5 \text{ mmol}}{5.262 \text{ g solution}} = 0.079\,34$ mol/kg solution

11-53. True mass $= m = \dfrac{(1.023)\left(1 - \dfrac{0.001\,2}{8.0}\right)}{\left(1 - \dfrac{0.001\,2}{1.33}\right)} = 1.023_8$ g

Failure to account for buoyancy introduces a systematic error of $100 \times (1.023_8 - 1.023) / 1.023 = 0.08\%$ in the calculated molarity of HCl. The true mass is higher than the measured mass of Tris, so the calculated HCl molarity is too low.

11-54. The mmoles of HgO in $0.194\,7$ g $= 0.898\,9$, which will make 1.798 mmol of OH^- by reaction with Br^- plus H_2O. HCl molarity $= 1.798$ mmol/17.98 mL $= 0.100\,0$ M.

11-55. 30 mL of 0.05 M $OH^- = 1.5$ mmol $= 0.31$ g of potassium acid phthalate.

11-56. (a) From a graph of weight percent vs pressure,
HCl $= 20.254\%$ when $P = 746$ Torr.

(b) We need $0.100\,00$ mole of HCl $= 3.646\,1$ g
$\dfrac{3.646\,1 \text{ g HCl}}{0.202\,54 \text{ g HCl/g solution}} = 18.001_9$ g of solution.
The mass required (weighed in air) is

$$m' = \dfrac{(18.001_9)\left(1 - \dfrac{0.001\,2}{1.096}\right)}{\left(1 - \dfrac{0.001\,2}{8.0}\right)} = 17.985 \text{ g}$$

11-57. (a) For a rectangular distribution of uncertainty in atomic mass, divide the uncertainty listed in the periodic table by $\sqrt{3}$ to find the standard uncertainty:

C: $12.010\,6 \pm 0.001\,0/\sqrt{3} = 12.010\,6 \pm 0.000\,5_8$

H: $1.007\,98 \pm 0.000\,14/\sqrt{3} = 1.007\,98 \pm 0.000\,08_1$

O: $15.999\,4 \pm 0.000\,4/\sqrt{3} = 15.999\,4 \pm 0.000\,2_3$

K: $39.098\,3 \pm 0.000\,1/\sqrt{3} = 39.098\,3 \pm 0.000\,05_8$

8C:	$8(12.010\,6 \pm 0.000\,5_8)$	$=$	$96.084\,8 \pm 0.004\,6$
5H:	$5(1.007\,98 \pm 0.000\,08_1)$	$=$	$5.039\,9 \pm 0.000\,40$
4O:	$4(15.999\,4 \pm 0.000\,2_3)$	$=$	$63.997\,6 \pm 0.000\,92$
1K:	$1(39.098\,3 \pm 0.000\,05_8)$	$=$	$39.098\,3 \pm 0.000\,05_8$

$C_8H_5O_4K$: $204.220\,6 \pm ?$

Uncertainty $= \sqrt{0.003\,7^2 + 0.000\,20^2 + 0.000\,69^2 + 0.000\,058^2} = 0.004\,7$

Answer: 204.221 ± 0.005 g/mol

(b) For a rectangular distribution, divide the stated uncertainty by $\sqrt{3}$ to find the standard uncertainty. Purity $= 1.000\,00 \pm 0.000\,05/\sqrt{3} = 1.000\,00 \pm 0.000\,03$

11-58. 5.00 mL of 0.033 6 M HCl $= 0.168\,0$ mmol. 6.34 mL of 0.010 0 M NaOH $= 0.063\,4$ mmol. HCl consumed by $NH_3 = 0.168\,0 - 0.063\,4 = 0.104\,6$ mmol $= 1.465$ mg of nitrogen. 256 µL of protein solution contains 9.702 mg protein. 1.465 mg of N/9.702 mg protein $= 15.1$ wt%.

11-59. (a) One mol HCl is required for one mol NH_3. 8.28 mL of 0.050 M HCl $= 0.414$ mmol HCl $= 0.414$ mmol $NH_3 = 0.414$ mmol N $= 5.80$ mg N

wt% N in protein $= 100 \times (5.80$ mg$/37.9$ mg$) = 15.3$ wt %.

(b) Mass of $B(OH)_3$ in trapping solution $= (5.00$ mL$)(0.040\,0$ g $B(OH)_3$/mL$) = 0.200$ g $B(OH)_3 = 3.234$ mmol $B(OH)_3$. If 0.414 mmol of NH_3 converts an equivalent amount of boric acid to borate, there would be a mixture of 0.414 mmol $B(OH)_4^-$ plus $(3.234 - 0.414) = 2.82$ mmol $B(OH)_3$.

$$pH = pK_a \text{ (boric acid)} + \log \frac{B(OH)_4^-}{B(OH)_3} = 9.237 + \log \frac{0.414 \text{ mmol}}{2.82 \text{ mmol}} = 8.40$$

(c) $pH = 8.40 = pK_a(NH_4^+) + \log \dfrac{mmol\ NH_3}{mmol\ NH_4^+} = 9.245 + \log \dfrac{mmol\ NH_3}{mmol\ NH_4^+}$

$\Rightarrow \log \dfrac{mmol\ NH_3}{mmol\ NH_4^+} = 8.40 - 9.245 \Rightarrow \dfrac{mmol\ NH_3}{mmol\ NH_4^+} = 10^{-0.845} = 0.143$

$Fraction\ protonated = \dfrac{mmol\ NH_4^+}{mmol\ NH_4^+ + mmol\ NH_3} =$

$\dfrac{mmol\ NH_4^+}{mmol\ NH_4^+ + 0.143\ mmol\ NH_4^+} = \dfrac{1.00}{1.143} = 87\%.$ Fraction unprotonated =

13%. (Reaction A does not go to completion because its equilibrium constant is ~1. However, formation of polyborates does protonate most of the NH_3.)

(d) $NH_3 + H^+ \rightleftharpoons NH_4^+$ $\qquad\qquad K = 1/K_a = 1/10^{-9.245} = 10^{9.245}$

$\dfrac{^+\ B(OH)_3 + H_2O \rightleftharpoons H^+ + B(OH)_4^-}{NH_3 + B(OH)_3 + H_2O \rightleftharpoons NH_4^+ + B(OH)_4^-}$ $\begin{array}{l} K_a = 10^{-9.237} \\ \hline K = (10^{9.245})(10^{-9.237}) = 1.02 \end{array}$

11-60. When an acid that is stronger than H_3O^+ is added to H_2O, it reacts to give H_3O^+ and is "leveled" to the strength of H_3O^+. Similarly, bases stronger than OH^- are leveled to the strength of OH^-.

11-61. Methanol and ethanol have nearly the same acidity as water. Both of the following equilibria are driven to the right because of the high concentration of H_2O:

$CH_3O^- + H_2O \rightarrow CH_3OH + OH^-$

$CH_3CH_2O^- + H_2O \rightarrow CH_3CH_2OH + OH^-$

11-62. (a) In acetic acid, strong acids are not leveled to the strength of $CH_3CO_2H_2^+$. Therefore, very weak bases can be titrated in acetic acid.

(b) If tetrabutylammonium hydroxide were added to an acetic acid solution, most of the hydroxide would react with acetic acid instead of analyte. However, OH^- will not react with pyridine, so this solvent would be suitable.

11-63. Sodium amide and phenyl lithium are stronger bases than OH^-. Each reacts with H_2O to give OH^-:

$NH_2^- + H_2O \rightarrow NH_3 + OH^-$

$C_6H_5^- + H_2O \rightarrow C_6H_6 + OH^-$

11-64. The reaction of pyridine with acid is [pyridine structure] $N + H^+ \rightleftharpoons$ [pyridinium structure] NH^+

Methanol is less polar than water. If methanol is added to the aqueous solution, the neutral pyridine molecule will tend to be favored over the protonated pyridinium cation. It will take a higher concentration of acid (a lower pH) to protonate pyridine in the mixed solvent. pK_a for pyridinium ion is lowered when methanol is added to the solution.

11-65. Titration reaction: $K^+HP^- + Na^+OH^- \rightarrow K^+Na^+P^{2-} + H_2O$

Begin with C_aV_a moles of K^+HP^- and add C_bV_b moles of NaOH

Fraction of titration $= \phi = \dfrac{C_bV_b}{C_aV_a}$

Charge balance: $[H^+] + [Na^+] + [K^+] = [HP^-] + 2[P^{2-}] + [OH^-]$

Substitutions: $[K^+] = \dfrac{C_aV_a}{V_a+V_b}$ $[Na^+] = \dfrac{C_bV_b}{V_a+V_b}$

$[HP^-] = \alpha_{HP^-}\dfrac{C_aV_a}{V_a+V_b}$ $[P^{2-}] = \alpha_{P^{2-}}\dfrac{C_aV_a}{V_a+V_b}$

Putting these expressions into the charge balance gives

$$[H^+] + \frac{C_bV_b}{V_a+V_b} + \frac{C_aV_a}{V_a+V_b} = \alpha_{HP^-}\frac{C_aV_a}{V_a+V_b} + 2\alpha_{P^{2-}}\frac{C_aV_a}{V_a+V_b} + [OH^-]$$

Multiply by V_a+V_b and collect terms:

$$[H^+]V_a + [H^+]V_b + C_bV_b + C_aV_a = \alpha_{HP^-}C_aV_a + 2\alpha_{P2^-}C_aV_a + [OH^-]V_a + [OH^-]V_b$$

$$V_a([H^+] + C_a - \alpha_{HP^-}C_a - 2\alpha_{P2^-}C_a - [OH^-]) = V_b([OH^-] - [H^+] - C_b)$$

$$\frac{V_b}{V_a} = \frac{\alpha_{HP^-}C_a + 2\alpha_{P2^-}C_a - C_a - [H^+] + [OH^-]}{C_b + [H^+] - [OH^-]}$$

Multiply both sides by $\dfrac{1/C_a}{1/C_b}$:

$$\phi = \frac{C_bV_b}{C_aV_a} = \frac{\alpha_{HP^-} + 2\alpha_{P2^-} - 1 - \dfrac{[H^+] - [OH^-]}{C_a}}{1 + \dfrac{[H^+] - [OH^-]}{C_b}}$$

11-66.

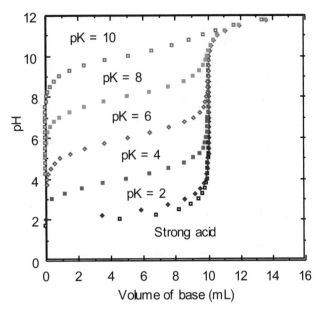

11-67.

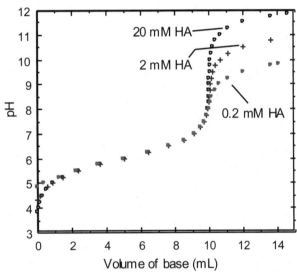

11-68.

	A	B	C	D	E	F	G
1	Effect of pKb in the titration of weak base with strong acid						
2							
3	Ca =	pH	[H+]	[OH-]	Alpha(BH+)	Phi	Va (mL)
4	0.1	2.00	1.00E-02	1.00E-12	9.90E-01	1.66E+00	16.557
5	Cb =	2.90	1.26E-03	7.94E-12	9.26E-01	1.00E+00	10.020
6	0.02	3.50	3.16E-04	3.16E-11	7.60E-01	7.78E-01	7.780
7	Vb =	4.00	1.00E-04	1.00E-10	5.00E-01	5.06E-01	5.055
8	50	4.50	3.16E-05	3.16E-10	2.40E-01	2.42E-01	2.419
9	K(BH+) =	6.00	1.00E-06	1.00E-08	9.90E-03	9.95E-03	0.100
10	1E-04	8.15	7.08E-09	1.41E-06	7.08E-05	5.17E-07	0.000
11	Kw =						
12	1E-14			E4 = C4/(C4+$A$10)			
13		C4 = 10^-B4		F4 = (E4+(C4-D4)/$A$6)/(1-(C4-D4)/$A$4)			
14		D4 = $A$12/C4		G4 = F4*$A$6*$A$8/$A$4			

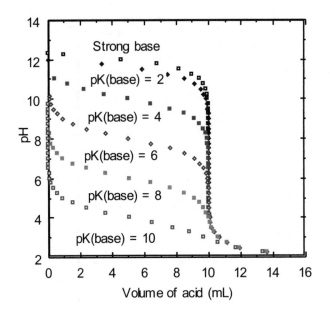

11-69. (a)

	A	B	C	D	E	F	G	H
1	Titrating weak acid with weak base							
2								
3	Cb =	pH	[H+]	[OH-]	Alpha(A-)	Alpha(BH+)	Phi	Vb (mL)
4	0.1	2.86	1.4E-03	7.2E-12	6.76E-02	1.00E+00	-1.4E-03	-0.01
5	Ca =	3.00	1.0E-03	1.0E-11	9.09E-02	1.00E+00	4.1E-02	0.41
6	0.02	4.00	1.0E-04	1.0E-10	5.00E-01	1.00E+00	4.9E-01	4.95
7	Va =	5.00	1.0E-05	1.0E-09	9.09E-01	9.99E-01	9.1E-01	9.09
8	50	6.00	1.0E-06	1.0E-08	9.90E-01	9.90E-01	1.0E+00	10.00
9	Ka =	7.00	1.0E-07	1.0E-07	9.99E-01	9.09E-01	1.1E+00	10.99
10	1E-04	8.00	1.0E-08	1.0E-06	1.00E+00	5.00E-01	2.0E+00	20.00
11	Kw =							
12	1E-14		A16 = $A$12/$A$14			D4 = $A$12/C4		
13	Kb =		C4 = 10^-B4			E4 = $A$10/(C4+$A$10)		
14	1E-06			F4 = C4/(C4+$A$16)				
15	K(BH+) =			G4 = (E4-(C4-D4)/$A$6)/(F4+(C4-D4)/$A$4)				
16	1E-08			H4 = G4*$A$6*$A$8/$A$4				

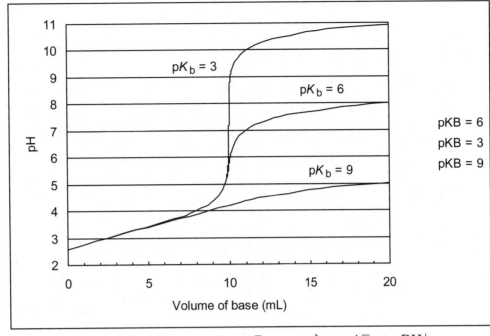

(b)   HA   +   B $\rightleftharpoons$ A$^-$ + BH$^+$

K_a = 1.75 × 10^{-5} K_b = 1.59 × 10^{-10}

V_a = 212 mL K_{BH^+} = 6.28 × 10^{-5}

C_a = 0.200 M V_b = 325 mL

                                C_b = 0.050 0 M

To find the equilibrium constant we write

HA $\rightleftharpoons$ A$^-$ + H$^+$ K_a

$\underline{H^+ + B \rightleftharpoons BH^+}$ $\underline{1/K_{BH^+}}$

HA + B $\rightleftharpoons$ A$^-$ + BH$^+$ $K = K_a/K_{BH^+} = 0.279$

A <u>pH of 4.16</u> gives V_b = 325.0 mL in the following spreadsheet:

	A	B	C	D	E	F	G	H
1	Mixing acetic acid and sodium benzoate							
2								
3	Cb =	pH	[H+]	[OH-]	Alpha(A-)	Alpha(BH+)	Phi	Vb (mL)
4	0.05	4.00	1.0E-04	1.0E-10	1.49E-01	6.14E-01	2.4E-01	204.28
5	Ca =	4.2	6.3E-05	1.6E-10	2.17E-01	5.01E-01	4.3E-01	365.98
6	0.2	4.1	7.9E-05	1.3E-10	1.81E-01	5.58E-01	3.2E-01	272.78
7	Va =	4.15	7.1E-05	1.4E-10	1.98E-01	5.30E-01	3.7E-01	315.79
8	212	4.1598	6.9E-05	1.4E-10	2.02E-01	5.24E-01	3.8E-01	325.03
9	Ka =							
10	1.750E-05		A16 = $A$12/$A$14					
11	Kw =		C4 = 10^-B4					
12	1.E-14		D4 = $A$12/C4					
13	Kb =		E4 = $A$10/(C4+$A$10)					
14	1.592E-10		F4 = C4/(C4+$A$16)					
15	K(BH+) =		G4 = (E4-(C4-D4)/$A$6)/(F4+(C4-D4)/$A$4)					
16	6.281E-05		H4 = G4*$A$6*$A$8/$A$4					

11-70.

	A	B	C	D	E	F	G	H
1	Titrating diprotic acid with strong base							
2								
3	Cb =	pH	[H+]	[OH-]	Alpha(HA-)	Alpha(A2-)	Phi	Vb (mL)
4	0.1	2.865	1.4E-03	7.3E-12	6.83E-02	5.00E-07	5.0E-05	0.000
5	Ca =	4.00	1.0E-04	1.0E-10	5.00E-01	5.00E-05	4.9E-01	4.946
6	0.02	6.00	1.0E-06	1.0E-08	9.80E-01	9.80E-03	1.0E+00	9.999
7	Va =	8.00	1.0E-08	1.0E-06	5.00E-01	5.00E-01	1.5E+00	15.000
8	50	10.0	1.0E-10	1.0E-04	9.90E-03	9.90E-01	2.0E+00	19.971
9	Kw =	12.0	1.0E-12	1.0E-02	1.00E-04	1.00E+00	2.8E+00	27.777
10	1E-14							
11	K1 =		C4 = 10^-B4			D4 = $A$12/C4		
12	1E-4		E4 = C4*$A$12/(C4^2+C4*$A$12+$A$12*$A$14)					
13	K2 =		F4 = $A$12*$A$14/(C4^2+C4*$A$12+$A$12*$A$14)					
14	1.E-08		G4 = (E4+2*F4-(C4-D4)/$A$6)/(1+(C4-D4)/$A$4)					
15			H4 = G4*$A$6*$A$8/$A$4					

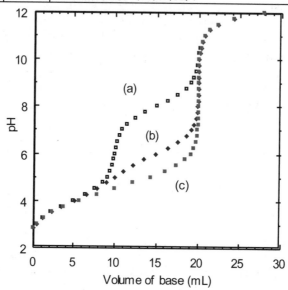

11-71.

	A	B	C	D	E	F	G	H
1	Titrating nicotine with strong acid							
2								
3	Cb =	pH	[H+]	[OH-]	Alpha(BH2)	Alpha(BH)	Phi	Va (mL)
4	0.1	1.75	1.8E-02	5.6E-13	9.62E-01	3.83E-02	2.6E+00	26.023
5	Ca =	2.00	1.0E-02	1.0E-12	9.34E-01	6.61E-02	2.3E+00	22.599
6	0.1	3.00	1.0E-03	1.0E-11	5.86E-01	4.14E-01	1.6E+00	16.117
7	Vb =	4.00	1.0E-04	1.0E-10	1.24E-01	8.76E-01	1.1E+00	11.258
8	10	5.00	1.0E-05	1.0E-09	1.39E-02	9.85E-01	1.0E+00	10.127
9	Kw =	6.00	1.0E-06	1.0E-08	1.39E-03	9.85E-01	9.9E-01	9.875
10	1.E-14	7.00	1.0E-07	1.0E-07	1.24E-04	8.76E-01	8.8E-01	8.764
11	KB1 =	8.00	1.0E-08	1.0E-06	5.86E-06	4.14E-01	4.1E-01	4.145
12	7.079E-7	9.00	1.0E-09	1.0E-05	9.34E-08	6.61E-02	6.6E-02	0.660
13	KB2 =	10.00	1.0E-10	1.0E-04	9.93E-10	7.03E-03	6.0E-03	0.060
14	1.41E-11	10.42	3.8E-11	2.6E-04	1.44E-10	2.68E-03	5.4E-05	0.001
15	KA1 =							
16	7.077E-4		C4 = 10^-B4			D4 = $A$12/C4		
17	KA2 =			E4 = C4*C4/(C4^2+C4*$A$16+$A$16*$A$18)				
18	1.413E-8			F4 = C4*$A$16/(C4^2+C4*$A$16+$A$16*$A$18)				
19				G4 = (F4+2*E4+(C4-D4)/$A$4)/(1-(C4-D4)/$A$6)				
20				H4 = G4*$A$4*$A$8/$A$6				

11-72.

	A	B	C	D	E	F	G	H	I
1	Titrating H_3A with NaOH								
2									
3	C_b =	pH	[H$^+$]	[OH$^-$]	Alpha(H$_2$A$^-$)	Alpha(HA^{2-})	Alpha(A^{3-})	Phi	V$_b$ (mL)
4	0.1	1.89	1.29E-02	7.76E-13	6.61E-01	5.50E-05	2.24E-12	1.50E-02	0.150
5	C_a =	2.00	1.00E-02	1.00E-12	7.15E-01	7.66E-05	4.02E-12	1.96E-01	1.958
6	0.02	3.00	1.00E-03	1.00E-11	9.61E-01	1.03E-03	5.40E-10	9.04E-01	9.037
7	V_a =	4.00	1.00E-04	1.00E-10	9.86E-01	1.06E-02	5.54E-08	1.00E+00	10.006
8	50	5.00	1.00E-05	1.00E-09	9.03E-01	9.67E-02	5.08E-06	1.10E+00	10.958
9	K_w =	6.00	1.00E-06	1.00E-08	4.83E-01	5.17E-01	2.71E-04	1.52E+00	15.176
10	1E-14	7.00	1.00E-07	1.00E-07	8.50E-02	9.10E-01	4.78E-03	1.92E+00	19.198
11	K_1 =	8.00	1.00E-08	1.00E-06	8.79E-03	9.42E-01	4.94E-02	2.04E+00	20.407
12	2.51E-02	9.00	1.00E-09	1.00E-05	6.12E-04	6.55E-01	3.44E-01	2.34E+00	23.441
13	K_2 =	10.00	1.00E-10	1.00E-04	1.49E-05	1.60E-01	8.40E-01	2.85E+00	28.478
14	1.07E-06	11.00	1.00E-11	1.00E-03	1.75E-07	1.87E-02	9.81E-01	3.06E+00	30.619
15	K_3 =	12.00	1.00E-12	1.00E-02	1.77E-09	1.90E-03	9.98E-01	3.89E+00	38.868
16	5.25E-10		C4 = 10^-B4						
17	pK_1 =		D4 = $A$10/C4						
18	1.60		E4 = $C4^2*$A$12/($C4^3+$C4^2*$A$12+$C4*$A$12*$A$14+$A$12*$A$14*$A$16)						
19	pK_2 =		F4 = $C4*$A$12*$A$14/($C4^3+$C4^2*$A$12+$C4*$A$12*$A$14+$A$12*$A$14*$A$16						
20	5.97		G4 = $A$12*$A$14*$A$16/($C4^3+$C4^2*$A$12+$C4*$A$12*$A$14+$A$12*$A$14*$A$16)						
21	pK_3 =		H4 = (E4+2*F4+3*G4-(C4-D4)/$A$6)/(1+(C4-D4)/$A$4)						
22	9.28		I4 = H4*$A$6*$A$8/$A$4						

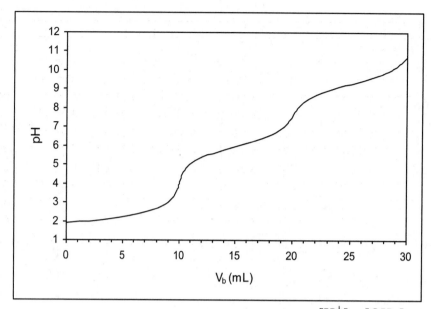

11-73. $\phi = \dfrac{C_a V_a}{C_b V_b} = \dfrac{\alpha_{BH^+} + 2\alpha_{BH_2^{2+}} + 3\alpha_{BH_3^{3+}} + 4\alpha_{BH_4^{4+}} + \dfrac{[H^+] - [OH^-]}{C_b}}{1 - \dfrac{[H^+] - [OH^-]}{C_a}}$

	A	B	C	D	E	F	G	H	I	J
1	Titrating Tetrabasic B with H^+									
2										
3	$C_b =$	pH	$[H^+]$	$[OH^-]$	$\alpha(BH^+)$	$\alpha(BH_2^{2+})$	$\alpha(BH_3^{3+})$	$\alpha(BH_4^{4+})$	Phi	V_b (mL)
4	0.02	10.9	1.3E-11	7.9E-04	3.1E-02	1.9E-06	4.6E-15	4.6E-25	-0.009	-0.090
5	$C_a =$	10.0	1.0E-10	1.0E-04	2.0E-01	1.0E-04	1.9E-12	1.5E-21	0.196	1.957
6	0.1	9.0	1.0E-09	1.0E-05	7.1E-01	3.6E-03	6.8E-10	5.4E-18	0.719	7.193
7	$V_b =$	8.0	1.0E-08	1.0E-06	9.2E-01	4.6E-02	8.8E-08	7.0E-15	1.009	10.094
8	50	7.0	1.0E-07	1.0E-07	6.6E-01	3.3E-01	6.3E-06	5.0E-12	1.330	13.303
9	$K_w =$	6.0	1.0E-06	1.0E-08	1.7E-01	8.3E-01	1.6E-04	1.3E-09	1.834	18.338
10	1E-14	5.0	1.0E-05	1.0E-09	2.0E-02	9.8E-01	1.9E-03	1.5E-07	1.983	19.830
11	$K_{a1} =$	4.0	1.0E-04	1.0E-10	2.0E-03	9.8E-01	1.9E-02	1.5E-05	2.024	20.238
12	1.26E-01	3.0	1.0E-03	1.0E-11	1.7E-04	8.4E-01	1.6E-01	1.3E-03	2.235	22.345
13	$K_{a2} =$	2.0	1.0E-02	1.0E-12	6.5E-06	3.3E-01	6.2E-01	5.0E-02	3.580	35.804
14	5.25E-03	1.7	2.0E-02	5.0E-13	1.9E-06	1.9E-01	7.0E-01	1.1E-01	4.902	49.022
15	$K_{a3} =$									
16	2.00E-07		C12 = 10^-A20							
17	$K_{a4} =$		C4 = 10^-B4		D4 = $A$10/C4					
18	3.98E-10		Denominator = ($C4^4+$C4^3*$A$12+$C4^2*$A$12*$A$14							
19	$pK_1 =$		+$C4*$A$12*$A$14*$A$16+$A$12*$A$14*$A$16*$A$18)							
20	0.90		E4 = $C4*$A$12*$A$14*$A$16/Denominator							
21	$pK_2 =$		F4 = $C4^2*$A$12*$A$14/Denominator							
22	2.28		G4 = $C4^3*$A$12//Denominator							
23	$pK_3 =$		H4 = $C4^4/Denominator							
24	6.70		I4 = (E4+2*F4+3*G4+4*H4+(C4-D4)/$A$4)/(1-(C4-D4)/$A$6)							
25	$pK_4 =$		J4 = I4*$A$4*$A$8/$A$6							
26	9.40									

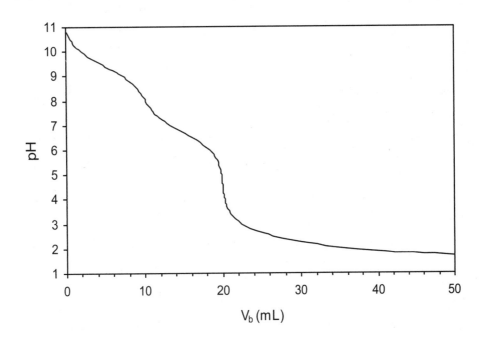

11-74. $A_{604} = \varepsilon_{In^-}[In^-](1.00) \Rightarrow [In^-] = \dfrac{0.118}{4.97 \times 10^4} = 2.37 \times 10^{-6}$ M

Since the indicator was diluted with KOH solution, the formal concentration of indicator is 0.700×10^{-5} M.

$[HIn] = 7.00 \times 10^{-6} - 2.37 \times 10^{-6} = 4.63 \times 10^{-6}$ M

$pH = pK_{In} + \log \dfrac{[In^-]}{[HIn]} = 7.95 + \log \dfrac{2.37}{4.63} = 7.66$

Call benzene-1,2,3-tricarboxylic acid H_3A, with $pK_1 = 2.88$, $pK_2 = 4.75$, and $pK_3 = 7.13$. Since the pH is 7.66, the main species is A^{3-} and the second main species is HA^{2-}. Enough KOH to react with H_3A and H_2A^- must have been added, and there is enough KOH to react with part of the HA^{2-}.

	HA^{2-}	+	OH^-	$\rightarrow$	A^{3-}	+	H_2O
Initial mmol:	1.00		x		—		
Final mmol:	$1.00 - x$		—		x		

$pH = pK_3 + \log \dfrac{[A^{3-}]}{[HA^{2-}]}$

$7.66 = 7.13 + \log \dfrac{x}{1.00 - x} \Rightarrow x = 0.77_2$ mmol of OH^-

The total KOH added is 2.77_2 mmol. The molarity is $\dfrac{2.77_2 \text{ mmol}}{20.0 \text{ mL}} = 0.139$ M.

11-75. The pH of the solution is 7.50, and the total concentration of indicator is 5.00×10^{-5} M. At pH 7.50, there is a negligible amount of H_2In, since $pK_1 = 1.00$. We can write

$[HIn^-] + [In^{2-}] = 5.0 \times 10^{-5}$

$pH = pK_2 + \log \dfrac{[In^{2-}]}{[HIn^-]}$

$7.50 = 7.95 + \log \dfrac{[In^{2-}]}{5.00 \times 10^{-5} - [In^{2-}]} \Rightarrow [In^{2-}] = 1.31 \times 10^{-5}$ M

$[HIn] = 3.69 \times 10^{-5}$ M

$A_{435} = \varepsilon_{435}[HIn^-] + \varepsilon_{435}[In^{2-}]$

$\quad\quad = (1.80 \times 10^4)(3.69 \times 10^{-5}) + (1.15 \times 10^4)(1.31 \times 10^{-5}) = 0.815$

CHAPTER 12
EDTA TITRATIONS

12-1. The chelate effect is the observation that multidentate ligands form more stable metal complexes than do similar, monodentate ligands.

12-2. $\alpha_{Y^{4-}}$ gives the fraction of all free EDTA in the form Y^{4-}.

(a) At pH 3.50:

$$\alpha_{Y^{4-}} = \frac{10^{-0.0}10^{-1.5}10^{-2.00}10^{-2.69}10^{-6.13}10^{-10.37}}{(10^{-3.50})^6 + (10^{-3.50})^5 10^{-0.0} +...+ 10^{-0.0}10^{-1.5}...10^{-10.37}} = 2.7 \times 10^{-10}$$

(b) At pH 10.50:

$$\alpha_{Y^{4-}} = \frac{10^{-0.0}10^{-1.5}10^{-2.00}10^{-2.69}10^{-6.13}10^{-10.37}}{(10^{-10.50})^6 + (10^{-10.50})^5 10^{-0.0} +...+ 10^{-0.0}10^{-1.5}...10^{-10.37}} = 0.57$$

12-3. (a) $K_f' = \alpha_{Y^{4-}} K_f = 0.041 \times 10^{8.79} = 2.5 \times 10^7$

(b) $$Mg^{2+} + ETDA \rightleftharpoons MgY^{2-}$$

$$\phantom{Mg^{2+}}\ x \ x \ 0.050 - x$$

$$\frac{0.050 - x}{x^2} = 2.5 \times 10^7 \Rightarrow [Mg^{2+}] = 4.5 \times 10^{-5}\ M$$

12-4. $[Ca^{2+}] = 10^{-9.00}\ M$, so essentially all calcium in solution is CaY^{2-}.

$$[CaY^{2-}] = \frac{1.95\ g}{(200.12\ g/mol)\ (0.500\ L)} = 0.019\ 49\ M$$

$$K_f' = (0.041)(10^{10.65}) = \frac{[CaY^{2-}]}{[EDTA]\ [Ca^{2+}]} = \frac{(1.949 \times 10^{-2})}{[EDTA]\ (10^{-9.00})}$$

$$\Rightarrow [EDTA] = 0.010\ 6\ M$$

Total EDTA needed = mol CaY^{2-} + mol free EDTA

$$= (0.019\ 49\ M)\ (0.500\ L)\ +\ (0.010\ 6\ M)\ (0.500\ L) = 0.015\ 0_4\ mol$$

$$= 5.60\ g\ Na_2EDTA \cdot 2\ H_2O$$

12-5.

Neutral H_5DTPA has 2 carboxylic acid protons and 3 ammonium protons. We are not given the pK_a values, but, by analogy with EDTA, we expect carboxyl pK_a values to be below ~3 and ammonium pK_a values to be above ~6. At pH 14, we expect all acidic protons of DTPA to be dissociated, so the predominant species will be $DTPA^{5-}$. At pH 3-4, nitrogen should be protonated, but carboxyl groups should be deprotonated. The predominant species is probably H_3DTPA^{2-}.

For HSO_4^-, $pK_a = 2.0$. At pH 14 and at pH 3, sulfate is in the form SO_4^{2-}.

At pH 14, $DTPA^{5-}$ is apparently a strong enough ligand to chelate Ba^{2+} and dissolve $BaSO_4(s)$. At pH 3-4, H_3DTPA^{2-} is not a strong enough ligand to dissolve $BaSO_4(s)$. An equivalent statement is that H^+ at a concentration of 10^{-3}—10^{-4} M competes with Ba^{2+} for binding sites on DTPA, but H^+ at a concentration of 10^{-14} M does not compete with Ba^{2+} for binding sites on DTPA.

Now that you have seen my reasoning, I'll provide some more information. The pK_a values for DTPA, beginning with the fully protonated H_8DTPA^{3+}, are

H_8DTPA^{3+}	$pK_1 = -0.1$	CO_2H		H_4DTPA^-	$pK_5 = 2.7$	CO_2H
H_7DTPA^{2+}	$pK_2 = 0.7$	CO_2H		H_3DTPA^{2-}	$pK_6 = 4.3$	NH^+
H_6DTPA^+	$pK_3 = 1.6$	CO_2H		H_2DTPA^{3-}	$pK_7 = 8.6$	NH^+
H_5DTPA	$pK_4 = 2.0$	CO_2H		$HDTPA^{4-}$	$pK_8 = 10.5$	NH^+

As pH is lowered from 14, the three nitrogen atoms are 50% protonated at pH 10.5, 8.6, and 4.3. The third nitrogen atom is not quite fully protonated at pH 3-4. The predominant species is H_3DTPA^{2-}, as I guessed correctly. The species H_4DTPA^- and H_2DTPA^{3-} are also present to some extent in the pH range 3-4.

12-6. (a) mmol EDTA = mmol M^{n+}

$(V_e)(0.050\,0\text{ M}) = (100.0\text{ mL})(0.050\,0\text{ M}) \Rightarrow V_e = 100.0\text{ mL}$

(b) $[M^{n+}] = \underbrace{\left(\frac{1}{2}\right)}_{\substack{\text{fraction}\\\text{remaining}}} \cdot \underbrace{(0.050\,0\text{ M})}_{\substack{\text{original}\\\text{concentration}}} \cdot \underbrace{\left(\frac{100\text{ mL}}{150\text{ mL}}\right)}_{\substack{\text{dilution}\\\text{factor}}} = 0.016\,7\text{ M}$

(c) 0.041 (Table 12-1)

(d) $K_f' = (0.041)(10^{12.00}) = 4.1 \times 10^{10}$

(e) $[MY^{n-4}] = (0.050\,0\text{ M})\left(\frac{100\text{ mL}}{200\text{ mL}}\right) = 0.025\,0\text{ M}$

$\dfrac{[MY^{n-4}]}{[M^{n+}][EDTA]} = \dfrac{0.025\,0 - x}{x^2} = 4.1 \times 10^{10} \Rightarrow x = [M^{n+}] = 7.8 \times 10^{-7}\text{ M}$

(f) $[EDTA] = (0.050\,0\text{ M})\left(\frac{10.0\text{ mL}}{210.0\text{ mL}}\right) = 2.38 \times 10^{-3}\text{ M}$

$[MY^{n-4}] = (0.050\,0\text{ M})\left(\frac{100.0\text{ mL}}{210.0\text{ mL}}\right) = 2.38 \times 10^{-2}\text{ M}$

$\dfrac{[MY^{n-4}]}{[M^{n+}][EDTA]} = \dfrac{(2.38 \times 10^{-2})}{[M^{n+}](2.38 \times 10^{-3})} = 4.1 \times 10^{10} \Rightarrow [M^{n+}] = 2.4 \times 10^{-10}\text{ M}$

12-7. $Co^{2+} + EDTA \rightleftharpoons CoY^{2-}$ $\alpha_{Y^{4-}} K_f = (1.8 \times 10^{-5})(10^{16.45}) = 5.1 \times 10^{11}$

$$V_e = (25.00)\left(\frac{0.020\,26\ M}{0.038\,55\ M}\right) = 13.14\ mL$$

(a) <u>12.00 mL</u>: $[Co^{2+}] = \left(\dfrac{13.14\ mL - 12.00\ mL}{13.14\ mL}\right)(0.020\,26\ M)\left(\dfrac{25.00\ mL}{37.00\ mL}\right)$

$$= 1.19 \times 10^{-3}\ M \Rightarrow pCo^{2+} = 2.93$$

(b) <u>V_e</u>: Formal concentration of $CoY^{2-} = \left(\dfrac{25.00\ mL}{38.14\ mL}\right)(0.020\,26\ M)$

$$= 1.33 \times 10^{-2}\ M$$

$$
\begin{array}{ccccc}
Co^{2+} & + & EDTA & \rightleftharpoons & CoY^{2-} \\
x & & x & & 1.33 \times 10^{-2} - x
\end{array}
$$

$$\frac{1.33 \times 10^{-2} - x}{x^2} = \alpha_{Y^{4-}} K_f \Rightarrow x = 1.6 \times 10^{-7}\ M \Rightarrow pCo^{2+} = 6.79$$

(c) <u>14.00 mL</u>: Formal concentration of CoY^{2-} is $\left(\dfrac{25.00\ mL}{39.00\ mL}\right)(0.020\,26\ M)$

$$= 1.30 \times 10^{-2}\ M$$

Formal concentration of EDTA is $\left(\dfrac{14.0\ mL - 13.14\ mL}{39.00\ mL}\right)(0.038\,55\ M) = 8.50 \times 10^{-4}\ M$

$$[Co^{2+}] = \frac{[CoY^{2-}]}{[EDTA]\,K_f'} = \frac{1.30 \times 10^{-2}}{8.50 \times 10^{-4}\,(5.1 \times 10^{11})} = 3.0 \times 10^{-11}\ M$$

$$\Rightarrow pCo^{2+} = 10.52$$

12-8. Titration reaction: $Mn^{2+} + EDTA \rightleftharpoons MnY^{2-}$

$$K_f' = \alpha_{Y^{4-}} K_f = \ = (4.2 \times 10^{-3})(10^{13.89}) = 3.3 \times 10^{11}$$

The equivalence point is 50.0 mL.   Sample calculations:

<u>20.0 mL</u>: The fraction of Mn^{2+} that has reacted is 2/5 and the fraction remaining is 3/5.

$$[Mn^{2+}] = \left(\frac{30.0\ mL}{50.0\ mL}\right)(0.020\,0\ M)\left(\frac{25.0\ mL}{45.0\ mL}\right) = 6.67 \times 10^{-3}\ M$$

$$\Rightarrow pMn^{2+} = 2.18$$

<u>50.0 mL</u>: The formal concentration of MnY^{2-} is

$$[MnY^{2-}] = \left(\frac{25.0\ mL}{75.0\ mL}\right)(0.020\,0\ M) = 0.006\,67\ M$$

$$
\begin{array}{ccccc}
Mn^{2+} & + & EDTA & \rightleftharpoons & MnY^{2-} \\
x & & x & & 0.006\,67 - x
\end{array}
$$

$$\frac{0.006\,67 - x}{x^2} = \alpha_{Y^{4-}} K_f \Rightarrow x = 1.4 \times 10^{-7} \Rightarrow pMn^{2+} = 6.85$$

60.0 mL: There are 10.0 mL of excess EDTA.

$$[EDTA] = \left(\frac{10.0 \text{ mL}}{85.0 \text{ mL}}\right)(0.010\,0 \text{ M}) = 1.176 \times 10^{-3} \text{ M}$$

$$[MnY^{2-}] = \left(\frac{25.0 \text{ mL}}{85.0 \text{ mL}}\right)(0.020\,0 \text{ M}) = 5.88 \times 10^{-3} \text{ M}$$

$$[Mn^{2+}] = \frac{[MnY^{2-}]}{[EDTA]K_f'} = 1.5 \times 10^{-11} \Rightarrow pMn^{2+} = 10.82$$

Volume (mL)	pMn^{2+}	Volume	pMn^{2+}	Volume	pMn^{2+}
0	1.70	49.0	3.87	50.1	8.82
20.0	2.18	49.9	4.87	55.0	10.51
40.0	2.81	50.0	6.85	60.0	10.82

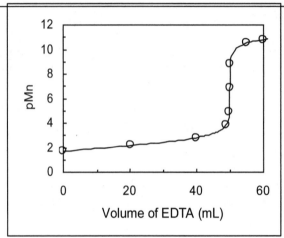

12-9.   Titration reaction: $Ca^{2+} + EDTA \rightleftharpoons CaY^{2-}$

$$K_f' = \alpha_{Y^{4-}} K_f = (0.30)(10^{10.65}) = 1.3_4 \times 10^{10}$$

The equivalence point is 50.0 mL.   Sample calculations:

20.0 mL: The fraction of EDTA consumed is 2/5.

$$[EDTA] = \left(\frac{30.0 \text{ mL}}{50.0 \text{ mL}}\right)(0.020\,0 \text{ M})\left(\frac{25.0 \text{ mL}}{45.0 \text{ mL}}\right) = 0.006\,67 \text{ M}$$

$$[CaY^{2-}] = \left(\frac{20.0 \text{ mL}}{50.0 \text{ mL}}\right)(0.020\,0 \text{ M})\left(\frac{25.0 \text{ mL}}{45.0 \text{ mL}}\right) = 0.004\,44 \text{ M}$$

$$[Ca^{2+}] = \frac{[CaY^{2-}]}{[EDTA]K_f'} = 4.9_7 \times 10^{-11} \Rightarrow pCa^{2+} = 10.30$$

50.0 mL: The formal concentration of CaY^{2-} is

$$[CaY^{2-}] = \left(\frac{25.0 \text{ mL}}{75.0 \text{ mL}}\right)(0.020\,0 \text{ M}) = 0.006\,67 \text{ M}$$

$$\begin{array}{ccccc} Ca^{2+} & + & EDTA & \rightleftharpoons & CaY^{2-} \\ x & & x & & 0.006\,67 - x \end{array}$$

$$\frac{0.006\,67 - x}{x^2} = \alpha_{Y^{4-}} K_f \Rightarrow x = 7.05 \times 10^{-7} \text{ M} \Rightarrow pCa^{2+} = 6.15$$

50.1 mL: There is an excess of 0.1 mL of Ca^{2+}.

$$[Ca^{2+}] = \left(\frac{0.1 \text{ mL}}{75.1 \text{ mL}}\right)(0.010\,0 \text{ M}) = 1.33 \times 10^{-5} \text{ M} \Rightarrow pCa^{2+} = 4.88$$

Volume (mL)	pCa^{2+}	Volume	pCa^{2+}	Volume	pCa^{2+}
0	(∞)	49.0	8.44	50.1	4.88
20.0	10.30	49.9	7.43	55.0	3.20
40.0	9.52	50.0	6.15	60.0	2.93

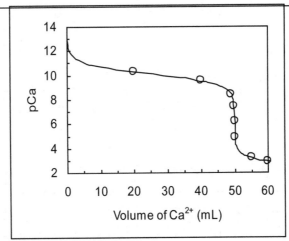

12-10. There is more VO^{2+} than EDTA in this solution.

$$[VO^{2+}] = \left(\frac{0.10 \text{ mL}}{29.9 \text{ mL}}\right)(0.010\,0 \text{ M}) = 3.34 \times 10^{-5} \text{ M}$$

$$[VOY^{2-}] = \left(\frac{9.90 \text{ mL}}{29.90 \text{ mL}}\right)(0.010\,0 \text{ M}) = 3.31 \times 10^{-3} \text{ M}$$

K_f for $VOY^{2-} = 10^{18.7}$; pK_6 for $H_6Y^{2+} = 10.37$; pH = 4.00

$$[Y^{4-}] = \frac{[VOY^{2-}]}{[VO^{2+}] K_f} = 1.98 \times 10^{-17} \text{ M}$$

$$[HY^{3-}] = \frac{[H^+][Y^{4-}]}{K_6} = 4.6 \times 10^{-11} \text{ M}$$

12-11.

	A	B	C	D	E	F	G
1	Titration of V_M mL of C_M M Cu^{2+} with C(ligand) M EDTA						
2							
3	$C_M =$	pM	M	Phi	V(EDTA)		
4	0.001	3.0	1.00E-03	0.000	0.000		
5	$V_M =$	4.0	1.00E-04	0.891	0.891		
6	10	5.0	1.00E-05	0.989	0.989		
7	C(ligand) =	6.0	1.00E-06	0.999	0.999		
8	0.01	7.0	1.00E-07	1.000	1.000		
9	$K_f^{'} =$	8.0	1.00E-08	1.000	1.000		
10	1.75E+12	9.0	1.00E-09	1.001	1.001		
11	$\alpha(Y^{4-})=$	10.0	1.00E-10	1.006	1.006		
12	2.90E-07	11.0	1.00E-11	1.057	1.057		
13	$K_f =$	12.0	1.00E-12	1.572	1.572		
14	6.0256E+18	12.3	5.01E-13	2.142	2.142		
15							
16	A10 = A12*A14						
17	C4 = 10^-B4						
18	D4 = (1+$A$10*C4-(C4+C4*C4*$A$10)/$A$4)/(C4*$A$10+(C4+C4*C4*$A$10)/$A$8)						
19	E4 = D4*$A$4*$A$6/$A$8						

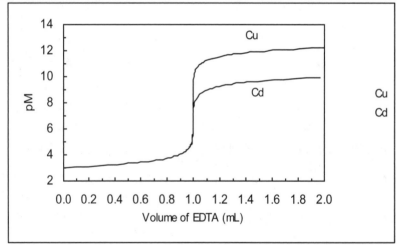

12-12. The spreadsheet below gives representative calculations for the pH 7.

	A	B	C	D	E	F
1	Titration of 10 mL of 1 mM Ca^{2+} with 1 mM EDTA vs pH					
2	pH 7					
3	C$_M$ =	pM	M	Phi	V(ligand)	
4	0.001	3.000	1.00E-03	0.000	0.000	
5	V$_M$ =	3.250	5.62E-04	0.280	2.801	
6	10	3.500	3.16E-04	0.520	5.196	
7	C(ligand) =	3.750	1.78E-04	0.698	6.982	
8	0.001	4.000	1.00E-04	0.819	8.186	
9	K$_f'$ =	4.500	3.16E-05	0.940	9.404	
10	1.70E+07	5.000	1.00E-05	0.986	9.859	
11	α(Y^{4-}) =	5.500	3.16E-06	1.012	10.121	
12	3.80E-04	6.000	1.00E-06	1.057	10.567	
13	K$_f$ =	6.500	3.16E-07	1.185	11.855	
14	4.4668E+10	7.000	1.00E-07	1.589	15.887	
15	A10 = A12*A14					
16	C4 = 10^-B4					
17	D4 = (1+A$10*C4-(C4+C4*C4*A$10)/A$4)/(C4*A$10+(C4+C4*C4*A$10)/A$8)					
18	E4 = D4*A$4*A$6/A$8					

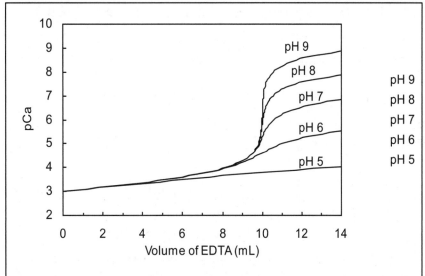

The only change in the spreadsheet for different values of pH is the value of α_Y4-
in cell A12.

12-13.

	A	B	C	D	E	F	G
1	Titration of EDTA with metal						
2							
3	C_M =	pM	M	Phi	V_M		
4	0.08	14.640	2.29E-15	0.004	0.100		
5	V(ligand) =	12.844	1.43E-13	0.200	5.004		
6	50	12.418	3.82E-13	0.400	10.007		
7	C(ligand) =	12.066	8.59E-13	0.600	15.004		
8	0.04	11.640	2.29E-12	0.800	20.003		
9	K_f' =	10.860	1.38E-11	0.960	24.005		
10	1.75E+12	6.910	1.23E-07	1.000	25.000		
11	$\alpha(Y^{4-})$	2.978	1.05E-03	1.040	25.999		
12	2.90E-07	2.301	5.00E-03	1.200	30.000		
13	K_f =						
14	6.0256E+18						
15							
16	A10 = A12*A14						
17	C4 = 10^-B4						
18	D4 = (C4*$A$10+(C4+C4*C4*$A$10)/$A$8)/(1+$A$10*C4-(C4+C4*C4*$A$10)/$A$4)						
19	E4 = D4*$A$8*$A$6/$A$4						

12-14. An auxiliary complexing agent forms a weak complex with analyte ion, thereby keeping it in solution without interfering with the EDTA titration. For example, NH_3 keeps Zn^{2+} in solution at high pH, but is easily displaced by EDTA.

12-15. (a) $\beta_2 = K_1K_2 = \beta_1K_2 \Rightarrow K_2 = \beta_2/\beta_1 = 10^{3.63}/10^{2.23} = 10^{1.40} = 25$

(b) $\alpha_{Cu^{2+}} = \dfrac{1}{1+\beta_1[L]+\beta_2[L]^2} = \dfrac{1}{1+10^{2.23}(0.100)+10^{3.63}(0.100)^2} = 0.016$

12-16. $Cu^{2+} + Y^{4-} \rightleftharpoons CuY^{2-}$ $\qquad K_f = 10^{18.78} = 6.0_3 \times 10^{18}$

$\alpha_{Y^{4-}} = 0.81$ at pH 11.00 (Table 12-1)

For Cu^{2+} and NH_3, Appendix I gives $\log\beta_1 = 3.99$, $\log\beta_2 = 7.33$, $\log\beta_3 = 10.06$, and $\log\beta_4 = 12.03$. Therefore, $\beta_1 = 9.8 \times 10^3$, $\beta_2 = 2.1 \times 10^7$, $\beta_3 = 1.15 \times 10^{10}$, and $\beta_4 = 1.07 \times 10^{12}$.

$$\alpha_{Cu^{2+}} = \dfrac{1}{1+\beta_1(1.00)+\beta_2(1.00)^2+\beta_3(1.00)^3+\beta_4(1.00)^4} = 9.2_3 \times 10^{-13}$$

$K_f' = \alpha_{Y^{4-}} K_f = 4.8_8 \times 10^{18}$

$K_f'' = \alpha_{Y^{4-}} \alpha_{Cu^{2+}} K_f = 4.5_1 \times 10^6$

Equivalence point = 50.00 mL

(a) At 0 mL, the total concentration of copper is $C_{Cu^{2+}} = 0.001\,00$ M and
$[Cu^{2+}] = \alpha_{Cu^{2+}} C_{Cu^{2+}} = 9.2_3 \times 10^{-16}$ M $\Rightarrow$ $pCu^{2+} = 15.03$

(b)   1.00 mL: $C_{Cu^{2+}} = \left(\dfrac{49.00\ mL}{50.00\ mL}\right)(0.001\ 00\ M)\left(\dfrac{50.00\ mL}{51.00\ mL}\right) = 9.61 \times 10^{-4}\ M$

$\qquad\qquad\qquad\quad$ fraction $\qquad$ original $\qquad$ dilution
$\qquad\qquad\qquad$ remaining $\quad$ concentration $\quad$ factor

$\qquad$ $[Cu^{2+}] = \alpha_{Cu^{2+}}\ C_{Cu^{2+}} = 8.8_7 \times 10^{-16}\ M \;\Rightarrow\; pCu^{2+} = 15.05$

(c)   45.00 mL: $C_{Cu^{2+}} = \left(\dfrac{5.00\ mL}{50.00\ mL}\right)(0.001\ 00)\left(\dfrac{50.00\ mL}{95.00\ mL}\right) = 5.26 \times 10^{-5}\ M$

$\qquad$ $[Cu^{2+}] = \alpha_{Cu^{2+}}\ C_{Cu^{2+}} = 5.0_4 \times 10^{-17}\ M \;\Rightarrow\; pCu^{2+} = 16.30$

(d)   Equivalence point:

$$C_{Cu^{2+}} \;+\; EDTA \quad\rightleftharpoons\quad CuY^{2-}$$

$\qquad\qquad$ x $\qquad\qquad\quad$ x $\qquad\qquad\quad$ $\left(\dfrac{50.00\ mL}{100.00\ mL}\right)(0.001\ 00) - x$

$$\dfrac{0.000\ 500 - x}{x^2} = K_f'' = 4.5_1 \times 10^6 \;\Rightarrow\; x = C_{Cu^{2+}} = 1.04 \times 10^{-5}\ M$$

$\qquad$ $[Cu^{2+}] = \alpha_{Cu^{2+}}\ C_{Cu^{2+}} = 9.6_2 \times 10^{-18}\ M \;\Rightarrow\; pCu^{2+} = 17.02$

(e)   Past the equivalence point at 55.00 mL:

$\qquad$ $[EDTA] = \left(\dfrac{5.00\ mL}{105.00\ mL}\right)(0.001\ 00\ M) = 4.76 \times 10^{-5}\ M$

$\qquad$ $[CuY^{2-}] = \left(\dfrac{50.00\ mL}{105.00\ mL}\right)(0.001\ 00\ M) = 4.76 \times 10^{-4}\ M$

$\qquad$ $K_f' = \dfrac{[CuY^{2-}]}{[Cu^{2+}][EDTA]} = \dfrac{(4.76 \times 10^{-4})}{[Cu^{2+}]\,(4.76 \times 10^{-5})}$

$\qquad$ $\Rightarrow\; [Cu^{2+}] = 2.05 \times 10^{-18}\ M \;\Rightarrow\; pCu^{2+} = 17.69$

12-17. (a) $\quad \alpha_{ML} = \dfrac{[ML]}{C_M} = \dfrac{\beta_1[M][L]}{[M]\{1 + \beta_1[L] + \beta_2[L]^2\}} = \dfrac{\beta_1[L]}{1 + \beta_1[L] + \beta_2[L]^2}$

$\qquad\qquad$ $\alpha_{ML_2} = \dfrac{[ML_2]}{C_M} = \dfrac{\beta_2[M][L]^2}{[M]\{1 + \beta_1[L] + \beta_2[L]^2\}} = \dfrac{\beta_2[L]^2}{1 + \beta_1[L] + \beta_2[L]^2}$

$\qquad$ (b) $\quad$ For $[L] = 0.100\ M$, $\beta_1 = 1.7 \times 10^2$, and $\beta_2 = 4.3 \times 10^3$, we get $\alpha_{ML} = 0.28$ and $\alpha_{ML_2} = 0.70$.

12-18. Let T = transferrin

$\qquad$ (a) $\quad Fe^{3+} + T \overset{K_1}{\rightleftharpoons} FeT \qquad\qquad\qquad K_1 = \dfrac{[FeT]}{[Fe^{3+}][T]}$

$\qquad\qquad\qquad Fe^{3+} + FeT \overset{K_2}{\rightleftharpoons} Fe_2T \qquad\qquad K_2 = \dfrac{[Fe_2T]}{[Fe^{3+}][FeT]}$

(b) $\quad K_1 = \dfrac{[Fe_aT] + [Fe_bT]}{[Fe^{3+}][T]} = \dfrac{[Fe_aT]}{[Fe^{3+}][T]} + \dfrac{[Fe_bT]}{[Fe^{3+}][T]} = k_{1a} + k_{1b}$

$\quad\quad \dfrac{1}{K_2} = \dfrac{[Fe^{3+}]([Fe_aT] + [Fe_bT])}{[Fe_2T]} = \dfrac{[Fe^{3+}][Fe_aT]}{[Fe_2T]} + \dfrac{[Fe^{3+}][Fe_bT]}{[Fe_2T]} = \dfrac{1}{k_{2b}} + \dfrac{1}{k_{2a}}$

(c) $\quad k_{1a}\, k_{2b} = \dfrac{[Fe_aT]}{[Fe^{3+}][T]} \dfrac{[Fe_2T]}{[Fe^{3+}]\,[Fe_aT]} = \dfrac{[Fe_bT]}{[Fe^{3+}][T]} \dfrac{[Fe_2T]}{[Fe^{3+}]\,[Fe_bT]} = k_{1b}\, k_{2a}$

(d) Substituting from Eq. (A) into Eq. (C) gives

$$19.44 = \dfrac{[FeT]^2}{(1 - [FeT] - [Fe_2T])\,[Fe_2T]} \qquad\qquad\qquad (D)$$

Substituting from Eq. (B) into Eq. (D) gives

$$19.44 = \dfrac{(0.8 - 2[Fe_2T])^2}{\{1 - (0.8 - 2[Fe_2T]) - [Fe_2T]\}\,[Fe_2T]} \quad \overset{\text{solve}}{\underset{\substack{\text{quadratic}\\\text{equation}}}{\Rightarrow}} \quad [Fe_2T] = 0.0773$$

Using $[Fe_2T] = 0.0773$ in Equations (A) and (B) gives $[FeT] = 0.645$ and

$[T] = 0.2773$. Now we also know that $\dfrac{k_{1a}}{k_{1b}} = \dfrac{[Fe_aT]}{[Fe_bT]} = 6.0$, which tells us

that $[Fe_aT] = \left(\dfrac{6.0}{7.0}\right)[FeT] = 0.5532$ and $[Fe_bT] = \left(\dfrac{1.0}{7.0}\right)[FeT] = 0.0922$

Final result: $[T] = 0.27_7$, $[Fe_aT] = 0.55_3$, $[Fe_bT] = 0.09_2$, $[Fe_2T] = 0.07_7$

12-19. In place of Equation 12-8, we write

$$M_{\text{free}} + EDTA \rightleftharpoons M(EDTA) \qquad\qquad K_f'' = \dfrac{[M(EDTA)]}{[M]_{\text{free}}[EDTA]}$$

where $[M]_{\text{free}}$ is the concentration of all metal not bound to EDTA. $[EDTA]$ is the concentration of all EDTA not bound to metal. The mass balances are

Metal: $\qquad [M]_{\text{free}} + [M(EDTA)] = \dfrac{C_M V_M}{V_M + V_{EDTA}}$

EDTA: $\qquad [EDTA] + [M(EDTA)] = \dfrac{C_{EDTA} V_{EDTA}}{V_M + V_{EDTA}}$

These equations have the same form as the first three equations in Section 12-4, with K_f replaced by K_f'', $[M]$ replaced by $[M]_{\text{free}}$, and $[L]$ replaced by $[EDTA]$. The derivation therefore leads to Equation 12-11, with K_f replaced by K_f'', $[M]$ replaced by $[M]_{\text{free}}$, and C_L replaced by C_{EDTA}.

12-20. (a)

	A	B	C	D	E	F
1	Titration of 50 mL of 0.001 M Zn^{2+} with 0.001 M EDTA/pH 10 with NH_3					
2						
3	$C_M =$	pM	M	$[M]_{tot}$	ϕ	V_{EDTA}
4	0.001	8.115	7.67E-09	4.29E-04	0.400	19.9814
5	$V_M =$	12.014	9.68E-13	5.41E-08	1.000	50.0000
6	50	15.278	5.27E-16	2.95E-11	1.200	59.9965
7	$C_{EDTA} =$					
8	0.001					
9	$K_f'' =$	A10 = A12*A16*10^A14				
10	1.70E+11	A12 = 1/(1+A20*A18+B20*A18^2+C20*A18^3+D20*A18^4)				
11	$\alpha(Zn^{2+}) =$					
12	1.79E-05	C4 = 10^-B4				
13	$\log K_f =$	D4 = C4/\$A\$12				
14	16.5	E4 = (1+\$A\$10*D4-(D4+D4^2*\$A\$10)/\$A\$4)/				
15	$\alpha(Y^{4-}) =$	(D4*\$A\$10+(D4+D4^2*\$A\$10)/\$A\$8)				
16	0.30	F4 = E4*\$A\$4*\$A\$6/\$A\$8				
17	$[NH_3] =$					
18	0.1					
19	$\beta_1 =$	$\beta_2 =$	$\beta_3 =$	$\beta_4 =$		
20	1.51E+02	2.69E+04	5.50E+06	5.01E+08		

(b)

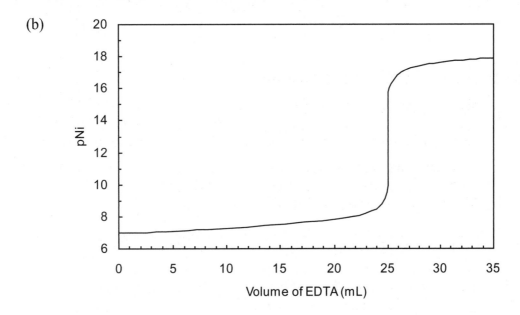

	A	B	C	D	E	F
1	Titration of 50 mL of 0.05 M Ni^{2+} with 0.1 M EDTA/pH 11/0.1 M Oxalate					
2						
3	C_M =	pM	M	$[M]_{tot}$	ϕ	V_{EDTA}
4	0.005	6.97	1.07E-07	4.94E-03	0.008	0.210
5	V_M =	7.00	1.00E-07	4.61E-03	0.054	1.342
6	50	7.20	6.31E-08	2.91E-03	0.324	8.106
7	C_{EDTA} =	7.50	3.16E-08	1.46E-03	0.618	15.461
8	0.01	8.00	1.00E-08	4.61E-04	0.868	21.696
9	K_f'' =	8.40	3.98E-09	1.83E-04	0.946	23.649
10	4.42E+13	8.80	1.58E-09	7.30E-05	0.978	24.456
11	$\alpha(Ni^{2+})$ =	9.50	3.16E-10	1.46E-05	0.996	24.891
12	2.17E-05	10.50	3.16E-11	1.46E-06	1.000	24.989
13	log K_f =	12.80	1.58E-13	7.30E-09	1.000	25.000
14	18.4	14.00	1.00E-14	4.61E-10	1.000	25.001
15	$\alpha(Y^{4-})$ =	15.00	1.00E-15	4.61E-11	1.000	25.012
16	0.81	16.00	1.00E-16	4.61E-12	1.005	25.123
17	$[Oxalate^{2-}]$ =	17.00	1.00E-17	4.61E-13	1.049	26.229
18	0.1	17.40	3.98E-18	1.83E-13	1.123	28.086
19	β_1 =	17.60	2.51E-18	1.16E-13	1.196	29.892
20	1.45E+05	17.80	1.58E-18	7.30E-14	1.310	32.753
21	β_2 =	17.90	1.26E-18	5.80E-14	1.390	34.760
22	3.16E+06	18.00	1.00E-18	4.61E-14	1.491	37.287
23						
24	A10 = A16*A12*10^A14					
25	A12 = 1/(1+A20*A18+A22*A18^2)					
26	C4 = 10^-B4					
27	D4 = C4/$A$12					
28	E4 = (1+$A$10*D4-(D4+D4*D4*$A$10)/$A$4)/					
29	(D4*$A$10+(D4+D4*D4*$A$10)/$A$8)					
30	F4 = E4*$A$4*$A$6/$A$8					

12-21. $\quad [L] + [ML] + 2[ML_2] = \dfrac{C_L V_L}{V_M + V_L}$

$$[L] + \alpha_{ML}\frac{C_M V_M}{V_M + V_L} + 2\alpha_{ML_2}\frac{C_M V_M}{V_M + V_L} = \frac{C_L V_L}{V_M + V_L}$$

Multiply both sides by $V_M + V_L$:

$$[L]V_M + [L]V_L + \alpha_{ML}C_M V_M + 2\alpha_{ML_2}C_M V_M = C_L V_L$$

Collect terms $\quad V_L([L] - C_L) = V_M(-[L] - \alpha_{ML}C_M - 2\alpha_{ML_2}C_M)$

$$\frac{V_L}{V_M} = \frac{[L] + \alpha_{ML}C_M + 2\alpha_{ML_2}C_M}{C_L - [L]}$$

Divide the denominator by C_L and divide the numerator by C_M to obtain ϕ, the fraction of the way to the equivalence point:

$$\phi = \frac{C_L V_L}{C_M V_M} = \frac{\dfrac{[L]}{C_M} + \alpha_{ML} + 2\alpha_{ML_2}}{1 - \dfrac{[L]}{C_L}}$$

12-22.

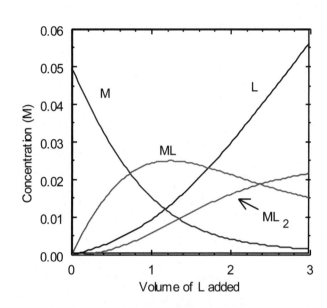

	A	B	C	D	E	F	G	H	I	J	K
1	Copper-acetate complexes ML and ML$_2$										
2											
3	C_M =	pL	[L]	α_M	α_{ML}	α_{ML2}	ϕ	V(ligand)	[M]	[ML]	[ML$_2$]
4	0.05	4.00	0.0001	0.983	0.017	0.000	0.019	0.019	0.0491	0.0008	0.0000
5	V_M =	3.00	0.0010	0.852	0.145	0.004	0.172	0.172	0.0419	0.0071	0.0002
6	10	2.80	0.0016	0.781	0.210	0.008	0.260	0.260	0.0381	0.0103	0.0004
7	C(ligand)	2.60	0.0025	0.688	0.294	0.019	0.383	0.383	0.0331	0.0141	0.0009
8	0.5	2.40	0.0040	0.573	0.388	0.039	0.550	0.550	0.0272	0.0184	0.0019
9	β_1 =	2.20	0.0063	0.446	0.478	0.076	0.766	0.766	0.0207	0.0222	0.0035
10	170	2.00	0.0100	0.319	0.543	0.137	1.039	1.039	0.0145	0.0246	0.0062
11	β_2 =	1.90	0.0126	0.262	0.560	0.178	1.199	1.199	0.0117	0.0250	0.0080
12	4300	1.80	0.0158	0.209	0.564	0.226	1.377	1.377	0.0092	0.0248	0.0099
13		1.70	0.0200	0.164	0.556	0.280	1.579	1.579	0.0071	0.0240	0.0121
14		1.60	0.0251	0.125	0.535	0.340	1.808	1.808	0.0053	0.0226	0.0144
15		1.50	0.0316	0.094	0.504	0.403	2.073	2.073	0.0039	0.0209	0.0167
16		1.40	0.0398	0.069	0.464	0.467	2.385	2.385	0.0028	0.0187	0.0189
17		1.30	0.0501	0.049	0.419	0.532	2.761	2.761	0.0019	0.0164	0.0208
18		1.25	0.0562	0.041	0.396	0.563	2.981	2.981	0.0016	0.0152	0.0217
19											
20	C4 = 10^-B4						H4 = G4*$A$4*$A$6/$A$8				
21	D4 = 1/(1+$A$10*C4+$A$12*C4*C4)						I4 = D4*$A$4*$A$6/($A$6+H4)				
22	E4 = $A$10*C4/(1+$A$10*C4+$A$12*C4*C4)						J4 = E4*$A$4*$A$6/($A$6+H4)				
23	F4 = $A$12*C4*C4/(1+$A$10*C4+$A$12*C4*C4)						K4 = F4*$A$4*$A$6/($A$6+H4)				
24	G4 = (C4/$A$4+E4+2*F4)/(1-C4/$A$8)										

12-23. Only a small amount of indicator is employed. Most of the Mg^{2+} is not bound to the indicator. Free Mg^{2+} reacts with EDTA before MgIn reacts. Therefore, the concentration of MgIn is constant until all of Mg^{2+} has been consumed. Only when MgIn begins to react does the color change.

12-24. 1.   With metal ion indicators

2.   With a mercury electrode

3.   With an ion-selective electrode

4.   With a glass electrode

12-25. HIn^{2-}, wine red, blue

12-26. Buffer (i) (pH 6–7) will give a yellow $\rightarrow$ blue color change that will be easier to observe than the violet $\rightarrow$ blue change expected with the other buffers.

12-27. A back titration is necessary if the analyte precipitates in the absence of EDTA, if it reacts too slowly with EDTA, or if it blocks the indicator.

12-28. In a displacement titration, analyte displaces a metal ion from a complex. The displaced metal ion is then titrated. An example is the liberation of Ni^{2+} from $Ni(CN)_4^{2-}$ by the analyte Ag^+. The liberated Ni^{2+} is then titrated by EDTA to find out how much Ag^+ was present.

12-29. The Mg^{2+} in a solution of Mg^{2+} and Fe^{3+} can be titrated by EDTA if the Fe^{3+} is masked with CN^- to form $Fe(CN)_6^{3-}$, which does not react with EDTA.

12-30. Hardness refers to the total concentration of alkaline earth cations in water, which normally means $[Ca^{2+}] + [Mg^{2+}]$. Hardness gets its name from the reaction of these cations with soap to form insoluble curds. Temporary hardness, due to $Ca(HCO_3)_2$, is lost by precipitation of $CaCO_3(s)$ upon heating. Permanent hardness derived from other salts, such as $CaSO_4$, is not affected by heat.

12-31. $(50.0 \text{ mL})(0.010\,0 \text{ mmol/mL}) = 0.500 \text{ mmol } Ca^{2+}$, which requires 0.500 mmol EDTA $= 10.0 \text{ mL EDTA}$.

0.500 mmol Al^{3+} requires the same amount of EDTA, 10.0 mL.

12-32. mmol EDTA = mmol Ni^{2+} + mmol Zn^{2+}

   1.250   = x +   0.250 $\Rightarrow$ 1.000 mmol Ni^{2+} in 50.0 mL = 0.020 0 M

12-33. The formula mass of $MgSO_4$ is 120.37. The 50.0 mL aliquot contains

$$\left(\frac{50.0 \text{ mL}}{500 \text{ mL}}\right)\left(\frac{0.450 \text{ g}}{120.37 \text{ g/mol}}\right) = 0.373 8 \text{ mmol of } Mg^{2+}$$

37.6 mL of EDTA reacts with this much Mg^{2+}, so the EDTA solution contains 0.373 8 mmol / 37.6 mL = 9.943×10^{-3} mmol/mL. The formula mass of $CaCO_3$ is 100.09. 1.00 mL of EDTA will react with 9.943×10^{-3} mmol of $CaCO_3$ = 0.995 mg.

12-34. 30.10 mL Ni^{2+} reacted with 39.35 mL 0.013 07 M EDTA, so Ni^{2+} molarity is

$$[Ni^{2+}] = \frac{(39.35 \text{ mL})(0.013 07 \text{ mol/L})}{30.10 \text{ mL}} = 0.017 09 \text{ M}.$$

25.00 mL Ni^{2+} contains 0.427 2 mmol Ni^{2+}. 10.15 mL EDTA = 0.132 7 mmol EDTA. The Ni^{2+} which must have reacted with CN^- was 0.427 2 – 0.132 7 = 0.294 5 mmol. Cyanide reacting with Ni^{2+} must have been (4)(0.294 5 mmol) = 1.178 mmol. Original $[CN^-]$ = 1.178 mmol/12.73 mL = 0.092 54 M.

12-35. For 1.00 mL of unknown:

   25.00 mL of EDTA = 0.968 0 mmol

   $-$

   23.54 mL of Zn^{2+} = 0.500 7 mmol

   ―――――――――――――――――――

   Co^{2+} + Ni^{2+} = 0.467 3 mmol

For 2.000 mL of unknown:

   25.00 mL of EDTA = 0.968 0 mmol

   $-$

   25.63 mL of Zn^{2+} = 0.545 2 mmol

   ―――――――――――――――――――

   Ni^{2+} in 2.000 mL   = 0.422 8 mmol

Co^{2+} in 2.000 mL of unknown = 2 (0.467 3) – 0.422 8 = 0.511 8 mmol. The Co^{2+} will react with 0.511 8 mmol of EDTA, leaving 0.968 0 – 0.511 8 = 0.456 2 mmol EDTA.

$$\text{mL Zn needed} = \frac{0.456 2 \text{ mmol}}{0.021 27 \text{ mmol/mL}} = 21.45 \text{ mL}$$

12-36.

$$\text{Total EDTA} = (25.0 \text{ mL})(0.045\,2 \text{ M}) = 1.130 \text{ mmol}$$

$$\underline{-\quad \text{Mg}^{2+} \text{ required} = (12.4 \text{ mL})(0.012\,3 \text{ M}) = 0.153 \text{ mmol}}$$

$$\text{Ni}^{2+} + \text{Zn}^{2+} = 0.977 \text{ mmol}$$

$\text{Zn}^{2+} = $ EDTA displaced by 2,3-dimercapto-1-propanol

$$= (29.2 \text{ mL})(0.012\,3 \text{ M}) = 0.359 \text{ mmol}$$

$$\Rightarrow \text{Ni}^{2+} = 0.977 - 0.359 = 0.618 \text{ mmol}; \quad [\text{Ni}^{2+}] = \frac{0.618 \text{ mmol}}{50.0 \text{ mL}} = 0.012\,4 \text{ M}$$

$$[\text{Zn}^{2+}] = \frac{0.359 \text{ mmol}}{50.0 \text{ mL}} = 0.007\,18 \text{ M}$$

12-37. The precipitation reaction is $\text{Cu}^{2+} + \text{S}^{2-} \rightarrow \text{CuS}\,(s)$.

$$\text{Total Cu}^{2+} \text{ used} = (25.00 \text{ mL})(0.043\,32 \text{ M}) = 1.083\,0 \text{ mmol}$$

$$\underline{-\quad \text{Excess Cu}^{2+} = (12.11 \text{ mL})(0.039\,27 \text{ M}) = 0.475\,6 \text{ mmol}}$$

$$\text{mmol of S}^{2-} = 0.607\,4 \text{ mmol}$$

$$[\text{S}^{2-}] = 0.607\,4 \text{ mmol}/25.00 \text{ mL} = 0.024\,30 \text{ M}$$

12-38.

mmol Bi in reaction $= (25.00 \text{ mL})(0.086\,40 \text{ M}) = 2.160 \text{ mmol}$

EDTA required $= (14.24 \text{ mL})(0.043\,7 \text{ M}) = 0.622 \text{ mmol}$

mmol Bi that reacted with Cs $= 2.160 - 0.622 = 1.538 \text{ mmol}$

Since 2 mol Bi react with 3 mol Cs to give $\text{Cs}_3\text{Bi}_2\text{I}_9$,

$$\text{mmol Cs}^+ \text{ in unknown} = \frac{3}{2}(1.538) = 2.307 \text{ mmol}$$

$$[\text{Cs}^+] = \frac{2.307 \text{ mmol}}{25.00 \text{ mL}} = 0.092\,28 \text{ M}.$$

12-39. Total standard $\text{Ba}^{2+} + \text{Zn}^{2+}$ added to the sulfate was $(5.000 \text{ mL})(0.014\,63 \text{ M}$ $\text{BaCl}_2) + (1.000 \text{ mL})(0.010\,00 \text{ M ZnCl}_2) = 0.083\,15 \text{ mmol}$. Total EDTA required was $(2.39 \text{ mL})(0.009\,63 \text{ M}) = 0.023\,0_2 \text{ mmol}$. Therefore, the original solid must have contained $0.083\,15 - 0.023\,0_2 = 0.060\,1_3 \text{ mmol}$ sulfur (which made $0.060\,1_3 \text{ mmol}$ sulfate that precipitated $0.060\,1_3 \text{ mmol Ba}^{2+}$). The mass of sulfur was $(0.060\,1_3 \text{ mmol})(32.066 \text{ mg/mmol}) = 1.92_8 \text{ mg}$. wt% S $= 100 \times$ $(1.92_8 \text{ mg S}/5.89 \text{ mg sphalerite}) = 32.7 \text{ wt\%}$. Theoretical wt% S in pure ZnS $=$ $100 \times (32.066 \text{ g S} / 97.46 \text{ g ZnS}) = 32.90 \text{ wt\%}$.

CHAPTER 13
ADVANCED TOPICS IN EQUILIBRIUM

13-1. As pH is lowered, $[H^+]$ increases. H^+ reacts with basic anions to increase the solubility of their salts. Dissolution of minerals such as galena and cerussite increases the concentration of Pb^{2+} in the environment.

Galena: $PbS(s) + H^+ \rightleftharpoons Pb^{2+} + HS^-$

Cerussite: $PbCO_3(s) + H^+ \rightleftharpoons Pb^{2+} + HCO_3^-$

13-2. (a) Hydroxybenzene = HA with $pK_{HA} = 9.997$

Mixture contains 0.010 0 mol HA and 0.005 0 mol KOH in 1.00 L.

Chemical reactions:

$$HA \rightleftharpoons A^- + H^+ \qquad\qquad K_{HA} = \frac{[H^+][A^-]}{[HA]} = 10^{-9.997}$$

$$H_2O \rightleftharpoons H^+ + OH^- \qquad\qquad K_w = [H^+][OH^-] = 10^{-14.00}$$

Charge balance:

$$[H^+] + [K^+] = [OH^-] + [A^-]$$

Mass balances:

$$[K^+] = 0.005\ 0\ M$$

$$[HA] + [A^-] = 0.010\ 0\ M = F_A$$

We have 5 equations and 5 chemical species.

Fractional composition equations:

$$[HA] = \alpha_{HA}F_A = \frac{[H^+]F_A}{[H^+] + K_{HA}}$$

$$[A^-] = \alpha_{A^-}F_A = \frac{K_{HA}F_A}{[H^+] + K_{HA}}$$

Substitute concentration expressions into the charge balance:

$$[H^+] + [0.005\ 0] = K_w/[H^+] + \alpha_{A^-}F_A \qquad\qquad (A)$$

We could solve Equation A for $[H^+]$ by using the solution to a quadratic equation. Instead, we will use Solver in the following spreadsheet, with an initial guess of pH =10 in cell H9. Select Solver and choose Options. Set Precision to 1e-15 and click OK. In the Solver window, Set Target Cell <u>E12</u> Equal To Value of <u>0</u> By Changing Cells <u>H9</u>. Click Solve and Solver finds pH = 9.98 in cell H9, giving a net charge near 0 in cell E12.

	A	B	C	D	E	F	G	H	I
1	Mixture of 0.010 M HA and 0.005 M NaOH								
2									
3	F_A =	0.010		[K$^+$] =	0.005				
4	pK_{HA} =	9.997		pK_w =	14.000				
5	K_{HA} =	1.01E-10		K_w =	1.00E-14				
6									
7	Species in charge balance:						Other concentrations:		
8	[H$^+$] =	1.05E-10		[A$^-$] =	4.90E-03		[HA] =	5.10E-03	
9	[K$^+$] =	5.00E-03		[OH$^-$] =	9.56E-05		pH =	9.980	
10							↑ initial value is a guess		
11									
12	Positive charge minus negative charge				-3.00E-18		= B8+B9-E8-E9		
13	Formulas:								
14	B5 = 10^-B4			B8 = 10^-H9					
15	E5 = 10^-E4			B9 = E3					
16	E8 = B5*B3/(B8+B5)			E9 = E5/B8					
17	H8 = B8*B3/(B8+B5)								

(b) From previous knowledge, we would have said that there is enough KOH to neutralize half of the HA. Therefore, [HA] = [A$^-$].

pH = pK_a + log([A$^-$]/[HA]) = pK_a + log(1) = pK_a = 10.00. The systematic treatment of equilibrium in the spreadsheet gave pH = 9.98.

(c) If we dilute HA to 0.000 10 M and KOH to 0.000 050 in cells B3 and E3, then Solver finds a pH of 9.45. It makes sense that as the solution becomes more dilute, the pH must move toward 7.

13-3. Use effective equilibrium constants, K', defined as follows:

$$K_{HA} = \frac{[H^+]\gamma_{H^+}[A^-]\gamma_{A^-}}{[HA]\gamma_{HA}} \Rightarrow K'_{HA} = \frac{[H^+][A^-]}{[HA]} = K_{HA}\frac{\gamma_{HA}}{\gamma_{H^+}\gamma_{A^-}}$$

$$K_w = [H^+]\gamma_{H^+}[OH^-]\gamma_{OH^-} = 10^{-13.995}$$

$$K'_w = \frac{K_w}{\gamma_{H^+}\gamma_{OH^-}} = [H^+][OH^-] \qquad \Rightarrow [OH^-] = K'_w/[H^+]$$

$$pH = -\log([H^+]\gamma_{H^+})$$

In the spreadsheet, enable circular definitions in Excel Options (in the File menu in Excel 2010) by selecting Formulas. Check "Enable iterative calculation" and set Maximum Change = 1E-15. In Solver Options, set Precision = 1E-15. Use solver to find the pH in cell H13 that produces a net charge of 0 in cell E16. Solver finds pH = 9.95 and ionic strength = 0.005 0 in cell C17.

	A	B	C	D	E	F	G	H	I
1	Mixture of 0.010 M HA and 0.005 M NaOH with activity coefficients								
2									
3	F_A =	0.010		[K$^+$] =	0.005				
4	pK_{HA} =	9.997		pK_w =	13.995				
5	K_{HA}' =	1.17E-10		K_w' =	1.18E-14				
6									
7	Activity coefficients:								
8	H$^+$ =	0.93		A$^-$	0.93				
9	OH$^-$ =	0.93		HA	1.00				
10									
11	Species in charge balance:						Other concentrations:		
12	[H$^+$] =	1.22E-10		[A$^-$] =	4.90E-03		[HA] =	5.10E-03	
13	[K$^+$] =	0.005		[OH$^-$] =	9.67E-05		pH =	9.947	
14							↑ initial value is a guess		
15									
16	Positive charge minus negative charge =				2.29E-17		= B12+B13-E12-E13		
17	Ionic strength =		0.005000	= 0.5*(B12+B13+E12+E13)					
18									
19	Formulas:								
20	B5 = (10^-B4)*E9/(B8*E8)				E9 = 1		B13 = E3		
21	E5 = (10^-E4)/(B8*B9)								
22	B8 = B9 = E8 = 10^(-0.51*1^2*(SQRT($C$17)/(1+SQRT($C$17))-0.3*$C$17))								
23	B12 = (10^-H13)/B8								
24	E12 = B5*B3/(B12+B5))				E13 = E5/B12				
25	C18 = 0.5*(B12+B13+E12+E13)				H12 = B12*B3/(B12+B5)				

13-4. Abbreviating the protonated form of glycine as H_2G^+, we write

$$H_2G^+ \rightleftharpoons HG + H^+ \qquad K_1 = \frac{[HG]\gamma_{HG}[H^+]\gamma_{H^+}}{[H_2G^+]\gamma_{H_2G^+}} \qquad pK_1 = 2.350$$

$$HG \rightleftharpoons G^- + H^+ \qquad K_2 = \frac{[G^-]\gamma_{G^-}[H^+]\gamma_{H^+}}{[HG]\gamma_{HG}} \qquad pK_2 = 9.778$$

At $\mu = 0.1$ M, the activity coefficient of a monovalent ion is 0.78. The activity coefficient of a neutral molecule is 1. Putting these coefficients into the expressions for K_1 and K_2 gives

$$K_1' = \frac{[HG][H^+]}{[H_2G^+]} \text{ (at } \mu = 0.1 \text{ M)} = K_1\frac{\gamma_{H_2G^+}}{\gamma_{HG}\gamma_{H^+}} = 10^{-2.350}\frac{0.78}{(1)(0.78)} = 10^{-2.350}$$

$$K_2' = \frac{[G^-][H^+]}{[HG]} \text{ (at } \mu = 0.1 \text{ M)} = K_2\frac{\gamma_{HG}}{\gamma_{G^-}\gamma_{H^+}} = 10^{-9.778}\frac{1}{(0.78)(0.78)} = 10^{-9.562}$$

The predicted values are $pK_1' = 2.350$ and $pK_2' = 9.562$. Values from fitting the data in the spreadsheet are 2.312 and 9.625. The change from pK_1 to pK_1' is expected to be zero and it is observed to be –0.038. The change from pK_2 to pK_2'

is expected to be –0.216 and it is observed to be –0.153.

13-5.   Ethylenediamine = B from diprotic H_2B^{2+} $pK_1 = 6.848$ $pK_2 = 9.928$

Mixture contains 0.100 mol B and 0.035 mol HBr in 1.00 L.

Chemical reactions:

$H_2B^{2+} \rightleftharpoons HB^+ + H^+$ $\qquad\qquad$ $K_1 = 10^{-6.848}$

$HB^+ \rightleftharpoons B + H^+$ $\qquad\qquad\qquad$ $K_2 = 10^{-9.928}$

$H_2O \rightleftharpoons H^+ + OH^-$ $\qquad\qquad$ $K_w = 10^{-14.00}$

Charge balance:

$[H^+] + 2[H_2B^{2+}] + [HB^+] = [OH^-] + [Br^-]$

Mass balances:

$[Br^-] = 0.035\ M;$ $[H_2B^+] + [HB] + [B] = 0.100\ M = F_B$

We have 6 equations and 6 chemical species, so there is enough information.

Fractional composition equations:

$$[H_2B^{2+}] = \alpha_{H_2B^{2+}}\, F_B = \frac{[H^+]^2 F_B}{[H^+]^2 + [H^+]K_1 + K_1K_2}$$

$$[HB^+] = \alpha_{HB^+}\, F_B = \frac{K_1[H^+]F_B}{[H^+]^2 + [H^+]K_1 + K_1K_2}$$

$$[B] = \alpha_B\, F_B = \frac{K_1K_2 F_B}{[H^+]^2 + [H^+]K_1 + K_1K_2}$$

Substitute into charge balance:

$$[H^+] + 2\alpha_{H_2B^{2+}} F_B + \alpha_{HB^+} F_B = K_w/[H^+] + [0.035\ M] \qquad\qquad (A)$$

We solve Equation A for $[H^+]$ by using Solver in the following spreadsheet, with an initial guess of pH = 10 in cell H11. Select Solver and choose Options. Set Precision to 1E-16 and click OK. In the Solver window, Set Target Cell E14 Equal To Value of 0 By Changing Cells H11. Click Solve and Solver finds pH = 10.194 in cell H11, giving a net charge of ~10^{-17} in cell E14.

	A	B	C	D	E	F	G	H	I
1	Mixture of 0.100 M B and 0.035 M HBr								
2									
3	F_B =	0.100		[Br⁻] =	0.035				
4	pK_1 =	6.848		pK_w =	14.000				
5	pK_2 =	9.928		K_w =	1.00E-14				
6	K_1 =	1.42E-07							
7	K_2 =	1.18E-10							
8									
9	Concentrations:								
10	[H⁺] =	6.39E-11		[H₂B²⁺] =	1.58E-05		[B] =	6.49E-02	
11	[Br⁻] =	3.50E-02		[HB⁺] =	3.51E-02		pH =	10.194	
12	[OH⁻] =	1.56E-04					↑ initial value is a guess		
13									
14	Positive charge minus negative charge				-1.70E-17				
15	Formulas:								
16	B6 = 10^-B4			B7 = 10^-B5	E5 = 10^-E4				
17	B10 = 10^-H11			B11 = E3	B12 = E5/B10				
18	E10 = B10^2*$B$3/(B10^2+B10*$B$6+$B$6*$B$7)								
19	E11 = B10*$B$6*$B$3/(B10^2+B10*$B$6+$B$6*$B$7)								
20	H10 = $B$6*$B$7*$B$3/(B10^2+B10*$B$6+$B$6*$B$7)								

In your earlier life, you would have solved this problem by noting that 0.035 mol HBr converts 0.035 mol B into 0.035 mol HB⁺, leaving (0.100 – 0.035) mol B.

$$\text{pH} = pK_2 + \log\frac{[\text{B}]}{[\text{HB}^+]} = 9.928 + \log\frac{0.065}{0.035} = 10.197 \text{ (close to spreadsheet}$$

answer)

13-6. Benzene-1,2,3-tricarboxylic acid = H_3A with $pK_1 = 2.86$, $pK_2 = 4.30$, $pK_3 = 6.28$. Imidazole = HB from diprotic H_2B^+ with $pK_1 = 6.993$, $pK_2 = 14.5$.

Mixture contains 0.040 mol H_3A, 0.030 mol HB, and 0.035 mol NaOH in 1.00 L.

Charge balance:

$$[\text{H}^+] + [\text{H}_2\text{B}^+] + [\text{Na}^+] = [\text{OH}^-] + [\text{H}_2\text{A}^-] + 2[\text{HA}^{2-}] + 3[\text{A}^{3-}] + [\text{B}^-]$$

Substitute fractional composition equations into charge balance:
$$[\text{H}^+] + \alpha_{\text{H}_2\text{B}^+} F_B + [0.035]$$
$$= K_w/[\text{H}^+] + \alpha_{\text{H}_2\text{A}^-} F_A + 2\alpha_{\text{H}_2\text{A}^{2-}} F_A + 3\alpha_{\text{HA}^{3-}} F_A + \alpha_{\text{B}^-} F_B$$

We solve for [H⁺] with the following spreadsheet, with an initial guess of pH = 7 in cell H14. Select Solver and choose Options. Set Precision to 1e-16 and click OK. In Solver, Set Target Cell E16 Equal To Value of 0 By Changing Cells H14. Click Solve and Solver finds pH = 4.52 in cell H14, giving a net charge of ~10⁻¹⁶ in cell E16.

	A	B	C	D	E	F	G	H	I
1	Mixture of 0.040 M H_3A, 0.030 M HB, and 0.035 M NaOH								
2									
3	F_A =	0.040		F_B =	0.030		$[Na^+]$ =	0.035	
4	pK_1 =	2.86		pK_{H2B} =	6.993		pK_w =	14.000	
5	pK_2 =	4.30		pK_{HB} =	14.5		K_w =	1.00E-14	
6	pK_3 =	6.28		K_{H2B} =	1.02E-07				
7	K_1 =	1.4E-03		K_{HB} =	3.16E-15				
8	K_2 =	5.0E-05							
9	K_3 =	5.2E-07							
10									
11	Concentrations								
12	$[H^+]$ =	3.05E-05		$[A^{3-}]$ =	4.21E-04		$[OH^-]$ =	3.28E-10	
13	$[H_3A]$ =	3.27E-04		$[H_2B^+]$ =	2.99E-02		$[Na^+]$ =	0.035	
14	$[H_2A^-]$ =	1.48E-02		$[HB]$ =	9.98E-05		pH =	4.516	
15	$[HA^{2-}]$ =	2.44E-02		$[B^-]$ =	1.04E-14		↑ initial value is a guess		
16	Positive charge minus negative charge =				-2.42E-17				
17									
18	Formulas:								
19	B7 = 10^-B4			B8 = 10^-B5			B9 = 10^-B6		
20	E6 = 10^-E4			E7 = 10^-E5					
21	B12 = 10^-H14			H12 = H5/B12			H13 = H3		
22	B13 = $B$12^3*$B$3/($B$12^3+$B$12^2*$B$7+$B$12*$B$7*$B$8+$B$7*$B$8*$B$9)								
23	B14 = $B$12^2*$B$7*$B$3/($B$12^3+$B$12^2*$B$7+$B$12*$B$7*$B$8+$B$7*$B$8*$B$9)								
24	B15 = $B$12*$B$7*$B$8*$B$3/($B$12^3+$B$12^2*$B$7+$B$12*$B$7*$B$8+$B$7*$B$8*B								
25	E12 = B$7*B$8*B$9*B$3/(B12^3+B12^2*B$7+B12*B$7*B$8+B$7*B$8*B$9)								
26	E13 = $B$12^2*$E$3/($B$12^2+$B$12*$E$6+$E$6*$E$7)								
27	E14 = $B$12*$E$6*$E$3/($B$12^2+$B$12*$E$6+$E$6*$E$7)								
28	E15 = $E$6*$E$7*$E$3/($B$12^2+$B$12*$E$6+$E$6*$E$7)								

13-7.   Arginine = HA from H_3A^{2+} with $pK_1 = 1.823$, $pK_2 = 8.991$, $pK_3 = 12.1$.

Glutamic acid = H_2G from H_3G^+ with $pK_1 = 2.160$, $pK_2 = 4.30$, $pK_3 = 9.96$.

Mixture contains 0.020 mol arginine, 0.030 mol glutamic acid, and 0.005 mol KOH in 1.00 L. $F_A = 0.020$ M and $F_G = 0.030$ M.

Charge balance:

$$[H^+] + 2[H_3A^{2+}] + [H_2A^+] + [H_3G^+] + [K^+] = [OH^-] + [A^-] + [HG^-] + 2[G^{2-}]$$

Substitute fractional composition equations into charge balance:
$$[H^+] + 2\alpha_{H_3A^{2+}} F_A + \alpha_{H_2A^+} F_A + \alpha_{H_3G^+} F_G + [0.005]$$
$$= K_w/[H^+] + \alpha_{A^-} F_A + \alpha_{HG^-} F_G + 2\alpha_{G^{2-}} F_G$$

	A	B	C	D	E	F	G	H	I	J
1	Mixture of 0.020 M arginine, 0.030 M glutamic acid, and 0.005 M KOH									
2										
3	F_A =	0.020		F_B =	0.030		[K$^+$] =	0.005		
4	pK_1 =	1.823		pK_{H3G} =	2.160		pK_w =	14.00		
5	pK_2 =	8.991		pK_{H2G} =	4.30		K_w =	1.00E-14		
6	pK_3 =	12.1		pK_{HG} =	9.96					
7	K_1 =	1.5E-02		K_{H3G} =	6.9E-03					
8	K_2 =	1.0E-09		K_{H2G} =	5.0E-05					
9	K_3 =	7.9E-13		K_{HG} =	1.1E-10					
10										
11	Concentrations									
12	[H$_3$A^{2+}] =	1.32E-05		[H$_3$G$^+$] =	7.14E-06		[H$^+$] =	9.94E-06		
13	[H$_2$A$^+$] =	2.00E-02		[H$_2$G] =	4.96E-03		[OH$^-$] =	1.01E-09		
14	[HA] =	2.05E-06		[HG$^-$] =	2.50E-02		[K$^+$] =	0.005		
15	[A$^-$] =	1.64E-13		[G^{2-}] =	2.76E-07		pH =	5.003		
16							↑ initial value is a guess			
17	Positive charge minus negative charge =				-9.94E-17					
18					= 2*B12+B13+E12+H12+H14-B15-E14-2*E15-H13					
19	Formulas:									
20	B7 = 10^-B4			B8 = 10^-B5			B9 = 10^-B6			
21	E7 = 10^-E4			E8 = 10^-E5			E9 = 10^-E6			
22	H12 = 10^-H15			H13 = H5/H12			H14 = H3			
23	B12 = $H$12^3*$B$3/($H$12^3+$H$12^2*$B$7+$H$12*$B$7*$B$8+$B$7*$B$8*$B$9)									
24	B13 = $H$12^2*$B$7*$B$3/($H$12^3+$H$12^2*$B$7+$H$12*$B$7*$B$8+$B$7*$B$8*$B$9)									
25	B14 = $H$12*$B$7*$B$8*$B$3/($H$12^3+$H$12^2*$B$7+$H$12*$B$7*$B$8+$B$7*$B$8*$B$9)									
26	B15 = $B$7*$B$8*$B$9*$B$3/($H$12^3+$H$12^2*$B$7+$H$12*$B$7*$B$8+$B$7*$B$8*$B$9)									
27	E12 = $H$12^3*$E$3/($H$12^3+$H$12^2*$E$7+$H$12*$E$7*$E$8+$E$7*$E$8*$E$9)									
28	E13 = $H$12^2*$E$7*$E$3/($H$12^3+$H$12^2*$E$7+$H$12*$E$7*$E$8+$E$7*$E$8*$E$9)									
29	E14 = $H$12*$E$7*$E$8*$E$3/($H$12^3+$H$12^2*$E$7+$H$12*$E$7*$E$8+$E$7*$E$8*$E$9)									
30	E15 = $E$7*$E$8*$E$9*$E$3/($H$12^3+$H$12^2*$E$7+$H$12*$E$7*$E$8+$E$7*$E$8*$E$9)									

We solve for [H$^+$] with the spreadsheet, with an initial guess of pH = 3 in cell H15. Select Solver and choose Options. Set Precision to 1e-16 and click OK. In Solver, Set Target Cell E17 Equal To Value of 0 By Changing Cells H55. Click Solve and Solver finds pH = 5.00 in cell H15, giving a net charge of ~10^{-16} in cell E17.

13-8. $H_3A^{2+} \rightleftharpoons H_2A^+ + H^+$ $\qquad K_1 = \dfrac{[H_2A^+]\gamma_{H_2A^+}[H^+]\gamma_{H^+}}{[H_3A^{2+}]\gamma_{H_3A^{2+}}} = 10^{-1.823}$

$H_2A^+ \rightleftharpoons HA + H^+$ $\qquad K_2 = \dfrac{[HA]\gamma_{HA}[H^+]\gamma_{H^+}}{[H_2A^+]\gamma_{H_2A^+}} = 10^{-8.991}$

$HA \rightleftharpoons A^- + H^+$ $\qquad K_3 = \dfrac{[A^-]\gamma_{A^-}[H^+]\gamma_{H^+}}{[HA]\gamma_{HA}} = 10^{-12.1}$

$$K_1' = K_1\left(\frac{\gamma_{H_3A^{2+}}}{\gamma_{H_2A^+}\gamma_{H^+}}\right) = \frac{[H_2A^+][H^+]}{[H_3A^{2+}]} \qquad\qquad K = K_2\left(\frac{\gamma_{H_2A^+}}{\gamma_{HA}\gamma_{H^+}}\right) = \frac{[HA][H^+]}{[H_2A^+]}$$

$$K_3' = K_3\left(\frac{\gamma_{HA}}{\gamma_{A^-}\gamma_{H^+}}\right) = \frac{[A^-][H^+]}{[HA]}$$

$$H_3G^+ \;\rightleftharpoons\; H_2G + H^+ \qquad K_{H_3G} = \frac{[H_2G]\gamma_{H_2G}[H^+]\gamma_{H^+}}{[H_3G^+]\gamma_{H_3G^+}} = 10^{-2.160}$$

$$H_2G \;\rightleftharpoons\; HG^- + H^+ \qquad K_{H_2G} = \frac{[HG^-]\gamma_{HG^-}[H^+]\gamma_{H^+}}{[H_2G]\gamma_{H_2G}} = 10^{-4.30}$$

$$HG^- \;\rightleftharpoons\; G^{2-} + H^+ \qquad K_{HG} = \frac{[G^{2-}]\gamma_{G^{2-}}[H^+]\gamma_{H^+}}{[HG^-]\gamma_{HG^-}} = 10^{-9.96}$$

$$K_{H_3G}' = K_{H_3G}\left(\frac{\gamma_{H_3G^+}}{\gamma_{H_2G}\gamma_{H^+}}\right) = \frac{[H_2G][H^+]}{[H_3G^+]}$$

$$K_{H_2G}' = K_{H_2G}\left(\frac{\gamma_{H_2G}}{\gamma_{HG^-}\gamma_{H^+}}\right) = \frac{[HG^-][H^+]}{[H_2G]}$$

$$K_{HG}' = K_{HG}\left(\frac{\gamma_{HG^-}}{\gamma_{G^{2-}}\gamma_{H^+}}\right) = \frac{[G^{2-}][H^+]}{[HG^-]}$$

In the spreadsheet, $[H^+]$ in cell H18 is computed from $(10^{-pH})/\gamma_{H^+}$. Don't forget that activity coefficient! Enable circular definitions in Excel Options. Ionic strength is computed in cell E23 and activity coefficients are computed in cells A13:H15. From the activity coefficients, effective equilibrium constants are computed in cells H4:H10. $pH = 3$ is *guessed* in cell H21. From pH and the K' values, concentrations are computed in cells A18:H21. Solver is used to vary pH in cell H21 until the net charge in cell H23 is near 0. If Solver does not find an answer, try a different initial value for pH or increase Precision in the Options window of Solver. A value of 1e-15 generally works, but for some problems I need larger numbers, such as 1E-10. pH computed with activities is 4.94 and $\mu = 0.025\,1$ M. When we found pH without activities in the previous problem, the pH was 5.00.

	A	B	C	D	E	F	G	H	I
1	Mixture of 0.020 M arginine, 0.030 M glutamic acid, and 0.005 M KOH								
2	Solved with Davies activity coefficients								
3	F_A =	0.020		F_B =	0.030		$[K^+]$ =	0.005	
4	pK_1 =	1.823		pK_{H3G} =	2.160		K_1' =	1.1E-02	
5	pK_2 =	8.991		pK_{H2G} =	4.30		K_2' =	1.0E-09	
6	pK_3 =	12.1		pK_{HG} =	9.96		K_3' =	1.1E-12	
7	K_1 =	1.5E-02		K_{H3G} =	6.9E-03		K_{H3G}' =	6.9E-03	
8	K_2 =	1.0E-09		K_{H2G} =	5.0E-05		K_{H2G}' =	6.8E-05	
9	K_3 =	7.9E-13		K_{HG} =	1.1E-10		K_{HG}' =	2.01E-10	
10	pK_w =	13.995		K_w =	1.01E-14		K_w' =	1.37E-14	
11									
12	Davies activity coefficients								
13	H_3A^{2+} =	0.55		H_3G^+ =	0.86				
14	H_2A^+ =	0.86		HG^- =	0.86		H^+ =	0.86	
15	A^- =	0.86		G^{2-} =	0.55		OH^- =	0.86	
16									
17	Concentrations								
18	$[H_3A^{2+}]$ =	2.41E-05		$[H_3G^+]$ =	9.58E-06		$[H^+]$ =	1.34E-05	
19	$[H_2A^+]$ =	2.00E-02		$[H_2G]$ =	4.95E-03		$[OH^-]$ =	1.02E-09	
20	$[HA]$ =	1.52E-06		$[HG^-]$ =	2.50E-02		$[K^+]$ =	0.005	
21	$[A^-]$ =	1.22E-13		$[G^{2-}]$ =	3.76E-07		pH =	4.939	
22							↑ initial value is a guess		
23	Positive charge minus negative charge =				-1.30E-16				
24			Ionic strength =		0.0251	=0.5*(4*B18+B19+B21+E18			
25						+E20+4*E21+H18+H19+H20)			
26									
27	Formulas								
28	K_1' = B7*B13/(B14*H14)			K_2' = B8*B14/(H14)			K_3' = B9/(B15*H14)		
29	K_{H3B}' = E7*E13/H14			K_{H2B}' =E8/(E14*H14)			K_{HB}' = E9*E14/(E15*H14)		
30	K_w' = E10/(H14*H15)			[H+] = (10^-H21)/H14			[OH-] = H10/H18		
31	Activity coefficient = 10^(-0.51*charge^2*(SQRT($D$24)/(1+SQRT($D$24))-0.3*$D$24))								
32	Denom1 = ($H$18^3+$H$18^2*$H$4+$H$18*$H$4*$H$5+$H$4*$H$5*$H$6)								
33	$[H_3A^{2+}]$ = $H$18^3*$B$3/Denom1						$[H_2A^+]$ = $H$18^2*$H$4*$B$3/Denom1		
34	$[HA]$ = $H$18*$H$4*$H$5*$B$3/Denom1						$[A^-]$ = $H$4*$H$5*$H$6*$B$3/Denom1		
35	Denom2 = ($H$18^3+$H$18^2*$H$7+$H$18*$H$7*$H$8+$H$7*$H$8*$H$9)								
36	$[H_3G^+]$ = $H$18^3*$E$3/Denom2						$[H_2G]$ = $H$18^2*$H$7*$E$3/Denom2		
37	$[HG^-]$ = $H$18*$H$7*$H$8*$E$3/Denom2						$[G^{2-}]$ = $H$7*$H$8*$H$9*$E$3/Denom2		
38	E23 = 2*B18+B19+E18+H18+H20-B21-E20-2*E21-H19								
39	D25 = 0.5*(4*B18+B19+B21+E18+E20+4*E21+H18+H19+H20)								

13-9. (a) This is the same problem that was worked in Section 13-2, but with $[KH_2PO_4]$ = 0.008 695 m and $[Na_2HPO_4]$ = 0.030 43 m in cells B3 and B4 of the spreadsheet in Figure 13-3. Enable circular definitions in Excel Options. Use Solver to find the pH in cell H15 that reduces the charge balance in cell E18 to near 0. The pH is found to be 7.420 and the ionic strength in cell E19 is 0.100 m.

(b) To use the Debye-Hückel equation, we can compute activity coefficients with ion size parameters from Table 8-1 or we can just use activity coefficients from Table 8-1 for μ = 0.1 M:

	A	B	C	D	E	F	G	H
8	Activity coefficients from table in textbook:							
9	H^+ =	0.83		H_3P =	1.00	(fixed at 1)	HP^{2-} =	0.36
10	OH^- =	0.76		H_2P^- =	0.78		P^{3-} =	0.10

These activity coefficients produce a pH of 7.403 after executing Solver to reduce the net charge to 0 in cell E18.

13-10. EDTA = H_4A from hexaprotic H_6A^{2+} with pK_1 = 0.0, pK_2 = 1.5, pK_3 = 2.00, pK_4 = 2.69, pK_5 = 6.13, pK_6 = 10.37; Lysine = HL from triprotic H_3L^{2+} with pK_1 = 1.77, pK_2 = 9.07, pK_3 = 10.82

Mixture contains 0.040 mol H_4A, 0.030 mol HL, and 0.050 mol NaOH in 1.00 L.

Charge balance:
$$[H^+] + 2[H_6A^{2+}] + [H_5A^+] + 2[H_3L^{2+}] + [H_2L^+] + [Na^+]$$
$$= [OH^-] + [H_3A^-] + 2[H_2A^{2-}] + 3[HA^{3-}] + 4[A^{4-}] + [L^-]$$

Substitute fractional composition equations into charge balance:
$$[H^+] + 2\alpha_{H_6A^{2+}} F_A + \alpha_{H_5A^+} F_A + 2\alpha_{H_3L^{2+}} F_L + \alpha_{H_2L^+} F_L + [0.050]$$
$$= K_w/[H^+] + \alpha_{H_3A^-} F_A + 2\alpha_{H_2A^{2-}} F_A + 3\alpha_{HA^{3-}} F_A + 4\alpha_{A^{4-}} F_A + \alpha_{L^-} F_L$$

We solve this equation for pH with the following spreadsheet, with an initial guess of pH = 7 in cell H18. In Solver, Set Target Cell E19 Equal To Value of 0 By Changing Cells H18. Click OK and Solver finds pH = 4.44 in cell H18.

	A	B	C	D	E	F	G	H	I
1	Mixture of 0.040 M H_4A, 0.030 M HL, and 0.05 M NaOH								
2									
3	$F_A =$	0.040		$F_L =$	0.030		$[Na^+] =$	0.050	
4	$pK_1 =$	0.0		$pK_{L1} =$	1.77				
5	$pK_2 =$	1.5		$pK_{L2} =$	9.07				
6	$pK_3 =$	2.0		$pK_{L3} =$	10.82		$pK_w =$	14.00	
7	$pK_4 =$	2.69		$K_{L1} =$	1.70E-02		$K_w =$	1.00E-14	
8	$pK_5 =$	6.13		$K_{L2} =$	8.51E-10				
9	$pK_6 =$	10.37		$K_{L3} =$	1.51E-11				
10	$K_1 =$	1.00E+00		$K_3 =$	1.00E-02		$K_5 =$	7.41E-07	
11	$K_2 =$	3.16E-02		$K_4 =$	2.04E-03		$K_6 =$	4.27E-11	
12									
13	Species in charge balance:								
14	$[H^+] =$	3.62E-05		$[HA^{3-}] =$	7.88E-04		$[OH^-] =$	2.76E-10	
15	$[H_6A^{2+}] =$	1.03E-13		$[A^{4-}] =$	9.28E-10		$[Na^+] =$	0.050	
16	$[H_5A^+] =$	2.84E-09		$[H_3L^{2+}] =$	6.39E-05		$[H_4A] =$	2.48E-06	
17	$[H_3A^-] =$	6.84E-04		$[H_2L^+] =$	2.99E-02		$[HL] =$	7.03E-07	
18	$[H_2A^{2-}] =$	3.85E-02		$[L^-] =$	2.94E-13		pH =	4.441	← initial value
19	Positive charge minus negative charge				-9.77E-17				is a guess
20	Formulas:								
21	B10 = 10^-B4 with analogous formulas for K_2 - K_6								
22	E7 = 10-E4 with analogous formulas for K_{L2} and K_{L3}								
23	B14 = 10^-H18			H14 = H7/B14			H15 = H3		
24	Denom1 = $B$14^6+$B$14^5*$B$10+$B$14^4*$B$10*$B$11+$B$14^3*$B$10*$B$11*$E$10								
25	+$B$14^2*$B$10*$B$11*$E$10*$E$11+$B$14*$B$10*$B$11*$E$10*$E$11*$H$10								
26	+$B$10*$B$11*$E$10*$E$11*$H$10*$H$11								
27	Denom2 = $B$14^3+$B$14^2*$E$7+$B$14*$E$7*$E$8+$E$7*$E$8*$E$9								
28	B15 = $B$14^6*$B$3/Denom1								
29	B16 = $B$14^5*$B$10*$B$3/Denom1								
30	B17 = $B$14^3*$B$10*$B$11*$E$10*$B$3/Denom1								
31	B18 = $B$14^2*$B$10*$B$11*$E$10*$E$11*$B$3/Denom1								
32	E14 = $B$14*$B$10*$B$11*$E$10*$E$11*$H$10*$B$3/Denom1								
33	E15 = $B$10*$B$11*$E$10*$E$11*$H$10*$H$11*$B$3/Denom1								
34	H16 = $B$14^4*$B$10*$B$11*$B$3/Denom1								
35	E16 = $B$14^3*$E$3/Denom2								
36	E17 = $B$14^2*$E$7*$E$3/Denom2								
37	E18 = $E$7*$E$8*$E$9*$E$3/Denom2								
38	H17 = $B$14*$E$7*$E$8*$E$3/Denom2								
39	E19 = B14+2*B15+B16+2*E16+E17+H15-B17-2*B18-3*E14-4*E15-E18-H14								

13-11. (a) $Fe^{3+} + SCN^- \rightleftharpoons Fe(SCN)^{2+}$ $[Fe(SCN)^{2+}] = \beta_1'[Fe^{3+}][SCN^-]$

   $Fe^{3+} + 2SCN^- \rightleftharpoons Fe(SCN)_2^+$ $[Fe(SCN)_2^+] = \beta_2'[Fe^{3+}][SCN^-]^2$

   $Fe^{3+} + H_2O \rightleftharpoons FeOH^{2+} + H^+$ $[FeOH^{2+}] = K_a'[Fe^{3+}]/[H^+]$

   $H_2O \rightleftharpoons H^+ + OH^-$ $[OH^-] = K_w'/[H^+]$

   $$\beta_1' = \beta_1 \frac{\gamma_{Fe^{3+}}\gamma_{SCN^-}}{\gamma_{Fe(SCN)^{2+}}} \quad \beta_2' = \beta_2 \frac{\gamma_{Fe^{3+}}\gamma_{SCN^-}^2}{\gamma_{Fe(SCN)_2^+}} \quad K_a' = K_a \frac{\gamma_{Fe^{3+}}}{\gamma_{FeOH^{2+}}\gamma_{H^+}}$$

   $$K_w' = \frac{K_w}{\gamma_{H^+}\gamma_{OH^-}} \quad \beta_1 = 10^{3.03} \quad \beta_2 = 10^{4.6} \quad K_a = 10^{-2.195}$$

(b) Charge balance: $[OH^-] + [SCN^-] + [NO_3^-] =$

   $[H^+] + 3[Fe^{3+}] + 2[Fe(SCN)^{2+}] + [Fe(SCN)_2^+] + 2[FeOH^{2+}] + [Na^+]$

(c) Mass balances:

   Total iron $\equiv F_{Fe} = 5.0$ mM $= [Fe^{3+}] + [Fe(SCN)^{2+}] + [Fe(SCN)_2^+] + [FeOH^{2+}]$

   Total thiocyanate $\equiv F_{SCN} = 5.0$ μM $= [Fe(SCN)^{2+}] + 2[Fe(SCN)_2^+] + [SCN^-]$

   $[Na^+] = 5.0$ μM

   $[NO_3^-] = 3(5.0$ mM$) + 15.0$ mM $= 30.0$ mM

(d) $[Fe(SCN)^{2+}] + 2[Fe(SCN)_2^+] + [SCN^-] = F_{SCN}$

   $\beta_1'[Fe^{3+}][SCN^-] + 2\beta_2'[Fe^{3+}][SCN^-]^2 + [SCN^-] = F_{SCN}$

   $[Fe^{3+}](\beta_1'[SCN^-] + 2\beta_2'[SCN^-]^2) = F_{SCN} - [SCN^-]$

   $$[Fe^{3+}] = \frac{F_{SCN} - [SCN^-]}{\beta_1'[SCN^-] + 2\beta_2'[SCN^-]^2}$$

(e) The spreadsheet uses Solver to find all concentrations beginning with
   estimates for $[SCN^-]$ and $[H^+]$. Results are

   $[Fe^{3+}] = 4.20$ mM $[SCN^-] = 2.03$ μM $[H^+] = 15.8$ mM
   $[Fe(SCN)^{2+}] = 2.97$ μM $[Fe(SCN)_2^+] = 106$ pM $[FeOH^{2+}] = 0.802$ mM
   $[OH^-] = 0.920$ pM $\mu = 0.043\ 4$ M

(f) Hydrolysis of Fe(III) produces 0.000 8 M H^+ ($Fe^{3+}+H_2O \rightleftharpoons FeOH^{2+}+H^+$)

(g) $\dfrac{[Fe(SCN)^{2+}]}{\{[Fe^{3+}] + [FeOH^{2+}]\}[SCN^-]} = 293$ (The graph in the textbook gives 270.)

(h) With 0.200 M, KNO_3, $[K^+] = 0.2$ M in cell J5 and $[NO_3^-] = 0.23$ M in cell J4.

   $[K^+]$ has to be added to the charge balance in cell J22. Results are

   $[Fe^{3+}] = 4.45$ mM $[SCN^-] = 2.81$ μM $[H^+] = 15.6$ mM

$$[Fe(SCN)^{2+}] = 2.19 \ \mu M \quad [Fe(SCN)_2^+] = 68.2 \ pM \quad [FeOH^{2+}] = 0.546 \ mM$$

$$[OH^-] = 1.18 \ pM \quad\quad \mu = 0.244 \ M$$

$$\frac{[Fe(SCN)^{2+}]}{\{[Fe^{3+}] + [FeOH^{2+}]\}[SCN^-]} = 156 \quad \text{(The graph in the textbook gives 150.)}$$

	A	B	C	D	E	F	G	H	I	J
1	Iron thiocyanate equilibria with Davies equation for activity coefficients								F_{Fe} =	5.000E-03
2	1. *Estimate* values in cells B11 and B12								F_{SCN} =	5.000E-06
3	2. Use Solver to adjust B11 and B12 to minimize sum in cell J23								$[Na^+]$ =	5.000E-06
4									$[NO_3^-]$ =	3.000E-02
5									$[K^+]$ =	0.000E+00
6	Ionic strength		μ = 0.5*(D10^2*C10+D11^2*C11+D12^2*C12+D13^2*C13+							
7	μ =	0.04339	D14^2*C14+D15^2*C15+D16^2*C16+J3+J4+J5)							
8					Davies	Activity				
9	Species	pC	C (M)	Charge	log γ	coefficient, γ	F10 = 10^E10		Check:	
10	Fe^{3+}		4.195E-03	3	-7.32E-01	1.855E-01	C10 formula below		Total Fe =	
11	SCN^-	5.69241	2.030E-06	-1	-8.13E-02	8.293E-01	C11=10^-B11		5.000E-03	
12	H^+	1.80130	1.580E-02	1	-8.13E-02	8.293E-01	C12=10^-B12		=C10+C13+C14+C15	
13	$Fe(SCN)^{2+}$		2.969E-06	2	-3.25E-01	4.730E-01	C13=D19*C10*C11			
14	$Fe(SCN)_2^+$		1.060E-10	1	-8.13E-02	8.293E-01	C14=D20*C10*C11^2		Total SCN⁻ =	
15	$FeOH^{2+}$		8.016E-04	2	-3.25E-01	4.730E-01	C15=D21*C10/C12		5.000E-06	
16	OH^-		9.202E-13	-1	-8.13E-02	8.293E-01	C16=D22/C12		=C11+C13+2*C14	
17										
18			K' (with activity coefficients)				Mass and charge balances:			b_i
19	$p\beta_1$ =	-3.03	β_1' =	3.49E+02	b_1 = 0 = F_{Fe} - $[Fe^{3+}]$-$[FeSCN^{2+}]$-$[Fe(SCN)_2^+]$-$[FeOH^{2+}]$ =					-8.06E-12
20	$p\beta_2$ =	-4.6	β_2' =	6.13E+03	b_2 = 0 = F_{SCN} - $[FeSCN^{2+}]$ - 2$[Fe(SCN)_2^+]$ - $[SCN^-]$ =					0.00E+00
21	pK_a =	2.195	K_a' =	3.02E-03	b_3 = 0 = $[OH^-]$ + $[SCN^-]$ + $[NO_3^-]$ - $[H^+]$ - 3$[Fe^{3+}]$ -					
22	pK_w =	14.00	K_w' =	1.45E-14	-2$[FeSCN^{2+}]$-$[Fe(SCN)_2^+]$-2$[FeOH^{2+}]$-$[Na^+]$ =					2.871E-12
23									Σb_i^2 =	7.32E-23
24	Initial values:							J19 = J1-C10-C13-C14-C15		
25	pSCN = 7	pH = 2						J20 = J2-C13-2*C14-C11		
26							J22 = C16+C11+J4-C12-3*C10-2*C13-C14-2*C15-J3			
27								J23 = J19^2 + J20^2 +J22^2		
28							C10 = (J2-C11)/(D19*C11+2*D20*C11^2)			
29						E10 = -0.51*D10^2*(SQRT($B$7)/(1+SQRT($B$7)) - 0.3*$B$7)				
30	Optimize both pSCN and pH together for a few cycles							D19 = (10^-B19)*F10*F11/F13		
31	Then try to optimize just pSCN or just pH while holding the other constant							D20 = (10^-B20)*F10*F11^2/F14		
32	Continue optimization as long as Σb_i^2 keeps getting smaller							D21 = (10^-B21)*F10/(F15*F12)		
33								D22 = (10^-B22)/(F12*F16)		

Spreadsheet for no added KNO_3

	A	B	C	D	E	F	G	H	I	J
1	Iron thiocyanate equilibria with Davies equation for activity coefficients								F_{Fe} =	5.000E-03
2	1. *Estimate* values in cells B11 and B12								F_{SCN} =	5.000E-06
3	2. Use Solver to adjust B11 and B12 to minimize sum in cell J23								$[Na^+]$ =	5.000E-06
4									$[NO_3^-]$ =	2.300E-01
5									$[K^+]$ =	2.000E-01
6	Ionic strength		μ = 0.5*(D10^2*C10+D11^2*C11+D12^2*C12+D13^2*C13+							
7	μ =	0.24391	D14^2*C14+D15^2*C15+D16^2*C16+J3+J4+J5)							
8					Davies	Activity				
9	Species	pC	C (M)	Charge	log γ	coefficient, γ	F10 = 10^E10		Check:	
10	Fe^{3+}		4.452E-03	3	-1.18E+00	6.583E-02	C10 formula below		Total Fe =	
11	SCN^-	5.55092	2.812E-06	-1	-1.31E-01	7.391E-01	C11=10^-B11		5.000E-03	
12	H^+	1.80839	1.555E-02	1	-1.31E-01	7.391E-01	C12=10^-B12		=C10+C13+C14+C15	
13	$Fe(SCN)^{2+}$		2.187E-06	2	-5.25E-01	2.984E-01	C13=D19*C10*C11			
14	$Fe(SCN)_2^+$		6.821E-11	1	-1.31E-01	7.391E-01	C14=D20*C10*C11^2		Total SCN$^-$ =	
15	$FeOH^{2+}$		5.455E-04	2	-5.25E-01	2.984E-01	C15=D21*C10/C12		5.000E-06	
16	OH^-		1.178E-12	-1	-1.31E-01	7.391E-01	C16=D22/C12		=C11+C13+2*C14	
17										
18			K' (with activity coefficients)				Mass and charge balances:			b_i
19	$p\beta_1$ =	-3.03	β_1' =	1.75E+02	b_1 = 0 = F_{Fe} - $[Fe^{3+}]$-$[FeSCN^{2+}]$-$[Fe(SCN)_2^+]$-$[FeOH^{2+}]$ =					-1.74E-14
20	$p\beta_2$ =	-4.6	β_2' =	1.94E+03	b_2 = 0 = F_{SCN} - $[FeSCN^{2+}]$ - 2$[Fe(SCN)_2^+]$ - $[SCN^-]$ =					0.00E+00
21	pK_a =	2.195	K_a' =	1.90E-03	charge: b_3 = 0 = $[OH^-]$ + $[SCN^-]$ + $[NO_3^-]$ - $[H^+]$ - 3$[Fe^{3+}]$ -					
22	pK_w =	14.00	K_w' =	1.83E-14	-2$[FeSCN^{2+}]$-$[Fe(SCN)_2^+]$-2$[FeOH^{2+}]$-$[Na^+]$-$[K^+]$ =					5.256E-13
23									Σb_i^2 =	2.77E-25
24	Initial values:								J19 = J1-C10-C13-C14-C15	
25	pSCN = 7	pH = 2							J20 = J2-C13-2*C14-C11	
26							J22 = C16+C11+J4-C12-3*C10-2*C13-C14-2*C15-J3-J5			
27							J23 = J19^2 + J20^2 +J22^2			
28							C10 = (J2-C11)/(D19*C11+2*D20*C11^2)			
29							E10 = -0.51*D10^2*(SQRT($B$7)/(1+SQRT($B$7)) - 0.3*$B$7)			
30	Optimize both pSCN and pH together for a few cycles						D19 = (10^-B19)*F10*F11/F13			
31	Then try to optimize just pSCN or just pH while holding the other constant						D20 = (10^-B20)*F10*F11^2/F14			
32	Continue optimization as long as Σb_i^2 keeps getting smaller						D21 = (10^-B21)*F10/(F15*F12)			
33							D22 = (10^-B22)/(F12*F16)			

Spreadsheet for 0.20 M KNO_3

13-12. (a) La^{3+} + SO_4^{2-} $\rightleftharpoons$ $La(SO_4)^+$ $\qquad$ $[La(SO_4)^+] = \beta_1'[La^{3+}][SO_4^{2-}]$

$\qquad$ La^{3+} + 2SO_4^{2-} $\rightleftharpoons$ $La(SO_4)_2^-$ $\qquad$ $[La(SO_4)_2^-] = \beta_2'[La^{3+}][SO_4^{2-}]^2$

$\qquad$ La^{3+} + H_2O $\rightleftharpoons$ $LaOH^{2+}$ + H^+ $\qquad$ $[LaOH^{2+}] = K_a'[La^{3+}]/[H^+]$

$\qquad$ H_2O $\rightleftharpoons$ H^+ + OH^- $\qquad$ $[OH^-] = K_w'/[H^+]$

$$\beta_1' = \beta_1 \frac{\gamma_{La^{3+}}\gamma_{SO_4^{2-}}}{\gamma_{La(SO_4)^+}} \qquad \beta_2' = \beta_2 \frac{\gamma_{La^{3+}}\gamma_{SO_4^{2-}}^2}{\gamma_{La(SO_4)_2^-}} \qquad K_a' = K_a \frac{\gamma_{La^{3+}}}{\gamma_{LaOH^{2+}}\gamma_{H^+}}$$

$$K_w' = \frac{K_w}{\gamma_{H^+}\gamma_{OH^-}} \qquad \beta_1 = 10^{3.64} \qquad \beta_2 = 10^{5.3} \qquad K_a = 10^{-8.5}$$

Charge balance:

$[OH^-] + 2[SO_4^{2-}] + [La(SO_4)_2^-] = [H^+] + 3[La^{3+}] + [La(SO_4)^+] + 2[LaOH^{2+}]$

Mass balances:

Lanthanum $\equiv F_{La} = 2.0$ mM $= [La^{3+}] + [La(SO_4)^+] + [La(SO_4)_2^-] + [LaOH^{2+}]$

Sulfate $\equiv F_{SO_4} = 3.0$ mM $= [La(SO_4)^+] + 2[La(SO_4)_2^-] + [SO_4^{2-}]$

Express $[La^{3+}]$ in terms of $[SO_4^{2-}]$:

$[La(SO_4)^+] + 2[La(SO_4)_2^-] + [SO_4^{2-}] = F_{SO_4}$

$\beta_1'[La^{3+}][SO_4^{2-}] + 2\beta_2'[La^{3+}][SO_4^{2-}]^2 + [SO_4^{2-}] = F_{SO_4}$

$[La^{3+}](\beta_1'[SO_4^{2-}] + 2\beta_2'[SO_4^{2-}]^2) = F_{SO_4} - [SO_4^{2-}]$

$$[La^{3+}] = \frac{F_{SO_4} - [SO_4^{2-}]}{\beta_1'[SO_4^{2-}] + 2\beta_2'[SO_4^{2-}]^2}$$

The spreadsheet uses Solver to find all concentrations beginning with estimates for $[SO_4^{2-}]$ and $[H^+]$. Results are

$[La^{3+}] = 0.569$ mM $\qquad [SO_4^{2-}] = 1.50$ mM $\qquad [H^+] = 1.14\ \mu M$

$[La(SO_4)^+] = 1.36$ mM $\qquad [La(SO_4)_2^-] = 66.9\ \mu M \qquad [LaOH^{2+}] = 1.12\ \mu M$

$[OH^-] = 10.5$ nM $\qquad\qquad pH = -\log[H^+]\gamma_{H^+} = 5.98$

(b) If $La_2(SO_4)_3$ were a strong electrolyte, $\mu = \frac{1}{2}\{[La^{3+}]\cdot(+3)^2 + [SO_4^{2-}]\cdot(-2)^2\}$

$= \frac{1}{2}\{(2.0$ mM $\cdot 9) + (3.0$ mM $\cdot 4)\} = 15.0$ mM. The actual ionic strength in cell C17 is 6.3 mM.

(c) $[La^{3+}]/F_{La} = 0.285$

(d) We expected the solution to have a pH near neutral. pK_a for HSO_4^- is 1.99. Therefore, we did not expect very much HSO_4^- to be present. With a pH near 6, the fraction on sulfate that is protonated is $\sim 10^{-4}$.

(e) Evaluate the solubility product expression for $La(OH)_3$:

$[La^{3+}][OH^-]^3\gamma_{La^{3+}}\gamma_{OH^-}^3 = (5.69 \times 10^{-4})(1.05 \times 10^{-8})^3(0.47)(0.92)^3$

$= 2.4 \times 10^{-28} < K_{sp}$ for $La(OH)_3 = 2 \times 10^{-21}$

$La(OH)_3(s)$ does <u>not</u> precipitate.

	A	B	C	D	E	F	G	H	I	J
1	Lanthanum sulfate equilibria with Davies equation for activity coefficients								$F_{La} =$	2.000E-03
2	1. *Estimate* values in cells B11 and B12								$F_{SO4} =$	3.000E-03
3	2. Use Solver to adjust B11 and B12 to minimize sum in cell J23									
4										
5										
6	Ionic strength		μ = 0.5*(D10^2*C10+D11^2*C11+D12^2*C12+D13^2*C13+							
7	μ =	0.00629	D14^2*C14+D15^2*C15+D16^2*C16+J3+J4+J5)							
8					Davies	Activity				
9	Species	pC	C (M)	Charge	log γ	coefficient, γ	F10 = 10^E10		Check:	
10	La^{3+}		5.693E-04	3	-3.29E-01	0.469	C10 formula below		Total La =	
11	SO_4^{2-}	2.82289	1.504E-03	-2	-1.46E-01	0.714	C11=10^-B11		2.000E-03	
12	H^+	5.94333	1.139E-06	1	-3.65E-02	0.919	C12=10^-B12		=C10+C13+C14+C15	
13	$La(SO_4)^+$		1.363E-03	1	-3.65E-02	0.919	C13=D19*C10*C11			
14	$La(SO_4)_2^-$		6.691E-05	-1	-3.65E-02	0.919	C14=D20*C10*C11^2		Total SO_4^{2-} =	
15	$LaOH^{2+}$		1.129E-06	2	-1.46E-01	0.714	C15=D21*C10/C12		3.000E-03	
16	OH^-		1.050E-08	-1	-3.65E-02	0.919	C16=D22/C12		=C11+C13+2*C14	
17										
18			K' (with activity coefficients)				Mass and charge balances:			b_i
19	$p\beta_1$ =	-3.64	β_1' =	1.59E+03	$b_1 = 0 = F_{La} - [La^{3+}]-[LaSO_4^+]-[La(SO_4)_2^-]-[LaOH^{2+}]$ =					-6.31E-13
20	$p\beta_2$ =	-5.3	β_2' =	5.20E+04	$b_2 = 0 = F_{SO4} - [LaSO_4^+] - 2[La(SO_4)_2^-] - [SO_4^{2-}]$ =					0.00E+00
21	pK_a =	8.5	K_a' =	2.26E-09	$b_3 = 0 = [OH^-] + 2[SO_4^{2-}] + [La(SO_4)_2^-]$					
22	pK_w =	13.995	K_w' =	1.20E-14	$- [H^+] - 3[La^{3+}] - [LaSO_4^+] - 2[LaOH^{2+}]$ =					3.819E-13
23									Σb_i^2 =	5.44E-25
24	Initial values:							J19 = J1-C10-C13-C14-C15		
25	pSO_4 = 3	pH = 7						J20 = J2-C13-2*C14-C11		
26							J22 = C16+2*C11+C14-C12-3*C10-C13-2*C15			
27							J23 = J19^2 + J20^2 +J22^2			
28							C10 = (J2-C11)/(D19*C11+2*D20*C11^2)			
29							E10 = -0.51*D10^2*(SQRT($B$7)/(1+SQRT($B$7)) - 0.3*$B$7)			
30	Optimize both pSO_4 and pH together for a few cycles							D19 = (10^-B19)*F10*F11/F13		
31	Then optimize just pSO_4 or just pH while holding the other constant							D20 = (10^-B20)*F10*F11^2/F14		
32	Continue optimization as long as Σb_i^2 keeps getting smaller							D21 = (10^-B21)*F10/(F15*F12)		
33								D22 = (10^-B22)/(F12*F16)		

13-13.

$$AgCN(s) \rightleftharpoons Ag^+ + CN^- \qquad [Ag^+] = K_{sp}'/[CN^-]$$

$$HCN(aq) \rightleftharpoons CN^- + H^+ \qquad [HCN(aq)] = [CN^-][H^+]/K_{HCN}'$$

$$Ag^+ + H_2O \rightleftharpoons AgOH(aq) + H^+ \qquad [AgOH] = K_a'[Ag^+]/[H^+]$$

$$Ag^+ + CN^- + OH^- \rightleftharpoons Ag(OH)(CN)^-$$

$$[Ag(OH)(CN)^-] = K_{Ag}' [Ag^+][CN^-][OH^-]$$

$$Ag^+ + 2CN^- \rightleftharpoons Ag(CN)_2^- \qquad [Ag(CN)_2^-] = \beta_2'[Ag^+][CN^-]^2$$

$$Ag^+ + 3CN^- \rightleftharpoons Ag(CN)_3^{2-} \qquad [Ag(CN)_3^{2-}] = \beta_3'[Ag^+][CN^-]^3$$

$$H_2O \rightleftharpoons H^+ + OH^- \qquad [OH^-] = K_w'/[H^+]$$

$$pH = -\log([H^+]\gamma_{H^+}) \qquad [H^+] = (10^{-pH})/\gamma_{H^+}$$

$$K'_{sp} = \frac{K_{sp}}{\gamma_{Ag^+}\gamma_{CN^-}} \qquad K'_{HCN} = \frac{K_{HCN}}{\gamma_{CN^-}\gamma_{H^+}} \qquad K'_a = K_a \frac{\gamma_{Ag^+}}{\gamma_{H^+}}$$

$$K'_{Ag} = K_{Ag} \frac{\gamma_{Ag^+}\gamma_{CN^-}\gamma_{OH^-}}{\gamma_{Ag(OH)(CN)^-}} \qquad K'_w = \frac{K_w}{\gamma_{H^+}\gamma_{OH^-}}$$

$$\beta'_2 = \beta_2 \frac{\gamma_{Ag^+}\gamma^2_{CN^-}}{\gamma_{Ag(CN)_2^-}} \qquad \beta'_3 = \beta_3 \frac{\gamma_{Ag^+}\gamma^3_{CN^-}}{\gamma_{Ag(CN)_3^{2-}}}$$

$$K_{sp} = 10^{-15.66} \qquad K_{HCN} = 10^{-9.21} \qquad K_a = 10^{-12.0}$$

$$K_{Ag} = 10^{13.22} \qquad \beta_2 = 10^{20.48} \qquad \beta_3 = 10^{21.7}$$

Charge balance:

$$[OH^-] + [CN^-] + [Ag(OH)(CN)^-] + [Ag(CN)_2^-] + 2[Ag(CN)_3^{2-}]$$
$$= [H^+] + [Ag^+] + [K^+] + [Na^+] \qquad (A)$$

Mass balances:

$$[K^+] = 0.10 \text{ M} \qquad (B)$$

{total silver } + $[K^+]$ = {total cyanide}

$$[Ag^+] + [AgOH] + [Ag(OH)(CN)^-] + [Ag(CN)_2^-] + [Ag(CN)_3^{2-}] + [K^+]$$
$$= [CN^-] + [HCN] + [Ag(OH)(CN)^-] + 2[Ag(CN)_2^-] + 3[Ag(CN)_3^{2-}]$$

which simplifies to

$$[Ag^+] + [AgOH] - [Ag(CN)_2^-] - 2[Ag(CN)_3^{2-}] + [K^+] - [CN^-] - [HCN] = 0 \, (C)$$

In the spreadsheet, $[K^+] = 0.10$ M and pH = 12.00 are input in cells J2:J3. Ionic strength is computed in cell B5. Remember to set Excel Options to accept circular definitions. pCN and pNa are the independent variables to be adjusted by Solver in cells B8 and B17 to minimize the mass and charge balances in cell J23. The initial value pNa = 2 comes from ~10^{-2} M NaOH to make pH = 12. The initial value of pCN = 5 was obtained by trying a few integers by hand and inspecting the resulting charge and mass balance sums.

The spreadsheet shows that 99.95% of silver is in the form $Ag(CN)_2^-$.

Fraction of Ag in for the form $[Ag(CN)_2^-]$ =

$$[Ag(CN)_2^-] / \{[Ag^+] + [AgOH] + [Ag(OH)(CN)^-] + [Ag(CN)_2^-] + [Ag(CN)_3^{2-}]\}$$

	A	B	C	D	E	F	G	H	I	J
1	AgCN equilibria with fixed pH and added KCN									
2	1. *Estimate* values in celsl B8 and B17								$[K^+]$ =	0.1
3	2. Use Solver to adjust B8 to minimize sum in cell J21								pH =	12
4	Ionic strength		μ=0.5*(D8^2*C8+D9^2*C9+D10^2*C10+D11^2*C11+D12^2*C12+D13^2*C13							
5	μ =	0.112964	+D14^2*C14+D15^2*C15+D16^2*C16+D17^2*C17+J2) $\leftarrow$ remember $[K^+]$							
6					Davies	Activity		E8 = -0.51*D8^2*(SQRT($B$5)/		
7	Species	pC	C (M)	Charge	log γ	coefficient, γ		(1+SQRT($B$5)) - 0.3*$B$5)		
8	CN^-	5.820043	1.51E-06	-1	-1.110E-01	7.744E-01		C8 = 10^-B8		
9	H^+		1.29E-12	1	-1.110E-01	7.744E-01		C9 = 10^-J3/F9		
10	Ag^+		2.41E-10	1	-1.110E-01	7.744E-01		C10 = D19/C8		
11	HCN		1.90E-09	0	0.000E+00	1.000E+00		C11 = C8*C9/D20		
12	AgOH		1.87E-10	0	0.000E+00	1.000E+00		C12 = D21*C10/C9		
13	$Ag(OH)(CN)^-$		4.69E-05	-1	-1.110E-01	7.744E-01		C13 = D22*C10*C8*C16		
14	$Ag(CN)_2^-$		1.00E-01	-1	-1.110E-01	7.744E-01		C14 = D23*C10*C8^2		
15	$Ag(CN)_3^{2-}$		4.19E-06	-2	-4.440E-01	3.597E-01		C15 = D24*C10*C8^3		
16	OH^-		1.29E-02	-1	-1.110E-01	7.744E-01		C16 = D25/C9		
17	Na^+	1.887417	1.30E-02	1	-0.1110088	7.744E-01		C17 = 10^-B17		
18			K' (with activity coefficients)					Mass and charge balances:		b_i
19	pK_{sp} =	15.66	K_{sp}' =	3.65E-16			$b_1 = 0 = [Ag^+] +[K^+] + [AgOH] - [CN^-]$			
20	pK_{HCN} =	9.21	K_{HCN}' =	1.03E-09			$- [HCN] - [Ag(CN)_2^-] - 2[Ag(CN)_3^{2-}] =$			-7.34E-15
21	pK_a =	12.0	K_a' =	1.00E-12			$b_2 = 0 = [OH^-] + [CN^-] + [Ag(OH(CN)^-] + [Ag(CN)_2^-]$			
22	pK_{Ag} =	-13.22	K_{Ag}' =	9.95E+12			$+2 [Ag(CN)_3^{2-}] - [H^+] - [Ag^+] - [K^+] - [Na^+] =$			2.17E-14
23	$p\beta_2$ =	-20.48	β_2' =	1.81E+20					Σb_i^2 =	5.24E-28
24	$p\beta_3$ =	-21.70	β_3' =	5.01E+21			J20 = C10+J2+C12-C8-C11-C14-2*C15			
25	pK_w =	14.00	K_w' =	1.67E-14			J22 = C16+C8+C13+C14+2*C15-C9-C10-C17-J2			
26								Σb_i^2 =	J23 = J20^2 + J22^2	
27	Initial values:						D19 = (10^-B19)/(F10*F8)			
28	pCN = 5	pNa = 2					D20 = (10^-B20)/(F8*F9)			
29							D21 = (10^-B21)*F10/F9			
30							D22 = (10^-B22)*F10*F8*F16/F13			
31	Optimize both pCN and pNa together for a few cycles						D23 = (10^-B23)*F10*F8^2/F14			
32	Then optimize just pCN or just pNa while holding the other constant						D24 = 10^-B24*F10*F8^3/F15			
33	Continue optimization as long as Σb_i^2 keeps getting smaller						D25 = 10^-B25/(F9*F16)			

13-14.

$$Fe^{2+} + G^- \rightleftharpoons FeG^+ \qquad [FeG^+] = \beta_1'[Fe^{2+}][G^-]$$

$$Fe^{2+} + 2G^- \rightleftharpoons FeG_2(aq) \qquad [FeG_2(aq)] = \beta_2'[Fe^{2+}][G^-]^2$$

$$Fe^{2+} + 3G^- \rightleftharpoons FeG_3^- \qquad [FeG_3^-] = \beta_3'[Fe^{2+}][G^-]^3$$

$$Fe^{2+} + H_2O \rightleftharpoons FeOH^+ + H^+ \qquad [FeOH^+] = K_a'[Fe^{2+}]/[H^+]$$

$$HG \rightleftharpoons G^- + H^+ \qquad [HG] = [G^-][H^+]/K_2'$$

$$H_2G^+ \rightleftharpoons HG + H^+ \qquad [H_2G^+] = [HG][H^+]/K_1'$$

$$H_2O \rightleftharpoons H^+ + OH^- \qquad [OH^-] = K_w'/[H^+]$$

$$\beta_1' = \beta_1 \frac{\gamma_{Fe^{2+}}\gamma_{G^-}}{\gamma_{FeG^+}} \qquad \beta_2' = \beta_2 \frac{\gamma_{Fe^{2+}}\gamma_{G^-}^2}{\gamma_{FeG_2}} \qquad \beta_3' = \beta_3 \frac{\gamma_{Fe^{2+}}\gamma_{G^-}^3}{\gamma_{FeG_3^-}}$$

$$K_a' = K_a \frac{\gamma_{Fe^{2+}}}{\gamma_{H^+}} \qquad K_2' = K_2 \frac{\gamma_{HG}}{\gamma_{G^-}\gamma_{H^+}} \qquad K_1' = K_1 \frac{\gamma_{H_2G^+}}{\gamma_{HG}\gamma_{H^+}} \qquad K_w' = \frac{K_w}{\gamma_{H^+}\gamma_{OH^-}}$$

$$\beta_1 = 10^{4.31} \qquad\qquad \beta_2 = 10^{7.65} \qquad \beta_3 = 10^{8.87}$$

$$K_a = 10^{-9.4} \qquad\qquad K_1 = 10^{-2.350} \qquad K_2 = 10^{-9.778}$$

Charge balance:

$$[OH^-] + [G^-] + [FeG_3^-] + [Cl^-] =$$

$$[H^+] + 2[Fe^{2+}] + [FeOH^+] + [FeG^+] + [H_2G^+] \qquad (A)$$

Mass balances:

$$F_{Fe} = 0.050\ M = [Fe^{2+}] + [FeG^+] + [FeG_2] + [FeG_3^-] + [FeOH^+] \qquad (B)$$

$$F_G = 0.100\ M = [G^-] + [HG] + [H_2G^+] + [FeG^+] + 2[FeG_2] + 3[FeG_3^-] \quad (C)$$

There are 11 concentrations to be found and there are 7 equilibria. In the absence of other information, we would need to solve for 4 independent concentrations. However, $[H^+]$ is known from the pH, so we need to find 3 independent variables. The amount of HCl added to the solution to achieve the pH is not known, so $[Cl^-]$ is one of the independent variables to be found by Solver. For the other two independent variables, we select $[Fe^{2+}]$ and $[G^-]$ because all of other concentrations can be expressed in terms of $[Fe^{2+}]$, $[G^-]$, and $[H^+]$. In the spreadsheet, we had to try a few different initial values of pFe, pG, and pCl to find the set pFe = 3, pG = 3, and pCl = 2 that roughly satisfied the mass and charge balances. With these initial values, Solver was able to solve the problem in a single iteration.

The fraction of iron and glycine in each form is computed from the concentrations in the spreadsheet:

Iron		Glycine	
Fe^{2+}	3.49%	G^-	0.95%
FeG^+	37.40%	FeG^+	18.70%
FeG_2	57.95%	FeG_2	57.95%
FeG_3^-	0.92%	FeG_3^-	1.38%
$FeOH^+$	0.24%	HG	21.02%
		H_2G^+	0.00%

The chemistry: We dissolved FeG_2 and found that the principal species are FeG^+, FeG_2, and HG. The chemistry that requires 20.9 mmol HCl to be added to 50 mmol FeG_2 to obtain pH 8.5 is $FeG_2 \rightleftharpoons FeG^+ + G^-$ followed by $G^- + H^+ \rightleftharpoons HG$. The base G^- released when FeG_2 dissolves requires HCl for neutralization.

	A	B	C	D	E	F	G	H	I
1	Fe(II)-glycine equilibria with fixed pH							pH =	8.50
2	1. *Estimate* values in cells B8. B9, B18							F_{Fe} =	0.050
3	2. Use Solver to adjust B8,B9,B17 to minimize sum in cell I24							F_G =	0.100
4	Ionic strength		μ=0.5*(D8^2*C8+D9^2*C9+D10^2*C10+D11^2*C11+D12^2*C12+D13^2*C13						
5	μ =	0.024051	+D14^2*C14+D15^2*C15+D16^2*C16+D17^2*C17+D18^2*C18)						
6									check mass
7	Species	pC	C (M)	Charge	Davies log γ	Activity coefficient, γ			balances:
8	Fe^{2+}	2.758698	1.74E-03	2	-2.592E-01	5.506E-01	C8 = 10^-B8		Fe_{total} =
9	G^-	3.020251	9.54E-04	-1	-6.479E-02	8.614E-01	C9 = 10^-B9		5.00E-02
10	H^+	known	3.67E-09	1	-6.479E-02	8.614E-01	C10 = 10^-I1/F10		G_{total} =
11	FeG^+		1.87E-02	1	-6.479E-02	8.614E-01	C11 = D20*C8*C9		1.00E-01
12	FeG_2		2.90E-02	0	0.000E+00	1.000E+00	C12 = D21*C8*C9^2		
13	FeG_3^-		4.59E-04	-1	-6.479E-02	8.614E-01	C13 = D22*C8*C9^3		
14	$FeOH^+$		1.21E-04	1	-6.479E-02	8.614E-01	C14 =D23*C8/C10		
15	HG		2.10E-02	0	0.000E+00	1.000E+00	C15 = C9*C10/D25		
16	H_2G^+		1.28E-08	1	-6.479E-02	8.614E-01	C16 = C15*C10/D24		
17	OH^-	known	3.67E-06	-1	-0.0647941	8.614E-01	C17 = D26/C10		
18	Cl^-	1.680037	2.09E-02	-1	-0.0647941	8.614E-01	C18 = 10^-B18		
19			K' (with activity coefficients)			Mass and charge balances:			b_i
20	$pβ_1$ =	-4.31	$β_1'$ =	1.12E+04	b_1=0=F_{Fe}-[Fe^{2+}]-[FeG^+]-[FeG_2]-[FeG_3^-]-[$FeOH^+$]				-1.27E-10
21	$pβ_2$ =	-7.65	$β_2'$ =	1.82E+07	b_2=0=F_G-[G^-]-[HG]-[H_2G^+]-[FeG^+]-2[FeG_2]-3[FeG_3^-]				8.37E-11
22	$pβ_3$ =	-8.9	$β_3'$ =	3.03E+08	b_3 = 0 = [OH^-]+ [G^-]+ [FeG_3^-] + [Cl^-]				
23	pK_a =	9.40	K_a' =	2.54E-10	- [H^+] -2[Fe^{2+}] - [$FeOH^+$] - [FeG^+] - [H_2G^+] =				-2.41E-12
24	pK_1 =	2.35	K_1' =	6.02E-03				$Σb_i^2$ =	2.31E-20
25	pK_2 =	9.78	K_2' =	1.67E-10		I20 = I2-C8-C11-C12-C13-C14			
26	pK_w =	14.00	K_w' =	1.35E-14		I21 = I3-C9-C15-C16-C11-2*C12-3*C13			
27						I23 = C17+C9+C13+C18-C10-2*C8-C14-C11-C16			
28	Initial values:					I24 = I20^2 + I21^2 + I23^2			
29	pFe = 3	pG = 3	pCl = 2			E8 = -0.51*D8^2*(SQRT($B$5)/(1+SQRT($B$5))) - 0.3*$B$5)			
30						D20 = 10^-B20*F8*F9/F11			
31						D21 = 10^-B21*F8*F9^2/F12			
32						D22 =10^-B22*F8*F9^3/F13			
33						D23 = 10^-B23*F8/F10			
34	Optimize pFe, pG, and pCl together					D24 = 10^-B24*F15/(F9*F10)			
35	Then optimize one or two p values while holding other(s) constant					D25 = 10^-B25*F16/(F15*F10)			
36	Continue optimization as long as $Σb_i^2$ keeps getting smaller					D26 = 10^-B26/(F10*F17)			

13-15. (a) A range of initial values of pK_1 and pK_2, such as 6 and 6 or 10 and 10, converge to the correct solution. Even choosing a ridiculous value for pK'_w, such as 10, converges to the correct value after executing Solver more than once.

    (b) Fixing pK'_w at 13.797 and using Solver gives the optimized values of pK_1 and $pK_2 = 2.312$ and 9.630, which are hardly different from the values obtained when pK'_w is allowed to vary. However, inspection of the following curves shows that $\bar{n}_H$(measured) deviates systematically from $\bar{n}_H$(theoretical) at the end of the titration when $\bar{n}_H$(measured) should approach 0.

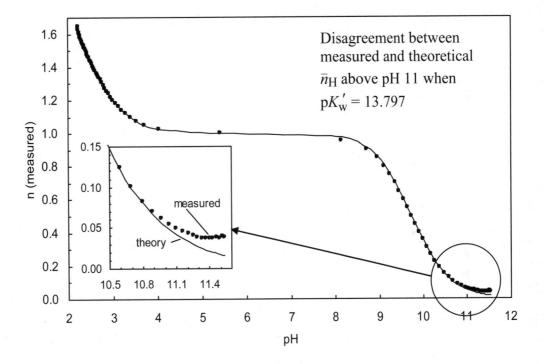

13-16. (a) $\quad \bar{n}_H = \dfrac{\text{moles of bound H}^+}{\text{total moles of weak acid}} = \dfrac{3[H_3A^{3+}] + 2[H_2A^{2+}] + [HA^+]}{[H_3A^{3+}] + [H_2A^{2+}] + [HA^+] + [A]}$

or $\quad \bar{n}_H F_{H_3A} = 3[H_3A^{3+}] + 2[H_2A^{2+}] + [HA^+]$ (A)

where $F_{H_3A} = [H_3A^{3+}] + [H_2A^{2+}] + [HA^+] + [A]$

charge balance:

$$[H^+] + [Na^+] + \underbrace{3[H_3A^{3+}] + 2[H_2A^{2+}] + [HA^+]}_{\displaystyle = \bar{n}_H F_{H_3A} \text{ from Eq. (A)}} = [OH^-] + [Cl^-]_{HCl} + \underbrace{[Cl^-]_{H_3A}}_{\displaystyle = 3F_{H_3A}}(B)$$

where $[Cl^-]_{HCl}$ is from HCl and $[Cl^-]_{H_3A}$ is from H_3A^{3+}.

Each mol of H_3A^{3+} brings $3Cl^-$, so $[Cl^-]_{H_3A} = 3F_{H_3A}$

We can rearrange Eq. (B) to solve for $\bar{n}_H$:

$$[H^+] + [Na^+] + \bar{n}_H F_{H_3A} = [OH^-] + [Cl^-]_{HCl} + 3F_{H_3A}$$

$$\bar{n}_H F_{H_3A} = 3F_{H_3A} + [OH^-] + [Cl^-]_{HCl} - [H^+] - [Na^+]$$

$$\bar{n}_H = 3 + \frac{[OH^-] + [Cl^-]_{HCl} - [H^+] - [Na^+]}{F_{H_3A}} \qquad (C)$$

Expression (C) is the same equation derived in the text, with $n = 3$.

The expression for $\bar{n}_H$(theoretical) is $\bar{n}_H$(theoretical) $= 3\alpha_{H_3A} + 2\alpha_{H_2A} + \alpha_{HA}$.

(b) Optimized values are $pK_w' = 13.819$, $pK_1 = 8.33$, $pK_2 = 9.48$, and $pK_3 = 10.19$ in cells B9:B12 in the following spreadsheet. These were obtained from initial guesses of $pK_w' = 13.797$, $pK_1 = 8$, $pK_2 = 9$, and $pK_3 = 10$, after executing Solver to minimize the sum of squared residuals in column K. The NIST database lists $pK_1 = 7.85$, $pK_2 = 9.13$, and $pK_3 = 10.03$ at $\mu = 0$ and $pK_1 = 8.42$, $pK_2 = 9.43$, and $pK_3 = 10.13$ at $\mu = 0.1$ M. Our observed values at $\mu = 0.1$ M are in reasonable agreement with the NIST values.

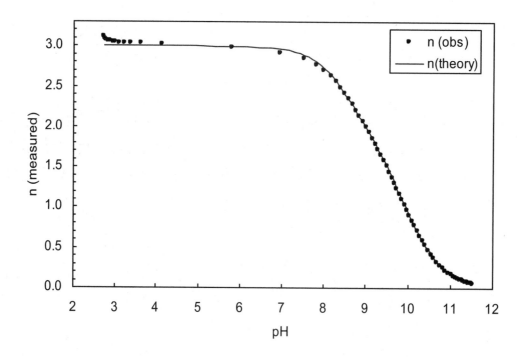

	A	B	C	D	E	F	G	H	I	J	K
1	Difference plot for tris(2-aminoethyl)amine										
2				C17 = 10^-B17/$B$8				H17 = 10^-$B$9/C17			
3	Titrant NaOH =	0.4905	C_b (M)	E17 = $B$7+($B$6-$B$3*A17-(C17-D17)*($B$4+A17))/$B$5							
4	Initial volume =	40	V_o (mL)	denom = $C17^3+$C17^2*$E$10+$C17*$E$10*$E$11+$E$10*$E$11*$E$12							
5	H_3A =	0.139	L (mmol)	F17 = $C17^3/denom							
6	HCl added =	0.115	A (mmol)	G17 = $C17^2*$E$10/denom							
7	Number of H^+=	3	n	H17 = $C17*$E$10*$E$11/denom							
8	Activity coeff =	0.78	γ_H	I17 = $E$10*$E$11*$E$12/denom							
9	pK_w' =	13.819		J17 = 3*F17+2*G17+H17							
10	pK_1 =	8.334		K_1 =	4.636E-09	= 10^-B10					
11	pK_2 =	9.483		K_2 =	3.289E-10	= 10^-B11					
12	pK_3 =	10.188		K_3 =	6.485E-11	= 10^-B12					
13	Σ(resid)^2 =	0.0510	= sum of column K								
14											
15	v	pH	[H^+] =	[OH] =	Measured	α_{H3A}	α_{H2A}	α_{HA}	α_A	Theoretical	(residuals)^2 =
16	mL NaOH		(10^{-pH})/γ_H	(10^{-pKw})/[H^+]	n_H					n_H	$(n_{meas} - n_{thoer})^2$
17	0.00	2.709	2.51E-03	6.06E-12	3.106	1.000	0.000	0.000	0.000	3.000	0.011303
18	0.02	2.743	2.32E-03	6.55E-12	3.090	1.000	0.000	0.000	0.000	3.000	0.008046
19	:										
20	0.34	8.158	8.91E-09	1.70E-06	2.628	0.649	0.338	0.012	0.000	2.637	0.000077
21	0.36	8.283	6.68E-09	2.27E-06	2.558	0.579	0.401	0.020	0.000	2.558	0.000001
22	:										
23	0.54	9.087	1.05E-09	1.45E-05	1.926	0.145	0.641	0.201	0.012	1.919	0.000045
24	0.56	9.158	8.91E-10	1.70E-05	1.856	0.121	0.630	0.232	0.017	1.855	0.000002
25	:										
26	0.78	9.864	1.75E-10	8.66E-05	1.100	0.010	0.277	0.520	0.192	1.106	0.000031
27	0.80	9.926	1.52E-10	9.98E-05	1.034	0.008	0.243	0.525	0.224	1.035	0.000001
28	:										
29	1.00	10.545	3.66E-11	4.15E-04	0.421	0.000	0.039	0.346	0.615	0.424	0.000011
30	1.02	10.615	3.11E-11	4.88E-04	0.372	0.000	0.030	0.314	0.656	0.375	0.000007
31	:										
32	1.38	11.496	4.09E-12	3.71E-03	0.062	0.000	0.001	0.059	0.940	0.061	0.000002
33	1.40	11.521	3.86E-12	3.93E-03	0.057	0.000	0.001	0.056	0.943	0.058	0.000000

(c) The fractions of each species are computed in columns F, G, H, and I beginning in row 17 in the spreadsheet. Results are shown in the graph.

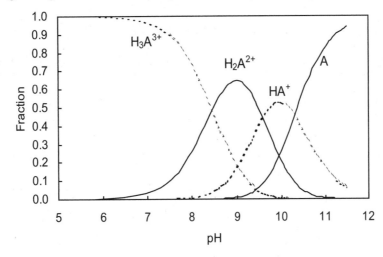

13-17. (a) $[Na^+] + [NaT^-] + [NaHT] = F_{Na} = F_{H_2T}$

Substitute $[NaT^-] = K_{NaT^-}[Na^+][T^{2-}]$ and $[NaHT] = K_{NaHT}[Na^+][HT^-]$:

$[Na^+] + K_{NaT^-}[Na^+][T^{2-}] + K_{NaHT}[Na^+][HT^-] = F_{H_2T}$

$[Na^+]\{1 + K_{NaT^-}[T^{2-}] + K_{NaHT}[HT^-]\} = F_{H_2T}$

$$[Na^+] = \frac{F_{H_2T}}{1 + K_{NaT^-}[T^{2-}] + K_{NaHT}[HT^-]} \qquad (A)$$

(b) $[H_2T] + [HT^-] + [T^{2-}] + [NaT^-] + [NaHT] = F_{H_2T}$

Make the following substitutions to find expressions in terms of $[T^{2-}]$, $[Na^+]$, and $[H^+]$:

$$[H_2T] = \frac{[H^+]^2}{K_1K_2}[T^{2-}]; \qquad [HT^-] = \frac{[H^+]}{K_2}[T^{2-}]; \qquad [NaT^-] = K_{NaT^-}$$

$[Na^+][T^{2-}]$

$$[NaHT] = K_{NaHT}[Na^+][HT^-] = K_{NaHT}[Na^+]\frac{[H^+]}{K_2}[T^{2-}]$$

Then put these expressions into the mass balance:

$$\frac{[H^+]^2}{K_1K_2}[T^{2-}] + \frac{[H^+]}{K_2}[T^{2-}] + [T^{2-}] + K_{NaT^-}[Na^+][T^{2-}] + K_{NaHT}[Na^+]\frac{[H^+]}{K_2}[T^{2-}] = F_{H_2T}$$

and solve for $[T^{2-}]$:

$$[T^{2-}] = \frac{F_{H_2T}}{\dfrac{[H^+]^2}{K_1K_2} + \dfrac{[H^+]}{K_2} + 1 + K_{NaT^-}[Na^+] + K_{NaHT}[Na^+]\dfrac{[H^+]}{K_2}} \qquad (B)$$

(c) To find $[HT^-]$, make the following substitutions in the mass balance for H_2T:

$$[H_2T] = \frac{[H^+]}{K_1}[HT^-]; \quad [T^{2-}] = \frac{K_2}{[H^+]}[HT^-]; \quad [NaHT] = K_{NaHT}[Na^+][HT^-]$$

$$[NaT^-] = K_{NaT^-}[Na^+][T^{2-}] = K_{NaT^-}[Na^+]\frac{K_2}{[H^+]}[HT^-]$$

Then put these expressions into the mass balance:

$$\frac{[H^+]}{K_1}[HT^-] + [HT^-] + \frac{K_2}{[H^+]}[HT^-] + K_{NaT^-}[Na^+]\frac{K_2}{[H^+]}[HT^-] + K_{NaHT}[Na^+][HT^-] = F_{H_2T}$$

and solve for $[HT^-]$:

$$[HT^-] = \frac{F_{H_2T}}{\dfrac{[H^+]}{K_1} + 1 + \dfrac{K_2}{[H^+]} + K_{NaT^-}[Na^+]\dfrac{K_2}{[H^+]} + K_{NaHT}[Na^+]} \qquad (C)$$

To find $[H_2T]$, make the following substitutions in the mass balance for H_2T:

$$[HT^-] = \frac{K_1}{[H^+]}[H_2T]; \qquad [T^{2-}] = \frac{K_1K_2}{[H^+]^2}[H_2T]$$

$$[NaT^-] = K_{NaT^-}[Na^+][T^{2-}] = K_{NaT^-}[Na^+]\frac{K_1K_2}{[H^+]^2}[H_2T]$$

$$[\text{NaHT}] = K_{\text{NaHT}}[\text{Na}^+][\text{HT}^-] = K_{\text{NaHT}}[\text{Na}^+]\frac{K_1}{[\text{H}^+]}[\text{H}_2\text{T}]$$

Then put these expressions into the mass balance:

$$[\text{H}_2\text{T}] + \frac{K_1}{[\text{H}^+]}[\text{H}_2\text{T}] + \frac{K_1 K_2}{[\text{H}^+]^2}[\text{H}_2\text{T}] + K_{\text{NaT}^-}[\text{Na}^+]\frac{K_1 K_2}{[\text{H}^+]^2}[\text{H}_2\text{T}] + K_{\text{NaHT}}[\text{Na}^+]\frac{K_1}{[\text{H}^+]}[\text{H}_2\text{T}]$$
$$= F_{\text{H}_2\text{T}}$$

and solve for $[\text{H}_2\text{T}]$:

$$[\text{H}_2\text{T}] = \frac{F_{\text{H}_2\text{T}}}{1 + \dfrac{K_1}{[\text{H}^+]} + \dfrac{K_1 K_2}{[\text{H}^+]^2} + K_{\text{NaT}^-}[\text{Na}^+]\dfrac{K_1 K_2}{[\text{H}^+]^2} + K_{\text{NaHT}}[\text{Na}^+]\dfrac{K_1}{[\text{H}^+]}} \qquad \text{(D)}$$

(d) Insert equations A, B, C, and D into the following spreadsheet to compute $[\text{Na}^+]$, $[\text{H}_2\text{T}]$, $[\text{HT}^-]$, and $[\text{T}^{2-}]$ in cells B12, H10, E11, and E12. Excel indicates a circular reference problem with these new formulas. In Excel 2007, click the Microsoft Office button at the upper left of the spreadsheet. Click on Excel Options at the bottom of the window. Select Formulas. In Calculation options, check Enable iterative calculation and set Maximum Change to 1e-16. Click OK. (In earlier versions of Excel, go to the Tools menu and choose Options. Select Calculation and choose Iteration. Set the maximum change to 1e-16 and click OK.) Guess a pH (such as 6) in cell H13. Select Solver and Set Target Cell E15 Equal To Value of 0 By Changing Cells H13. Click Solve and your spreadsheet should find the concentrations in the spreadsheet on the next page.

	A	B	C	D	E	F	G	H	I
1	Mixture of 0.020 M Na^+HT^-, 0.015 M PyH^+Cl^-, and 0.010 M KOH								
2	With some Na^+ ion pairs								
3	F_{H2T} =	0.020		F_{PyH+} =	0.015		$[K^+]$ =	0.010	
4	pK_1 =	3.036		pK_a =	5.20		K_w =	1.00E-14	
5	pK_2 =	4.366		K_a =	6.31E-06		K_{NaT-} =	8	
6	K_1 =	9.20E-04					K_{NaHT} =	1.6	
7	K_2 =	4.31E-05							
8									
9	Species in charge balance:						Other concentrations:		
10	$[H^+]$ =	5.45E-05		$[OH^-]$ =	1.84E-10		$[H_2T]$ =	5.93E-04	
11	$[PyH^+]$ =	1.34E-02		$[HT^-]$ =	1.00E-02		$[Py]$ =	1.56E-03	
12	$[Na^+]$ =	0.0185		$[T^{2-}]$ =	7.92E-03		$[NaHT]$ =	2.97E-04	
13	$[K^+]$ =	0.0100		$[Cl^-]$ =	0.0150		pH =	4.264	← initial value
14				$[NaT^-]$ =	1.17E-03				is a guess
15	Positive charge minus negative charge =				-3.69E-18				
16					E15 = B10+B11+B12+B13-E10-E11-2*E12-E13-E14				
17	Check: $[PH^+] + [P]$ =				0.01500				
18	Check: $[H_2T]+[HT^-]+[T^{2-}]+[NaT^-]+[NaHT]$ =				0.02000				
19	Check: $[Na^+]+[NaT^-]+[NaHT]$ =				0.02000				
20	Formulas:								
21	B6 = 10^-B4			B7 = 10^-B5		E5 = 10^-E4	E10 = H4/B10		
22	B10 = 10^-H13				B13 = H3		E13 = E3		
23	E11 = B3/(B10/B6 + 1 + B7/B10 + H5*B12*B7/B10 + H6*B12)								
24	E12 = B3/((B10^2/(B6*B7))+(B10/B7)+1+H5*B12+H6*B12*B10/B7)								
25	H10 = B3/(1 + B6/B10 + B6*B7/B10^2 + H5*B12*B6*B7/B10^2 + H6*B12*B6/B10)								
26	B12 = B3/(1 + H5*E12 + H6*E11)						B11 = B10*E3/(B10+E5)		
27	E14 = H5*B12*E12						H11 = E5*E3/(B10+E5)		
28	H12 = H6*B12*E11								
29									
30									
31	Na distribution (%)			H_2T distribution (%)					
32	$[Na^+]$ =	92.6		$[H_2T]$ =	3.0				
33	$[NaT^-]$ =	5.9		$[HT^-]$ =	50.1				
34	$[NaHT]$ =	1.5		$[T^{2-}]$ =	39.6				
35				$[NaT^-]$ =	5.9				
36				$[NaHT]$ =	1.5				

CHAPTER 14
FUNDAMENTALS OF ELECTROCHEMISTRY

14-1. Electric charge (coulombs) refers to the quantity of positive or negative particles. Current (amperes) is the quantity of charge moving past a point in a circuit each second. Electric potential (volts) measures the work that can be done by (or must be done to) each coulomb of charge as it moves from one point to another.

14-2. (a) $1/1.602\,176\,53 \times 10^{-19}$ C/electron $= 6.241\,509\,48 \times 10^{18}$ electrons/C

   (b) $F = 96\,485.338\,3$ C/mol

14-3. (a) $I =$ coulombs/s. Every mol of O_2 accepts 4 mol of e^-. 16 mol O_2/day $= 64$ mol e^-/day $= 7.41 \times 10^{-4}$ mol e^-/s $= 71._5$ C/s $= 71._5$ A

   (b) $I =$ Power/$E = 500$ W/115 V $= 4.35$ A. The resting human uses 16 times as much current as the refrigerator.

   (c) Power $= E \cdot I = (1.1$ V$)(71._5$ A$) = 79$ W

14-4. (a) $I = \dfrac{6.00 \text{ V}}{2.0 \times 10^3 \text{ W}} = 3.00$ mA $= 3.00 \times 10^{-3}$ C/s

   $\left(\dfrac{3.00 \times 10^{-3} \text{ C/s}}{9.649 \times 10^4 \text{ C/mol}}\right)(6.022 \times 10^{23} \text{ } e^-\text{/mole}) = 1.87 \times 10^{16} \text{ } e^-\text{/s}$

   (b) $P = E \cdot I = (6.00$ V$)(3.00 \times 10^{-3}$ A$) = 1.80 \times 10^{-2}$ W
   The power appears as heat in the resistor.
   Heat per electron $= \dfrac{1.80 \times 10^{-2} \text{ J/s}}{1.87 \times 10^{16} \text{ } e^-\text{/s}} = 9.63 \times 10^{-19}$ J/e^-

   (c) $(1.87 \times 10^{16} \text{ } e^-\text{/s})(1\,800 \text{ s}) = 3.37 \times 10^{19}$ electrons $= 5.60 \times 10^{-5}$ mol

   (d) $P = EI = E(E/R) = E^2/R \Rightarrow E = \sqrt{PR} = \sqrt{(100 \text{ W})(2.00 \times 10^3 \text{ W})} = 447$ V

14-5. (a) $I_2 + 2e^- \rightleftharpoons 2I^-$ (I_2 is the oxidant)

   (b) $2S_2O_3^{2-} \rightleftharpoons S_4O_6^{2-} + 2e^-$ ($S_2O_3^{2-}$ is the reductant)

   (c) 1.00 g $S_2O_3^{2-}/(112.13$ g/mol$) = 8.92$ mmol $S_2O_3^{2-} = 8.92$ mmol e^-
   $(8.92 \times 10^{-3}$ mol$)(9.649 \times 10^4$ C/mol$) = 861$ C

   (d) Current (A) $=$ coulombs/s $= 861$ C/60 s $= 14.3$ A

14-6. (a) Oxidation numbers of reactants: N (in NH_4^+)   Cl (in ClO_4^-) Al

   -3 $+7$ 0

   Oxidation numbers of products: N (in N_2)   Cl (in HCl)   Al (in Al_2O_3)

   0   -1 $+3$

   NH_4^+ and Al are reducing agents and ClO_4^- is the oxidizing agent.

(b) Formula mass of reactants = 6(FM NH_4ClO_4) + 10(FM Al) = 974.75

   Heat released per gram = 9 334 kJ/974.75 g = 9.576 kJ/g

14-7. In a galvanic cell, two half-reactions are physically separated from each other. At the anode, oxidation generates electrons that can flow through the electric circuit to reach the cathode, where a reduction occurs. The favorable free energy change for the net reaction provides the driving force for electrons to flow through the circuit. There must be a connector (such as a salt bridge) between the anode and cathode compartments to allow ions to flow to maintain electroneutrality.

14-8. (a) $Fe(s) \mid FeO(s) \mid KOH(aq) \mid Ag_2O(s) \mid Ag(s)$

   $FeO(s) + H_2O + 2e^- \rightleftharpoons Fe(s) + 2OH^-$

   $Ag_2O + H_2O + 2e^- \rightleftharpoons 2Ag(s) + 2OH^-$

(b) $Pb(s) \mid PbSO_4(s) \mid K_2SO_4(aq) \mid\mid H_2SO_4(aq) \mid PbSO_4(s) \mid PbO_2(s) \mid Pb(s)$

   $PbSO_4(s) + 2e^- \rightleftharpoons Pb(s) + SO_4^{2-}$

   $PbO_2 + 4H^+ + SO_4^{2-} + 2e^- \rightleftharpoons PbSO_4(s) + 2H_2O$

14-9. $Fe^{3+} + e^- \rightleftharpoons Fe^{2+}$

   $Cr_2O_7^{2-} + 14H^+ + 6e^- \rightleftharpoons 2Cr^{3+} + 7H_2O$

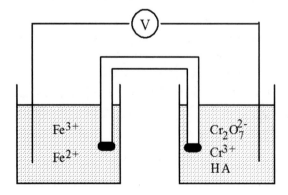

14-10. (a) $Zn^{2+} + 2e^- \rightleftharpoons Zn(s)$ $E° = -0.762$ V

$Cl_2(l) + 2e^- \rightleftharpoons 2Cl^-$ $E° \approx 1.4$ V

The Appendix lists standard reduction potentials for $Cl_2(g)$ and $Cl_2(aq)$, but not for $Cl_2(l)$. Both listed potentials are close to 1.4 V, so the potential for $Cl_2(l)$ is probably also close to 1.4 V. Electrons flow from the more negative electrode (Zn) through the circuit to the more positive electrode (C).

(b)   One mol of Cl_2 requires 2 mol of e^-.

Moles of Cl_2 consumed in 1.00 h $= \frac{1}{2}$ (mol of e^-/hr) $=$

$\left[\frac{1}{2}\left(1.00 \times 10^3 \frac{C}{s}\right)/(9.64 \times 10^4 \text{ C/mol})\right](3\,600 \text{ s/h}) = 18.7 \text{ mol } Cl_2 = 1.32 \text{ kg.}$

14-11. (a)   anode: $C_6Li \rightleftharpoons C_6 + Li^+ + e^-$

cathode: $2Li_{0.5}CoO_2 + Li^+ + e^- \rightleftharpoons 2LiCoO_2$

Formula mass of reactants $= 188.832$ g $(2Li_{0.5}CoO_2) + 79.032$ g $(LiC_6) = 267.863$ g

(b) $1 \text{ A} = 1 \frac{C}{s}$. $1 \text{ h} = 3\,600$ s; $1 \text{ A·h} = \left(1 \frac{C}{s}\right)(3\,600 \text{ s}) = 3\,600$ C

$(3\,600 \text{ C})/(96\,485 \text{ C/mol}) = 0.037\,311$ mol e^- in 1 A·h

(c)   0.267 863 kg of reactants $(2Li_{0.5}CoO_2 + LiC_6)$ provides 1 mol e^-.

$(1 \text{ mol } e^-)/(0.267\,863 \text{ kg}) = 3.733\,3 \text{ mol } e^-/\text{kg}$

$\left(3.733\,3 \frac{\text{mol } e^-}{\text{kg}}\right) \div \left(0.037\,311 \frac{\text{mol } e^-}{\text{A·h}}\right) = 100.1 \text{ A·h/kg}$

(d)   Power (W) $= E·I =$ volts $\times$ amperes. $(3.7 \text{ V})(100 \text{ A·h/kg}) = 370$ W·h/kg

14-12. Cl_2 is strongest because it has the most positive reduction potential.

14-13. (a)   Since it becomes harder to reduce Fe(III) to Fe(II) in the presence of CN^-, Fe(III) is stabilized more than Fe(II).

(b)   Since it becomes easier to reduce Fe(III) to Fe(II) in the presence of phenanthroline, Fe(II) is stabilized more than Fe(III) by phenanthroline.

14-14. $E°$ applies when activities of reactants and products are unity. E applies to whatever activities exist. At equilibrium, E goes to zero. $E°$ is constant.

14-15. (a) $Zn(s) \mid Zn^{2+}(0.1 \text{ M}) \parallel Cu^{2+}(0.1 \text{ M}) \mid Cu(s)$

right half-cell: $Cu^{2+} + 2e^- \rightleftharpoons Cu(s)$ $E_+^{\circ} = 0.339 \text{ V}$

left half-cell: $Zn^{2+} + 2e^- \rightleftharpoons Zn(s)$ $E_-^{\circ} = -0.762 \text{ V}$

$$E = \left(0.339 - \frac{0.059\,16}{2} \log \frac{1}{0.1}\right) - \left(-0.762 - \frac{0.059\,16}{2} \log \frac{1}{0.1}\right) = 1.101 \text{ V}$$

The Cu electrode is more positive than the Zn electrode. Electrons move from Zn to Cu. The net reaction is $Cu^{2+} + Zn(s) \rightleftharpoons Cu(s) + Zn^{2+}$.

(b)   Since Cu^{2+} ions are consumed in the right half-cell, Zn^{2+} ions must migrate from the left half-cell into the salt bridge to help balance charge. I hope you like Zn^{2+}, because that is what your body will take up.

14-16. $E = -0.238 - \dfrac{0.059\,16}{3} \log \dfrac{P_{AsH_3}}{[H^+]^3}$

$$E = -0.238 - \frac{0.059\,16}{3} \log \frac{1.0 \times 10^{-3}}{(10^{-3.00})^3} = -0.356 \text{ V}$$

14-17. (a) $Pt(s) \mid Br_2(l) \mid HBr(aq, 0.10 \text{ M}) \parallel Al(NO_3)_3(aq, 0.010 \text{ M}) \mid Al(s)$

(b)   right half-cell: $Al^{3+} + 3e^- \rightleftharpoons Al(s)$ $E_+^{\circ} = -1.677 \text{ V}$

left half-cell: $Br_2(l) + 2e^- \rightleftharpoons 2Br^-$ $E_-^{\circ} = 1.078 \text{ V}$

right half-cell: $E_+ = \left(-1.677 - \dfrac{0.059\,16}{3} \log \dfrac{1}{[0.010]}\right) = -1.716_4 \text{ V}$

left half-cell: $E_- = \left(1.078 - \dfrac{0.059\,16}{2} \log [0.10]^2\right) = 1.137_2 \text{ V}$

$E = E_+ - E_- = -1.716_4 - 1.137_2 = -2.854 \text{ V}$. The Pt electrode is more positive, so electrons flow from Al to Pt. Br_2 is reduced at the Pt electrode:

$$\tfrac{3}{2} Br_2(l) + Al(s) \rightleftharpoons 3Br^- + Al^{3+}$$

The species reacting at Pt is $Br_2(aq)$, which is in equilibrium with $Br_2(l)$.

(c)   14.3 mL of Br_2 = 44.6 g = 0.279 mol of Br_2. 12.0 g of Al = 0.445 mol of Al. The reaction requires 3/2 mol of Br_2 for every mol of Al. The Br_2 will be used up first.

(d)   0.231 mL of Br_2 = 0.721 g of Br_2 = 4.51×10^{-3} mol Br_2 = 9.02×10^{-3} mol e^- = 870 C.   Work = $E \cdot q$ = (1.50)(870) = 1.31 kJ.

(e) $I = \sqrt{P/R} = \sqrt{(1.00 \times 10^{-4})/(1.20 \times 10^3)} = 2.89 \times 10^{-4}$ A
$= 2.99 \times 10^{-9}$ mol e^-/s $= 9.97 \times 10^{-10}$ mol Al/s $= 2.69 \times 10^{-8}$ g/s

14-18. The activities of the solid reagents do not change until they are used up. The only aqueous species, OH^-, is created at the cathode and consumed in equal amounts at the anode, so its concentration remains constant in the cell. Therefore, none of the activities change during the life cycle of the cell until something is used up.

14-19. (a) $\frac{1}{2}O_2(g) + 2H^+ + 2e^- \rightleftharpoons H_2O(l)$ $\qquad E° = 1.229$ V at 25°C

$$E_{O_2} = \left(1.23 - \frac{0.059\ 16}{2} \log \frac{1}{P_{O_2}^{1/2}[H^+]^2}\right)$$

$$= \left(1.23 - \frac{0.059\ 16}{2} \log \frac{1}{(0.2)^{1/2}[0.5]^2}\right) = 1.201 \text{ V}$$

$2H^+ + 2e^- \rightleftharpoons H_2(g)$ $\qquad E° = 0$

$$E_{H_2} = \left(0 - \frac{0.059\ 16}{2} \log \frac{P_{H_2}}{[H^+]^2}\right)$$

$$= \left(0 - \frac{0.059\ 16}{2} \log \frac{(1.0)}{[0.5]^2}\right) = -0.018 \text{ V}$$

$E = E_{O_2} - E_{H_2} = 1.219$ V

(b) Power is the product of current × voltage ($P = IE$).

Current $= I = P/E = 20$ kW/220 V $= 90.9$ A.

e^- flow $= (90.9 \text{ C/s})/(96\ 485 \text{ C/mol}) = 0.942$ mmol/s

For each mol $H_2(g)$ consumed at the cathode, $2e^-$ are produced.

Therefore, $H_2(g)$ consumed if cell is 100% efficient $= \frac{1}{2}$ rate of e^- production

$= (0.942 \text{ mmol/s})/2 = 0.471$ mmol/s. If the cell is only 70% efficient, the consumption of $H_2(g)$ is $(0.471 \text{ mmol/s})/0.70 = 0.673$ mmol/s.

$H_2(g)$ consumption in 1 h $= (0.673 \text{ mmol/s})(3\ 600 \text{ s/h}) = 2.42$ mol/h

$\qquad = 4.88$ g/h

(c) $(20\ 000 \text{ W})\left(\dfrac{1 \text{ horsepower}}{745.7 \text{ W}}\right) = 26.8$ horsepower

14-20. (a)   Diagram of alkaline battery

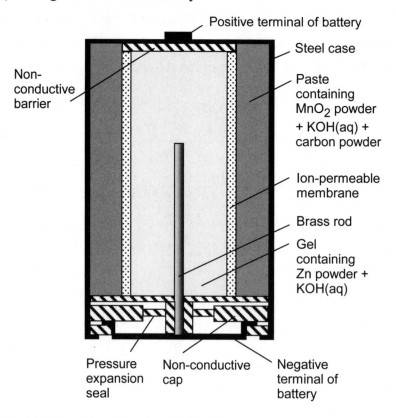

(b)   Half-reactions for the alkaline battery:

Cathode: $2MnO_2\ (s) + H_2O(l) + 2e^- \rightleftharpoons Mn_2O_3(s) + 2OH^-$ $E° = +0.147\ V$ (A)

Anode: $ZnO(s) + H_2O(l) + 2e^- \rightleftharpoons Zn(s) + 2OH^-$ $E° = -1.260\ V$ (B)

Net: $2MnO_2\ (s) + Zn(s) \rightleftharpoons Mn_2O_3(s) + ZnO(s)$ $E° = 1.407\ V$ (C)

(c)   When an alkaline battery runs down, its chemistry changes and $H_2(g)$ can be generated, causing the pressure expansion seal at the base to rupture. $KOH(aq)$ can then leak from the cell and react with CO_2 in the air to give a white, crystalline $K_2CO_3(s)$ crust on the outside of the battery.

(d)   The ZnO half-reaction (B) appears in Appendix H. The manganese oxide half-reaction (A) is the sum $(1) - (2) + (3) - (4)$ of four reactions found in the Appendix:

(1) $2MnO_2(s) + 8H^+ + 4e^- \rightleftharpoons 2Mn^{2+} + 4H_2O$ $E_1^\circ = 1.230$ V

(2) $Mn_2O_3(s) + 6H^+ + 2e^- \rightleftharpoons 2Mn^{2+} + 3H_2O$ $E_2^\circ = 1.485$ V

(3) $2H_2O + 2e^- \rightleftharpoons H_2(g) + 2OH^-$ $E_3^\circ = -0.828$ V

(4) $2H^+ + 2e^- \rightleftharpoons H_2(g)\}$ $E_4^\circ \equiv 0$

Net reaction (A): $2MnO_2(s) + H_2O(l) + 2e^- \rightleftharpoons Mn_2O_3(s) + 2OH^-$ $E_5^\circ = ?$

For each half-reaction, $\Delta G^\circ = -nNFE^\circ$, where nN is the moles of electrons in that half-reaction. ΔG° for the net reaction is $\Delta G_5^\circ = \Delta G_1^\circ - \Delta G_2^\circ + \Delta G_3^\circ - \Delta G_4^\circ$. When you multiply a half-reaction, you *do not* multiply E° because the work done *per electron* is unchanged.

$\Delta G_5^\circ = \Delta G_1^\circ - \Delta G_2^\circ + \Delta G_3^\circ - \Delta G_4^\circ$

$-2FE_5^\circ = [-4F(1.230)] - [-2F(1.485)] + [-2F(-0.828)] - [-2F(0)]$

$$E_5^\circ = \frac{-4(1.230) + 2(1.485) + 2(0.828)}{-2} = 0.147 \text{ V}$$

In (b), we see that E° for the net cell reaction of the battery is $+0.147 - (-1.260) = 1.407$ V

14-21. (a)   right half-cell: $E_+ = \left(0.222 - \dfrac{0.059\ 16}{2} \log [Cl^-]^2\right) = 0.281_2$ V

left half-cell: $E_- = \left(-0.350 - \dfrac{0.059\ 16}{2} \log [F^-]^2\right) = -0.290_8$ V

$E = E_+ - E_- = 0.281_2 - (-0.290_8) = 0.572$ V

(b)   Electrons flow from the left half-cell $(E = -0.290_8$ V) to the right half-cell $(E = 0.281_2$ V).

(c) $[Pb^{2+}] = K_{sp}$ (for PbF_2) / $[F^-]^2 = (3.6 \times 10^{-8}) / (0.10)^2 = 3.6 \times 10^{-6}$ M

$[Ag^+] = K_{sp}$ (for AgCl) / $[Cl^-] = (1.8 \times 10^{-10}) / (0.10) = 1.8 \times 10^{-9}$ M

right half-cell: $E_+ = \left(0.799 - \dfrac{0.059\ 16}{2} \log \dfrac{1}{[Ag^+]^2}\right) = 0.281_2$ V

left half-cell: $E_- = \left(-0.126 - \dfrac{0.059\ 16}{2} \log \dfrac{1}{[Pb^{2+}]}\right) = -0.287_0$ V

$E = E_+ - E_- = 0.281_2 - (-0.287_0) = 0.568$ V

The agreement between the two calculations is reasonable.

14-22. A hydrogen pressure of 727.2 Torr corresponds to (727.2 Torr)/(760 Torr/atm) = 0.956 8 atm.   (0.956 8 atm)(1.013 25 bar/atm) = 0.969 5 bar.

$$0.798\ 3 = E^{o}_{Ag^{+}\,|\,Ag} - 0.059\ 16\ \log \frac{[0.010\ 00]\ (0.914)}{(0.969\ 5)^{1/2}\ [0.010\ 00](0.898)}$$

$$\Rightarrow E^{o}_{Ag^{+}\,|\,Ag} = 0.799\ 2\ V$$

14-23. Balanced reaction: $HOBr + 2e^{-} + H^{+} \rightleftharpoons Br^{-} + H_2O$

$$HOBr \rightarrow \tfrac{1}{2}Br_2 \qquad\qquad \Delta G^{o}_1 = -1F(1.584)$$

$$\tfrac{1}{2}Br_2 \rightarrow Br^{-} \qquad\qquad \Delta G^{o}_2 = -1F(1.098)$$

$$\overline{\qquad HOBr \rightarrow Br^{-} \qquad\qquad \Delta G^{o}_3 = \Delta G^{o}_1 + \Delta G^{o}_2 = -2FE^{o}_3 }$$

$$E^{o}_3 = \frac{-1F(1.584) - 1F(1.098)}{-2F} = 1.341\ V$$

14-24.

$$2X^{+} + 2e^{-} \rightleftharpoons 2X(s) \qquad\qquad E^{o}_{+} = E^{o}_2$$

$$^{-}\ \underline{X^{3+} + 2e^{-} \rightleftharpoons X^{+}} \qquad\qquad \underline{E^{o}_{-} = E^{o}_1}$$

$$3X^{+} \rightleftharpoons X^{3+} + 2X(s) \qquad\qquad E^{o}_3 = E^{o}_2 - E^{o}_1$$

If $E^{o}_2 > E^{o}_1$, then $E^{o}_3 \geq 0$ and disproportionation is spontaneous.

14-25. right half-cell: $Cu^{2+} + 2e^{-} \rightleftharpoons Cu(s)$ $\qquad E^{o}_{+} = 0.339\ V$

left half-cell: $Ni^{2+} + 2e^{-} \rightleftharpoons Ni(s)$ $\qquad E^{o}_{-} = -0.236\ V$

The ionic strength of the right half-cell is 0.009 0 M, and the ionic strength of the left half-cell is 0.008 0 M. At $\mu = 0.009\ 0$ M, $\gamma_{Cu^{2+}} = 0.690$.

At $\mu = 0.008\ 0$ M, $\gamma_{Ni2+} = 0.705$.

$$E_{+} = E^{o}_{+} - \frac{0.059\ 16}{2}\ \log \frac{1}{[Cu^{2+}]\gamma_{Cu^{2+}}}$$

$$= 0.339 - \frac{0.059\ 16}{2}\ \log \frac{1}{(0.003\ 0)(0.690)} = 0.259_6\ V$$

$$E_{-} = E^{o}_{-} - \frac{0.059\ 16}{2}\ \log \frac{1}{[Ni^{2+}]\gamma_{Ni2+}}$$

$$= -0.236 - \frac{0.059\ 16}{2}\ \log \frac{1}{(0.002\ 0)(0.705)} = -0.320_3\ V$$

$$E = E_{+} - E_{-} = 0.580\ V$$

Electrons flow from Ni ($E = -0.320_3$ V) to Cu ($E = 0.259_6$ V).

14-26. (a) Anode: $PbSO_4(s) + 2e^- \rightleftharpoons Pb(s) + SO_4^{2-}$ $\qquad E° = -0.355$ V

Cathode: $PbO_2(s) + SO_4^{2-} + 4H^+ + 2e^- \rightleftharpoons PbSO_4(s) + 2H_2O$ $\;E° = 1.685$ V

Net reaction: $Pb(s) + PbO_2(s) + SO_4^{2-} + 4H^+ \rightleftharpoons 2PbSO_4(s) + 2H_2O$

$$E° = 1.685 - (-0.355) = 2.040 \text{ V}$$

(b) $Pb(s) \mid PbSO_4(s) \mid H_2SO_4(aq) \mid PbSO_4(s) \mid PbO_2(s) \mid Pb(s)$

(c) Reactions during charging:

Anode: $PbSO_4(s) + 2H_2O \rightleftharpoons PbO_2(s) + SO_4^{2-} + 4H^+ + 2e^-$

Cathode: $PbSO_4(s) + 2e^- \rightleftharpoons Pb(s) + SO_4^{2-}$

(d) $E_{\text{cathode}} = 1.685 - \dfrac{0.059\,16}{2} \log \dfrac{m_{H_2O}^2 \gamma_{H_2O}^2}{m_{SO_4^{2-}} \gamma_{SO_4^{2-}} m_{H^+}^4 \gamma_{H^+}^4}$

$E_{\text{anode}} = -0.355 - \dfrac{0.059\,16}{2} \log m_{SO_4^{2-}} \gamma_{SO_4^{2-}}$

$E_{\text{net reaction}} = E_{\text{cathode}} - E_{\text{anode}} =$

$$[1.685 - (-0.355)] - \left(\dfrac{0.059\,16}{2} \log \dfrac{m_{H_2O}^2 \gamma_{H_2O}^2}{m_{SO_4^{2-}} \gamma_{SO_4^{2-}} m_{H^+}^4 \gamma_{H^+}^4} \right)$$

$$- \left(-\dfrac{0.059\,16}{2} \log m_{SO_4^{2-}} \gamma_{SO_4^{2-}} \right)$$

$$E_{\text{net reaction}} = 2.040 - \dfrac{0.059\,16}{2} \log \dfrac{m_{H_2O}^2 \gamma_{H_2O}^2}{m_{SO_4^{2-}}^2 \gamma_{SO_4^{2-}}^2 m_{H^+}^4 \gamma_{H^+}^4}$$

(e) $E_{\text{net reaction}} = 2.040 - \dfrac{0.059\,16}{2} \log \dfrac{\mathcal{A}_{H_2O}^2}{m_{H^+}^4 m_{SO_4^{2-}}^2 \gamma_{H^+}^4 \gamma_{SO_4^{2-}}^2}$

But $(\gamma_{H^+}^2 \gamma_{SO_4^{2-}}) = \gamma_\pm^3$, so

$$E_{\text{net reaction}} = 2.040 - \dfrac{0.059\,16}{2} \log \dfrac{\mathcal{A}_{H_2O}^2}{m_{H^+}^4 \, m_{SO_4^{2-}}^2 \, \gamma_\pm^3}$$

$$E_{\text{net reaction}} = 2.040 - \dfrac{0.059\,16}{2} \log \dfrac{(0.66)^2}{(11.0)^4 (5.5)^2 (0.22)^3} = 2.159 \text{ V}$$

14-27. $2LiOH(s) + CO_2(g) \rightleftharpoons Li_2CO_3(s) + H_2O(l)$

LiOH weighs less than an equivalent number of moles of NaOH or KOH. The mass of material launched into space must be minimized, so LiOH is chosen.

14-28. $E^\circ = \dfrac{-\Delta G^\circ}{nF} = \dfrac{(+257 \times 10^3 \text{ J/mol})}{(2)(9.648\,5 \times 10^4 \text{ C/mol})} = 1.33 \text{ V}$

$K = 10^{nE^\circ/0.059\,16} = 1 \times 10^{45}$

14-29. (a)

$\begin{aligned} & 4[\text{Co}^{3+} + e^- \rightleftharpoons \text{Co}^{2+}] && E_+^\circ = 1.92 \text{ V} \\ & -2[\tfrac{1}{2}\text{O}_2 + 2\text{H}^+ + 2e^- \rightleftharpoons \text{H}_2\text{O}] && E_-^\circ = 1.229 \text{ V} \\ \hline & 4\text{Co}^{3+} + 2\text{H}_2\text{O} \rightleftharpoons 4\text{Co}^{2+} + \text{O}_2 + 4\text{H}^+ && E^\circ = 0.69_1 \text{ V} \end{aligned}$

$\Delta G^\circ = -4F\,E^\circ = -2.7 \times 10^5 \text{ J} \qquad K = 10^{4E^\circ/0.059\,16} = 10^{47}$

(b)

$\begin{aligned} & \text{Ag(S}_2\text{O}_3)_2^{3-} + e^- \rightleftharpoons \text{Ag}(s) + 2\text{S}_2\text{O}_3^{2-} && E_+^\circ = 0.017 \text{ V} \\ & {}^-\text{Fe(CN)}_6^{3-} + e^- \rightleftharpoons \text{Fe(CN)}_6^{4-} && E_-^\circ = 0.356 \text{ V} \\ \hline & \text{Ag(S}_2\text{O}_3)_2^{3-} + \text{Fe(CN)}_6^{4-} \rightleftharpoons \text{Ag}(s) + 2\text{S}_2\text{O}_3^{2-} + \text{Fe(CN)}_6^{3-} && E^\circ = -0.339 \text{ V} \end{aligned}$

$\Delta G^\circ = -1F E^\circ = 32.7 \text{ kJ} \qquad K = 10^{1E^\circ/0.059\,16} = 1.9 \times 10^{-6}$

14-30. (a)

$\begin{aligned} & 5\text{Ce}^{4+} + 5e^- \rightleftharpoons 5\text{Ce}^{3+} && E_+^\circ = 1.70 \text{ V} \\ & {}^-\text{MnO}_4^- + 8\text{H}^+ + 5e^- \rightleftharpoons \text{Mn}^{2+} + 4\text{H}_2\text{O} && E_-^\circ = 1.507 \text{ V} \\ \hline & 5\text{Ce}^{4+} + \text{Mn}^{2+} + 4\text{H}_2\text{O} \rightleftharpoons 5\text{Ce}^{3+} + \text{MnO}_4^- + 8\text{H}^+ && E^\circ = 0.19_3 \text{ V} \end{aligned}$

(b) $\Delta G^\circ = -5F E^\circ = -93._1 \text{ kJ} \qquad K = 10^{5E^\circ/0.059\,16} = 2 \times 10^{16}$

(c) $E = \left(1.70 - \dfrac{0.059\,16}{5} \log \dfrac{[\text{Ce}^{3+}]^5}{[\text{Ce}^{4+}]^5}\right) - \left(1.507 - \dfrac{0.059\,16}{5} \log \dfrac{[\text{Mn}^{2+}]}{[\text{MnO}_4^-][\text{H}^+]^8}\right)$

$= 1.52_{23} - 1.54_{25} = -0.02_0 \text{ V}$

(d) $\Delta G = -5F E = +10 \text{ kJ}$

(e) At equilibrium, $E = 0 \Rightarrow E^\circ = \dfrac{0.059\,16}{5} \log \dfrac{[\text{Ce}^{3+}]^5[\text{MnO}_4^-][\text{H}^+]^8}{[\text{Ce}^{4+}]^5[\text{Mn}^{2+}]}$

$\Rightarrow [\text{H}^+] = 0.62 \Rightarrow \text{pH} = 0.21$

14-31.

right half-cell: $\qquad\qquad \text{Sn}^{4+} + 2e^- \rightleftharpoons \text{Sn}^{2+}$

left half-cell: $\quad -2\text{VO}^{2+} + 4\text{H}^+ + 2e^- \rightleftharpoons 2\text{V}^{3+} + 2\text{H}_2\text{O}$

net reaction: $\quad \text{Sn}^{4+} + 2\text{V}^{3+} + 2\text{H}_2\text{O} \rightleftharpoons \text{Sn}^{2+} + 2\text{VO}^{2+} + 4\text{H}^+$

$E = E^\circ - \dfrac{0.059\,16}{2} \log \dfrac{[\text{VO}^{2+}]^2[\text{H}^+]^4[\text{Sn}^{2+}]}{[\text{V}^{3+}]^2[\text{Sn}^{4+}]}$

$-0.289 = E^\circ - \dfrac{0.059\,16}{2} \log \dfrac{(0.116)^2(1.57)^4(0.031\,8)}{(0.116)^2(0.031\,8)} \Rightarrow E^\circ = -0.266 \text{ V}$

$\Rightarrow K = 10^{2E^\circ/0.059\,16} = 1.0 \times 10^{-9}$

14-32.

$$Pd(OH)_2(s) + 2e^- \rightleftharpoons Pd(s) + 2OH^- \qquad E_+^o$$

$$\underline{-\ Pd^{2+} + 2e^- \rightleftharpoons Pd(s)} \qquad\qquad E_-^o = 0.915\ V$$

$$Pd(OH)_2 \overset{K_{sp}}{\rightleftharpoons} Pd^{2+} + 2OH^- \qquad E^o = E_+^o - 0.915$$

But $K_{sp} = 3 \times 10^{-28} \Rightarrow E^o = \dfrac{0.059\ 16}{2} \log K_{sp} = -0.814$

$-0.814 = E_1^o - 0.915 \Rightarrow E_1^o = 0.101\ V$

14-33.

$$Br_2(l) + 2e^- \rightleftharpoons 2Br^- \qquad\qquad E_+^o = 1.078\ V$$

$$\underline{-\ Br_2(aq) + 2e^- \rightleftharpoons 2Br^-} \qquad E_-^o = 1.098\ V$$

$$Br_2(l) \overset{K}{\rightleftharpoons} Br_2(aq) \qquad\qquad E^o = -0.020\ V$$

At equilibrium, $E = 0$. Therefore, $0 = -0.020 - \dfrac{0.059\ 16}{2} \log \dfrac{[Br_2(aq)]}{[Br_2(l)]}$

$$\Rightarrow K = \frac{[Br_2(aq)]}{[Br_2(l)]} = 0.21_1\ M.$$

That is, the solubility of Br_2 in water is $0.21_1\ M = 34\ g/L$.

14-34. (a)

$$Cl_2(g) + 2e- \rightleftharpoons 2Cl^- \qquad E^o = 1.360\ 4\ V \qquad (A)$$

$$\underline{-\ Cl_2(aq) + 2e- \rightleftharpoons 2Cl^-} \qquad E^o = 1.396\ V \qquad (B)$$

$$Cl_2(g) \rightleftharpoons Cl_2(aq) \qquad E^o = 1.360\ 4 - 1.396 = -0.035_6\ V$$

$$K_h = \frac{[Cl_2(aq)]}{P_{Cl_2}} = 10^{nE^{\circ}/0.059\ 16} = 10^{2(-0.035_6)/0.059\ 16} = 0.062_6\ M/bar$$

When $P_{Cl_2} = 1.00$ bar, $[Cl_2(aq)] = 0.063\ M$

E^o and the constant 0.059 16 V both apply at 298.15 K

(b) For reaction A, $dE^o/dT = -1.248$ mV/K. For an increase in temperature of 50 K (from 298.15 to 323.15 K), the change in E^o is

$$E^o(at\ 323.15\ K) = E^o(at\ 298.15\ K) + (dE^o/dT)\Delta T$$
$$= 1.360\ 4\ V + (-1.248 \times 10^{-3}\ V/K)(25\ K) = 1.329_2\ V$$

For reaction B, $dE^o/dT = -0.72$ mV/K, so

$$E^o(at\ 323.15\ K) = 1.396\ V + (-0.72 \times 10^{-3}\ V/K)(25\ K) = 1.378\ V$$

E^o(net reaction at 323.15 K) = $1.329_2 - 1.378 = -0.048_8$ V.

The number 0.059 16 V in the Nernst equation is $(RT \ln 10)/F$, which changes with temperature. At $T = 323.15$ K,

$(RT \ln 10)/F = (8.314\ 5\ J/[mol \cdot K])(323.15\ K)(\ln 10)/(96\ 485.336\ C/mol)$

$= 0.0641\ 2\ J/C = = 0.0641\ 2\ V$

$$K_h = \frac{[Cl_2(aq)]}{P_{Cl_2}} = 10^{nE^\circ/0.064\,12} = 10^{2(-0.048_8)/0.064\,12} = 0.030_1 \text{ M/bar}$$

The solubility of $Cl(g)$ decreases when the temperature is raised, which is typical of most gases.

14-35.

$FeY^- + e^- \rightleftharpoons FeY^{2-}$	E_+°
$^-\ FeY^- + e^- \rightleftharpoons Fe^{2+} + Y^{4-}$	$E_-^\circ = -0.730 \text{ V}$
$Fe^{2+} + Y^{4-} \rightleftharpoons FeY^{2-}$	$E^\circ = E_+^\circ + 0.730$

But $E^\circ = 0.059\,16 \ \log [K_f \text{ (for } FeY^{2-})] = 0.847 \text{ V} \Rightarrow E_+^\circ = E^\circ - E_-^\circ = 0.117 \text{ V}.$

14-36. $E^\circ(T) = E^\circ + \dfrac{dE^\circ}{dT} \Delta T$

$E^\circ(50°C) = -1.677 \text{ V} + (0.533 \times 10^{-3} \text{ V/K})(25 \text{ K}) = -1.664 \text{ V}$

14-37.

1. $2Cu(s) + 2I^- \rightleftharpoons 2CuI(s) + 2e^-$	$\Delta G_1^\circ = +2F(-0.185) = -35.7_0 \text{ kJ}$
2. $2Cu^{2+} + 4e^- \rightleftharpoons 2Cu(s)$	$\Delta G_2^\circ = -4F(0.339) = -130._{83} \text{ kJ}$
3. hydro $\rightleftharpoons$ quinone $+ 2H^+ + 2e^-$	$\Delta G_3^\circ = +2F(0.700) = 135._{08} \text{ kJ}$
4. (1)+(2)+(3): $2Cu^{2+} + 2I^- +$ hydro $\rightleftharpoons$	$\Delta G_4^\circ = \Delta G_1^\circ + \Delta G_2^\circ + \Delta G_3^\circ$
$\qquad 2CuI(s) +$ quinone $+ 2H^+$	$= -31._4 \text{ kJ}$

Since $2e^-$ are transferred in the net reaction, $E_4^\circ = \dfrac{-\Delta G_4^\circ}{2F} = +0.16_3 \text{ V}$

$K = 10^{2(0.163)/0.059\,16} = 3._2 \times 10^5.$

14-38. (a) $E(\text{right}) = E^\circ(\text{right}) - \dfrac{RT}{2F} \ln \dfrac{\mathcal{A}_{F^-}^2(\text{right})}{P_{O_2}^{1/2}(\text{right})}$

$E(\text{left}) = E^\circ(\text{left}) - \dfrac{RT}{2F} \ln \dfrac{\mathcal{A}_{F^-}^2(\text{left})}{P_{O_2}^{1/2}(\text{left})}$

Net reaction: reverse left half-reaction and add it to right half-reaction:

$MgF_2(s) + Al_2O_3(s) + \frac{1}{2} O_2(g) + 2e^- \rightleftharpoons MgAl_2O_4(s) + 2F^-$

$\qquad\qquad MgO(s) + 2F^- \rightleftharpoons MgF_2(s) + \frac{1}{2} O_2(g) + 2e^-$

net reaction: $\quad Al_2O_3(s) + MgO(s) \rightleftharpoons MgAl_2O_4(s)$

Nernst equation for net reaction:

$E(\text{right}) - E(\text{left}) = E^\circ(\text{right}) - E^\circ(\text{left}) - \dfrac{RT}{2F} \ln \dfrac{\mathcal{A}_{F^-}^2(\text{right})}{P_{O_2}^{1/2}(\text{right})} + \dfrac{RT}{2F} \ln \dfrac{\mathcal{A}_{F^-}^2(\text{left})}{P_{O_2}^{1/2}(\text{left})}$

The activities of F^- are the same on both sides and the activities of O_2 are also the same on both sides, so the ln terms cancel, leaving $E(\text{cell}) = E^\circ(\text{right}) - E^\circ(\text{left}) = E^\circ(\text{cell})$.

(b) $\Delta G^\circ = -nFE^\circ = -(2)(9.648\ 5 \times 10^4\ \text{C/mol})(0.152\ 9\ \text{J/C}) = -29.51\ \text{kJ/mol}$,

where we made use of the fact that a volt is equivalent to one joule/coulomb.

(c) $\Delta G^\circ = \Delta H^\circ - T\Delta S^\circ$

$-nFE^\circ = \Delta H^\circ - T\Delta S^\circ = -nF(0.122\ 3 + 3.06 \times 10^{-5}\ T)$

$\qquad\qquad = -nF(0.122\ 3\ \text{V}) - nF(3.06 \times 10^{-5}\ T)$

$\qquad\qquad = \underbrace{-nF(0.122\ 3\ \text{V})}_{\Delta H^\circ} - T\underbrace{\{nF(3.06 \times 10^{-5}\ \text{V/K})\}}_{\Delta S^\circ}$

$\Delta H^\circ = -nF(0.122\ 3\ \text{V}) = -(2)(9.648\ 5 \times 10^4\ \text{C/mol})(0.122\ 3\ \text{J/C})$

$\qquad = -23.60\ \text{kJ/mol}$

$\Delta S^\circ = nF(3.06 \times 10^{-5}\ \text{V/K})$

$\qquad = (2)(9.648\ 5 \times 10^4\ \text{C/mol})(3.06 \times 10^{-5}\ \text{V/K})$

$\qquad = 5.90\ \text{C·V/(K·mol)} = 5.90\ \text{J/(K·mol)}$, where we made use of the

conversion coulomb·volt = joule.

14-39. In the right half-cell, the reaction $Hg^{2+} + Y^{4-} \rightleftharpoons HgY^{2-}$ is at equilibrium, even though the net cell reaction $Hg^{2+} + H_2 \rightleftharpoons Hg(l) + 2H^+$ is not at equilibrium.

14-40. (a) $\begin{array}{ll} AgCl(s) + e^- \rightleftharpoons Ag(s) + Cl^- & E^\circ_+ = 0.222\ \text{V} \\ {}^-\ \ H^+ + e^- \rightleftharpoons \frac{1}{2}\ H_2(g) & E^\circ_- = 0\ \text{V} \\ \hline AgCl(s) + \frac{1}{2}H_2(g) \rightleftharpoons Ag(s) + H^+ + Cl^- & E^\circ = 0.222\ \text{V} \end{array}$

$E = 0.222 - 0.059\ 16 \log \dfrac{[H^+][Cl^-]}{\sqrt{P_{H_2}}}$

(b) $0.485 = 0.222 - 0.059\ 16 \log \dfrac{(10^{-3.60})[Cl^-]}{\sqrt{1.00}} \Rightarrow [Cl^-] = 0.14_3\ \text{M}$

14-41. (a) Left: quinone $+ 2H^+ + 2e^- \rightleftharpoons$ hydroquinone $\qquad E^\circ_- = 0.700\ \text{V}$

Right: $Hg_2Cl_2 + 2e^- \rightleftharpoons 2Hg(l) + 2Cl^-$ $\qquad E^\circ_+ = 0.268\ \text{V}$

$E(\text{left}) = 0.700 - \dfrac{0.059\ 16}{2} \log \dfrac{[\text{hydroquinone}]}{[\text{quinone}][H^+]^2}$

$E(\text{right}) = 0.268 - \dfrac{0.059\ 16}{2} \log [Cl^-]^2$

(b) $E(\text{cell}) = E(\text{right}) - E(\text{left})$

$$= \left(0.268 - \frac{0.059\,16}{2}\log[Cl^-]^2\right) - \left(0.700 - \frac{0.059\,16}{2}\log\frac{[\text{hydroquinone}]}{[\text{quinone}][H^+]^2}\right)$$

$$E(\text{cell}) = -0.432 - \frac{0.059\,16}{2}\log\frac{[\text{quinone}][H^+]^2[Cl^-]^2}{[\text{hydroquinone}]}$$

Setting $[Cl^-] = 0.50$ M and noting [quinone] = [hydroquinone], we find

$$E(\text{cell}) = -0.432 - 0.059\,16\log(0.50) - 0.059\,16\log[H^+]$$

$$E(\text{cell}) = -0.414 + 0.059\,16\,\text{pH} \quad (A = -0.414, B = 0.059\,16\text{ V per pH unit})$$

(c) $E(\text{cell}) = -0.414 + 0.059\,16\,(4.50) = -0.148$

Since $E < 0$, electrons flow from right to left (Hg → Pt) through the meter.

14-42. $H^+ (1.00 \text{ M}) + e^- \rightleftharpoons \frac{1}{2} H_2 (g, 1.00 \text{ bar})$ $E_+^\circ = 0$ V

$$\underline{\ H^+ (x \text{ M}) + e^- \ \rightleftharpoons \ \frac{1}{2} H_2 (g, 1.00 \text{ bar}) \qquad E_-^\circ = \ 0 \text{ V}}$$

$$\underline{H^+ (1.00 \text{ M}) \qquad \rightleftharpoons \qquad H^+ (x \text{ M}) \qquad\qquad E^\circ = 0 \text{ V}}$$

$$E(\text{cell}) = 0.490 = E_+ - E_- = \{0\} - \left(0 - 0.059\,16\log\frac{1}{[H^+]}\right)$$

$$\Rightarrow [H^+] = 5.2 \times 10^{-9} \text{ M}$$

$$K_b = \frac{[RNH_3^+][OH^-]}{[RNH_2]} = \frac{(0.050)(K_w/[H^+])}{0.10} = 9.6 \times 10^{-7}$$

14-43. $M^{2+} + 2e^- \rightleftharpoons M(s)$ $E_+^\circ = -0.266$ V

$$\underline{\ 2H^+ + 2e^- \ \rightleftharpoons \ H_2 (g, 0.50 \text{ bar}) \qquad\qquad E_-^\circ = \ 0 \text{ V}}$$

$$\underline{H_2(g) + M^{2+} \rightleftharpoons 2H^+ + M(s) \qquad\qquad E^\circ = -0.266 \text{ V}}$$

$$E = \left(-0.266 - 0.059\,16\log\frac{1}{[M^{2+}]}\right) - \left(0 - 0.059\,16\log\frac{P_{H_2}}{[H^+]^2}\right)$$

$$E = -0.266 - \frac{0.059\,16}{2}\log\frac{[H^+]^2}{P_{H_2}[M^{2+}]}$$

$[H^+]$ in the left half-cell is found by considering the titration of 28.0 mL of the tetraprotic pyrophosphoric acid (abbreviated H_4P) with 72.0 mL of KOH.

28.0 mL of 0.010 0 M H_4P = 0.280 mmol

72.0 mL of 0.010 0 M KOH = 0.720 mmol

First, 0.280 mmol OH^- consumes 0.280 mmol of H_4P, giving 0.280 mmol of H_3P^- and (0.720 − 0.280 =) 0.440 mmol of OH^-. Then 0.280 mmol OH^- consumes 0.280 mmol of H_3P^-, giving 0.280 mmol of H_2P^{2-} and (0.440 − 0.280 =) 0.160 mmol of OH^-. Finally, 0.160 mmol of OH^- reacts with 0.280 mmol of H_2P^{2-} to create 0.160 mmol of HP^{3-}, leaving 0.120 mmol of unreacted H_2P^{2-}.

$$\text{H}_4\text{P} \quad + \quad \text{OH}^- \quad \rightarrow \quad \text{H}_2\text{P}^{2-} \quad + \quad \text{HP}^{3-}$$

$$\text{0.280 mmol} \qquad \text{0.720 mmol} \qquad \text{0.120 mmol} \qquad \text{0.160 mmol}$$

$$\text{pH} = pK_3 + \log \frac{[\text{HP}^{3-}]}{[\text{H}_2\text{P}^{2-}]} = 6.70 + \log \frac{0.160}{0.120} = 6.82 \Rightarrow [\text{H}^+] = 1.5_0 \times 10^{-7}\text{ M}$$

Putting the known values of $[\text{H}^+]$ and P_{H_2} into the Nernst equation gives

$$-0.246 = -0.266 - \frac{0.059\,16}{2} \log \frac{[\text{H}^+]^2}{P_{\text{H}_2}[\text{M}^{2+}]}$$

$$= -0.266 - \frac{0.059\,16}{2} \log \frac{[1.5_0 \times 10^{-7}]^2}{0.50\,[\text{M}^{2+}]} \quad \Rightarrow [\text{M}^{2+}] = 2.1_3 \times 10^{-13}\text{ M}$$

In the right half-cell we have the equilibrium

$$\text{M}^{2+} \quad + \quad \text{F}_{\text{EDTA}} \quad \rightleftharpoons \quad \text{MY}^{2-}$$

initial mmol/mL	$\frac{0.280}{100}$	$\frac{0.720}{100}$	—
final mmol/mL	small	$\frac{0.440}{100}$	$\frac{0.280}{100}$

$$K_f = \frac{[\text{MY}^{2-}]}{[\text{M}^{2+}]\,\alpha_{\text{Y}4^-}\text{F}_{\text{EDTA}}} = \frac{0.280/100}{(2.1_3 \times 10^{-13})(0.004\,2)(0.440/100)} = 7.1 \times 10^{14}$$

14-44. right half-cell: $\quad \text{Pb}^{2+}(\text{right}) + 2e^- \rightleftharpoons \text{Pb}(s) \qquad E_+^\circ = -0.126\text{ V}$

left half-cell: $\quad \underline{^-\ \text{Pb}^{2+}(\text{left}) + 2e^- \rightleftharpoons \text{Pb}(s)} \qquad \underline{E_-^\circ = -0.126\text{ V}}$

$$\text{Pb}^{2+}(\text{right}) \rightleftharpoons \text{Pb}^{2+}(\text{left}) \qquad\qquad E^\circ = 0$$

$$E_{\text{cell}} = E_{\text{right}} - E_{\text{left}}$$

$$= \left(-0.126 - 0.059\,16 \log \frac{1}{[\text{Pb}^{2+}(\text{right})]}\right) - \left(-0.126 - 0.059\,16 \log \frac{1}{[\text{Pb}^{2+}(\text{left})]}\right)$$

Nernst equation for net cell reaction:

$$E_{\text{cell}} = -0.001\,8 = -\frac{0.059\,16}{2} \log \frac{[\text{Pb}^{2+}(\text{left})]}{[\text{Pb}^{2+}(\text{right})]} \Rightarrow \frac{[\text{Pb}^{2+}(\text{left})]}{[\text{Pb}^{2+}(\text{right})]} = 1.15$$

For each half-cell, we can write $[\text{CO}_3^{2-}] = K_{\text{sp}}$ (for PbCO_3) / $[\text{Pb}^{2+}]$

$$\frac{[\text{CO}_3^{2-}(\text{left})]}{[\text{CO}_3^{2-}(\text{right})]} = \frac{K_{\text{sp}}\,(\text{for PbCO}_3)/[\text{Pb}^{2+}(\text{left})]}{K_{\text{sp}}\,(\text{for PbCO}_3)/[\text{Pb}^{2+}\,(\text{right})]} = \frac{1}{1.15} = 0.87$$

In each compartment the Ca^{2+} concentration is equal to the total concentration of all carbonate species (since PbCO_3 is much less soluble than CaCO_3). Let the fraction of all carbonate species in the form CO_3^{2-} be $\alpha_{\text{CO}_3^{2-}}$

(i.e., $[\text{CO}_3^{2-}] = \alpha_{\text{CO}_3^{2-}}$ [total carbonate]). We can say that $[\text{Ca}^{2+}] = $ [total carbonate] $ = [\text{CO}_3^{2-}]\,/\,\alpha_{\text{CO}_3^{2-}}$. The value of $\alpha_{\text{CO}_3^{2-}}$ is the same in both compartments, since the pH is the same. Now we can write

$$\frac{K_{sp}(\text{calcite})}{K_{sp}(\text{aragonite})} = \frac{[Ca^{2+}(\text{left})][CO_3^{2-}(\text{left})]}{[Ca^{2+}(\text{right})][CO_3^{2-}(\text{right})]} = \frac{[CO_3^{2-}(\text{left})]^2/\alpha_{CO_3^{2-}}}{[CO_3^{2-}(\text{right})]^2/\alpha_{CO_3^{2-}}}$$

$$= (0.87)^2 = 0.76.$$

14-45. $Cu^{2+} + 2e^- \rightleftharpoons Cu(s)$ $E_+^\circ = 0.339$ V

$-$ $Ni^{2+} + 2e^- \rightleftharpoons Ni(s)$ $E_-^\circ = -0.236$ V

$\overline{ Cu^{2+} + Ni(s) \rightleftharpoons Cu(s) + Ni^{2+} }$ $\overline{E^\circ = 0.575 \text{ V}}$

The ionic strength of the right half-cell is 0.10 M and $\gamma_{Cu^{2+}} = 0.405$.

The ionic strength of the left half-cell is 0.010 M and $\gamma_{Ni^{2+}} = 0.675$.

$E_{\text{cell}} = E_{\text{right}} - E_{\text{left}}$

$$= \left(0.339 - 0.059\,16 \log \frac{1}{[Cu^{2+}]\,\gamma_{Cu^{2+}}}\right) - \left(-0.236 - 0.059\,16 \log \frac{1}{[Ni^{2+}]\,\gamma_{Ni^{2+}}}\right)$$

$$E_{\text{cell}} = 0.575 - \frac{0.059\,16}{2} - 0.059\,16 \log \frac{[Ni^{2+}]\gamma_{Ni^{2+}}}{[Cu^{2+}]\gamma_{Cu^{2+}}}$$

$$0.512 = 0.575 - \frac{0.059\,16}{2} \log \frac{(0.002\,5)(0.675)}{[Cu^{2+}](0.405)} \Rightarrow [Cu^{2+}] = 3.09 \times 10^{-5} \text{ M}$$

$$K_{sp} = [Cu^{2+}]\gamma_{Cu^{2+}}[IO_3^-]^2\gamma_{IO_3^-}^2) = (3.09 \times 10^{-5})(0.405)(0.10)^2(0.775)^2 = 7.5 \times 10^{-8}$$

14-46. $E^{\circ\prime}$ is the effective reduction potential for a half-reaction at pH 7, instead of pH 0. Since living systems tend to have a pH much closer to 7 than to 0, $E^{\circ\prime}$ provides a better indication of redox behavior in an organism.

14-47. (a) $E = 0.731 - \dfrac{0.059\,16}{2} \log \dfrac{P_{C_2H_4}}{P_{C_2H_2}[H^+]^2}$

(b) $E = \underbrace{0.731 + 0.059\,16 \log [H^+]}_{\text{This is } E^{\circ\prime} \text{ when pH} = 7} - \dfrac{0.059\,16}{2} \log \dfrac{P_{C_2H_4}}{P_{C_2H_2}}$

(c) $E^{\circ\prime} = 0.731 + 0.059\,16 \log (10^{-7.00}) = 0.317$ V

14-48. $E = E^\circ - \dfrac{0.059\,16}{2} \log \dfrac{[HCN]^2}{P_{(CN)_2}[H^+]^2}$

Substituting $[HCN] = \dfrac{[H^+]\,F_{HCN}}{[H^+] + K_a}$ into the Nernst equation gives

$E = 0.373 - \dfrac{0.059\,16}{2} \log \dfrac{[H^+]^2\,F_{HCN}^2}{([H^+] + K_a)^2\,P_{(CN)_2}[H^+]^2}$

$$E = \underbrace{0.373 + 0.059\,16 \log ([H^+] + K_a)}_{\text{This is } E^{\circ\prime} \text{ when pH} = 7} - \frac{0.059\,16}{2} \log \frac{F_{HCN}^2}{P_{(CN)_2}}.$$

Inserting $K_a = 6.2 \times 10^{-10}$ for HCN and $[H^+] = 10^{-7.00}$ gives

$$E^{\circ\prime} = 0.373 + 0.059\,16 \log (10^{-7.00} + 6.2 \times 10^{-10}) = -0.041 \text{ V}.$$

14-49. $H_2C_2O_4 + 2H^+ + 2e^- \rightleftharpoons 2HCO_2H$

$$E = 0.204 - \frac{0.059\,16}{2} \log \frac{[HCO_2H]^2}{[H_2C_2O_4][H^+]^2}$$

But $[HCO_2H] = \dfrac{[H^+]\, F_{HCO_2H}}{[H^+] + K_a}$ and $[H_2C_2O_4] = \dfrac{[H^+]^2\, F_{H_2C_2O_4}}{[H^+]^2 + K_1[H^+] + K_1K_2}.$

Putting these expressions into the Nernst equation gives

$$E = 0.204 - \frac{0.059\,16}{2} \log \frac{[H^+]^2\, F_{HCO_2H}^2 ([H^+]^2 + K_1[H^+] + K_1K_2)}{([H^+] + K_a)^2\, [H^+]^2\, F_{H_2C_2O_4}\, [H^+]^2}$$

$$E = 0.204 - \underbrace{\frac{0.059\,16}{2} \log \frac{[H^+]^2 + K_1[H^+] + K_1K_2}{([H^+] + K_a)^2\, [H^+]^2}}_{\text{This is } E^{\circ\prime} \text{ when pH} = 7} - \frac{0.059\,16}{2} \log \frac{F_{HCO_2H}^2}{F_{H_2C_2O_4}}.$$

Putting in $[H^+] = 10^{-7.00}$ M, $K_a = 1.80 \times 10^{-4}$, $K_1 = 5.62 \times 10^{-2}$, and $K_2 = 5.42 \times 10^{-5}$ gives $E^{\circ\prime} = -0.268$ V.

14-50. $E = E^\circ - 0.059\,16 \log \dfrac{[H_2Red^-]}{[HOx]}$

But $[HOx] = \dfrac{[H^+]\, F_{HOx}}{[H^+] + K_a}$ and $[H_2Red^-] = \dfrac{[H^+]^2\, F_{H_2Red^-}}{[H^+]^2 + [H^+]K_1 + K_1K_2}.$

Putting these values into the Nernst equation gives

$$E = E^\circ - 0.059\,16 \log \frac{[H^+]^2\, F_{H_2Red^-} ([H^+] + K_a)}{[H^+]\, F_{HOx} ([H^+]^2 + [H^+]K_1 + K_1K_2)}$$

$$E = E^\circ - \underbrace{0.059\,16 \log \frac{[H^+]([H^+] + K_a)}{[H^+]^2 + [H^+]K_1 + K_1K_2}}_{E^{\circ\prime}} - 0.059\,16 \log \frac{F_{H_2Red^-}}{F_{HOx}}.$$

Since $E^{\circ\prime} = 0.062$ V, we find $E^\circ = -0.036$ V.

14-51. $E = E^\circ - \dfrac{0.059\ 16}{2} \log \dfrac{[HNO_2]}{[NO_3^-][H^+]^3}$

But $[HNO_2] = \dfrac{[H^+]F_{HNO_2}}{[H^+] + K_a}$ and $[NO_3^-] = F_{NO_3^-}$.

Putting these values into the Nernst equation gives

$$E = \underbrace{E^\circ - \frac{0.059\ 16}{2} \log \frac{1}{([H^+] + K_a)[H^+]^2}}_{E^{\circ\prime}} - \frac{0.059\ 16}{2} \log \frac{F_{HNO_2}}{F_{NO_3^-}}$$

$E^{\circ\prime} = 0.433 = 0.940 - \dfrac{0.059\ 16}{2} \log \dfrac{1}{(10^{-7} + K_a)(10^{-7})^2} \Rightarrow K_a = 7.2 \times 10^{-4}.$

14-52.

$$
\begin{array}{lr}
H_2PO_4^- + H^+ + 2e^- \rightleftharpoons HPO_3^{2-} + H_2O & E_+^\circ \\
-\ HPO_4^{2-} + 2H^+ + 2e^- \rightleftharpoons HPO_3^{2-} + H_2O & E_-^\circ = -0.234\ V \\
\hline
H_2PO_4^- \rightleftharpoons HPO_4^{2-} + H^+ & E^\circ = E_+^\circ - E_-^\circ
\end{array}
$$

$E^\circ = \dfrac{0.059\ 16}{2} \log K_{a2}$ (for H_3PO_4) $= \dfrac{0.059\ 16}{2} \log(6.34 \times 10^{-8}) = -0.213\ V$

$E_+^\circ = E^\circ - 0.234\ V = -0.447\ V$

14-53. (a) $A = 0.500 =$

$(1.12 \times 10^4\ M^{-1}cm^{-1})[Ox](1.00\ cm) + (3.82 \times 10^3\ M^{-1}cm^{-1})[Red](1.00\ cm)$

But $[Ox] = 5.70 \times 10^{-5} - [Red]$. Combining these two equations gives

$[Ox] = 3.82 \times 10^{-5}\ M$ and $[Red] = 1.88 \times 10^{-5}\ M$.

(b) $[S^-] = [Ox] = 3.82 \times 10^{-5}\ M$ and $[S] = Red = 1.88 \times 10^{-5}\ M$.

(c)

$$
\begin{array}{ll}
\quad S + e^- \rightleftharpoons S^- & E_+^{\circ\prime} = ? \\
-\ Ox + e^- \rightleftharpoons Red & E_-^{\circ\prime} = -0.128\ V \\
\hline
Red + S \rightleftharpoons Ox + S^- & E^{\circ\prime} = 0.059\ 16 \log \dfrac{[Ox][S^-]}{[Red][S]} = 0.036\ V
\end{array}
$$

$E_+^{\circ\prime} = E^{\circ\prime} + E_-^{\circ\prime} = -0.092\ V$

CHAPTER 15
ELECTRODES AND POTENTIOMETRY

15-1. (a) $AgCl(s) + e^- \rightleftharpoons Ag(s) + Cl^-$

$Hg_2Cl_2(s) + 2e^- \rightleftharpoons 2Hg(l) + 2Cl^-$

(b) $E = E_+ - E_- = 0.241 - 0.197 = 0.044$ V

15-2. Diagram for (a):

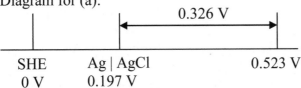

SHE   Ag | AgCl   0.523 V

0 V   0.197 V

(a) 0.326 V   (b) 0.086 V   (c) 0.019 V   (d) –0.021 V   (e) 0.021 V

15-3. $E = E_+ - E_-$

$$E = \left(0.771 - 0.059\,16 \log \frac{[Fe^{2+}]}{[Fe^{3+}]}\right) - (0.241) = 0.684 \text{ V}$$

15-4. $E = E° - 0.059\,16 \log \mathcal{A}_{Cl^-}$

$0.280 = 0.268 - 0.059\,16 \log \mathcal{A}_{Cl^-} \Rightarrow \mathcal{A}_{Cl^-} = 0.627$

15-5. For the saturated Ag-AgCl electrode, we can write: $E = E° - 0.059\,16 \log \mathcal{A}_{Cl^-}$.
Putting in $E = 0.197$ and $E° = 0.222$ V gives $\mathcal{A}_{Cl^-} = 2.6_5$. For the S.C.E., we
can write: $E = E° - 0.059\,16 \log \mathcal{A}_{Cl^-} = 0.268 \text{ V} - 0.059\,16 \log 2.6_5$
$= 0.243$ V.

15-6. (a) $Cu^{2+} + 2e^- \rightleftharpoons Cu(s)$ $E° = 0.339$ V

(b) $E_+ = 0.339 - \dfrac{0.059\,16}{2} \log \dfrac{1}{[Cu^{2+}]} = 0.309$ V

(c) $E = E_+ - E_- = 0.309 - 0.241 = 0.068$ V

15-7. A silver electrode serves as an indicator for Ag^+ by virtue of the equilibrium
$Ag^+ + e^- \rightleftharpoons Ag(s)$ that occurs at its surface. If the solution is saturated with
silver halide, then $[Ag^+]$ is affected by changes in halide concentration.
Therefore, the electrode is also an indicator for halide.

15-8. $V_e = 20.0$ mL. $Ag^+ + e^- \rightleftharpoons Ag(s) \Rightarrow E_+ = 0.799 - 0.059\,16 \log \dfrac{1}{[Ag^+]}$

0.1 mL: $\quad [Ag^+] = \underbrace{\left(\dfrac{19.9\ \text{mL}}{20.0\ \text{mL}}\right)}_{\substack{\text{Fraction} \\ \text{remaining}}} \underbrace{(0.050\,0\ \text{M})}_{\substack{\text{Original} \\ \text{concentration}}} \underbrace{\left(\dfrac{10.0\ \text{mL}}{10.1\ \text{mL}}\right)}_{\substack{\text{Dilution} \\ \text{factor}}} = 0.049\,3\ \text{M}$

$$E = E_+ - E_- = \left(0.799 - 0.059\,16 \log \dfrac{1}{0.049\,3}\right) - 0.241 = 0.481\ \text{V}$$

30.0 mL: This is 10.0 mL past $V_e \Rightarrow [Br^-] = \left(\dfrac{10.0}{40.0}\right)(0.025\,0\ \text{M}) = 0.006\,25\ \text{M}$

$$[Ag^+] = K_{sp}/[Br^-] = (5.0 \times 10^{-13})/0.006\,25 = 8.0 \times 10^{-11}\ \text{M}$$

$$E = E_+ - E_- = \left\{0.799 - 0.059\,16 \log \dfrac{1}{8.0 \times 10^{-11}}\right\} - 0.241 = -0.039\ \text{V}.$$

15-9. The reaction in the right half-cell is $Hg^{2+} + 2e^- \rightleftharpoons Hg(l)$

$$E = E_+ - E_-$$

$$-0.027 = \left(0.852 - \dfrac{0.059\,16}{2} \log \dfrac{1}{[Hg^{2+}]}\right) - (0.241)$$

$$\Rightarrow [Hg^{2+}] = 2.7 \times 10^{-22}\ \text{M}.$$

The cell contains 5.00 mmol EDTA (in all forms) and 1.00 mmol Hg(II) in 100 mL. 1.00 mmol EDTA reacts with 1.00 mmol Hg(II), leaving 4.00 mmol EDTA.

$$K_f = \dfrac{[HgY^{2-}]}{[Hg^{2+}][Y^{4-}]} = \dfrac{[HgY^{2-}]}{[Hg^{2+}]\alpha_{Y4-}\ [EDTA]}$$

$$K_f = \dfrac{(1.00\ \text{mmol}/100\ \text{mL})}{(2.7 \times 10^{-22})(0.30)(4.00\ \text{mmol}/100\ \text{mL})} = 3._1 \times 10^{21}$$

15-10. (a) $Fe^{3+} + e^- \rightleftharpoons Fe^{2+} \qquad\qquad E° = 0.771\ \text{V}$

(b) $E = E_+ - E_-$

$$-0.126 = 0.771 - 0.059\,16 \log \dfrac{[Fe^{2+}]}{[Fe^{3+}]} - 0.241 \Rightarrow \dfrac{[Fe^{2+}]}{[Fe^{3+}]} = 1._2 \times 10^{11}$$

(c) $\dfrac{K_f(FeEDTA^-)}{K_f(FeEDTA^{2-})} = \dfrac{[FeEDTA^-]}{[Fe^{3+}][EDTA^{4-}]} \div \dfrac{[FeEDTA^{2-}]}{[Fe^{2+}][EDTA^{4-}]}$

$$= \left(\dfrac{[FeEDTA^-]}{[FeEDTA^{2-}]}\right)\left(\dfrac{[Fe^{2+}]}{[Fe^{3+}]}\right) = \left(\dfrac{1.00 \times 10^{-3}}{2.00 \times 10^{-3}}\right)(1._2 \times 10^{11}) = 6 \times 10^{10}$$

15-11. $E = E_+ - E_- = -0.429 - 0.059\,16 \log \dfrac{[CN^-]^2}{[Cu(CN)_2^-]} - 0.197$

Putting in $E = -0.440$ V and $[Cu(CN)_2^-] = 1.00$ mM gives $[CN^-] = 0.847$ mM.

$pH = pK_a(HCN) + \log \dfrac{[CN^-]}{[HCN]} = 9.21 + \log \dfrac{8.47 \times 10^{-4}}{1.00 \times 10^{-3} - 8.47 \times 10^{-4}} = 9.95_4$

Now we use the pH to see how much HA reacted with KOH:

	HA	+	OH⁻	→	A⁻	+	H₂O
initial mmol	10.0		x		—		
final mmol	$10.0 - x$		—		x		

$pH = pK_a(HA) + \log \dfrac{[A^-]}{[HA]}$

$9.95_4 = 9.50 + \log \dfrac{x}{10.0 - x} \Rightarrow x = 7.4_0$ mmol of OH⁻

$[KOH] = \dfrac{7.4_0 \text{ mmol}}{25.0 \text{ mL}} = 0.29_6$ M

15-12. Junction potential arises because different ions diffuse at different rates across a liquid junction, leading to a separation of charge. The resulting electric field retards fast-moving ions and accelerates the slow-moving ions until a steady-state junction potential is reached. This limits the accuracy of a potentiometric measurement, because we do not know what part of a measured cell voltage is due to the process of interest and what is due to the junction potential. The cell in Figure 14-4 has no junction potential because there are no liquid junctions.

15-13. H^+ has greater mobility than K^+. The HCl side of the HCl | KCl junction will be negative because H^+ diffuses into the KCl region faster than K^+ diffuses into the HCl region. K^+ has a greater mobility than Na^+, so this junction has the opposite sign. The HCl | KCl voltage is larger, because the difference in mobility between H^+ and K^+ is greater than the difference in mobility between K^+ and Na^+.

15-14. Relative mobilities:

$K^+ \rightarrow 7.62 \qquad NO_3^- \rightarrow 7.40$

$5.19 \leftarrow Na^+ \qquad 7.91 \leftarrow Cl^-$

Both the cation and anion diffusion cause negative charge to build up on the <u>left</u>.

15-15. The junction potentials are dominated by the high mobility of H^+ and OH^-. The high mobility of H^+ makes the right side of 0.1 M HCl | 0.1 M KCl become positive. The high mobility of OH^- makes the right side of 0.1 M NaOH | 0.1 M KCl become negative. There is only a small junction potential at 0.1 M NaOH | KCl (saturated) because diffusion is dominated by the high concentrations of K^+ and Cl^-, which have nearly equal mobility.

15-16. Both half-cells contain saturated KCl. Therefore it makes sense to use saturated KCl in the salt bridge. The Ag | AgCl compartment also contains a very low concentration of Ag^+ and the calomel compartment contains a very low concentration of Hg_2^{2+}. With $[Cl^-] \approx 4.2$ M,

$$[Ag^+] \approx K_{sp}(AgCl)/[Cl^-] = (1.8 \times 10^{-10})/(4.2) \approx 10^{-11} \text{ M}$$

$$[Hg_2^{2+}] \approx K_{sp}(Hg_2Cl_2)/[Cl^-]^2 = (1.2 \times 10^{-18})/(4.2)^2 \approx 10^{-19} \text{ M}$$

The Ag^+ and Hg_2^{2+} concentrations are too low to affect the junction potential. Both junctions contain saturated KCl on both sides of each junction. The junction potential should therefore be very close to 0 at both junctions.

15-17. Velocity = mobility × field = $(36.30 \times 10^{-8} \text{ m}^2/(s \cdot V)) \times (7\,800 \text{ V/m}) = 2.83 \times 10^{-3}$ m s^{-1} for H^+ and $(7.40 \times 10^{-8})(7\,800) = 5.77 \times 10^{-4}$ m s^{-1} for NO_3^-. To cover 0.120 m will require $(0.120 \text{ m})/(2.83 \times 10^{-3} \text{ m s}^{-1}) = 42.4$ s for H^+ and $(0.120)/(5.77 \times 10^{-4}) = 208$ s for NO_3^-.

15-18. (a) $E° = 0.799$ V $\Rightarrow K = 10^{0.799/0.059\,16} = 3._{20} \times 10^{13}$

   (b) $K' = 10^{0.801/0.059\,16} = 3._{46} \times 10^{13}$. $K'/K = 1.08$. The increase is 8%.

   (c) $K = 10^{0.100/0.059\,16} = 49.0$. $K' = 10^{0.102/0.059\,16} = 53.0$
   $K'/K = 1.08$. The change is still 8%.

15-19. Both half-cell reactions are the same (AgCl + $e^- \rightleftharpoons$ Ag + Cl^-) and the concentration of Cl^- is the same on both sides. In principle, the voltage of the cell would be 0 if there were no junction potential. The measured voltage can be attributed to the junction potential. In practice, if both sides contained 0.1 M HCl (or 0.1 M KCl), the two electrodes would probably produce a small voltage because no two real cells are identical. This voltage can be measured and subtracted from the voltage measured with the HCl | KCl junction.

15-20. **(a)** In phase α, we have 0.1 M H^+ ($u = 36.3 \times 10^{-8}$ m^2 s^{-1} V^{-1}) and 0.1 M Cl$^-$

($u = 7.91 \times 10^{-8}$ m^2 s^{-1} V^{-1}). In phase β, we have 0.1 M K^+

($u = 7.62 \times 10^{-8}$ m^2 s^{-1} V^{-1}) and 0.1 M Cl$^-$ ($u = 7.91 \times 10^{-8}$ m^2 s^{-1} V^{-1}).

Substituting into the Henderson equation gives

$$E_j = \frac{(36.3 \times 10^{-8})[0 - 0.1] + (7.62 \times 10^{-8})[0.1 - 0] - (7.91 \times 10^{-8})[0.1 - 0.1]}{(36.3 \times 10^{-8})[0 - 0.1] + (7.62 \times 10^{-8})[0.1 - 0] + (7.91 \times 10^{-8})[0.1 - 0.1]} \times$$

$$0.059\,16 \log \frac{(36.3 \times 10^{-8})(0.1) + (7.91 \times 10^{-8})(0.1)}{(7.62 \times 10^{-8})(0.1) + (7.91 \times 10^{-8})(0.1)} = 26.9 \text{ mV}$$

(b)

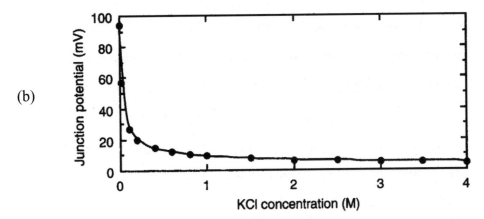

(c)

[HCl]	y M HCl | 1mM KCl	y M HCl | 4 M KCl
10^{-4} M	9.1 mV	4.6 mV
10^{-3} M	26.9 mV	3.6 mV
10^{-2} M	57.3 mV	3.0 mV
10^{-1} M	93.6 mV	4.7 mV

15-21. The electrode should be calibrated at 37° using two buffers bracketing the pH of the blood. It would be reasonable to use the MOPSO and HEPES buffers in Table 15-3 that are recommended for use with physiologic fluids. The pH of these standards at 37°C is 6.695 and 7.370. The standards should be thermostatted to 37° during calibration, and the blood should also be at 37° during the measurement.

15-22. Uncertainty in pH of standard buffers, junction potential, junction potential drift, sodium or acid errors at extreme pH values, equilibration time, hydration of glass, temperature of measurement and calibration, and cleaning of electrode

15-23. The error measured in the graph is –0.33 pH units. The electrode will indicate $11.00 - 0.33 = 10.67$.

15-24. Saturated potassium hydrogen tartrate and 0.05 m potassium hydrogen phthalate

15-25. If the alkaline solution has a high concentration of Na^+ (as in NaOH), the Na^+ cation competes with H^+ for cation exchange sites on the glass surface. The glass responds as if H^+ were present, and the apparent pH is lower than the actual pH.

15-26. The junction potential changes from -6.4 mV to -0.2 mV. A change of $6.4 - 0.2$ $= +6.2$ mV appears to be a pH change of $+6.2/59.16 = +0.10$ pH units.

15-27. (a) $(4.63)(59.16 \text{ mV}) = 274$ mV. The factor 59.16 mV is the value of $(RT \ln 10)/F$ at 298.15 K.

(b) At 310.15 K (37°C), $(RT \ln 10)/F$

$= (8.3145 \text{ J mol}^{-1} \text{ K}^{-1})(310.15 \text{ K})(\ln 10)/(96\,485 \text{ C mol}^{-1}) = 61.54$ mV

$(4.63)(61.54 \text{ mV}) = 285$ mV.

15-28. pH of 0.025 m KH_2PO_4/0.025 m Na_2HPO_4 at 20°C = 6.881

pH of 0.05 m potassium hydrogen phthalate at 20°C = 4.002

$$\frac{E_{unknown} - E_{S1}}{pH_{unknown} - pH_{S1}} = \frac{E_{S2} - E_{S1}}{pH_{S2} - pH_{S1}}$$

$$\frac{E_{unknown} - (-18.3 \text{ mV})}{pH_{unknown} - 6.881} = \frac{(+146.3 \text{ mV}) - (-18.3 \text{ mV})}{4.002 - 6.881} = -57.17_3 \text{ mV/pH unit}$$

$$pH_{unknown} = \frac{E_{unknown} - (-18.3 \text{ mV})}{-57.17_3 \text{ mV/pH unit}} + 6.881 = 5.686$$

Observed slope $= -57.17_3$ mV/pH unit

Theoretical slope $= -\dfrac{RT \ln 10}{F}$

$$= -\frac{[8.3145 \text{ V·C/(mol·K)}][293.15 \text{ K}] \ln 10}{9.64853 \times 10^4 \text{ C/mol}} = -0.05817 \text{ V}$$

$$\beta = \frac{\text{observed slope}}{\text{theoretical slope}} = \frac{-57.17_3 \text{ mV}}{-58.17 \text{ mV}} = 0.983$$

15-29. (a) There is negligible change in the concentrations of the buffer species when we mix the acid $H_2PO_4^-$ with its conjugate base, HPO_4^{2-}. The ionic strength of 0.0250 m KH_2PO_4 (a 1:1 electrolyte) is 0.0250 m. The ionic strength of 0.0250 m Na_2HPO_4 (a 2:1 electrolyte) is 0.0750 m. The total ionic strength is 0.100 m.

(b) $\quad K_2 = \dfrac{[H^+]\gamma_{H^+}[HPO_4^{2-}]\gamma_{HPO_4^{2-}}}{[H_2PO_4^-]\gamma_{H_2PO_4^-}}$

But $K_2 = 10^{-7.198}$ and $[H^+]\gamma_{H^+} = 10^{-pH} = 10^{-6.865}$

Therefore, $\dfrac{\gamma_{HPO_4^{2-}}}{\gamma_{H_2PO_4^-}} = \dfrac{K_2[H_2PO_4^-]}{[H^+]\gamma_{H^+}[HPO_4^{2-}]} = \dfrac{10^{-7.198}[0.025\,0]}{10^{-6.865}[0.025\,0]} = 0.464_5$

(We can use molality or any other units for concentrations because they cancel in the numerator and denominator.)

(c) To get a pH of 7.000, we need to increase the concentration of base (HPO_4^{2-}) and decrease the concentration of acid ($H_2PO_4^-$). To maintain a constant ionic strength, we must decrease KH_2PO_4 three times as much as we increase Na_2HPO_4, because Na_2HPO_4 contributes three times as much as KH_2PO_4 to the ionic strength. So let's increase Na_2HPO_4 by x and decrease KH_2PO_4 by $3x$.

$$K_2 = \dfrac{[H^+]\gamma_{H^+}[HPO_4^{2-}]\gamma_{HPO_4^{2-}}}{[H_2PO_4^-]\gamma_{H_2PO_4^-}}$$

$$\Rightarrow 10^{-7.198} = \dfrac{10^{-7.000}[0.025\,0 + x]}{[0.025\,0 - 3x]}(0.464_5) \Rightarrow x = 0.001\,8\ m.$$

The new concentrations should be $Na_2HPO_4 = 0.026\,8\ m$ and $KH_2PO_4 = 0.019\,6\ m$.

15-30. (a)

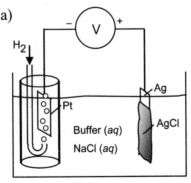 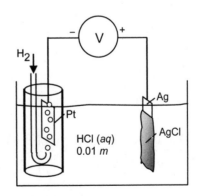

(b) $\quad \log \gamma_{Cl} = \dfrac{-Az^2\sqrt{\mu}}{1 + B\alpha\sqrt{\mu}} = \dfrac{(-0.511)(-1)^2\sqrt{0.1}}{1 + 1.5\sqrt{0.1}} \Rightarrow \gamma_{Cl} = 0.777$

$-\log \mathcal{A}_H\gamma_{Cl} = p(\mathcal{A}_H\gamma_{Cl})^o = 6.972$ from the graph at zero added NaCl

$-\log \mathcal{A}_H(0.777) = 6.972 \Rightarrow \mathcal{A}_H = (10^{-6.972})/0.777 = 1.373 \times 10^{-7}$

$pH = -\log(1.373 \times 10^{-7}) = 6.862$ (value in table = 6.865)

15-31. (a) Analyte ions equilibrate between the outer solution and ligand L in the ion-selective membrane. This equilibrium creates a slight charge imbalance (an electric potential difference) across the interface between the membrane and the analyte solution because other ions in the membrane and outer solution are not free to cross the interface. Changes in analyte ion concentration in the outer solution change the potential difference across the outer boundary of the ion-selective membrane.

(b) Compound electrodes contain a conventional electrode surrounded by a membrane that isolates (or generates) the analyte to which the electrode responds.

15-32. The selectivity coefficient $K_{A,X}^{Pot}$ tells us the relative response of an ion-selective electrode to the ion of interest (A) and an interfering ion (X). The smaller $K_{A,X}^{Pot}$, the more selective is the electrode (smaller response to the interfering ion).

15-33. A mobile molecule dissolved in the membrane liquid phase binds tightly to the ion of interest and weakly to interfering ions.

15-34. A metal ion buffer maintains the desired (small) concentration of metal ion from a large reservoir of metal complex (ML) and free ligand (L). If you just tried to dissolve 10^{-8} M metal ion in most solutions or containers, the metal would probably bind to the container wall or to an impurity in the solution and be lost.

15-35. Electrodes respond to *activity*. If the ionic strength is constant, the activity coefficient of analyte will be constant in all standards and unknowns. In this case, the calibration curve can be written directly in terms of concentration.

15-36. (a) $-0.230 = \text{constant} - 0.059\,16 \log(1.00 \times 10^{-3}) \Rightarrow \text{constant} = -0.407$ V

(b) $-0.300 = -0.407 - 0.059\,16 \log x \Rightarrow x = 1.5_5 \times 10^{-2}$ M

(c) $-0.230 = \text{constant} - 0.059\,16 \log(1.00 \times 10^{-3})$

$\underline{-0.300 = \text{constant} - 0.059\,16 \log\ x}$

subtract: $0.070 = -0.059\,16 \log \dfrac{1.00 \times 10^{-3}}{x} \Rightarrow x = 1.5_2 \times 10^{-2}$ M

15-37. $E_1 = \text{constant} + \dfrac{0.059\,16}{2} \log\,[1.00 \times 10^{-4}]$

$E_2 = \text{constant} + \dfrac{0.059\,16}{2} \log\,[1.00 \times 10^{-3}]$

$\Delta E = E_2 - E_1 = \dfrac{0.059\,16}{2} \log\dfrac{1.00 \times 10^{-3}}{1.00 \times 10^{-4}} = +0.029\,6 \text{ V}$

✱ 15-38. $[\text{F}^-]_{\text{Providence}} = 1.00 \text{ mg F}^-/\text{L} = 5.26 \times 10^{-5} \text{ M}$

$E_{\text{Providence}} = \text{constant} - 0.059\,16 \log\,[5.26 \times 10^{-5}]$

$E_{\text{Foxboro}} = \text{constant} - 0.059\,16 \log\,[\text{F}^-]_{\text{Foxboro}}$

$\Delta E = E_{\text{Foxboro}} - E_{\text{Providence}} = 0.040\,0 \text{ V}$

$= -0.059\,16 \log\dfrac{[\text{F}^-]_{\text{Foxboro}}}{5.26 \times 10^{-5}} \Rightarrow [\text{F}^-]_{\text{Foxboro}} = 1.11 \times 10^{-5} \text{ M} = 0.211 \text{ mg/L}$

15-39. K^+ has the largest selectivity coefficient of Group 1 ions and therefore interferes the most. Sr^{2+} and Ba^{2+} are the worst of the Group 2 ions. Since $\log K_{\text{Li}^+,\text{K}^+}^{\text{Pot}} \approx -2$, there must be 100 times more K^+ than Li^+ to give equal response.

15-40. $\dfrac{[\text{ML}]}{[\text{M}][\text{L}]} = 4.0 \times 10^8 = \dfrac{0.030 \text{ M}}{[\text{M}](0.020 \text{ M})} \Rightarrow [\text{M}] = 3.8 \times 10^{-9} \text{ M}$

15-41. (a) The least squares parameters are

$E = 51.10\,(\pm0.24) + 28.14\,(\pm0.08_5)\log\,[\text{Ca}^{2+}] \quad (s_y = 0.2_7)$

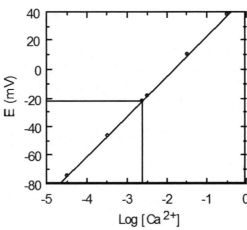

(b) The slope is $0.028\,14 \text{ V} = \beta(0.059\,16 \text{ V})/2 \Rightarrow \beta = 0.951$.

(c)   If we use Equation 4-27 in a spreadsheet, we find log $[Ca^{2+}] =$ $-2.615_3(\pm0.007_2)$ using $s_y = 0.3$ and $k = 4$.

From Table 3-1, we can write that if $F = 10^x$, $e_F/F = (\ln 10)e_x$.

In this problem, $F = [Ca^{2+}] = 10^{-2.6153(\pm0.0072)}$

$e_F/F = (\ln 10)(0.007_2) = 0.016_6$

$e_F = (0.016_6)\,F = 4.0 \times 10^{-5} \Rightarrow F = 2.43(\pm0.04) \times 10^{-3}$ M.

15-42.   At pH 7.2 the effect of H^+ will be negligible because $[H^+] \ll [Li^+]$:

-0.333 V $=$ constant $+ 0.059\,16 \log [3.44 \times 10^{-4}] \Rightarrow$ constant $= -0.128$ V.

At pH 1.1 ($[H^+] = 0.079$ M), we must include interference by H^+:

$E = -0.128 + 0.059\,16 \log [3.44 \times 10^{-4} + (4 \times 10^{-4})(0.079)] = -0.331$ V.

15-43.   The function to plot on the y-axis is $(V_o + V_s)\,10^{E/S}$, where $S = -(\beta RT \ln 10)/nF$. (The minus sign comes from the equation for the response of the electrode, which has a minus sign in front of the log term.) Putting in $\beta = 0.933$, $R = 8.314\,5$ J/(mol·K), $F = 96\,485$ C/mol, $T = 303.15$ K, and $n = 2$ gives $S = -0.028\,06$ J/C $=$ 0.028 06 V. (You can get the relation of J/C = V from the equation $\Delta G = -nFE$, in which the units are J = (mol)(C/mol)(V).)

V_s (mL)	E (V)	y
0	0.079 0	0.084 1
0.100	0.072 4	0.144 9
0.200	0.065 3	0.259 9
0.300	0.058 8	0.443 8
0.800	0.050 9	0.856 5

The graph has a slope of $m = 0.989$ and an intercept of $b = 0.080\,9$, giving an x-intercept of $-b/m = -0.081_8$ mL. The concentration of original unknown is

$$c_X = -\frac{(x\text{-intercept})\,c_s}{V_o} = -\frac{(-0.081_8 \text{ mL})(0.020\,0 \text{ M})}{55.0 \text{ mL}} = 3.0 \times 10^{-5} \text{ M}.$$

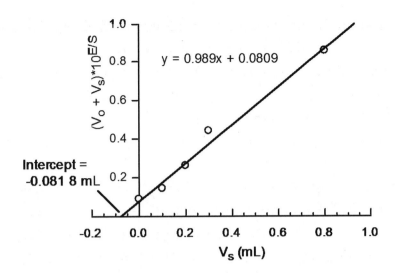

15-44. (a) The spreadsheet and graph are shown below. The x-intercept is at -3.65 mL with a standard deviation in cell B26 of $s = 0.484$ mL. The intercept gives us the concentration of ammonia nitrogen in the volume $V_0 = 101.0$ mL:

$$x\text{-intercept} = -3.65 \text{ mL} = -\frac{V_0 c_X}{c_S}$$

$$\Rightarrow c_X = \frac{(x\text{-intercept})c_S}{V_0} = \frac{(3.65 \text{ mL})(10.0 \text{ ppm})}{101.0 \text{ mL}} = 0.361_4 \text{ ppm}$$

The concentration of ammonia nitrogen in the original 100.0 mL of seawater, which had been diluted from 100.0 to 101.0 mL, is

$\frac{101.0 \text{ mL}}{100.0 \text{ mL}} (0.3614 \text{ ppm}) = 0.365$ ppm. The 95% confidence interval is equal to Student's t times the standard deviation:

$$95\% \text{ confidence interval} = \pm t \cdot s = \pm(3.18)(0.484 \text{ mL}) = \pm 1.54 \text{ mL}$$

where t is for 95% confidence and $5 - 2 = 3$ degrees of freedom because there are 5 data points on the standard addition curve. You can find t in the table of Student's t or you can compute it with the statement "=TINV(0.05,3)" in cell G24 in the spreadsheet. The confidence interval of ± 1.54 mL corresponds to a relative uncertainty of $100 \times \frac{1.54 \text{ mL}}{3.65 \text{ mL}} = 42\%$. The absolute uncertainty is $(0.042)(0.365 \text{ ppm}) = 0.15$ ppm. The concentration of ammonia nitrogen in the seawater can be expressed as 0.36 ± 0.15 ppm.

	A	B	C	D	E	F	G	H
1	Standard Addition: Ammonia in Seawater							
2								
3	$V_0 =$	101	mL					
4	$c_S =$	10	ppm					
5	$s =$	0.0566	V					
6								
7	x			y				
8	Added standard	E = cell	$V_0 + V_s$	$(V_0 + V_s)10^{E/s}$				
9	(mL)	voltage (V)	(mL)	(mL)				
10	0.00	-0.0844	101.0	3.26				
11	10.00	-0.0581	111.0	10.44				
12	20.00	-0.0469	121.0	17.95				
13	30.00	-0.0394	131.0	26.37				
14	40.00	-0.0347	141.0	34.37				
15								
16		LINEST output:			Highlight cells B17:D19			
17	m	0.7815	2.8502	b	Type '=LINEST(D10:D14,A10:A14,TRUE,TRUE)			
18	u_m	0.0137	0.3364	u_b	Press CTRL+SHIFT+ENTER (on PC)			
19	R^2	0.9991	0.4343	s_y	Press COMMAND+RETURN (on Mac)			
20								
21	x-intercept = -b/m =	-3.647	mL			To find 95% confidence interval		
22	n =	5	B22 = COUNT(A10:A14)			we need Student's t for		
23	Mean y =	18.479	B23 = AVERAGE(D10:D14)			3 degrees of freedom		
24	$\Sigma(x_i - \text{mean } x)^2 =$	1000	B24 = DEVSQ(A10:A14)			TINV(0.05,3) =	3.182446	
25	Std uncertainty of					t*s =	1.541109	
26	x-intercept =	0.484	mL					
27	B26 = (C19/ABS(B17))*SQRT((1/B22) + B23^2/(B17^2*B24))							

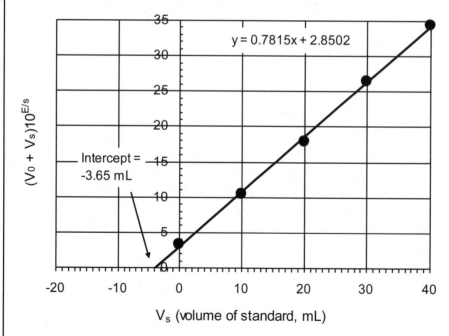

$$y = 0.7815x + 2.8502$$

Intercept = -3.65 mL

(b)   Added standards should increase analytical signal by a factor of 1.5 to 3. In this experiment, analytical signal is $(V_0 + V_S)10^{E/S} = 3.26$ mL for unknown and 34.37 mL for final standard. The final signal is 10 times as great as the initial signal, which is ~3 times more than the recommended limit.

15-45. For the first line of data, with $A = Na^+$ and $X = Mg^{2+}$

$$\log K_{A,X}^{Pot} = \frac{z_A F(E_X - E_A)}{RT \ln 10} + \log\left(\frac{\mathcal{A}_A}{\mathcal{A}_X^{z_A/z_X}}\right)$$

$$= \frac{(+1)(96\,485\ C/mol)(-0.385\ V)}{(8.3145\ V \cdot C/[mol \cdot K])(294.65\ K)\ln 10} + \log\left(\frac{10^{-3}}{(10^{-3})^{1/2}}\right) = -8.09.$$

For the second line of data, $\log K_{A,X}^{Pot} = -8.15$.

The first and second lines should give the same selectivity coefficient. The difference is experimental error.

For the third line of data, with $A = Na^+$ and $X = K^+$:

$$\log K_{A,X}^{Pot} = \frac{(+1)(96\,485\ C/mol)(-0.285\ V)}{(8.3145\ V \cdot C/[mol \cdot K])(294.65\ K)\ln 10} + \log\left(\frac{10^{-3}}{(10^{-3})^{1/1}}\right) = -4.87.$$

For the fourth line of data, $\log K_{A,X}^{Pot} = -4.87$.

15-46. For Na^+: % error in $\mathcal{A}_{H^+} = \dfrac{(10^{-8.6})^{1/1}(10^{-2.0})}{(10^{-8.0})^{1/1}} \times 100 = 0.25\%$

For Ca^{2+}: % error in $\mathcal{A}_{H^+} = \dfrac{(10^{-7.8})^{2/1}(10^{-2.0})}{(10^{-8.0})^{2/1}} \times 100 = 2.5\%$

15-47. $E = \underset{A}{\text{constant}} + \underset{B}{\frac{\beta(0.059\,16)}{2}} \log ([Ca^{2+}] + K_{Ca^{2+},Mg^{2+}}^{Pot} [Mg^{2+}])$

For the first two solutions we can write

$$-52.6\ mV = A + B\ \log(1.00 \times 10^{-6}) = A - 6\,B$$

$$+16.1\ mV = A + B\ \log(2.43 \times 10^{-4}) = A - 3.614\,B.$$

Subtraction gives $68.7\ mV = 2.386\,B \Rightarrow B = 28.80\ mV$.

Putting this value of B back into the first equation gives $A = 120.2\ mV$.

The third set of data now gives the selectivity coefficient:

$$-38.0\ mV = 120.2 + 28.80 \log [10^{-6} + K_{Ca^{2+},Mg^{2+}}^{Pot} (3.68 \times 10^{-3})]$$

$$\Rightarrow K_{Ca^{2+},Mg^{2+}}^{Pot} = 6.0 \times 10^{-4}$$

$$E = 120.2 + 28.80 \log ([Ca^{2+}] + 6.0 \times 10^{-4} [Mg^{2+}]).$$

15-48. There is a large excess of EDTA in the buffer. We expect essentially all lead to be in the form PbY^{2-} (where Y = EDTA).

$$[PbY^{2-}] = \frac{1.0}{101.0}(0.10\ M) = 9.9 \times 10^{-4}\ M$$

$$\text{Total EDTA} = \frac{100.0}{101.0}(0.050\ M) = 0.049_5\ M$$

$$\text{Free EDTA} = 0.049_5\ M - 9.9 \times 10^{-4}\ M = \underbrace{0.048_5\ M}_{\substack{\text{EDTA bound} \\ \text{to } Pb^{2+}}}$$

$$Pb^{2+} + Y^{4-} \rightleftharpoons PbY^{2-} \qquad K_f' = \alpha_{Y^{4-}}K_f = (1.46 \times 10^{-8})(10^{18.0}) = 1.4_6 \times 10^{10}$$

$$K_f' = \frac{[PbY^{2-}]}{[Pb^{2+}][EDTA]}$$

$$\Rightarrow [Pb^{2+}] = \frac{[PbY^{2-}]}{K_f'[EDTA]} = \frac{9.9 \times 10^{-4}}{(1.4_6 \times 10^{10})(0.048_5)} = 1.4 \times 10^{-12}\ M$$

15-49. $[Hg^{2+}]$ in the buffers is computed from equilibrium constants for the solubility of HgX_2 and formation of complex ions such as HgX_3^-. Since the data for $HgCl_2$ are not in line with the data for $Hg(NO_3)_2$ and $HgBr_2$, equilibrium constants used for the $HgCl_2$ system could be in error. Whenever we make a buffer by mixing *calculated* quantities of reagents, we are at the mercy of the quality of tabulated equilibrium constants.

15-50. (a) $\text{slope} = 29.58\ mV = \dfrac{E_2 - E_1}{\log \mathcal{A}_2 - \log \mathcal{A}_1} = \dfrac{(-25.90) - 2.06}{\log \mathcal{A}_2 - (-3.000)}$

$$\Rightarrow \mathcal{A}_2 = 1.13 \times 10^{-4}$$

(b) $\quad Ca^{2+} \qquad + \qquad A^{3-} \qquad\qquad \rightleftharpoons \qquad CaA^-$

$\quad\ \ 5.00 \times 10^{-4} - x \qquad (0.998)[(5.00 \times 10^{-4}) - x] \qquad\qquad x$

But $\mathcal{A}_{Ca^{2+}} = 1.13 \times 10^{-4} = (5.00 \times 10^{-4} - x)\,\underset{\substack{\uparrow \\ \gamma \text{ from Table 8-1}}}{(0.405)}$

$$\Rightarrow\ x = 2.2 \times 10^{-4}\ M$$

$$K_f = \frac{[CaA^-]\gamma_{CaA^-}}{[Ca^{2+}]\gamma_{Ca^{2+}}[A^{3-}]\gamma_{A^{3-}}}$$

$$K_f = \frac{(2.20 \times 10^{-4})(0.79)}{(1.13 \times 10^{-4})[(0.998)(5 \times 10^{-4} - 2.20 \times 10^{-4})](0.115)}$$

$$K_f = 4.8 \times 10^4$$

15-51. Analyte adsorbed on the surface of the gate changes the electric potential of the gate. This, in turn, changes the current between the source and drain. The potential that must be applied by the external circuit to restore the current to its initial value is a measure of the change in gate potential. Following the Nernst equation, there is close to a 59 mV change in gate potential for each factor-of-10 change in activity of univalent analyte at 25°C. The key to ion-specific response is to have a chemical on the gate that selectively binds one analyte.

15-52. DNA is randomly cleaved into many small fragments. Many copies ($\sim 10^5$-10^6) of one strand of one fragment are attached to one bead placed into one well of the device. Different beads in each well carry different fragments of DNA. The enzyme and primer required for DNA replication are added to the fragments on each bead. A solution containing a single nucleotide triphosphate is then added to all of the wells. If the next required nucleotide for chain extension was added to DNA in a particular well, it is incorporated and liberates some H^+ from the reaction with the existing strand of DNA. The H^+-sensitive tantalum oxide at the base of the well regulates the current in a field effect transistor below each well. If the right base was added to extend the DNA in a particular well, H^+ is released and there is a voltage change in the transistor associated with that well. If the wrong base was added, no reaction occurs in that well and no signal is recorded by the transistor. After a few seconds, the wells are washed and fresh solution with a different nucleotide base is added. The sequence is repeated to see which wells respond to which added base. The purpose of each field effect transistor is to respond when the next required base is added to each well.

CHAPTER 16
REDOX TITRATIONS

16-1. (a) $Ce^{4+} + Fe^{2+} \rightarrow Ce^{3+} + Fe^{3+}$

(b) $Fe^{3+} + e^- \rightleftharpoons Fe^{2+}$ $E^o = 0.767$ V

$Ce^{4+} + e^- \rightleftharpoons Ce^{3+}$ $E^o = 1.70$ V

(c) $E = \left(0.767 - 0.059\,16 \log \dfrac{[Fe^{2+}]}{[Fe^{3+}]} \right) - (0.241)$ (A)

$E = \left(1.70 - 0.059\,16 \log \dfrac{[Ce^{3+}]}{[Ce^{4+}]} \right) - (0.241)$ (B)

(d) <u>10.0 mL</u>:   Use Eq. (A) with $[Fe^{2+}]/[Fe^{3+}] = 40.0/10.0$, since
$V_e = 50.0$ mL $\Rightarrow E = 0.490$ V.

    <u>25.0 mL</u>: $[Fe^{2+}]/[Fe^{3+}] = 25.0/25.0 \Rightarrow E = 0.526$ V

    <u>49.0 mL</u>: $[Fe^{2+}]/[Fe^{3+}] = 1.0/49.0 \Rightarrow E = 0.626$ V

    <u>50.0 mL</u>:   This is V_e, where $[Ce^{3+}] = [Fe^{3+}]$ and $[Ce^{4+}] = [Fe^{2+}]$.
Eq. 16-11 gives $E_+ = 1.23$ V and $E = 0.99$ V.

    <u>51.0 mL</u>:   Use Eq. (B) with $[Ce^{3+}]/[Ce^{4+}] = 50.0/1.0 \Rightarrow E = 1.36$ V.

    <u>60.0 mL</u>: $[Ce^{3+}]/[Ce^{4+}] = 50.0/10.0 \Rightarrow E = 1.42$ V

    <u>100.0 mL</u>: $[Ce^{3+}]/[Ce^{4+}] = 50.0/50.0 \Rightarrow E = 1.46$ V

16-2. (a) $Ce^{4+} + Cu^+ \rightarrow Ce^{3+} + Cu^{2+}$

(b) $Ce^{4+} + e^- \rightleftharpoons Ce^{3+}$ $E^o = 1.70$ V

$Cu^{2+} + e^- \rightleftharpoons Cu^+$ $E^o = 0.161$ V

(c) $E = \left(1.70 - 0.059\,16 \log \dfrac{[Ce^{3+}]}{[Ce^{4+}]} \right) - (0.197)$ (A)

$E = \left(0.161 - 0.059\,16 \log \dfrac{[Cu^+]}{[Cu^{2+}]} \right) - (0.197)$ (B)

(d) <u>1.00 mL</u>:   Use Eq. (A) with $[Ce^{3+}]/[Ce^{4+}] = 1.00/24.0$, since
$V_e = 25.0$ mL $\Rightarrow E = 1.58$ V.

    <u>12.5 mL</u>: $[Ce^{3+}]/[Ce^{4+}] = 12.5/12.5 \Rightarrow E = 1.50$ V

    <u>24.5 mL</u>: $[Ce^{3+}]/[Ce^{4+}] = 24.5/0.5 \Rightarrow E = 1.40$ V

    <u>25.0 mL</u>: $E_+ = 1.70 - 0.059\,16 \log \dfrac{[Ce^{3+}]}{[Ce^{4+}]}$

$$E_+ = 0.161 - 0.059\,16 \log \dfrac{[Cu^+]}{[Cu^{2+}]}$$

$$2E_+ = 1.86_1 - 0.059\,16 \log \dfrac{[Ce^{3+}][Cu^+]}{[Ce^{4+}][Cu^{2+}]}$$

222

At the equivalence point, $[Ce^{3+}] = [Cu^{2+}]$ and $[Ce^{4+}] = [Cu^+]$.
Therefore, the log term above is zero and $E_+ = 1.86_1/2 = 0.930$ V.

$$E = 0.930 - 0.197 = 0.733 \text{ V}$$

25.5 mL: Use Eq. (B) with $[Cu^+]/[Cu^{2+}] = 0.5/25.0 \Rightarrow E = 0.065$ V.

30.0 mL: $[Cu^+]/[Cu^{2+}] = 5.0/25.0 \Rightarrow E = 0.005$ V

50.0 mL: $[Cu^+]/[Cu^{2+}] = 25.0/25.0 \Rightarrow E = -0.036$ V

16-3. (a) $Sn^{2+} + Tl^{3+} \rightarrow Sn^{4+} + Tl^+$

(b) $Sn^{4+} + 2e^- \rightleftharpoons Sn^{2+} \quad E° = 0.139$ V

$Tl^{3+} + 2e^- \rightleftharpoons Tl^+ \quad E° = 0.77$ V

(c) $E = \left(0.139 - \dfrac{0.059\,16}{2} \log \dfrac{[Sn^{2+}]}{[Sn^{4+}]}\right) - (0.241)$ (A)

$E = \left(0.77 - \dfrac{0.059\,16}{2} \log \dfrac{[Tl^+]}{[Tl^{3+}]}\right) - (0.241)$ (B)

(d) 1.00 mL: Use Eq. (A) with $[Sn^{2+}]/[Sn^{4+}] = 4.00/1.00$, since
$V_e = 5.00$ mL $\Rightarrow E = -0.120$ V.

2.50 mL: $[Sn^{2+}]/[Sn^{4+}] = 2.50/2.50 \Rightarrow E = -0.102$ V

4.90 mL: $[Sn^{2+}]/[Sn^{4+}] = 0.10/4.90 \Rightarrow E = -0.052$ V

5.00 mL: $E_+ = 0.139 - \dfrac{0.059\,16}{2} \log \dfrac{[Sn^{2+}]}{[Sn^{4+}]}$

$\underline{E_+ = 0.77 - \dfrac{0.059\,16}{2} \log \dfrac{[Tl^+]}{[Tl^{3+}]}}$

$2E_+ = 0.90_9 - \dfrac{0.059\,16}{2} \log \dfrac{[Sn^{2+}][Tl^+]}{[Sn^{4+}][Tl^{3+}]}$

At the equivalence point, $[Sn^{4+}] = [Tl^+]$ and $[Sn^{2+}] = [Tl^{3+}]$.
Therefore, the log term above is zero and $E_+ = 0.90_9/2 = 0.45_4$ V.

$$E = 0.45_4 - 0.241 = 0.21 \text{ V}$$

5.10 mL: Use Eq. (B) with $[Tl^+]/[Tl^{3+}] = 5.00/0.10 \Rightarrow E = 0.48$ V

10.0 mL: Use Eq. (B) with $[Tl^+]/[Tl^{3+}] = 5.00/5.00 \Rightarrow E = 0.53$ V

16-4. (a) $2Fe^{3+} + \text{ascorbic acid} + H_2O \rightarrow 2Fe^{2+} + \text{dehydroascorbic acid} + 2H^+$

(b) The equivalence volume is 10.0 mL.
At 5.0 mL, half of the Fe^{3+} is titrated and the ratio $[Fe^{2+}]/[Fe^{3+}]$ is 5.0/5.0:

$Fe^{3+} + e^- \rightleftharpoons Fe^{2+} \qquad\qquad E° = 0.767$ V

$E = E_+ - E_- = \left(0.767 - 0.059\,16 \log \dfrac{[Fe^{2+}]}{[Fe^{3+}]}\right) - (0.197)$

$$= \left(0.767 - 0.059\,16 \log \frac{5.0}{5.0}\right) - (0.197) = 0.570$$

<u>10.0 mL</u> is the equivalence point. We multiply the ascorbic acid Nernst equation by 2 and add it to the iron Nernst equation:

$$E_+ = 0.767 - 0.059\,16 \log \frac{[Fe^{2+}]}{[Fe^{3+}]}$$

$$2E_+ = 2\left(0.390 - \frac{0.059\,16}{2} \log \frac{[\text{ascorbic acid}]}{[\text{dehydro}][H^+]^2}\right)$$

$$3E_+ = 1.547 - 0.059\,16 \log \frac{[Fe^{2+}][\text{ascorbic acid}]}{[Fe^{3+}][\text{dehydro}][H^+]^2}$$

At the equivalence point, the stoichiometry of the titration reaction tells us that $[Fe^{2+}] = 2[\text{dehydroascorbic acid}]$ and $[Fe^{3+}] = 2[\text{ascorbic acid}]$. Inserting these equalities into the log term just shown gives

$$3E_+ = 1.547 - 0.059\,16 \log \frac{2[\text{dehydro}][\text{ascorbic acid}]}{2[\text{ascorbic acid}][\text{dehydro}][H+]^2}$$

$$3E_+ = 1.547 - 0.059\,16 \log \frac{1}{[H+]^2}$$

Using $[H^+] = 10^{-0.30}$ gives $E_+ = 0.504$ V and $E = 0.504 - 0.197 = 0.307$ V.

At <u>15.0 mL</u>, the ratio [dehydro]/[ascorbic acid] is 10.0/5.0:

dehydroascorbic acid $+ 2H^+ + 2e^- \rightleftharpoons$ ascorbic acid $+ H_2O$ $E° = 0.390$ V

$$E = E_+ - E_- = \left(0.390 - \frac{0.059\,16}{2} \log \frac{[\text{ascorbic acid}]}{[\text{dehydro}][H^+]^2}\right) - (0.197)$$

$$= \left(0.390 - \frac{0.059\,16}{2} \log \frac{[5.0]}{[10.0][\,10^{-0.30}]^2}\right) - (0.197) = 0.184 \text{ V}$$

16-5. (a) Titration reaction: $Sn^{2+} + 2Fe^{3+} \rightarrow Sn^{4+} + 2Fe^{2+}$ $V_e = 25.0$ mL

(b) $Fe^{3+} + e^- \rightleftharpoons Fe^{2+}$ $E° = 0.732$ V

$Sn^{4+} + 2e^- \rightleftharpoons Sn^{2+}$ $E° = 0.139$ V

(c) $E = \left(0.732 - 0.059\,16 \log \frac{[Fe^{2+}]}{[Fe^{3+}]}\right) - (0.241)$ (A)

$E = \left(0.139 - \frac{0.059\,16}{2} \log \frac{[Sn^{2+}]}{[Sn^{4+}]}\right) - (0.241)$ (B)

(d) Representative calculations:

<u>1.0 mL</u>: $E_+ = 0.139 - \frac{0.059\,16}{2} \log \frac{[Sn^{2+}]}{[Sn^{4+}]}$

initial mol $Sn^{2+} = (25.0 \text{ mL})(0.050\,0 \frac{\text{mmol}}{\text{mL}}) = 1.25$ mmol

mol Fe^{3+} added $= (1.0 \text{ mL})(0.100 \frac{\text{mmol}}{\text{mL}}) = 0.10$ mmol

Note that 1 mol Fe^{3+} consumes ½ mol Sn^{2+} and gives ½ mol Sn^{4+}

$$[Sn^{4+}] = \frac{½\,(0.10\text{ mmol})}{26.0\text{ mL}} = 1.9_2 \times 10^{-3}\text{ M}$$

$$[Sn^{2+}] = \frac{1.25 - ½\,(0.10)\text{ mmol}}{26.0\text{ mL}} = 4.62 \times 10^{-2}\text{ M}$$

$$E_+ = 0.139 - \frac{0.059\,16}{2}\log\frac{[Sn^{2+}]}{[Sn^{4+}]}$$

$$E_+ = 0.139 - \frac{0.059\,16}{2}\log\frac{4.62 \times 10^{-2}]}{[1.9_2 \times 10^{-3}]} = 0.098\text{ V}$$

$$E = E_+ - E_- = 0.098 - 0.241 = -0.143\text{ V}$$

25.0 mL: At the equivalence point, we add the two indicator electrode Nernst equations. To make the factor in front of the log term the same in both equations, we can multiply the $Sn^{4+}\,|\,Sn^{2+}$ equation by 2:

$$E_+ = 0.139 - \frac{0.059\,16}{2}\log\frac{[Sn^{2+}]}{[Sn^{4+}]}$$

$$2E_+ = 0.278 - 0.059\,16\log\frac{[Sn^{2+}]}{[Sn^{4+}]}$$

$$E_+ = 0.732 - 0.059\,16\log\frac{[Fe^{2+}]}{[Fe^{3+}]}$$

Now add the last two equations to get

$$3E_+ = 1.010 - 0.059\,16\log\left(\frac{[Sn^{2+}][Fe^{2+}]}{[Sn^{4+}][Fe^{3+}]}\right)$$

But at the equivalence point, $2[Sn^{4+}] = [Fe^{2+}]$ and $2[Sn^{2+}] = [Fe^{3+}]$. Substituting these identities into the log term gives

$$3E_+ = 1.010 - 0.059\,16\log\left(\frac{[Sn^{2+}]2[Sn^{4+}]}{[Sn^{4+}]2[Sn^{2+}]}\right)$$

So the log quotient in the log term is 1 and the logarithm is 0. Therefore, $E_+ = 1.010/3 = 0.337$ V and $E = E_+ - E_- = 0.337 - 0.241 = 0.096$ V.

26.0 mL: $E_+ = 0.732 - 0.059\,16\log\dfrac{[Fe^{2+}]}{[Fe^{3+}]}$

There is 1.0 mL of Fe^{3+} beyond the equivalence point.

$$[Fe^{3+}] = \frac{(1.0\text{ mL})\,(0.100\text{ M})}{51.0\text{ mL}} = 1.9_6 \times 10^{-3}\text{ M}$$

The first 25.0 mL of Fe^{3+} were converted to Fe^{2+}, so

$$[Fe^{2+}] = \frac{(25.0\text{ mL})\,(0.100\text{ M})}{51.0\text{ mL}} = 4.90 \times 10^{-2}\text{ M}$$

$$E_+ = 0.732 - 0.059\,16\log\frac{[Fe^{2+}]}{[Fe^{3+}]}$$

$$E_+ = 0.732 - 0.059\,16 \log \frac{[4.90 \times 10^{-2}]}{[1.9_6 \times 10^{-3}]} = 0.649 \text{ V}$$

$$E = E_+ - E_- = 0.649 - 0.241 = 0.408 \text{ V}$$

mL	E (V)	mL	E (V)	mL	E (V)
1.0	–0.143	24.0	–0.061	26.0	0.408
12.5	–0.102	25.0	0.096	30.0	0.450

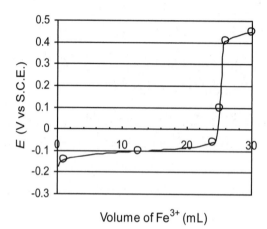

Volume of Fe^{3+} (mL)

16-6. Diphenylamine sulfonic acid: colorless $\rightarrow$ red -violet

Diphenylbenzidine sulfonic acid: colorless $\rightarrow$ violet

tris (2,2'-bipyridine) iron: red $\rightarrow$ pale blue

Ferroin: red $\rightarrow$ pale blue

16-7. The reduction potentials are

$$Sn^{4+} + 2e^- \rightleftharpoons Sn^{2+} \qquad\qquad E° = 0.139 \text{ V}$$

$$Mn(EDTA)^- + e^- \rightleftharpoons Mn(EDTA)^{2-} \qquad E° = 0.825 \text{ V}$$

The end point (versus S.H.E.) is between 0.139 and 0.825 V. Tris(2,2'-bipyridine) iron has too high a reduction potential (1.120 V) to be useful for this titration.

16-8. Preoxidation and prereduction refer to adjusting the oxidation state of analyte to a suitable value for a titration. The preoxidation or prereduction agent must be destroyed so it does not interfere with the titration by reacting with titrant.

16-9. $2S_2O_8^{2-} + 2H_2O \xrightarrow{\text{boiling}} 4SO_4^{2-} + O_2 + 4H^+$

$Ag^{3+} + H_2O \xrightarrow{\text{boiling}} Ag^+ + \frac{1}{2}O_2 + 2H^+$

$2H_2O_2 \xrightarrow{\text{boiling}} O_2 + 2H_2O$

16-10. A Jones reductor is a column packed with zinc granules coated with zinc amalgam. Prereduction is accomplished by passing analyte solution through the column.

16-11. Cr^{3+} and TiO^{2+} would interfere if they were reduced to Cr^{2+} and Ti^{3+}. In the Jones reductor, Zn is a strong enough reductant to react with Cr^{3+} and TiO^{2+}.

$E° = -0.764$ for the $Zn^{2+}|Zn$ couple

$E° = -0.42$ for the $Cr^{3+}|Cr^{2+}$ couple

$E° = 0.1$ for the $TiO^{2+}|Ti^{3+}$ couple

In the Walden reductor, Ag is not strong enough to reduce Cr^{3+} and TiO^{2+}:

$E° = 0.222$ for the $AgCl|Ag$ couple

16-12. A weighed amount of the solid mixture is added to a solution containing excess standard Fe^{2+} plus phosphoric acid. Each mol of $(NH_4)_2S_2O_8$ oxidizes 2 mol of Fe^{2+} to Fe^{3+}. Excess Fe^{2+} is then titrated with standard $KMnO_4$ to find out how much Fe^{2+} was consumed by the $(NH_4)_2S_2O_8$. The phosphoric acid masks the yellow color of Fe^{3+}, making the end point easier to see.

16-13. (a) $MnO_4^- + 8H^+ + 5e^- \rightleftharpoons Mn^{2+} + 4H_2O$

(b) $MnO_4^- + 4H^+ + 3e^- \rightleftharpoons MnO_2(s) + 2H_2O$

(c) $MnO_4^- + e^- \rightleftharpoons MnO_4^{2-}$

16-14. $3MnO_4^- + 5Mo^{3+} + 4H^+ \rightarrow 3Mn^{2+} + 5MoO_2^{2+} + 2H_2O$

$(16.43 - 0.04) = 16.39$ mL of $0.010\,33$ M $KMnO_4 = 0.169\,3$ mmol of MnO_4^-, which will react with $(5/3)(0.169\,3) = 0.282\,2$ mmol of Mo^{3+}.

$[Mo^{3+}] = 0.282\,2$ mmol/25.00 mL $= 0.011\,29$ M (= original $[MoO_4^{2-}]$).

16-15. $2MnO_4^- + 5H_2O_2 + 6H^+ \rightarrow 2Mn^{2+} + 5O_2 + 8H_2O$

$(27.66 - 0.04) = 27.62$ mL of $0.021\,23$ M $KMnO_4 = 0.586\,3_7$ mmol of MnO_4^-, which reacts with $(5/2)(0.586\,3_7) = 1.465\,9$ mmol of H_2O_2, which came from 25.00 mL of diluted solution $\Rightarrow$ $[H_2O_2] = 1.465\,9$ mmol/25.00 mL $= 0.058\,64$ M in the dilute solution. The original solution was 10 times more concentrated $= 0.586\,4$ M.

16-16. (a) Scheme 1:

$$2\,[8H^+ + MnO_4^- + 5e^- \rightarrow Mn^{2+} + 4H_2O]$$
$$\quad\quad\quad\;\; \overset{+7}{}\quad\quad\quad\quad\quad\overset{+2}{}$$

$$5\,[H_2O_2 \rightarrow O_2 + 2e^- + 2H^+]$$
$$\quad\;\; \overset{-1}{}\quad\quad\overset{0}{}$$

$$6H^+ + 2MnO_4^- + 5H_2O_2 \rightarrow 2Mn^{2+} + 5O_2 + 8H_2O$$

Scheme 2:

$$2\,[MnO_4^- \rightarrow Mn^{2+} + 2O_2 + 3e^-]$$
$$\quad\;\; \overset{+7\;-2}{}\quad\quad\overset{+2}{}\quad\quad\overset{0}{}$$

$$3\,[H_2O_2 + 2H^+ + 2e^- \rightarrow 2H_2O]$$
$$\quad\;\; \overset{-1}{}\quad\quad\quad\quad\quad\overset{-2}{}$$

$$6H^+ + 2MnO_4^- + 3H_2O_2 \rightarrow 2Mn^{2+} + 4O_2 + 6H_2O$$

(b) $\dfrac{1.023 \text{ g NaBO}_3 \cdot 4H_2O}{153.86 \text{ g/mol}} = 6.649 \text{ mmol NaBO}_3$

One tenth of this quantity was titrated $= 0.664\,9$ mmol $NaBO_3$, producing $0.664\,9$ mmol H_2O_2 by the reaction $BO_3^- + 2H_2O \rightarrow H_2O_2 + H_2BO_3^-$.

In Scheme 1, $2MnO_4^-$ react with $5H_2O_2$

$\Rightarrow 0.664\,9$ mmol H_2O_2 requires $\dfrac{2}{5}(0.664\,9) = 0.266\,0$ mmol MnO_4^-

$\dfrac{0.266\,0 \text{ mmol MnO}_4^-}{0.010\,46 \text{ mmol KMnO}_4/\text{ml}} = 25.43 \text{ mL KMnO}_4 \text{ required}$

In Scheme 2, $2MnO_4^-$ react with $3H_2O_2$

$\Rightarrow 0.664\,9$ mmol H_2O_2 requires $\dfrac{2}{3}(0.664\,9) = 0.443\,3$ mmol MnO_4^-

$\dfrac{0.443\,3 \text{ mmol MnO}_4^-}{0.010\,46 \text{ mmol KMnO}_4/\text{ml}} = 42.38 \text{ mL KMnO}_4 \text{ required}$

16-17. $2MnO_4^- + 5H_2C_2O_4 + 6H^+ \rightarrow 2Mn^{2+} + 10CO_2 + 8H_2O$

18.04 mL of 0.006 363 M $KMnO_4 = 0.114\,8$ mmol of MnO_4^-, which reacts with $(5/2)(0.114\,8) = 0.287\,0$ mmol of $H_2C_2O_4$, which came from $(2/3)(0.287\,0) = 0.191\,3$ mmol of La^{3+}. $[La^{3+}] = 0.191\,3$ mmol$/50.00$ mL $= 3.826$ mM.

16-18. $C_3H_8O_3 \quad + \quad 3H_2O \;\rightleftharpoons\; 3HCO_2H \;+\; 8e^- + 8H^+$
glycerol   formic acid
(average oxidation   (oxidation
number of C $= -2/3$)   number of C $= +2$)

$$8Ce^{4+} + 8e^- \rightleftharpoons 8Ce^{3+}$$

$$C_3H_8O_3 + 8Ce^{4+} + 3H_2O \rightleftharpoons 3HCO_2H + 8Ce^{3+} + 8H^+$$

One mole of glycerol requires eight moles of Ce^{4+}.

50.0 mL of 0.083 7 M Ce^{4+} = 4.185 mmol

12.11 mL of 0.044 8 M Fe^{2+} = 0.543 mmol

Ce^{4+} reacting with glycerol = 3.642 mmol

glycerol = (1/8) (3.642) = 0.455_2 mmol = 41.9 mg

$\Rightarrow$ original solution = 41.9 wt% glycerol

16-19. Ferrous ammonium sulfate is $Fe(NH_4)_2(SO_4)_2 \cdot 6H_2O$

50.00 mL of 0.118 6 M Ce^{4+} = 5.930 mmol Ce^{4+}

31.13 mL of 0.042 89 M Fe^{2+} = <u>1.335 mmol Fe^{2+}</u>

4.595 mmol Ce^{4+} consumed by NO_2^-

Since two moles of Ce^{4+} react with one mole of NO_2^-, there must have been

1/2 (4.595) = 2.298 mmol of $NaNO_2$ = 0.158 5 g in 25.0 mL. In 500.0 mL,

there would be $\left(\dfrac{500.0}{25.0}\right)(0.158\ 5)$ = 3.170 g = 78.67% of the 4.030 g sample.

16-20. Step 2 gives the total Cr content of the crystal, since each Cr^{x+} ion in any
oxidation state is oxidized and reacts with $3Fe^{2+}$.

Step 2: $\dfrac{(0.703\ mL)(2.786\ mM)}{0.156\ 6\ g\ of\ crystal} = \dfrac{12.51\ \mu mol\ Fe^{2+}}{g\ of\ crystal}$

$\dfrac{\frac{1}{3}(12.51)\ \mu mol\ Cr}{g\ of\ crystal} = \dfrac{4.169\ \mu mol\ Cr}{g\ of\ crystal}$

Step 1 tells us how much Cr^{x+} is oxidized above the +3 state. Each Cr^{x+} reacts
with $(x-3)\ Fe^{2+}$.

Step 1: $\dfrac{(0.498\ mL)(2.786\ mM)}{0.437\ 5\ g\ of\ crystal} = \dfrac{3.171\ \mu mol\ Fe^{2+}}{g\ of\ crystal}$

Since one gram of crystal contains 4.169 µmol of Cr that reacts with 3.171 µm of
Fe^{2+}, the average oxidation state of Cr is $3 + \frac{3.171}{4.169}$ = +3.761.

Total Cr (from Step 2) = 4.169 µmol Cr per gram = 217 µg per gram of crystal.

16-21. (a) Theoretical molarity = (3.214 g/L)/(158.034 g/mol) = $0.020\ 33_7$ M.

(b) 25.00 mL of $0.020\ 33_7$ M $KMnO_4$ = $0.508\ 43$ mmol. Two moles of MnO_4^-

react with five moles of H_3AsO_3, which comes from $\frac{5}{4}$ moles of As_4O_6.

The moles of As_4O_6 needed to react with $0.508\ 43$ mmol of MnO_4^- =

$(^1/_2)(^5/_4)(0.508\ 43$ mmol) = $0.317\ 77_7$ mmol = $0.125\ 74_4$ g of As_4O_6.

(c) $\dfrac{0.508\,4_3 \text{ mmol KMnO}_4}{0.125\,74 \text{ g As}_2\text{O}_3} = \dfrac{x \text{ mmol KMnO}_4}{0.146\,8 \text{ g As}_4\text{O}_6} \Rightarrow x = 0.593\,6_1 \text{ mmol}$

KMnO_4 in $(29.98 - 0.03) = 29.95$ mL $\Rightarrow [\text{KMnO}_4] = 0.019\,82$ M.

16-22. I^- reacts with I_2 to give I_3^-. This reaction increases the solubility of I_2 and decreases its volatility.

16-23. Standard triiodide can be prepared from a weighed amount of KIO_3 with acid plus excess iodide. Alternatively, triiodide solution can be standardized by reaction with standard $\text{S}_2\text{O}_3^{2-}$ prepared from anhydrous $\text{Na}_2\text{S}_2\text{O}_3$.

16-24. Starch is not added until just before the end point in iodometry, so it does not irreversibly bind to I_2, which is present during the whole titration.

16-25. $\text{S}_4\text{O}_6^{2-} + 2e^- \rightleftharpoons 2\text{S}_2\text{O}_3^{2-}$ or $\text{S}_4\text{O}_6^{2-} + 4\text{H}^+ + 2e^- \rightleftharpoons 2\text{H}_2\text{SO}_3$

$E°$ for the second half-reaction above is 0.57 V. $E°$ for the half-reaction $\tfrac{1}{2}\text{O}_2(g) + 2\text{H}^+ + 2e^- \rightleftharpoons \text{H}_2\text{O}$ is 1.23 V. O_2 is a stronger oxidant than tetrathionate.

16-26. (a) 50.00 mL contains exactly 1/10 of the $\text{KIO}_3 = 0.102\,2$ g $= 0.477\,5_7$ mmol KIO_3. Each mol of iodate makes 3 mol of triiodide, so $\text{I}_3^- = 3(0.477\,5_7)$ $= 1.43\,2_7$ mmol.

(b) Two moles of thiosulfate react with one mole of I_3^-. Therefore, there must have been $2(1.43\,2_7) = 2.86\,5_4$ mmol of thiosulfate in 37.66 mL, so the concentration is $(2.86\,5_4 \text{ mmol})/(37.66 \text{ mL}) = 0.076\,08_7$ M.

(c) 50.00 mL of KIO_3 make $1.43\,2_7$ mmol I_3^-. The unreacted I_3^- requires 14.22 mL of sodium thiosulfate $= (14.22 \text{ mL})(0.076\,08_7 \text{ M}) = 1.082_0$ mmol, which reacts with $\tfrac{1}{2}(1.082_0 \text{ mmol}) = 0.541_0$ mmol I_3^-. The ascorbic acid must have consumed the difference $= 1.432_7 - 0.541_0 = 0.891_7$ mmol I_3^-. Each mole of ascorbic acid consumes one mole of I_3^-, so mol ascorbic acid $= 0.891_7$ mmol, which has a mass of $(0.891_7 \times 10^{-3} \text{ mol})(176.13 \text{ g/mol}) = 0.157_1$ g. Ascorbic acid in the unknown $= 100 \times (0.157_1 \text{ g})/(1.223 \text{ g}) = 12.8$ wt%.

(d) Starch should not be added until just before the end point because I_3^- is present throughout the titration and will irreversibly bind to starch if the starch is added too early.

16-27. $2Cu^{2+} + 5I^- \rightarrow 2CuI(s) + I_3^-$ $\qquad$ $I_3^- + 2S_2O_3^{2-} \rightarrow 3I^- + S_4O_6^{2-}$

23.33 mL of 0.046 68 M $Na_2S_2O_3$ = 1.089$_0$ mmol $S_2O_3^{2-}$ = 0.544 5 mmol I_3^-, which came from 1.089$_0$ mmol Cu^{2+} = 69.20 mg Cu. This much Cu comes from 1/5 of the original solid, which therefore contained 346.0 mg Cu = 11.43 wt%. There is a great deal of I_3^- present at the start of the titration, so starch should not be added until just before the end point.

16-28. (a) 4.0 mL of the 297.6 mL sample were displaced by $MnSO_4$ and alkali solution. After mixing, the resulting 297.6 mL volume contains 293.6 mL of the creek water, or 293.6/297.6 = 98.66%.

(b) After adding 2.0 mL of H_2SO_4, 2.0 mL of solution are displaced from the bottle, which now contains 291.6 mL of creek water in a volume of 297.6 mL. The fraction of creek water in the bottle is 291.6/297.6 = 97.98%.

(c) 200.0 mL that are titrated contain (0.979 8)(200.0 mL) = 196.0 mL of creek water.

(d) $1O_2$ makes $4Mn(OH)_3$, which makes $2I_3^-$.

(e) Thiosulfate required = (14.05 mL)(0.010 22 mM) = 0.143 6 mmol $S_2O_3^{2-}$
2 moles of $S_2O_3^{2-}$ react with 1 mole of I_3^-, so mmol I_3^- produced by O_2 = ½(0.143 6 mmol) = 0.071 80 mmol O_2 = 2.297 mg O_2.
Dissolved O_2 concentration (mg/L) = 2.297 mg/0.196 0 L = 11.7 mg/L

(f) Fraction of saturation = (11.7 mg/L)/(14.6 mg/L) = 80%

(g) $2HNO_2 + 2H^+ + 3I^- \rightarrow 2NO + I_3^- + 2H_2O$
(NO could go on to react with even more I^-)

16-29. $H_2S + I_3^- \rightarrow S(s) + 3I^- + 2H^+$ $\qquad$ $I_3^- + 2S_2O_3^{2-} \rightarrow 3I^- + S_4O_6^{2-}$

25.00 mL of 0.010 44 M I_3^- = 0.261 0$_0$ mmol I_3^-
14.44 mL of 0.009 336 M $Na_2S_2O_3$ = 0.134 8$_1$ mmol $Na_2S_2O_3$, which would have reacted with 0.067 40$_6$ mmol I_3^-. Therefore, the quantity of I_3^- that reacted with H_2S was 0.261 0$_0$ − 0.067 40$_6$ = 0.193 5$_9$ mmol. Since 1 mol of I_3^- reacts with 1 mol of H_2S, the H_2S concentration was 0.193 5$_9$ mmol/25.00 mL = 0.007 744 M. I_3^- is present at the start of the titration, so starch should not be added until just before the end point.

16-30. (a) $I_2(aq) + 2e^- \rightleftharpoons 2I^-$ $\qquad\qquad\qquad$ $E° = 0.620$ V

$\qquad\qquad$ $3I^- \rightleftharpoons I_3^- + 2e^-$ $\qquad\qquad\qquad$ $\underline{E° = -0.535 \text{ V}}$

$\qquad$ $I_2(aq) + I^- \rightleftharpoons I_3^-$ $\qquad\qquad\qquad\qquad$ $E° = 0.085$ V

$\qquad\qquad$ $K = 10^{2(0.085)/0.059\ 16} = 7 \times 10^2$

(b) $\quad I_2(s) + 2e^- \rightleftharpoons 2I^-$ $\qquad\qquad\qquad$ $E° = 0.535$ V

$\qquad\qquad$ $3I^- \rightleftharpoons I_3^- + 2e^-$ $\qquad\qquad\qquad$ $\underline{E° = -0.535 \text{ V}}$

$\qquad$ $I_2(s) + I^- \rightleftharpoons I_3^-$ $\qquad\qquad\qquad\qquad$ $E° = 0.000$ V

$\qquad\qquad$ $K = 10^{2(-0.000)/0.059\ 16} = 1.0$

(c) $\quad I_2(s) + 2e^- \rightleftharpoons 2I^-$ $\qquad\qquad\qquad$ $E° = 0.535$ V

$\qquad\qquad$ $2I^- \rightleftharpoons I_2(aq) + 2e^-$ $\qquad\qquad\qquad$ $\underline{E° = -0.620 \text{ V}}$

$\qquad$ $I_2(s) \rightleftharpoons I_2(aq)$ $\qquad\qquad\qquad\qquad$ $E° = -0.085$ V

$\qquad\qquad$ $K = [I_2(aq)] = 10^{2(-0.085)/0.059\ 16} = 1.3 \times 10^{-3}$ M $= 0.34$ g of I_2/L

16-31. Each mole of NH_3 liberated in the Kjeldahl digestion reacts with 1 mole of H^+ in the standard H_2SO_4 solution. Six moles of H^+ left (3 moles of H_2SO_4) after reaction with NH_3 will react with 1 mole of iodate by Reaction 16-18 to release 3 moles of I_3^-. Two moles of thiosulfate react with 1 mole of I_3^- in Reaction 16-19. Therefore each mole of thiosulfate corresponds to $\frac{1}{2}$ mol of residual H_2SO_4.

$$\text{mol } NH_3 = 2 \text{ (initial mol } H_2SO_4 - \text{final mol } H_2SO_4)$$
$$\text{mol } NH_3 = 2 \text{ (initial mol } H_2SO_4 - \tfrac{1}{2} \times \text{mol thiosulfate)}$$

16-32. (a) $IO_3^- + 8I^- + 6H^+ \rightleftharpoons 3I_3^- + 3H_2O$. The stock solution contained $\{0.804\ 3$ g KIO_3 (FM 214.00) $+ 6.0$ g KI (FM 166.00)$\}$ / 100 mL, which translates into 0.037 58 M KIO_3 plus 0.36 M KI, giving a mole ratio $KI/KIO_3 = 18$, which is a good excess of the 8:1 ratio required in the reaction. 5.00 mL of the stock solution contain $0.187\ 9_2$ mmol KIO_3 plus 1.5 mmol KI. 1.0 mL of 6.0 M H_2SO_4 contains 6 mmol H_2SO_4, which is a large excess for the reaction. Neither KI nor H_2SO_4 need to be measured accurately.

(b) $I_3^- + SO_3^{2-} + H_2O \rightarrow 3I^- + SO_4^{2-} + 2H^+$

(c) $0.187\,9_2$ mmol KIO_3 delivered to the wine generates $3 \times 0.187\,9_2 = 0.563\,7_6$ mmol I_3^-. The excess, unreacted I_3^- required 12.86 mL of 0.048 18 M $Na_2S_2O_3 = 0.619\,5_9$ mmol $Na_2S_2O_3$. Each mole of unreacted I_3^- requires 2 moles of $Na_2S_2O_3$, so there must have been $(0.619\,5_9)/2 = 0.309\,8$ mmol I_3^- left over from the reaction with sulfite. Therefore, the I_3^- that reacted with sulfite was $(0.563\,7_6 - 0.309\,8) = 0.254\,0$ mmol I_3^-. One mole of I_3^- reacts with 1 mole of sulfite, so there must have been 0.254 0 mmol SO_3^{2-} in 50.0 mL of wine. $[SO_3^{2-}] = 0.254\,0$ mmol/50.0 mL $= 5.07\,9 \times 10^{-3}$ M. With a formula mass of 80.06 for sulfite, the sulfite content is 406.6 mg/L.

(d) $s_{pooled} = \sqrt{\dfrac{2.2^2\,(3-1) + 2.1^2\,(3-1)}{3 + 3 - 2}} = 2.15$

$t_{calculated} = \dfrac{|277.7 - 273.2|}{2.15}\sqrt{\dfrac{3 \cdot 3}{3 + 3}} = 2.56$

$t_{table} = 2.776$ for 95% confidence and $3 + 3 - 2 = 4$ degrees of freedom $t_{calculated} < t_{table}$, so the difference is not significant at 95% confidence level.

16-33. 25.00 mL of 0.020 00 M $KBrO_3 = 0.500\,0$ mmol of BrO_3^-, which generates 1.500 mmol of Br_2. One mole of excess Br_2 generates one mole of I_2 (from I^-) and one mole of I_2 consumes 2 moles of $S_2O_3^{2-}$. Since mmol of $S_2O_3^{2-} = (8.83)(0.051\,13)$ $= 0.451\,5$ mmol, $I_2 = 0.225\,7$ mmol and Br_2 consumed by reaction with 8-hydroxyquinoline $= 1.500 - 0.225\,7 = 1.274$ mmol. But one mole of 8-hydroxyquinoline consumes 2 moles of Br_2, so 8-hydroxyquinoline $= 0.637\,1$ mmol and $Al^{3+} = 0.6371/3 = 0.212\,4$ mmol $= 5.730$ mg.

16-34. (a) $YBa_2Cu_3O_7$ contains 1 Cu^{3+} and 2 Cu^{2+}. $YBa_2Cu_3O_{6.5}$ contains no Cu^{3+} and 3 Cu^{2+}. The moles of Cu^{3+} in the formula $YBa_2Cu_3O_{7-z}$ are therefore $1 - 2z$. The moles of superconductor in 1 g of superconductor are $(1\ g)/[(666.246 - 15.999\,4\ z)g/mol]$. The difference between experiments B and A is $5.68 - 4.55 = 1.13$ mmol $S_2O_3^{2-}$/g of superconductor. Since 1 mol of thiosulfate is equivalent to 1 mol of Cu^{3+}, there are 1.13 mmol Cu^{3+}/g of superconductor.

$$\frac{mol\ Cu^{3+}}{mol\ superconductor} = 1 - 2z = \frac{1.13\ 10^{-3}\ mol\ Cu^{3+}}{\left(\dfrac{1\ g\ superconductor}{(666.246 - 15.999\,4\ z)\ g/mol}\right)}$$

Solving this equation gives $z = 0.125$. The formula is $YBa_2Cu_3O_{6.875}$.

(b) $1 - 2z = \dfrac{[5.68(\pm0.05) - 4.55(\pm0.10)]\ 10^{-3}}{\left(\dfrac{1}{666.246 - 15.999\,4\,z}\right)}$

$1 - 2z = \dfrac{1.13\ (\pm0.112)\ 10^{-3}}{\left(\dfrac{1}{666.246 - 15.999\,4\,z}\right)}$

$1 - 2z = 0.752\,86\ (\pm0.074\,49) - 0.018\,079\ (\pm0.001\,789)\,z$

$0.247\,124\ (\pm0.074\,488) = 1.981\,92\ (\pm0.001\,79)\,z$

$z = 0.125 \pm 0.038.$ The formula is $YBa_2Cu_3O_{6.875\ \pm0.038}.$

16-35.   A superconductor containing unknown quantities of Cu(I), Cu(II), Cu(III), and peroxide (O_2^{2-}) is dissolved in a known excess of Cu(I) in oxygen-free HCl solution. Possible reactions are

$$Cu^{3+} + Cu^+ \rightarrow 2Cu^{2+}$$
$$H_2O_2 + 2Cu^+ + 2H^+ \rightarrow 2H_2O + 2Cu^{2+}$$

Unreacted Cu(I) is then measured by coulometry to find out how much Cu(I) was consumed by the dissolving superconductor. The amount of Cu(I) consumed is equal to the moles of Cu^{3+} plus 2 times the moles of O_2^{2-} in the superconductor. The coulometry is done under Ar to prevent oxidation of Cu(I) by O_2 from the air. If the superconductor contained Cu(I) (but no Cu(III) or peroxide), then the amount of Cu(I) found by coulometry would be greater than the known amount used in the original solution.

16-36.   (a)   Initial Fe^{2+} in 5.000 mL = (5.000 mL)(0.100 0 M) = 0.500 0 mmol.
    $K_2Cr_2O_7$ required for titration of unreacted Fe^{2+}

   = (3.22 8 mL)(0.015 93 M $K_2Cr_2O_7$) = 0.0514 2 mmol.

But 1 mmol $K_2Cr_2O_7$ reacts with 6 mmol Fe^{2+} by the reaction

$K_2Cr_2O_7 + 6Fe^{2+} + 14H^+ \rightarrow 2Cr^{3+} + 6\,Fe^{3+} + 2K^+ + 14H_2O.$

Therefore, Fe^{2+} left after reaction with $Li_{1+y}CoO_2$

   = (0.0514 2 mmol)(6 mmol Fe^{2+}/mmol $K_2Cr_2O_7$) = 0.308 5 mmol.

Fe^{2+} consumed by Co^{3+} = (0.500 0 − 0.308 5) = 0.191 5 mmol.

1 mol Fe^{2+} is consumed by 1 mol Co^{3+}, so Co^{3+} in 25.00 mg solid sample = 0.191 5 mmol.

   (b)   Co in 25.00 mg solid = (0.564 g Co/g solid)(25.00 g solid) = 14.10 mg
        Co in 25.00 mg solid = (14.10 mg)/(58.933 g/mol) = 0.239 3 mmol
        From (a), we know that Co^{3+} = 0.191 5 mmol, so

   Co^{2+} = 0.239 3 − 0.191 5 = 0.047 8 mmol.

$$\text{Co oxidation state} = \frac{(0.047\ 8\ \text{mmol})(2+) + (0.191\ 5\ \text{mmol})(3+)}{0.239\ 2\ \text{mmol}} = 2.80$$

(c) If Co has an average oxidation number of +2.80 and O has an oxidation number of –2, Li must contribute a charge of $4 - 2.80 = 1.20$. Therefore, the formula is $Li_{1.20}CoO_2$ and $y = 0.20$.

(d) Theoretical weight percent for metals in $Li_{1.20}CoO_2$:

Formula mass $= 1.20(6.941) + 1(58.933) + 2(15.9994) = 99.261$

wt% Li $= 100 \times 1.20(6.941)/99.261 = 8.39\%$

wt% Co $= 100 \times 58.933/99.261 = 59.37\%$

$\dfrac{\text{wt\% Li}}{\text{wt\% Co}} = 0.141\ 3$

The observed quotient wt% Li/wt% Co is $0.138\ 8 \pm. 0.000\ 6$, which is not exactly equal to the quotient computed from the oxidation number. The difference represents experimental error between the two methods used find the stoichiometry.

16-37. Denote the average oxidation number of Bi as $3 + b$ and the average oxidation number of Cu as $2 + c$.

$$Bi_2^{3+b}Sr_2^{2+}Ca^{2+}Cu_2^{2+c}O_x$$

Positive charge $= 6 + 2b + 4 + 2 + 4 + 2c = 16 + 2b + 2c$

The charge must be balanced by $O^{2-} \Rightarrow x = 8 + b + c$.

The formula mass of the superconductor is $760.37 + 15.999\ 4(8 + b + c)$.

One gram contains $1/[760.37 + 15.999\ 4(8 + b + c)]$ moles.

(a) <u>Experiment A:</u> Initial $Cu^+ = 0.200\ 0$ mmol; final $Cu^+ = 0.108\ 5$ mmol.

Therefore, 102.3 mg of superconductor consumed 0.091 5 mmol Cu^+.

$2 \times$ mmol Bi^{5+} + mmol Cu^{3+} in 102.3 mg of superconductor $= 0.091\ 5$.

<u>Experiment B:</u> Initial $Fe^{2+} = 0.100\ 0$ mmol; final $Fe^{2+} = 0.057\ 7$ mmol.

Therefore, 94.6 mg of superconductor consumed 0.042 3 mmol Fe^{2+}.

$2 \times$ mmol Bi^{5+} in 94.6 mg of superconductor $= 0.042\ 3$.

Normalizing to 1 gram of superconductor gives

Expt A: 2(mmol Bi^{5+}) + mmol Cu^{3+} in 1 g of superconductor $= 0.894\ 43$

Expt B: 2(mmol Bi^{5+}) in 1 g of superconductor $= 0.447\ 15$

It is easier not to get lost in the arithmetic if we suppose that the oxidized bismuth is Bi^{4+} and equate one mole of Bi^{5+} to two moles of Bi^{4+}.

Therefore, we can rewrite the two previous equations as

mmol Bi^{4+} + mmol Cu^{3+} in 1 g of superconductor $= 0.894\ 43$ (1)

mmol Bi^{4+} in 1 g of superconductor $= 0.447\,15$ (2)

Subtracting (2) from (1) gives

mmol Cu^{3+} in 1 g of superconductor $= 0.447\,28$ (3)

Equations (2) and (3) tell us that the stoichiometric relationship in the formula of the superconductor is $b/c = 0.447\,15/0.447\,28 = 0.999\,7$.

Since 1 g of superconductor contains $0.447\,28$ mmol Cu^{3+}, we can say

$$\frac{\text{mol } Cu^{3+}}{\text{mol solid}} = 2c$$

$$\frac{\text{mol } Cu^{3+} / \text{mol solid}}{\text{gram solid} / \text{mol solid}} = \frac{2c}{760.37 + 15.999\,4\,(8 + b + c)}$$

$$\frac{\text{mol } Cu^{3+}}{\text{gram solid}} = \frac{2c}{760.37 + 15.999\,4\,(8 + b + c)} = 4.472\,8 \times 10^{-4}$$ (4)

Substituting $b = 0.999\,7c$ in the denominator of (4) allows us to solve for c:

$$\frac{2c}{760.37 + 15.999\,4(8 + 1.999\,7c)} = 4.472\,8 \times 10^{-4} \Rightarrow c = 0.200_1$$

$$\Rightarrow b = 0.999\,7c = 0.200_0$$

The average oxidation numbers are $Bi^{3.200_0+}$ and $Cu^{2.200_1+}$ and the formula of the compound is $Bi_2Sr_2CaCu_2O_{8.400_1}$, since the oxygen stoichiometry derived at the beginning of the solution is $x = 8 + b + c$.

(b) Propagation of error:

Expt A: 102.3 ($\pm$0.2) mg compound consumed 0.091 5 ($\pm$0.000 7) mmol Cu^+

Expt B: 94.6 ($\pm$0.2) mg compound consumed 0.042 3 ($\pm$0.000 7) mmol Fe^{2+}

Normalizing to 1 gram of superconductor gives

Expt A: mmol Bi^{4+} + mmol Cu^{3+} in 1 g of superconductor

$$= \frac{0.091\,5\,(\pm0.000\,7)}{0.102\,3\,(\pm0.000\,2)} = 0.894\,43\,(\pm0.007\,06)\,\frac{\text{mmol}}{\text{gram}}$$

Expt B: mmol Bi^{4+} in 1 g of superconductor

$$= \frac{0.042\,3\,(\pm0.000\,7)}{0.094\,6\,(\pm0.000\,2)} = 0.447\,15\,(\pm0.007\,46)\,\frac{\text{mmol}}{\text{gram}}$$

$$\frac{\text{mmol } Cu^{3+}}{\text{g superconductor}} = 0.894\,43\,(\pm0.007\,06) - 0.447\,15\,(\pm0.007\,46)$$

$$= 0.447\,28\,(\pm0.010\,27)$$

$$\frac{b}{c} = \frac{0.447\,15\,(\pm0.007\,46)}{0.447\,28\,(\pm0.010\,27)} = 0.999\,7\,(\pm0.028\,4)$$

$$\frac{2c}{760.37 + 15.999\,4(8 + [1.999\,7(\pm0.028\,4)]c)} = 4.472\,8\,(\pm0.102\,7) \times 10^{-4}$$

$$[4\,471.47\,(\pm102.7)]\,c = 888.365 + [31.999\,4\,(\pm0.445)]\,c$$

$$\Rightarrow c = 0.200\,1\,(\pm0.004\,6)$$

The relative uncertainty in b just given as $0.007\,46/0.447\,15$ is smaller than the relative uncertainty in c, which is $0.010\,27/0.447\,28$.

$$\text{Uncertainty in } b = \frac{0.007\,46/0.447\,15}{0.010\,27/0.447\,28} \text{ (uncertainty in } c\text{)}$$

$$= \frac{0.007\,46/0.447\,15}{0.010\,27/0.447\,28} (\pm 0.004\,6) = \pm 0.003\,3$$

$$\Rightarrow b = 0.200\,0 \, (\pm 0.003\,3)$$

The average oxidation numbers are $Bi^{+3.200\,0}(\pm 0.003\,3)$ and $Cu^{+2.200\,1}(\pm 0.004\,6)$ and the formula of the compound is $Bi_2Sr_2CaCu_2O_{8.400\,1}(\pm 0.005\,7)$.

CHAPTER 17
ELECTROANALYTICAL TECHNIQUES

17-1. We observe that the silver electrode requires ~0.5 V more negative potential than the platinum electrode for reduction of H_3O^+ to H_2. The extra voltage needed to liberate H_2 at the silver surface is the overpotential required to overcome the activation energy for the reaction. In Table 17-1, we see that the difference in overpotential between Pt and Ag is ~0.5 V for a current density of 100 A/m^2.

17-2. charge (C) $= q = nNF$, where, n = unit charges per e$^-$, N = mol e$^-$, and F = Faraday constant.

$q = (1 \text{ charge/e}^-)(0.100 \text{ mol})(96\,485 \text{ C/mol}) = 9.648 \times 10^3$ C

1 ampere = 1 C/s, so the time required for 9.648×10^3 C is

$(9.648 \times 10^3 \text{ C})/(1.00 \text{ C/s}) = 9.648 \times 10^3 \text{ s} = 2.68 \text{ h}$

17-3. $\Delta G° \quad = \quad -n \quad \cdot \quad N \quad \cdot \quad F \quad \cdot \quad E°$

where n = unit charges per molecule, N = moles, F = Faraday constant, and $E°$ = standard voltage. For the reaction $H_2O(l) \rightleftharpoons H_2(g) + \frac{1}{2}O_2(g)$, $n = 2e^-$ per mole of H_2 or per half mole of O_2, N = 1 mol of reaction as written, which means 1 mole of H_2O, 1 mole of H_2, and $\frac{1}{2}$ mol of O_2.

$E° = -\Delta G°/nNF = -237.13 \times 10^3 \text{ J}/[(2)(1 \text{ mol})(964\,85 \text{ C/mol})]$

$= -1.228\,8 \text{ J/C} = -1.228\,8 \text{ V}$

"Standard" means that reactants and products are in their standard states (1 bar for gases, pure liquid for water, unit activity for H$^+$ and OH$^-$).

17-4. (a) $E = E(\text{cathode}) - E(\text{anode})$

$= \left\{ E° (\text{cathode}) - 0.059\,16 \log P_{H_2}^{1/2} [OH^-] \right\}$

$\quad - \left\{ E° (\text{anode}) - 0.059\,16 \log [Br^-] \right\}$

(remember to write both reactions as reductions)

$= \{ E° -0.828 - 0.059\,16 \log (1.0)^{1/2} [0.10] \}$

$\quad - \{ 1.078 - 0.059\,16 \log [0.10] \} = -1.906 \text{ V}$

(b) Ohmic potential $= I \cdot R = (0.100 \text{ A})(2.0 \text{ }\Omega) = 0.20 \text{ V}$

(c) $E = E(\text{cathode}) - E(\text{anode}) - I \cdot R - \text{Overpotentials}$

$= -1.906 - 0.20 - (0.20 + 0.40) = -2.71 \text{ V}$

(d) $E(\text{cathode}) = E° (\text{cathode}) - 0.059\,16 \log P_{H_2}^{1/2} [OH^-]_s$

$= -0.828 - 0.059\,16 \log (1.0)^{1/2} [1.0] = -0.828 \text{ V}$

$$E(\text{anode}) = E^\circ (\text{anode}) - 0.059\,16 \log [\text{Br}^-]_s$$
$$= 1.078 - 0.059\,16 \log [0.010] = 1.196 \text{ V}$$
$$E = E(\text{cathode}) - E(\text{anode}) - I \cdot R - \text{Overpotentials}$$
$$= -0.828 - 1.196 - 0.20 - (0.20 + 0.40) = -2.82 \text{ V}$$

17-5. (a) V_2 is the voltage between the working and reference electrodes, which is held constant. Working: Reference: Auxiliary:

$$\longrightarrow\!\circ \qquad \longrightarrow \qquad \longrightarrow\!\dashv$$

(b) The objective is to measure the potential difference between the reference electrode and a point in solution in an electrochemical cell. Negligible current flows between the opening of the Luggin capillary and the reference electrode in the reservoir connected to the Luggin capillary. There is negligible ohmic loss between the opening of the capillary and the surface of the reference electrode. There is also negligible overpotential and negligible concentration polarization at the reference electrode because there is negligible current at the reference electrode. With negligible potential losses between the reference electrode and the opening of the Luggin capillary, the potential inside the capillary is the potential of the reference electrode. The potential outside the capillary is that of the electrochemical system.

17-6. (a) Copper electrode: $Cu(s) \rightleftharpoons Cu^{2+} + 2e^-$

$$E_+ = 0.339 - \frac{0.059\,16}{2} \log[Cu^{2+}] = 0.339 \text{ V if } [Cu^{2+}] = 1.0 \text{ M}$$

Silver-silver chloride electrode: $E_- = 0.197 \text{ V}$

Predicted equilibrium voltage $= E_+ - E_- = 0.339 - 0.197 = 0.142 \text{ V}$

(observed $= 0.109$ mV)

The difference between the observed value of 109 mV and the predicted 142 mV is due to neglect of the activity coefficient of Cu^{2+} and to the difference between the reference electrode potential with saturated KCl or 3 M KCl.

(b) Increasing the imposed potential difference above 1.000 V would increase the current flowing between the Cu electrodes. The overpotential at each electrode increases as current density (current per unit area) increases. The anode intercept would rise above 122 mV and the cathode intercept would go below 85 mV.

17-7. (a) For every mole of Hg produced, one mole of electrons flows.

1.00 mL Hg = 13.53 g Hg = 0.067 45 mol Hg = 0.067 45 mol e^-.

(0.067 45 mol) (96 485 C/mol) = 6 508 C.

Work = $q \cdot E$ = (6 508 C) (1.02 V) = 6.64×10^3 J.

(b) The power is 0.209 J/min = 0.003 48 J/s.

$P = I^2 R \Rightarrow I = \sqrt{P/R} = \sqrt{(0.003\ 48\ \text{W})/(100\ \Omega)}$ = 5.902 mA.

In 1 h the total charge flowing through the circuit is

$(5.902 \times 10^{-3}$ C/s)· (3 600 s) = 21.25 C/(96 485 C/mol)

= 2.202×10^{-4} mol of e^-/h = 1.101×10^{-4} mol of Cd/h

= 0.012 4 g Cd/h.

17-8. Hydroxide generated at the cathode and Cl^- in the anode compartment cannot cross the membrane. Na^+ from seawater crosses from the anode to the cathode to preserve charge balance. Therefore, NaOH can be formed free from Cl^-.

17-9. $E = E(\text{cathode}) - E(\text{anode}) - IR - \text{overpotentials}$

Suppose that the open-circuit voltage of each cell is $E(\text{cathode}) - E(\text{anode})$ = 2.2 V when no current flows. Ohmic loss and overpotentials for the two half-reactions decrease the output of the cell by 0.2 V, giving a net cell voltage of 2.0 V when the cell is delivering current. The cell can be recharged at very low current flow by applying just over 2.2 V in the opposite direction to reverse the cell chemistry. To charge at a significant rate requires additional voltage to overcome ohmic loss and overpotentials. The recharge requires ~0.2 V more than open-circuit voltage, or ~2.4 V. Electrical losses [IR, overpotentials, and concentration polarization in the terms $E(\text{cathode})$ and $E(\text{anode})$] always decrease the magnitude of the voltage that can be delivered by a cell and increase the magnitude of the voltage required to reverse the spontaneous cell reaction.

17-10. $Pb(\text{lactate})_2 + 2H_2O \longrightarrow PbO_2(s) + 2\ \text{lactate}^- + 4H^+ + 2e^-$

$\quad\quad\quad Pb^{2+} \quad\quad\quad\quad\quad\quad\quad Pb^{4+}$

Lead lactate is oxidized to PbO_2 at the anode.

The mass of lead lactate (FM 385.3) giving 0.111 1 g of PbO_2

(FM = 239.2) is (385.3/239.2)(0.111 1 g) = 0.179 0 g.

wt% Pb = $\dfrac{0.179\ 0\ \text{g}}{0.326\ 8\ \text{g}} \times 100$ = 54.77%

17-11. Cathode: $Sn^{2+} + 2e^- \rightleftharpoons Sn(s)$ $E^\circ = -0.141$ V

$$E(\text{cathode, vs S.H.E.}) = -0.141 - \frac{0.059\ 16}{2} \log \frac{1}{1.0 \times 10^{-8}} = -0.378 \text{ V}$$

$E(\text{cathode, vs S.C.E.}) = -0.378 - 0.241 = -0.619$ V

The voltage will be more negative if concentration polarization occurs.
Concentration polarization means that $[Sn^{2+}]_s < 1.0 \times 10^{-8}$ M.

17-12. When 99.99% of Cd(II) is reduced, the formal concentration will be 1.0×10^{-5} M, and the predominant form is $Cd(NH_3)_4^{2+}$.

$$\beta_4 = \frac{[Cd(NH_3)_4^{2+}]}{[Cd^{2+}][NH_3]^4} = \frac{(1.0 \times 10^{-5})}{[Cd^{2+}](1.0)^4} \Rightarrow [Cd^{2+}] = 2.8 \times 10^{-12} \text{ M}$$

$Cd^{2+} + 2e^- \rightleftharpoons Cd(s)$ $E^\circ = -0.402$

$$E(\text{cathode}) = -0.402 - \frac{0.059\ 16}{2} \log \frac{1}{[Cd^{2+}]} = -0.744 \text{ V}$$

17-13. Ni deposited = (0.479 8 g – 0.477 5 g) = 2.3 mg = 39.19 μmol Ni which would require $(39.19 \times 10^{-6}$ mol Ni$)(2e^-/\text{Ni})(96\ 485$ C/mol$) = 7.562$ C. Percentage of current going to reduction of $Ni^{2+} = 100 \times \frac{7.562 \text{ C}}{8.082 \text{ C}} = 94\%$. The remainder went into reduction of H^+ to H_2.

17-14. When excess Br_2 appears in the solution, current flows at a low applied potential difference (0.25 V) in the detector circuit by virtue of the reactions

anode: $2Br^- \rightarrow Br_2 + 2e^-$

cathode: $Br_2 + 2e^- \rightarrow 2Br^-$

17-15. A mediator shuttles electrons between analyte and the electrode. After being oxidized or reduced by analyte, the mediator is regenerated at the electrode.

17-16. (a) 0.005 C/s $\times 0.1$ s $= 0.000\ 5$ C

$$\frac{0.000\ 5 \text{ C}}{96\ 485 \text{ C/mol}} = 5._2 \times 10^{-9} \text{ mol } e^-$$

(b) A 0.01 M solution of a two-electron reductant delivers 0.02 moles of electrons/liter.

$$\frac{5._2 \times 10^{-9} \text{ moles}}{0.02 \text{ moles/liter}} = 2._6 \times 10^{-7} \text{ L} = 0.000\ 2_6 \text{ mL} = 0.2_6 \text{ μL}$$

17-17. (a) $\text{mol } e^- = \dfrac{I \cdot t}{nF} = \dfrac{(5.32 \times 10^{-3} \text{ C/s})(964 \text{ s})}{(1 \text{ charge/}e^-)(96\,485 \text{ C/mol})} = 5.32 \times 10^{-5} \text{ mol}$

(b) One mol e^- reacts with ½ mol Br_2, which reacts with ½ mol cyclohexene, so 5.32×10^{-5} mol e^- produces 2.66×10^{-5} mol cyclohexene.

(c) 2.66×10^{-5} mol/5.00×10^{-3} L $= 5.32 \times 10^{-3}$ M

17-18. $2I^- \rightarrow I_2 + 2e^- \Rightarrow$ one mole of I_2 is created when two moles of electrons flow.

$\text{mol } e^- = \dfrac{I \cdot t}{nF} = \dfrac{(52.6 \times 10^{-3} \text{ C/s})(812 \text{ s})}{(1 \text{ charge/}e^-)(96\,485 \text{ C/mol})} = 0.442\,7 \text{ mmol of } e^- = 0.221\,3$

mmol of I_2. Therefore, there must have been 0.221 3 mmol of H_2S (FM 34.08) $=$ 7.542 mg of H_2S/50.00 mL $= 7.542 \times 10^3$ µg of H_2S/50.00 mL $= 151$ µg/mL.

17-19. D-fructose $\rightarrow$ 5-keto-D-fructose $+ 2H^+ + 2e^-$

$n = 2$ electrons per mol of fructose

2.00 nmol fructose will liberate 4.00 nmol electrons

$q = (4.00 \text{ nmol } e^-)(96\,485 \text{ C/mol}) = 0.386$ mC

The asymptotic current in the figure is near 0.38_7 mC.

17-20. (a) When one B atom is substituted for one C atom, the crystal lattice contains one less valence electron, making it *p*-type.

(b) Each ferrocene loses one electron. To convert charge to moles, divide by the Faraday constant: $\dfrac{23 \times 10^{-6} \text{ C}}{(0.38 \text{ cm}^2)(96\,485 \text{ C/mol})} = 6.2_7 \times 10^{-10} \dfrac{\text{mol}}{\text{cm}^2}$. If there are two ferrocene groups attached to each N atom, the surface density of N atoms is half of the surface density of ferrocene groups $= 3.1_4 \times 10^{-10} \dfrac{\text{mol}}{\text{cm}^2}$.

Surface density of carbon atoms $= \dfrac{1.7 \times 10^{15} \text{ atoms/cm}^2}{6.022 \times 10^{23} \text{ atoms/mol}} = 2.8_2 \times 10^{-9} \dfrac{\text{mol}}{\text{cm}^2}$. Fraction of C atoms substituted by N $= \dfrac{3.1_4 \times 10^{-10} \text{ mol/cm}^2}{2.8_2 \times 10^{-9} \text{ mol/cm}^2} \approx 11\%$.

17-21. (a) $C_6H_5N{=}NC_6H_5 + 4H^+ + 4e^- \rightarrow 2C_6H_5NH_2$

Electron flow $=$

$\left(4 \dfrac{\text{electrons}}{C_6H_5N{=}NC_6H_5}\right)\left(25.9 \dfrac{\text{nmol}}{\text{s}}\right)\left(96\,485 \dfrac{\text{C}}{\text{mol}}\right) = 1.00 \times 10^{-2}$ C/s

current density $= \dfrac{1.00 \times 10^{-2} \text{ A}}{1.00 \times 10^{-4} \text{ m}^2} = 1.00 \times 10^2$ A/m²

$\Rightarrow$ overpotential $= 0.85$ V

(b) $E(\text{cathode}) = 0.100 - 0.059\ 16 \log \dfrac{[Ti^{3+}]_s}{[TiO^{2+}]_s[H^+]^2}$

$= 0.100 - 0.059\ 16 \log \dfrac{[0.10]}{[0.050][0.10]^2} = -0.036\ V$

(c) $O_2 + 4H^+ + 4e^- \rightleftharpoons 2H_2O$ $\qquad\qquad\qquad E^\circ = 1.229\ V$

$E(\text{anode}) = 1.229 - \dfrac{0.059\ 16}{4} \log \dfrac{1}{P_{O_2}[H^+]^4}$

$= 1.229 - \dfrac{0.059\ 16}{4} \log \dfrac{1}{(0.20)[0.10]^4} = 1.160\ V$

(d) $E = E(\text{cathode}) - E(\text{anode}) - I \cdot R - \text{Overpotential}$

$= -0.036 - 1.160 - (1.00 \times 10^{-2}\ A)(52.4\ \Omega) - 0.85 = -2.57\ V$

17-22. $F = \dfrac{\text{coulombs}}{\text{mol}} = \dfrac{I \cdot t}{\text{mol}}$

$= \dfrac{[0.203\ 639\ 0(\pm 0.000\ 000\ 4)\ A][18\ 000.075\ (\pm 0.010)\ s]}{[4.097\ 900\ (\pm 0.000\ 003)\ g]/[107.868\ 2\ (\pm 0.000\ 2)g/mol]}$

$= \dfrac{[0.203\ 639\ 0(\pm 1.96\ 10^{-4}\ \%)][18\ 000.075\ (\pm 5.56\ 10^{-5}\ \%)]}{[4.097\ 900\ (\pm 7.32\ 10^{-5}\ \%)]/[107.868\ 2\ (\pm 1.85\ 10^{-4}\ \%)]}$

$= 9.648\ 667 \times 10^4\ (\pm 2.85 \times 10^{-4}\ \%) = 964\ 86.6_7 \pm 0.2_8\ C/mol$

17-23. (a) $H_2SO_3 \overset{pK_1 = 1.86}{\rightleftharpoons} HSO_3^- \overset{pK_2 = 7.17}{\rightleftharpoons} SO_3^{2-}$

H_2SO_3 is predominant below pH 1.86. HSO_3^- dominates between pH 1.86 and 7.17. SO_3^{2-} is dominant above pH 7.17.

(b) Cathode: $\quad H_2O + e^- \rightarrow \frac{1}{2}H_2(g) + OH^-$

Anode: $\qquad 3I^- \rightarrow I_3^- + 2e^-$

(c) $I_3^- + HSO_3^- + H_2O \rightarrow 3I^- + SO_4^{2-} + 3H^+$

$$I_3^- + 2S_2O_3^{2-} \rightleftharpoons 3I^- + \underset{\substack{| \\ O^-}}{\overset{\substack{O \\ \|}}{O=S}} - S - S - \underset{\substack{| \\ O^-}}{\overset{\substack{O \\ \|}}{S=O}}$$

$\qquad\qquad$ Thiosulfate $\qquad\qquad\qquad$ Tetrathionate

(d) In Step 3, I_3^- was generated by a current of 10.0 mA ($=10.0 \times 10^{-3}$ C/s) for 4.00 min ($= 240$ s).

charge $= I \cdot t = (10.0 \times 10^{-3}$ C/s$)(240$ s$) = 2.40$ C

mol e$^- = I/F = (2.40$ C$)/(96\ 485$ C/mol$) = 24.8_7$ μmol e$^-$

The anode reaction generates ½ mol I_3^- for 1 mol e^-. Therefore, 24.8_7 µmol e^- will generate ½(24.8_7) = 12.4_4 µmol I_3^-.

In Step 5, 0.500 mL of 0.050 7 M thiosulfate = 25.3_5 µmol $S_2O_3^{2-}$. But 2 mol $S_2O_3^{2-}$ consume 1 mol I_3^-. Therefore, 25.3_5 µmol $S_2O_3^{2-}$ consume ½(25.3_5) = 12.6_8 µmol I_3^-.

We added excess $S_2O_3^{2-}$ in Step 5 and consumed the excess in Step 6. In Step 6, we had to generate I_3^- at 10.0 mA for 131 s to react with excess $S_2O_3^{2-}$.

charge = $I \cdot t$ = $(10.0 \times 10^{-3}$ C/s$)(131$ s$)$ = 1.31 C

mol e^- = I/F = $(1.31$ C$)/(96\ 485$ C/mol$)$ = 13.5_8 µmol e^-

mol I_3^- = ½(13.5_8) = 6.79 µmol I_3^-

Here is where we are so far:

Step 3: 12.4_4 µmol I_3^- were generated.

Step 4: x µmol I_3^- were consumed by sulfite in wine.

Step 5: We added enough $S_2O_3^{2-}$ to consume 12.6_8 µmol I_3^-.

Step 6: We had to generate 6.79 µmol I_3^- to consume excess $S_2O_3^{2-}$ from

Step 5. Therefore, I_3^- left after Step 4 = 12.6_8 – 6.79 = 5.8_9 µmol I_3^-.

We began with 12.4_4 µmol I_3^- and 5.89 µmol I_3^- were left after reaction with sulfite in wine. Therefore, sulfite in wine consumed 12.4_4 – 5.8_9 = 6.5_5 µmol I_3^-. But 1 mol I_3^- reacts with 1 mol sulfite. Therefore, the wine contained 6.5_5 µmol sulfite in the 2.00 mL injected for analysis.

The wine sample prepared in Step 1 consisted of 9.00 mL wine diluted to 10.00 mL. Therefore, the original wine contained 10.00/9.00 of the amount found in the analysis. That is, 2.000 mL of pure wine contains $(10.00/9.00)(6.5_5$ µmol sulfite$)$ = 7.2_8 µmol sulfite.

$$\text{sulfite in wine} = \frac{7.2_8 \text{ µmol sulfite}}{2.00 \text{ mL}} = 3.64 \text{ mM}$$

This problem left out a description of the blank titration that should be done in a real analysis. There are components in wine in addition to sulfite that could react with I_3^-. For the blank titration, 1 M formaldehyde is added to the wine to bind all sulfite. The sulfite-formaldehyde adduct is not decomposed in 2 M NaOH and does not react with I_3^-. The blank titration consists of taking this formaldehyde/wine solution through the entire procedure. We subtract I_3^- consumed by the blank from I_3^- consumed by the wine without formaldehyde.

17-24. (a) Balance carbon: $B = c$;   balance halogen: $C = x$;   balance nitrogen: $D = n$

Balance oxygen: $o + A = 2B \Rightarrow o + A = 2c \Rightarrow A = 2c - o$

Balance hydrogen: $h + 2A = 3D + E \Rightarrow h + 2(2c - o) = 3n + E$

$\Rightarrow E = h + 4c - 2o - 3n$

Charge balance, $F = E - C = h + 4c - 2o - 3n - c = h - c/2 + o - 3n$

(b) To consume F e^- requires $F/4$ O_2, because each O_2 consumes $4e^-$.

(c) $F = (9.43 \times 10^{-3}\ C)/(9.648\ 5 \times 10^4\ C/mol) = 9.77_4 \times 10^{-8}$ mol e^-

mol $O_2 = F/4 = 2.44_3 \times 10^{-8}$ mol

(d) The mass of O_2 in (c) is $(2.44_3 \times 10^{-8}\ mol)(32.00\ g/mol) = 7.81_9 \times 10^{-7}$ g. This much O_2 was required to react with 13.5 µL of sample. The mass of O_2 that would react with 1 L of sample is $(7.81_9 \times 10^{-7}\ g)/(13.5 \times 10^{-6}\ L) =$ 0.057 9 g/L = 57.9 mg/L.

(e) The balanced equation oxidation half-reaction is

$C_9H_6NO_2ClBr_2 + 16H_2O \rightarrow 9CO_2 + 3X^- + NH_3 + 35H^+ + 32e^-$.

The observed number of electrons in the reaction was $9.77_4 \times 10^{-8}$ mol e^-, so there must have been $(9.77_4 \times 10^{-8}$ mol $e^-)/(32$ mol $e^-/mol\ C_9H_6NO_2ClBr_2) =$ $3.05_4 \times 10^{-9}$ mol $C_9H_6NO_2ClBr_2$ in 13.5 µL. The molarity of $C_9H_6NO_2ClBr_2$ is $(3.05_4 \times 10^{-9}$ mo$)/(13.5 \times 10^{-6}\ L) = 2.26 \times 10^{-4}$ M.

17-25. The Clark electrode measures dissolved oxygen by reducing it to H_2O at a gold tip on a platinum electrode held at –0.75 V with respect to Ag|AgCl. The opening of the body of the electrode is filled with a 10- to 40-µm-long plug of silicone rubber that is permeable to O_2. Current is proportional to the concentration of dissolved O_2 in the external medium. The electrode needs to be calibrated in solutions of known O_2 concentration.

17-26. **(a)** The glucose monitor has a test strip with two carbon indicator electrodes and a silver-silver chloride reference electrode. Indicator electrode 1 is coated with glucose oxidase and a mediator. When a drop of blood is placed on the test strip, glucose from the blood is oxidized near indicator electrode 1 by mediator to gluconolactone and the mediator is reduced. With a potential of +0.2 V (with respect to the Ag|AgCl electrode) on the indicator electrode, reduced mediator is re-oxidized at the indicator electrode. The current between indicator electrode 1 and the reference electrode is proportional to the rate of oxidation of the mediator, which is proportional to the concentration of glucose plus any interfering species in the blood. Indicator electrode 2 has mediator, but no glucose oxidase. Current measured between indicator electrode 2 and the reference electrode is proportional to the concentration of interfering species in the blood. The difference between the two currents is proportional to the concentration of glucose in the blood.

(b) In the absence of a mediator, the rate of oxidation of glucose depends on the concentration of O_2 in the blood. If $[O_2]$ is low, the current will be low and the monitor will give an incorrect, low reading for the glucose concentration. A mediator such as 1,1′-dimethylferrocene can replace O_2 in the glucose oxidation and be subsequently reduced at the indicator electrode. The concentration of mediator is constant and high enough so that variations in electrode current are due mainly to variations in glucose concentration. Also, by lowering the required electrode potential for oxidation of the mediator, there is less possible interference by other species in the blood.

(c) Glucose oxidase is replaced by glucose dehydrogenase, which does not use O_2 as a reactant. The enzyme oxidizes glucose and reduces the PQQ cofactor to $PQQH_2$. $PQQH_2$ is oxidized back to PQQ by a nearby Os^{3+} bound to the polymer chain. A nearby Os^{2+} can exchange electrons with the Os^{3+}. By moving from Os to Os, electrons eventually reach the carbon electrode. The coulometric sensor measures the total number of electrons needed to oxidize all of the glucose in the small blood sample.

(d) Amperometry measures current during the enzyme-catalyzed oxidation of glucose. Current is proportional to the rate of the oxidation reaction. The rates of most chemical reactions increase with increasing temperature. Therefore, the current will increase with increasing temperature of the blood sample. Coulometry measures the total number of electrons released in the oxidation. Glucose releases 2 electrons per molecule, regardless of temperature. The coulometric signal should have no temperature dependence.

(e) 1.00 g glucose/L = 5.55 mM glucose. A volume of 0.300×10^{-6} L contains 1.665 nmol glucose. Each mole of glucose releases $2e^-$ and $2H^+$ during oxidation. Therefore, 2×1.665 nmol = 3.33 nmol e^- are released. The charge is $Q = (3.33 \times 10^{-9} \text{ mol})(96\,485 \text{ C/mol}) = 321 \text{ }\mu\text{C}$.

17-27. At low potential ($< \sim 0$ V), the chemistry is $Fe(CN)_6^{3-} + e^- \rightarrow Fe(CN)_6^{4-}$. At high potential ($> \sim 0.5$ V), the chemistry is $Fe(CN)_6^{4-} \rightarrow Fe(CN)_6^{3-} + e^-$. The current density at each plateau depends on the rate at which reactant reaches the electrode which, in turn, depends on the bulk concentration of the reactant. In all cases, the bulk concentration of $Fe(CN)_6^{3-}$ is 10 mM, so the low voltage plateau is at the same current density for all solutions. The bulk concentration of $Fe(CN)_6^{4-}$ varies from 20 to 60 mM, so each curve reaches a different current density at high potential. If the rotation speed were decreased, the thickness of the diffusion layer would increase and analyte would require a longer time to transit the diffusion layer to the electrode. The magnitude of the positive current density and the magnitude of the negative current density on both ends of the curves would decrease.

17-28. ω is the rotation rate in radians per second. We need to convert rpm (revolutions per minute) to radians per second.

$$\left(2.00 \times 10^3 \frac{\text{revolutions}}{\text{min}}\right)\left(\frac{1 \text{ min}}{60 \text{ s}}\right)\left(\frac{2\pi \text{ radians}}{\text{revolution}}\right) = 209 \text{ rad/s} = 209 \text{ s}^{-1}$$
(because radian is a dimensionless unit)

$\delta = 1.61 D^{1/3} \nu^{1/6} \omega^{-1/2}$
$= 1.61(2.5 \times 10^{-9} \text{ m}^2/\text{s})^{1/3}(1.1 \times 10^{-6} \text{ m}^2/\text{s})^{1/6}(209 \text{ rad/s})^{-1/2} = 1.5_3 \times 10^{-5} \text{ m}$

To calculate current density, we need to express the concentration of the species reacting at the electrode in mol/m^3 instead of mol/L. Since 1 L is the volume of a 10-cm cube, there are 1 000 L in 1 m^3. The concentration of $K_4Fe(CN)_6$ is 50.0 mM $= \left(0.050\,0 \frac{\text{mol}}{\text{L}}\right)\left(1\,000 \frac{\text{L}}{\text{m}^3}\right) = 50.0 \text{ mol/m}^3$.

Current density $= 0.62 n F D^{2/3} \nu^{-1/6} \omega^{1/2} C_0$

$= 0.62(1)\left(96\,485 \frac{\text{C}}{\text{mol}}\right)\left(2.5 \times 10^{-9} \frac{\text{m}^2}{\text{s}}\right)^{2/3}\left(1.1 \times 10^{-6} \frac{\text{m}^2}{\text{s}}\right)^{-1/6}\left(209 \frac{1}{\text{s}}\right)^{1/2}\left(50.0 \frac{\text{mol}}{\text{m}^3}\right)$

$= 7.8_4 \times 10^2 \frac{\text{C}}{\text{m}^2 \cdot \text{s}} = 7.8_4 \times 10^2 \frac{\text{A}}{\text{m}^2}$

17-29. (a)

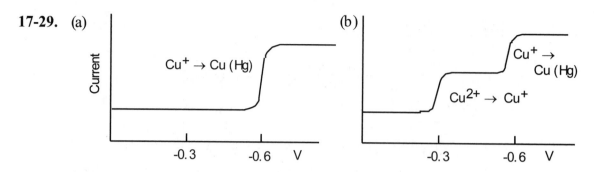

(c) The potential for the reaction Cu(I) → Cu(Hg) will change if Pt is used, since the product obviously cannot be copper amalgam.

17-30. (a) Charging current arises from charging or discharging of the electric double layer at the electrode-solution interface. Faradaic current arises from oxidation or reduction reactions.

(b) Charging current decays more rapidly than faradaic current. By 1 s after a potential step, the charging current decays to near zero and the faradaic current is still significant. The ratio of the desired signal (faradaic current) to the undesired background (charging current) is larger at 1 s than it was at earlier times. If we wait too long, both signals become too small to measure.

(c) In square wave voltammetry, an anodic pulse follows each cathodic pulse; the signal is the difference between the two. The anodic pulse oxidizes the product of each cathodic pulse, thereby replenishing electroactive species at the electrode surface for the next pulse. Analyte concentration at the electrode surface is therefore greater in square wave voltammetry.

17-31. Electrons flowing in 3.4 min =

$$\frac{(14 \times 10^{-6}\ \text{C/s})(60\ \text{s/min})(3.4\ \text{min})}{96\,485\ \text{C/mol}} = 2.9_6 \times 10^{-8}\ \text{mol e}^-$$

For the reaction $Cd^{2+} + 2e^- \rightarrow Cd(\text{in Hg})$,

$$\text{moles of } Cd^{2+} = \tfrac{1}{2}\ \text{moles of e}^- = 1.4_8 \times 10^{-8}\ \text{mol}$$

moles of Cd^{2+} in 25 mL of 0.50 mM solution $= 1.25 \times 10^{-5}$ mol

$$\text{percentage of } Cd^{2+} \text{ reduced} = \frac{1.4_8 \times 10^{-8}}{1.25 \times 10^{-5}} \times 100 = 0.11_8\%$$

17-32. $\dfrac{[X]_i}{[S]_f + [X]_f} = \dfrac{I_X}{I_{S+X}}$

$$\dfrac{x(mM)}{3.00\left(\dfrac{2.00}{52.00}\right) + x\left(\dfrac{50.0}{52.0}\right)} = \dfrac{0.37\ \mu A}{0.80\ \mu A} \Rightarrow x = 0.096\ mM$$

17-33. In anodic stripping voltammetry, analyte is reduced and concentrated at the working electrode at a controlled potential for a constant time. The potential is then ramped in a positive direction to reoxidize the analyte, during which time current is measured. The height of the oxidation wave is proportional to the original concentration of analyte. Stripping is the most sensitive voltammetric technique because analyte is concentrated from a dilute solution. The longer the period of concentration, the more sensitive is the analysis.

17-34. (a) Concentration (deposition) stage: $Cu^{2+} + 2e^- \rightarrow Cu(s)$

(b) Stripping stage: $Cu(s) \rightarrow Cu^{2+} + 2e^-$

(c) All solutions were made up to the same volume, so $[X]_i = [X]_f \equiv x$. Prepare a graph of I vs. $[S]_f$ using data measured from the figure in the problem. The intercept is at -313 ppb, so the original concentration of Cu^{2+} is 313 ppb.

Added standard (ppb)	Current (μA)
0	0.59_9
100	0.77_4
200	0.94_3
300	1.12_8
400	1.31_4
500	1.54_4

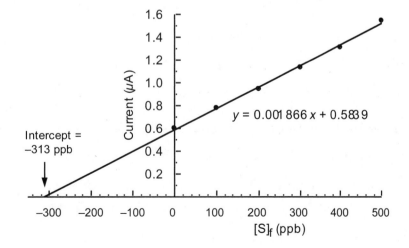

17-35.

	A	B	C	D	E
1	Standard Addition Constant Volume Least-Squares Spreadsheet				
2	x	y			
3	Added Fe(III)	Relative			
4	(pM)	peak height			
5	0	1.00			
6	50	1.56			
7	100	1.98			
8	B10:C12 = LINEST(B5:B7,A5:A7,TRUE,TRUE)				
9		LINEST output:			
10	m	0.0098	1.0233	b	
11	u_m	0.0008	0.0522	u_b	
12	R^2	0.9932	0.0572	s_y	
13	x-intercept = -b/m =	-104.422			
14	n =	3	B14 = COUNT(A6:A7)		
15	Mean y =	1.513	B15 = AVERAGE(B5:B7)		
16	$\Sigma(x_i - \text{mean } x)^2 =$	5000	B16 = DEVSQ(A5:A7)		
17	Std uncertainty of				
18	x-intercept (u_x) =	13.174			
19	B18 = (C12/ABS(B10))*SQRT((1/B14) + B15^2/(B10^2*B16))				

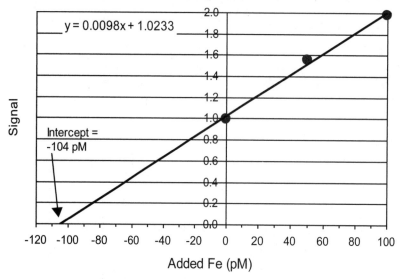

y = 0.0098x + 1.0233

Intercept = -104 pM

Cells B13 and B18 of the spreadsheet tell us that [Fe(III)] = 104 ± 13 pM.

17-36. In the oxidation step, neutral poly(3-octylthiophene) is oxidized to make it positively charged. One ClO_4^- molecule diffuses into the PVC membrane to neutralize each positive charge of poly(3-octylthiophene). The number of ClO_4^- molecules diffusing into the PVC membrane in a fixed oxidation time is proportional to the bulk concentration of ClO_4^-. The rate of poly(3-octylthiophene) oxidation cannot exceed the rate at which ClO_4^- diffuses into the PVC. In the cathodic stripping step, electrons are added back to oxidized poly(3-octyl-thiophene) to reduce the charge back to 0. One ClO_4^- diffuses out of the PVC membrane for each electron added to poly(3-octylthiophene) to maintain charge balance. The number of electrons in the cathodic faradaic stripping current equals the number of ClO_4^- molecules that were taken up by the PVC which, in turn, was proportional to the ClO_4^- concentration in bulk solution.

17-37. In the first cathodic scan from 0 to −1.0 V, the reduction is
$$RNO_2 + 4e^- + 4H^+ \rightarrow RNHOH + H_2O$$
In the reverse anodic scan from −1.0 to + 0.9 V, the oxidation is

Peak B: $RNHOH \rightarrow RNO + 2H^+ + 2e^-$

In the second cathodic scan from +0.9 to −0.4 V, the new peak C appears. This peak could be assigned to

Peak C: $RNO + 2H^+ + 2e^- \rightarrow RNHOH$

Peak C was not seen in the first cathodic scan because there was no RNO present in the initial solution. RNO was created by the first anodic scan.

17-38.

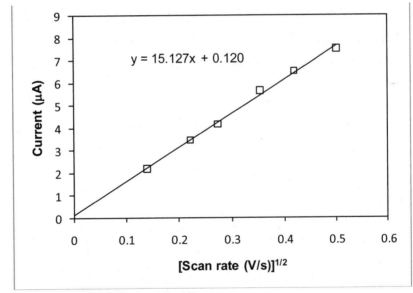

$$I_p = (2.69 \times 10^8)n^{3/2}ACD^{1/2}v^{1/2}$$
$$\text{slope} = (2.69 \times 10^8)n^{3/2}ACD^{1/2}$$
$$\Rightarrow D = \frac{\text{slope}^2}{(2.69 \times 10^8)^2 n^3 A^2 C^2}$$
$$= \frac{(15.1 \times 10^{-6}\ A/\sqrt{V/s})^2}{(2.69 \times 10^8)^2 1^3 (0.020\ 1 \times 10^{-4}\ m^2)^2 (1.00^{-3}\ M)^2} = 7.8 \times 10^{-10}\ m^2/s$$

17-39. Microelectrodes fit into small places, are useful in nonaqueous solution (because of small ohmic losses), and allow rapid voltage scans (because of small capacitance), which permits the study of short-lived species. Low capacitance gives low background charging current, which increases sensitivity to analyte by orders of magnitude.

17-40. The Nafion membrane permits neutral and cationic species to pass through to the electrode, but excludes anions. It reduces the background signal from ascorbate anion, which would otherwise swamp the signal from dopamine.

17-41. (a) Upper plateau: $Fe(CN)_6^{4-} \rightarrow Fe(CN)_6^{3-} + e^-$

Lower plateau: $Fe(CN)_6^{3-} + e^- \rightarrow Fe(CN)_6^{4-}$

(b) $r = \dfrac{I_{limit}}{4nFDC} = \dfrac{18.5 \text{ nA}}{4 \cdot 1 \ (96\ 485 \text{ C/mol})(9.2 \times 10^{-10} \text{ m}^2/\text{s})(0.010 \text{ M})}$

Converting concentration in mol/L to concentration in mol/m^3 gives

$= \dfrac{18.5 \times 10^{-9} \text{ C/s}}{4 \cdot 1 \ (96\ 485 \text{ C/mol})(9.2 \times 10^{-10} \text{ m}^2/\text{s})(0.010 \text{ mol}/10^{-3} \text{ m}^3)} = 5.2 \ \mu\text{m}$

Note that the diffusion coefficient cited in Problem 17-28 is almost 3 times greater than the value cited in the present problem. The higher diffusion coefficient gives an electrode radius of $r = 1.9 \ \mu\text{m}$.

17-42. $ROH + SO_2 + B \rightarrow BH^+ + ROSO_2^-$

$H_2O + I_2 + ROSO_2^- + 2B \rightarrow ROSO_3^- + 2BH^+I^-$

One mole of H_2O plus one mole of I_2 are required for the oxidation of the alkyl sulfite to an alkyl sulfate in the second reaction.

17-43. The bipotentiometric detector maintains a constant current (~10 μA) between two detector electrodes, while measuring the voltage needed to sustain that current. Before the equivalence point, the solution contains I^-, but little I_2. To maintain a current of 10 μA, the cathode potential must be negative enough to reduce some component of the solvent system (perhaps $CH_3OH + e^- \rightleftharpoons CH_3O^- + \frac{1}{2}H_2(g)$). At the equivalence point, excess I_2 suddenly appears and current can be carried at low voltage by the reactions below. The abrupt voltage drop marks the end point.

Cathode: $I_3^- + 2e^- \rightarrow 3I^-$

Anode: $3I^- \rightarrow I_3^- + 2e^-$

CHAPTER 18
FUNDAMENTALS OF SPECTROPHOTOMETRY

18-1. (a) double (b) halve (c) double

18-2. (a) $E = h\nu = hc/\lambda = (6.626\,2 \times 10^{-34}\text{ J s})(2.997\,9 \times 10^8\text{ m s}^{-1})/(650 \times 10^{-9}\text{ m})$
$= 3.06 \times 10^{-19}$ J/photon $= 184$ kJ/mol

(b) For $\lambda = 400$ nm, $E = 299$ kJ/mol.

18-3. $\nu = c/\lambda = 2.997\,9 \times 10^8$ m s$^{-1}/562 \times 10^{-9}$ m $= 5.33 \times 10^{14}$ Hz
$\tilde{\nu} = 1/\lambda = 1.78\ 10^6$ m^{-1} (1 m/100 cm) $= 1.78\ 10^4$ cm^{-1}
$E = h\nu = (6.626\,2 \times 10^{-34}\text{ J s})(5.33 \times 10^{14}\text{ s}^{-1}) = 3.53 \times 10^{-19}$ J/photon
$(3.53 \times 10^{-19}\text{ J/photon})(6.022 \times 10^{23}\text{ photon/mol}) = 213$ kJ/mol

18-4. Microwave energies correspond to molecular rotation energies. Infrared energies correspond to vibrational energies. Visible light can promote electrons to excited states (in colored compounds). Ultraviolet light also promotes electrons and can even break chemical bonds.

18-5. From the definition of index of refraction, we can write
$c_{\text{vacuum}} = n \cdot c_{\text{air}}$
$\lambda_{\text{vacuum}} \cdot \nu = n \cdot \lambda_{\text{air}} \cdot \nu$
$\lambda_{\text{air}} = \lambda_{\text{vacuum}}/n = \lambda_{\text{vacuum}}/1.000\,292\,6$
$\nu = c/\lambda_{\text{vacuum}} = 5.088\,491\,0$ and $5.083\,335\,8 \times 10^{14}$ Hz
$\lambda_{\text{air}} = \lambda_{\text{vacuum}}/n = 588.985\,54$ and $589.582\,86$ nm
$\tilde{\nu}_{\text{air}} = 1/\lambda_{\text{air}} = 1.697\,834\,5$ and $1.696\,114\,4\ 10^4$ cm^{-1}

18-6. (a) The light source is the mercury vapor lamp.

(b) The detector is a photodiode.

(c) The monochromator is the filter that admits 254 nm radiation to the detector and rejects other wavelengths from the Hg lamp.

(d) The optical pathlength, b, is the sum of the lengths of the two side legs and the bottom leg of the instrument.

(e) This instrument measures absorption of 254 nm radiation. Absorbance is proportional to the concentration of O_3.

18-7. Transmittance (T) is the fraction of incident light that is transmitted by a substance: $T = P/P_0$, where P_0 is incident irradiance and P is transmitted irradiance. Absorbance is logarithmically related to transmittance: $A = -\log T$. When all light is transmitted, absorbance is zero. When no light is transmitted, absorbance is infinite. Absorbance is proportional to concentration. Molar absorptivity is the constant of proportionality between absorbance at a particular wavelength and the product cb, where c is concentration and b is pathlength.

18-8. An absorption spectrum is a graph of absorbance vs. wavelength.

18-9. The color of transmitted light is the complement of the color that is absorbed. If blue-green light is absorbed, red light is transmitted.

18-10.

Curve	Absorption peak (nm)	Predicted color (Table 18-1)	Observed color
A	760	green	green
B	700	green	blue-green
C	600	blue	blue
D	530	violet	violet
E	500	red or purple red	red
F	410	green-yellow	yellow

18-11. If absorbance is too high, too little light reaches the detector for accurate measurement. If absorbance is too low, there is too little difference between sample and reference for accurate measurement.

18-12. $\varepsilon = A/bc = 0.822/[(1.00\ \text{cm})(2.31 \times 10^{-5}\ \text{M})] = 3.56 \times 10^4\ \text{M}^{-1}\ \text{cm}^{-1}$

18-13. Violet blue, according to Table 18-1.

18-14. [Fe] in reference cell $= \left(\dfrac{10.0\ \text{mL}}{50.0\ \text{mL}}\right)(6.80 \times 10^{-4}) = 1.36 \times 10^{-4}\ \text{M}$. Setting the absorbances of sample and reference equal to each other gives $\varepsilon_s b_s c_s = \varepsilon_r b_r c_r$. But $\varepsilon_s = \varepsilon_r$, so $(2.48\ \text{cm})c_s = (1.00\ \text{cm})(1.36 \times 10^{-4}\ \text{M}) \Rightarrow c_s = 5.48 \times 10^{-5}\ \text{M}$. This is a 1/4 dilution of runoff, so [Fe] in runoff $= 2.19 \times 10^{-4}\ \text{M}$.

18-15. (a) $A = -\log T = -\log 0.244 = 0.612_6$

(b) $PV = nRT$. Molar concentration $= \dfrac{\text{moles}}{\text{volume}} = \dfrac{n}{V} = \dfrac{P}{RT}$.

To convert 30.3 μbar $= 30.3 \times 10^{-6}$ bar to mol/L, we write

$$\frac{n}{V} = \frac{P}{RT} = \frac{30.3 \times 10^{-6} \text{ bar}}{\left(0.083\,145\,\dfrac{\text{L} \cdot \text{bar}}{\text{mol} \cdot \text{K}}\right)(298 \text{ K})} = 1.22_3 \times 10^{-6} \text{ M}$$

(c) Molar absorptivity $= \varepsilon = \dfrac{A}{bc}$

Pathlength $b = 3.00$ cm; concentration $c = 1.22_3 \times 10^{-6}$ M.

$$\varepsilon = \frac{A}{bc} = \frac{0.612_6}{(3.00 \text{ cm})(1.22_3 \times 10^{-6} \text{ M})} = 1.67 \times 10^5 \text{ M}^{-1} \text{ cm}^{-1}$$

18-16. (a) Measured from graph:

$\sigma \approx 1.3 \times 10^{-20}$ cm^2 at 325 nm $\qquad$ $\sigma \approx 3.5 \times 10^{-19}$ cm^2 at 300 nm

at 325 nm: $T = e^{-(8 \times 10^{18} \text{ cm}^{-3})(1.3 \times 10^{-20} \text{ cm}^2)(1 \text{ cm})} = 0.90$

$\qquad\qquad A = -\log T = 0.045$

at 300 nm: $T = e^{-(8 \times 10^{18} \text{ cm}^{-3})(3.5 \times 10^{-19} \text{ cm}^2)(1 \text{ cm})} = 0.061$

$\qquad\qquad A = -\log T = 1.22$

(b) $T = e^{-n\sigma b}$ $\qquad 0.14 = e^{-(8 \times 10^{18} \text{ cm}^{-3})\sigma(1 \text{ cm})}$

$\qquad\qquad \Rightarrow \sigma = 2.4_{576} \times 10^{-19}$ cm^2

If n is decreased by 1%, $T = e^{-(7.92 \times 10^{18} \text{ cm}^{-3})(2.457\,6 \times 10^{-19} \text{ cm}^2)(1 \text{ cm})}$

$\qquad = 0.142\,8$

Increase in transmittance is $\dfrac{0.142\,8 - 0.14}{0.14} = 2.0\%$.

Note that the fractional increase in transmittance is greater than the fractional decrease in ozone concentration.

(c) $T_{\text{winter}} = e^{-(290 \text{ D.U.})(2.69 \times 10^{16} \text{ molecules/cm}^3/\text{D.U.})(2.5 \times 10^{-19} \text{ cm}^2)(1 \text{ cm})}$

$\qquad\qquad = 0.142$

$T_{\text{summer}} = e^{-(350)(2.69 \times 10^{16})(2.5 \times 10^{-19})(1)} = 0.095$

Fractional increase in transmittance is $(0.142 - 0.095) / (0.095) = 49\%$.

18-17. Neocuproine reacts with Cu(I) and prevents it from forming a complex with ferrozine that would give a false positive result in the analysis of iron.

18-18. (a) $c = A/\varepsilon b = 0.427/[(6\,130\ \text{M}^{-1}\ \text{cm}^{-1})(1.000\ \text{cm})] = 6.97 \times 10^{-5}\ \text{M}$

(b) The sample had been diluted $\times\ 10 \Rightarrow 6.97 \times 10^{-4}\ \text{M}$.

(c) $\dfrac{x\ \text{g}}{(292.16\ \text{g/mol})(5.00 \times 10^{-3}\ \text{L})} = 6.97 \times 10^{-4}\ \text{M} \Rightarrow x = 1.02\ \text{mg}$

18-19. Yes

18-20. (a) $\varepsilon = \dfrac{A}{cb} = \dfrac{0.267 - 0.019}{(3.15 \times 10^{-6}\ \text{M})(1.000\ \text{cm})} = 7.87 \times 10^4\ \text{M}^{-1}\ \text{cm}^{-1}$

(b) $c = \dfrac{A}{\varepsilon b} = \dfrac{0.175 - 0.019}{(7.87 \times 10^4\ \text{M}^{-1}\ \text{cm}^{-1})(1.000\ \text{cm})} = 1.98 \times 10^{-6}\ \text{M}$

18-21. (a) $\varepsilon_{280\ \text{nm}}\ (\text{M}^{-1}\ \text{cm}^{-1}) \approx 5\,500\ n_{\text{Trp}} + 1\,490\ n_{\text{Tyr}} + 125\ n_{\text{S-S}}$

$= 5\,500\ (8) + 1\,490\ (26) + 125\ (19) = 8.512 \times 10^4\ \text{M}^{-1}\ \text{cm}^{-1}$

(b) 1 wt% protein $\approx$ 10 g/L if density = 1.00 g/mL

molarity = (10 g/L)/(79 550 g/mol) = $1.257 \times 10^{-4}\ \text{M}$

Absorbance of 1 wt% = εbc

$= (8.512 \times 10^4\ \text{M}^{-1}\ \text{cm}^{-1})(1.000\ \text{cm})(1.257 \times 10^{-4}\ \text{M}) = 10.7$

(c) A 1 wt% solution is predicted to have A = 10.7. A solution with A = 1.50 has a concentration of (1.50/10.7)(1 wt%) = 0.140 wt% .

Since 1 wt% = 10 mg/mL, 0.140 wt% = (0.140)(10 mg/mL) = 1.40 mg/mL

18-22. (a) The absorbance due to the colored product from nitrite added to sample C is $0.967 - 0.622 = 0.345$. The concentration of colored product due to added nitrite in sample C is $\dfrac{(7.50 \times 10^{-3}\ \text{M})(10.0 \times 10^{-6}\ \text{L})}{0.054\ \text{L}} = 1.389 \times 10^{-6}\ \text{M}$.

$\varepsilon = A/bc = 0.345/[(1.389 \times 10^{-6})(5.00)] = 4.97 \times 10^4\ \text{M}^{-1}\ \text{cm}^{-1}$

(b) 7.50×10^{-8} mol of nitrite (from 10.0 µL added to sample C) gives $A = 0.345$. In sample B, x mole of nitrite in food extract gives $A = 0.622 - 0.153 = 0.469$.

$\dfrac{x\ \text{mol}}{7.50 \times 10^{-8}\ \text{mol}} = \dfrac{0.469}{0.345} \Rightarrow x = 1.020 \times 10^{-7}\ \text{mol NO}_2^- = 4.69\ \text{µg}$

18-23. (a) Mass of iron required $= (500 \text{ mL})(500 \text{ µg Fe/mL}) = 2.50 \times 10^5 \text{ µg Fe} =$
250 mg Fe $= 0.250$ g Fe

Mass fraction of Fe in $Fe(H_3NCH_2CH_2NH_3)(SO_4)_2 \cdot 4H_2O =$

$$= \frac{55.845 \text{ g Fe}}{382.15 \text{ g reagent}} = 0.146\,13_4$$

Mass of $Fe(H_3NCH_2CH_2NH_3)(SO_4)_2 \cdot 4H_2O$ required to provide 0.250 g Fe

$$= \frac{0.250 \text{ g Fe}}{0.146\,13_4 \text{ g Fe/g reagent}} = 1.711 \text{ g}$$

(b) g Fe weighed out $= (0.146\,13_4 \text{ g Fe/g reagent})(1.627 \text{ g reagent})$

$$= 0.237_8 \text{ g Fe}$$

$$[Fe] = \frac{0.237_8 \text{ g Fe}}{500.0 \text{ mL}} = 0.4755 \text{ mg Fe/mL} = 475.5 \text{ µg Fe/mL}$$

(c) To prepare ~ 1 µg Fe/mL, dilute 1.000 mL of stock solution from (b) up to
500.0 mL with 0.1 M H_2SO_4

$$[Fe] = \frac{475.5 \text{ µg Fe}}{500.0 \text{ mL}} = 0.950_9 \text{ µg Fe/mL}$$

In a similar manner, dilute 2.000, 3.00, 4.00, and 5.00 mL of stock solution
from (b) up to 500.0 mL with 0.1 M H_2SO_4 to obtain [Fe] = 1.902 µg Fe/mL,
2.85_3 µg Fe/mL, 3.80 µg Fe/mL, and 4.76 µg Fe/mL.

(d) First make a 1/10 dilution by diluting 10 mL of ~500 µg Fe/mL stock
solution up to 50 mL with 0.1 M H_2SO_4. This solution contains ~50 µg
Fe/mL. To prepare ~ 1 µg Fe/mL, dilute 1.000 mL of ~50 µg Fe/mL stock
solution up to 50.0 mL with 0.1 M H_2SO_4. In a similar manner, dilute 2.000,
3.00, 4.00, and 5.00 mL of ~50 µg Fe/mL stock solution up to 50.0 mL with
0.1 M H_2SO_4 to obtain [Fe] $\approx$ 2, 3, 4, and 5 µg Fe/mL.

18-24. (a) Mass of iron required $= (500 \text{ mL})(1\,000 \text{ µg Fe/mL}) = 5.00 \times 10^5 \text{ µg Fe} =$
500 mg Fe $= 0.500$ g Fe

Mass fraction of Fe in $Fe(NH_4)_2(SO_4)_2 \cdot 6H_2O =$

$$= \frac{55.845 \text{ g Fe}}{392.15 \text{ g reagent}} = 0.142\,40_7$$

Mass of $Fe(NH_4)_2(SO_4)_2 \cdot 6H_2O$ required to provide 0.500 g Fe

$$= \frac{0.500 \text{ g Fe}}{0.142\,40_7 \text{ g Fe/g reagent}} = 3.511 \text{ g}$$

(b) g Fe weighed out = $(0.142\,40_7$ g Fe/g reagent$)(3.627$ g reagent$)$

$$= 0.516\,5_1 \text{ g Fe}$$

$$[\text{Fe}] = \frac{0.516\,5_1 \text{ g Fe}}{500.0 \text{ mL}} = 1.033_0 \text{ mg Fe/mL} = 1\,033._0 \text{ μg Fe/mL}$$

(c) The table below lists two serial dilutions to prepare each desired standard

First dilution volume (mL) of 1 000 μg/mL to dilute to 250 mL	Resulting [Fe] (μg/mL)	Second dilution volume of first dilution to dilute to 250 mL	Resulting [Fe] (μg/mL)
10	$\left(\dfrac{10}{250}\right)(1\,033.0$ μg Fe/mL$)$ $= 41.32$ μg Fe/mL	5	$\left(\dfrac{5}{250}\right)(41.32$ μg Fe/mL$)$ $= 0.826$ μg Fe/mL
10	41.32 μg Fe/mL	10	$\left(\dfrac{10}{250}\right)(41.32$ μg Fe/mL$)$ $= 1.653$ μg Fe/mL
10	41.32 μg Fe/mL	15 (= 10 + 5)	$\left(\dfrac{15}{250}\right)(41.32$ μg Fe/mL$)$ $= 2.479$ μg Fe/mL
15 (= 10 + 5)	$\left(\dfrac{15}{250}\right)(1\,033.0$ μg Fe/mL$)$ $= 61.98$ μg Fe/mL	15 (= 10 + 5)	$\left(\dfrac{15}{250}\right)(61.98$ μg Fe/mL$)$ $= 3.719$ μg Fe/mL
15 (= 10 + 5)	61.98 μg Fe/mL	(20) (= 10 + 10)	$\left(\dfrac{20}{250}\right)(61.98$ μg Fe/mL$)$ $= 4.958$ μg Fe/mL
20 (= 10 + 10)	$\left(\dfrac{20}{250}\right)(1\,033.0$ μg Fe/mL$)$ $= 82.64$ μg Fe/mL	(20) (= 10 + 10)	$\left(\dfrac{20}{250}\right)(82.64$ μg Fe/mL$)$ $= 6.611$ μg Fe/mL
20 (= 10 + 10)	82.64 μg Fe/mL	(25) (= 10 + 10 + 5)	$\left(\dfrac{25}{250}\right)(82.64$ μg Fe/mL$)$ $= 8.264$ μg Fe/mL
20 (= 10 + 10)	82.64 μg Fe/mL	(30) (= 10 + 10 + 10)	$\left(\dfrac{30}{250}\right)(82.64$ μg Fe/mL$)$ $= 9.917$ μg Fe/mL

18-25. Uncertainty in 3-mL pipet = ± 0.01 mL = 0.333%.

Uncertainty in delivery from two 3-mL pipets = $\pm \sqrt{0.01^2 + 0.01^2} = \pm 0.014$ mL

Volume delivered from two 3-mL pipets = 6.00 ± 0.014 mL = $6.00 \pm 0.236\%$ mL

Uncertainty in volume of 1-L flask = ± 0.30 mL = $\pm 0.030\%$

Fe stock concentration = 1.086 ± 0.002 g Fe/L = $1.086 \pm 0.184\%$ g Fe/L

After diluting two 3-mL volumes of stock solution to 1 L, the concentration is

$$[\text{Fe}] = \left(\frac{6.00 \pm 0.236\% \text{ mL}}{1\,000.0 \pm 0.03\% \text{ mL}}\right)(1.086 \pm 0.184\% \text{ g/L})$$

$$= 0.006\,516 \pm \sqrt{0.236\%^2 + 0.03\%^2 + 0.184\%^2} \text{ g/L}$$

$$= 0.006\,516 \pm 0.301\% \text{ g/L} = 0.006\,516 \pm 0.000\,020 \text{ g/L}$$

$$= 6.516 \pm 0.020 \text{ mg/L} = 6.516 \pm 0.020 \text{ µg/mL} = 6.516 \pm 0.30\% \text{ µg Fe/mL}$$

18-26. (a) Prior to the equivalence point, all added Fe(III) binds to the protein to form a red complex whose absorbance is measured in the figure. After the equivalence point, there are no more binding sites available on the protein. The slight increase in absorbance arises from the color of the iron titrant.

(b) 163×10^{-6} L of 1.43×10^{-3} M Fe(III) = 2.33×10^{-7} mol Fe(III)

(c) 1.17×10^{-7} mol transferrin in 2.00×10^{-3} L $\Rightarrow 5.83 \times 10^{-5}$ M transferrin

18-27. Theoretical equivalence point =

$$\frac{\left(2\,\dfrac{\text{mol Ga}}{\text{mol transferrin}}\right)\left(\dfrac{0.003\,57 \text{ g transferrin}}{81\,000 \text{ g transferrin/mol transferrin}}\right)}{0.006\,64\,\dfrac{\text{mol Ga}}{\text{L}}} = 13.3 \text{ µL}$$

Observed end point $\approx$ intersection of lines taken from first 6 points and last 4 points in the following graph = 12.2 µL, corresponding to $\dfrac{12.2}{13.3} = 91.7\%$ of 2 Ga/transferrin = 1.83 Ga/transferrin. In the absence of oxalate, there is no evidence for specific binding of Ga to the protein, since the slope of the curve is small and does not change near 1 or 2 Ga/transferrin.

18-28. (a) In the graph below, least-squares lines were put through the first 6 points and the last 3 points. Their intersection is the estimated end point of 21.4 µL. The moles of TCNQ in the titration are $(0.700 \text{ mL})(1.00 \times 10^{-4} \text{ M TCNQ}) = 70.0$ nmol TCNQ. This many moles of Au(0) are present in 21.4 µL of nanoparticle solution. The mass of nanoparticles in 21.4 µL is $(21.4 \text{ µL})(1.43 \text{ g/L}) = 30.6$ µg nanoparticles. To find the moles of Au(0) in 1.00g of nanoparticles, set up a proportion:

$$\frac{70.0 \times 10^{-9} \text{ mol Au(0)}}{30.6 \times 10^{-6} \text{ g nanoparticles}} = \frac{x \text{ mol Au(0)}}{1.00 \text{ g nanoparticles}} \Rightarrow x = 2.29 \text{ mmol Au(0)}$$

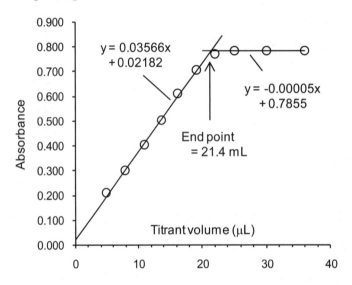

(b) 1.00 g of nanoparticles is estimated to contain 0.25 g $C_{12}H_{25}S$ (FM 201.40), which is 1.24 mmol $C_{12}H_{25}S$.

(c) From (a), the mass of Au(0) in 1.00 g is $(2.29 \text{ mmol Au(0)})(196.97 \text{ g/mol}) = 0.451$ g. The mass of Au(I) is estimated as the difference $1.00 - 0.451 - 0.25 = 0.299$ g $= 1.52$ mmol Au(I). The calculated mole ratio Au(I):$C_{12}H_{25}S$ is $1.52/1.24 = 1.23$. Ideally, this mole ratio should be 1.00.

18-29.   (a)   Both titrations show that 1 mol of streptavidin binds 4 mol of biotin

                 Addition of BF to SA: Addition of SA to BF:

                 Equivalence point = 3.99 [BF]:[SA]   Equivalence point = 0.250 [SA]:[BF]

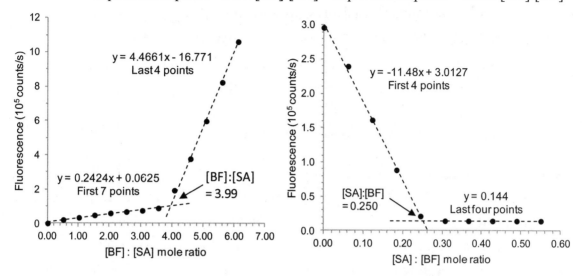

(b)   When adding BF to SA, initial fluorescence is 0 with no BF present. As long as added BF binds to SA, fluorescence increases slowly because BF is not highly fluorescent when bound to SA. When [BF]:[SA] > 4, fluorescence increases rapidly because highly fluorescent BF is present in solution.

When adding SA to BF, initial fluorescence of free BF is high. Fluorescence then decreases rapidly as SA is added until enough SA has been added to bind all BF ([SA]:[BF] = 0.25). Beyond this point, fluorescence is low and constant because all BF is bound to SA.

18-30. $\dfrac{570 \text{ kW·h}}{2 \text{ kW·h/kg coal}} = 285 \text{ kg coal}$ $\dfrac{0.6 \text{ kg C}}{\text{kg coal}} \times 285 \text{ kg coal} = 171 \text{ kg C}$

$\dfrac{44 \text{ kg } CO_2}{12 \text{ kg C}} \times 171 \text{ kg C} = 630 \text{ kg } CO_2$

$\dfrac{0.02 \text{ kg S}}{\text{kg coal}} \times 285 \text{ kg coal} = 5.7 \text{ kg S}$ $\dfrac{64 \text{ kg } SO_2}{32 \text{ kg S}} \times 5.7 \text{ kg S} = 11 \text{ kg } SO_2$

18-31. $n \rightarrow \pi^*(T_1)$:

$$E = h\nu = h\frac{c}{\lambda} = (6.626\,1 \times 10^{-34}\,\text{J·s}) \frac{2.997\,9 \times 10^8\,\text{s}^{-1}}{397 \times 10^{-9}\,\text{m}} = 5.00 \times 10^{-19}\,\text{J}$$

To convert to J/mol, multiply by Avogadro's number:

5.00×10^{-19} J/molecule $\times\ 6.022 \times 10^{23}$ molecules/mol $=\ 301$ kJ/mol.

$n \rightarrow \pi^*(S_1)$:

$$E = (6.626\,1 \times 10^{-34}\,\text{J·s}) \frac{2.997\,9 \times 10^8\,\text{s}^{-1}}{355 \times 10^{-9}\,\text{m}} = 5.60 \times 10^{-19}\,\text{J} = 337\,\text{kJ/mol.}$$

The difference between the T_1 and S_1 states is $337 - 301 = 36$ kJ/mol.

18-32. Fluorescence is emission of light with no change in the electronic spin state of the molecule (for example, singlet $\rightarrow$ singlet). In phosphorescence, the electronic spin does change during emission (for example, triplet $\rightarrow$ singlet). Phosphorescence is less probable, so molecules spend more time in the excited state prior to phosphorescence than to fluorescence. That is, phosphorescence has a longer lifetime than fluorescence. Phosphorescence also comes at lower energy (longer wavelength) than fluorescence, because the triplet excited state is at lower energy than the singlet excited state.

18-33. Luminescence is light given off after a molecule absorbs light. Chemiluminscence is light given off by a molecule created in an excited state in a chemical reaction.

18-34. In Rayleigh scattering, electrons in molecules oscillate at the frequency of incoming radiation and emit that same frequency in all directions. The time scale is $\sim 10^{-15}$ s for visible light. In Raman scattering, molecules extract a quantum of vibrational energy from incoming light and scatter light with less energy than the incoming light. Again, the time scale is $\sim 10^{-15}$ s for visible light. Fluorescence occurs in 10^{-8} to 10^{-4} s, which is 10^7 to 10^{11} times longer than scattering.

18-35. Phosphorescence is emitted at longer wavelength than fluorescence. Absorption is at shortest wavelength.

18-36. In an excitation spectrum, the exciting wavelength (λ_{ex}) is varied while the detector wavelength (λ_{em}) is fixed. In an emission spectrum, λ_{ex} is held constant and λ_{em} is varied. The excitation spectrum resembles an absorption spectrum because emission intensity is proportional to absorption of the exciting radiation.

18-37. The spreadsheet computes relative fluorescence given by

$$\text{relative intensity} = \frac{I}{k'P_0} = 10^{-\varepsilon_{ex}b_1 c}(1 - 10^{-\varepsilon_{ex}b_2 c})10^{-\varepsilon_{em}b_3 c}.$$

	A	B	C
1	Anthracene fluorescence response		
2			
3	$\varepsilon_{ex} =$	9000	$M^{-1} \, cm^{-1}$
4	$\varepsilon_{em} =$	50	$M^{-1} \, cm^{-1}$
5			
6	$b_1 =$	0.3	cm
7	$b_2 =$	0.4	cm
8	$b_3 =$	0.5	cm
9			
10		Relative	
11		fluorescence	
12	c (M)	$I/k'P_o$	Intensity/c
13	1.E-08	8.288E-05	8288
14	2.E-08	0.0001658	8288
15	4.E-08	0.0003314	8286
16	6.E-08	0.000497	8284
17	8.E-08	0.0006626	8282
18	1.E-07	0.0008281	8281
19	2.E-07	0.0016544	8272
20	4.E-07	0.0033019	8255
21	6.E-07	0.0049426	8238
22	8.E-07	0.0065764	8221
23	1.E-06	0.0082034	8203
24	2.E-06	0.0162369	8118
25	4.E-06	0.0318052	7951
26	6.E-06	0.0467266	7788
27	8.E-06	0.0610222	7628
28	1.E-05	0.0747124	7471
29	2.E-05	0.1347552	6738
30	4.E-05	0.2195664	5489
31	6.E-05	0.2689283	4482
32	8.E-05	0.2934519	3668
33	1.E-04	0.300872	3009
34	2.E-04	0.2307768	1154
35	4.E-04	0.0783318	196
36	6.E-04	0.0230136	38
37	8.E-04	0.0065982	8
38	1.E-03	0.0018832	2

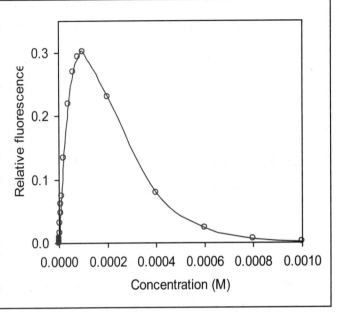

Fluorescence increases with concentration and then decreases because of self-absorption. Most of the absorption occurs at the excitation wavelength, and a little comes at the emission wavelength. Column C of the spreadsheet gives the relative fluorescence intensity divided by concentration. If intensity were proportional to concentration, then column C would be constant. We see that it is constant at low concentration, and falls by ~5% at ~5 µM. The calibration curve in the text goes up to 0.6 µM, which is in the linear range.

18-38. The equation for standard addition is

$$I_{S+X}\left(\frac{V}{V_o}\right) = \underbrace{I_X + \frac{I_X}{[X]_i}}_{\text{Function to plot on } y\text{-axis}} [S]_i \underbrace{\left(\frac{V_s}{V_o}\right)}_{\text{Function to plot on } x\text{-axis}}$$

where V_o is the volume of unknown in the cuvet (2.00 mL), V_s is the volume of standard added (0 to 40 µL), V is the total volume of unknown plus added standard, $[S]_i$ is the initial concentration of standard (1.40 µg Se/mL), $[X]_i$ is the initial concentration of unknown in the 2.00-mL solution, I_X is the fluorescence intensity from the unknown, and I_{S+X} is the fluorescence intensity from the unknown plus standard addition. We make a graph of $I_{S+X} V/V_o$ versus $[S]_i V_s/V_o$ in the following spreadsheet.

	A	B	C	D	E	F
1	Standard Addition Variable Volume Least-Squares Spreadsheet					
2						
3	Vo (mL) =	Vs (mL) =		x		y
4	2.00	volume				
5	[S]i (µg/mL) =	standard	Total volume	x-axis function	I(s+x) =	y-axis function
6	1.40	added	V = Vo + Vs	Si*Vs/Vo	signal	I(s+x)*V/Vo
7		0.0000	2.000	0.00000	41.4	41.4
8		0.0100	2.010	0.00700	49.2	49.4
9		0.0200	2.020	0.01400	56.4	57.0
10		0.0300	2.030	0.02100	63.8	64.8
11		0.0400	2.040	0.02800	70.3	71.7
12						
13	B16:D18 = LINEST(F7:F11,D7:D11,TRUE,TRUE)					
14						
15		LINEST output:				
16	m	1084.6	41.670	b		
17	u_m	14.9	0.255	u_b		
18	R²	0.9994	0.330	s_y		
19	x-intercept					
20	= -b/m =	-0.03842				
21						
22	n =		5	B22 = COUNT(B7:B11)		
23	Mean y =		56.8546	B23 = AVERAGE(F7:F11)		
24	Σ(xi - mean x)² =		0.00049	B24 = DEVSQ(D7:D11)		
25						
26	Std uncertainty of					
27	x-intercept (ux) =	0.000732				
28	B27 =(C18/ABS(B16))*SQRT((1/B22) + B23^2/(B16^2*B24))					

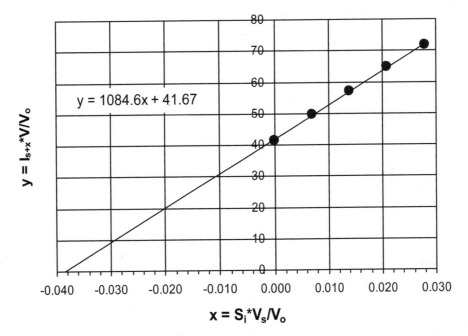

The x-intercept is found from the equation of the straight line by setting $y = 0$:

$0 = 1084.6x + 41.67 \Rightarrow x = -0.038\ 42$. The concentration of Se in the unknown

is $0.038\ 42$ µg/mL. All Se from 0.108 g of Brazil nuts was dissolved in 10.0 mL

of solvent, which contained $(10.0\ \text{mL})(0.038\ 42\ \text{µg/mL}) = 0.384\ 2$ µg Se. The

wt% Se in the nuts is $100 \times (0.384\ 2 \times 10^{-6}\ \text{g}/0.108\ \text{g}) = 3.56 \times 10^{-4}$ wt%.

The standard deviation of the x-intercept is

$$\begin{array}{c} \text{Standard deviation} \\ \text{of } x\text{-intercept} \end{array} = \frac{s_y}{|m|} \sqrt{\frac{1}{n} + \frac{\bar{y}^2}{m^2\ \Sigma(x_i - \bar{x})^2}}$$

where s_y is the standard deviation of y, m is the slope, n is the number of data

points $(= 5)$, $\bar{y}$ is the mean value of y for the 5 points, x_i are individual values of x,

and $\bar{x}$ is the mean value of x for the 5 points. Cell B27 of the spreadsheet gives a

standard deviation of $0.000\ 732$ for the x-intercept.

The relative uncertainty in the intercept is $0.000\ 732/0.038\ 42 = 1.91\%$. This is

the relative uncertainty in wt% Se if other sources of error are insignificant.

Uncertainty in wt% $= (0.0191)(3.56 \times 10^{-4}\ \text{wt%}) = 6.8 \times 10^{-6}$ wt%. Answer $=$

$3.56\ (\pm 0.06_8) \times 10^{-4}$ wt%.

The confidence interval is $\pm t \times$ (standard deviation) where t is Student's t (Table

4-2) for $n - 2 = 3$ degrees of freedom. The 95% confidence interval is

$\pm(3.182)(0.06_8 \times 10^{-4}\ \text{wt%}) = \pm 0.22 \times 10^{-4}$ wt%. The value $t = 3.182$ was taken

from Table 4-2 for 3 degrees of freedom. Answer $= 3.56\ (\pm 0.22) \times 10^{-4}$ wt%.

CHAPTER 19
APPLICATIONS OF SPECTROPHOTOMETRY

19-1. Putting $b = 0.100$ cm into the determinants gives

$$[X] = \frac{\begin{vmatrix} 0.233 & 387 \\ 0.200 & 642 \end{vmatrix}}{\begin{vmatrix} 1640 & 387 \\ 399 & 642 \end{vmatrix}} = 8.03 \times 10^{-5} \text{ M} \qquad [Y] = \frac{\begin{vmatrix} 1640 & 0.233 \\ 399 & 0.200 \end{vmatrix}}{\begin{vmatrix} 1640 & 387 \\ 399 & 642 \end{vmatrix}} = 2.62 \times 10^{-4} \text{ M}$$

The spreadsheet solution looks like this, with answers in column F:

	A	B	C	D	E	F	G
1	Analysis of a mixture by spreadsheet matrix operations						
2							
3	Wavelength	Coefficient Matrix		Absorbance		Concentrations	
4				of unknown		in mixture	
5	272	1640	387	0.233		8.034E-05	<-[X]
6	327	399	642	0.2		2.616E-04	<-[Y]
7		K		A		C	

2.

	A	B	C	D	E	F	G	H
1	Analysis of a Mixture When You Have More Data Points than Components of the Mixture							
2				Measured			Calculated	
3				Absorbance			Absorb-	
4	Wave-	Absorbance of Standard		of Mixture	Molar Absorptivity		ance	
5	length	MnO4	Cr2O7	Am	MnO4	Cr2O7	Acalc	[Acalc-Am]^2
6	266	0.042	0.410	0.766	420.0	4100.0	0.7650	1.017E-06
7	288	0.082	0.283	0.571	820.0	2830.0	0.5723	1.763E-06
8	320	0.168	0.158	0.422	1680.0	1580.0	0.4217	1.132E-07
9	350	0.125	0.318	0.672	1250.0	3180.0	0.6706	2.050E-06
10	360	0.056	0.181	0.366	560.0	1810.0	0.3690	9.106E-06
11							sum=	1.405E-05
12	Standards		Concentrations in the mixture					
13	[Mn](M)=		(to be found by Solver)					
14	1.00E-04		[MnO4]=	8.356E-05				
15	[Cr](M) =		[Cr2O7] =	1.780E-04				
16	1.00E-04							
17	Pathlength		E6 = B6/($A$19*$A$14)					
18	(cm) =		F6 = C6/($A$19*$A$16)					
19	1.000		G6 = E6*$A$19*$D$14+F6*$A$19*$D$15					
20			H6 = (G6-D6)^2					

19-3. If the spectra of two compounds with a constant total concentration cross at any wavelength, all mixtures with the same total concentration will go through that same point, called an isosbestic point. The appearance of isosbestic points in a chemical reaction is good evidence that we are observing one main species being converted to one other major species.

19-4. As VO^{2+} is added (traces 1-9), the peak at 439 decreases and a new one near 485 nm develops, with an isosbestic point at 457 nm. When VO^{2+}/xylenol orange > 1, the peak near 485 nm decreases and a new one at 566 nm grows in, with an isosbestic point at 528 nm. This sequence is logically interpreted by the sequence

$$M + L \rightarrow ML$$
$$ 439 \text{ nm} 485 \text{ nm}$$

$$ML + M \rightarrow M_2L$$
$$485 \text{ nm} 566 \text{ nm}$$

where M is vanadyl ion and L is xylenol orange. The structure of xylenol orange in Table 11-3 shows that it has metal-binding groups on both ends of the molecule, and could form an M_2L complex.

19-5. Convert T to A ($= -\log T$) and then convert A to ε ($= A/bc = A/[(0.005)(0.01)]$)

| | Absorbance | | | ε (M^{-1} cm^{-1}) | |
	2 022	1 993 cm^{-1}		2 022	1 993 cm^{-1}
A	0.508 6	0.098 54	A	10 170	1 971
B	0.011 44	0.699 0	B	228.8	13 980

For the mixture, $A_{2022} = -\log(0.340) = 0.468\,5$ and $A_{1993} = -\log(0.383) = 0.416\,8$. Equation 19-6 gives [A] $= 9.11 \times 10^{-3}$ M and [B] $= 4.68 \times 10^{-3}$ M.

19-6.

	A	B	C	D	E	F	G	H
1	Solving Simultaneous Linear Equations with Excel Matrix Operations							
2								
3	Wavelength	Coefficient Matrix			Absorbance		Concentrations	
4		TB	STB	MTB	of unknown		in mixture	
5	455	4800	11100	18900	0.412		1.2194E-05	<-[TB]
6	485	7350	11200	11800	0.350		9.2953E-06	<-[STB]
7	545	36400	13900	4450	0.632		1.3243E-05	<-[MTB]
8			K		A		C	
9								
10	1. Highlight block of blank cells required for solution (G5:G7)							
11	2. Type the formula "= MMULT(MINVERSE(B5:D7),E5:E7)"							
12	3. Press CONTROL+SHIFT+ENTER on a PC or COMMAND+RETURN on a Mac							
13	4. The answer appears in cells G5:G7							

19-7.

	A	B	C	D	E	F	G	H	I
1	Solving 4 Simultaneous Linear Equations with Excel Matrix Operations								
2	Wave-								
3	length	Coefficient Matrix			Ethyl-	Absorbance		Conc. in	
4	(μm)	p-xylene	m-xylene	o-xylene	benzene	of unknown		mixture	
5	12.5	1.5020	0.0514	0	0.0408	0.1013		0.0627	p-xylene
6	13.0	0.0261	1.1516	0	0.0820	0.09943		0.0795	m-xylene
7	13.4	0.0342	0.0355	2.532	0.2933	0.2194		0.0759	o-xylene
8	14.3	0.0340	0.0684	0	0.3470	0.03396		0.0761	Ethylbz
9			**K**				**A**		**C**
10									
11	1. Highlight block of blank cells required for solution (H5:H8)								
12	2. Type the formula "= MMULT(MINVERSE(B5:E8),F5:F8)"								
13	3. Press CONTROL+SHIFT+ENTER on a PC or COMMAND+RETURN on a Mac								
14	4. The answer appears in cells H5:H8								

19-8. (a) Cells D15:D16 in the spreadsheet give $[In^-]$ = 3.28 μM and $[HIn]$ = 6.91 μM. The fact that they add up to 10.19 instead of 10.00 μM represents 2% experimental error in the quality of the procedure and measurements.

	A	B	C	D	E	F	G	H
1	Analysis of a Mixture When You Have More Data Points than Components of the Mixture							
2				Measured				
3				Absorbance	Molar Absorptivity		Calculated	
4	Wave-	Absorbance of Standards		of Mixture	$(M^{-1}\,cm^{-1})$		Absorbance	
5	length nm)	Blue In^-	Yellow HIn	A_m	In-	HIn	A_{calc}	$[A_{calc}-A_m]$^2
6	400	0.182	0.274	0.248	18200.0	27400.0	0.2490	9.819E-07
7	432	0.076	0.344	0.265	7600.0	34400.0	0.2626	5.533E-06
8	450	0.046	0.323	0.237	4600.0	32300.0	0.2383	1.707E-06
9	550	0.356	0.027	0.131	35600.0	2700.0	0.1353	1.830E-05
10	570	0.570	0.011	0.193	57000.0	1100.0	0.1943	1.742E-06
11	616	0.811	0.005	0.272	81100.0	500.0	0.2691	8.302E-06
12							sum =	3.656E-05
13	Standards		Concentrations in the mixture					
14	[In-](M)=		(to be found by Solver)					
15	1.0E-05		[In-] =	3.2757E-06	M			
16	[HIn](M)=		[HIn] =	6.9114E-06	M			
17	1.0E-05							
18	Pathlength		E6 = B6/($A$19*$A$14)					
19	(cm) =		F6 = C6/($A$19*$A$16)					
20	1.0		G6 = E6*$A$19*$D$14+F6*$A$19*$D$15					
21			H6 =(G6-D6)^2					

(b) $pH = pK_{HIn} + \log \dfrac{[In^-]}{[HIn]} = 7.10 + \log \dfrac{3.28\ \mu M}{6.91\ \mu M} = 6.78$

(c)   Absorbance and absorptivity appear in rows 7 and 11 of the spreadsheet, with $b = 1$ cm. In Equations 19-6, let X = In⁻, Y = HIn, $\lambda' = 432$ nm, and λ'' = 432 nm.

$$[\text{In}^-] = \frac{\begin{vmatrix} 0.265 & 34\,400 \\ 0.272 & 500 \end{vmatrix}}{\begin{vmatrix} 7\,600 & 34\,400 \\ 81\,100 & 500 \end{vmatrix}} = 3.31 \times 10^{-6}\,\text{M}$$

$$[\text{HIn}] = \frac{\begin{vmatrix} 7\,600 & 0.265 \\ 81\,100 & 0.272 \end{vmatrix}}{\begin{vmatrix} 7\,600 & 34\,400 \\ 81\,100 & 500 \end{vmatrix}} = 6.97 \times 10^{-6}\,\text{M}$$

The spreadsheet solution looks like this, with answers in column F:

	A	B	C	D	E	F	G
1	Solving Simultaneous Linear Equations with Excel Matrix Operations						
2							
3	Wavelength	Coefficient Matrix		Absorbance		Concentrations	
4	(nm)			of unknown		in mixture (M)	
5	432	7600	34400	0.265		3.31E-06	<-[X]
6	616	81100	500	0.272		6.97E-06	<-[Y]
7		K		A		C	
8							
9	1. Enter matrix of coefficients εb in cells B5:C6						
10	2. Enter absorbance of unknown at each wavelength (cells D5:D6)						
11	3. Highlight block of blank cells required for solution (F5 and F6)						
12	4. Type the formula "= MMULT(MINVERSE(B5:C6),D5:D6)"						
13	5. Press CONTROL+SHIFT+ENTER on a PC or COMMAND+RETURN on a Mac						
14	6. Behold! The answer appears in cells F5 and F6						

The calculation in (a) uses six wavelengths to find [In⁻] and [HIn], whereas (c) uses just two wavelengths. We expect that (a) should be more accurate. Notice that the sum [In⁻] + [HIn] in (a) is 10.19 μM and the sum in (c) is 10.28 μM. The answer from (a) is higher than the theoretical by 2% and the answer from (c) is higher than the theoretical by 3%.

19-9. The quantity of HIn is small compared to aniline and sulfanilic acid. Calling aniline B and sulfanilic acid HA, we can write

	B	+	HA	$\overset{K}{\rightleftharpoons}$	BH⁺	+	A⁻
Initial mmol:	2.00		1.500		—		—
Final mmol:	$2.00 - x$		$1.500 - x$		x		x

$$K = \frac{K_a K_b}{K_w} = \frac{(10^{-3.232})(K_w/10^{-4.601})}{K_w} = 23.39$$

$$\frac{x^2}{(2.00 - x)(1.500 - x)} = 23.39 \Rightarrow x = 1.372 \text{ mmol}$$

$$pH = pK_{BH^+} + \log\frac{[B]}{[BH^+]} = 4.601 + \log\frac{2.00 - 1.372}{1.372} = 4.26$$

For HIn we can write:

absorbance $= 0.110 = (2.26 \times 10^4)(5.00)\,[HIn] + (1.53 \times 10^4)\,(5.00)[In^-]$.
Substituting $[HIn] = 1.23 \times 10^{-6} - [In^-]$ gives $[In^-] = 7.94 \times 10^{-7}$ and
$[HIn] = 4.36 \times 10^{-7}$. The Henderson-Hasselbalch equation for HIn is therefore

$$pH = pK_{HIn} + \log\frac{[In^-]}{[HIn]} \Rightarrow 4.26 = pK_{HIn} + \log\frac{7.94 \times 10^{-7}}{4.36 \times 10^{-7}} \Rightarrow pK_{HIn} = 4.00.$$

19-10. (a) $A_{620} = \varepsilon_{620}^{HIn^-} b[HIn^-] + \varepsilon_{620}^{In^{2-}} b[In^{2-}]$

$A_{434} = \varepsilon_{434}^{HIn^-} b[HIn^-] + \varepsilon_{434}^{In^{2-}} b[In^{2-}]$

The solution of these two equations is given by Equation 19-6 in the text:

$$[HIn^-] = \frac{1}{D}\left(A_{620}\varepsilon_{434}^{In^{2-}}b - A_{434}\varepsilon_{620}^{In^{2-}}b\right)$$

$$[In^{2-}] = \frac{1}{D}\left(A_{434}\varepsilon_{620}^{HIn^-}b - A_{620}\varepsilon_{434}^{HIn^-}b\right)$$

where $D = b^2\left(\varepsilon_{620}^{HIn^-}\varepsilon_{434}^{In^{2-}} - \varepsilon_{620}^{In^{2-}}\varepsilon_{434}^{HIn^-}\right)$

Dividing the expression for $[In^{2-}]$ by the expression for $[HIn^-]$ gives

$$\frac{[In^{2-}]}{[HIn^-]} = \frac{A_{434}\varepsilon_{620}^{HIn^-} - A_{620}\varepsilon_{434}^{HIn^-}}{A_{620}\varepsilon_{434}^{In^{2-}} - A_{434}\varepsilon_{620}^{In^{2-}}}$$

Dividing numerator and denominator on the right side by A_{434} gives

$$\frac{[In^{2-}]}{[HIn^-]} = \frac{\varepsilon_{620}^{HIn^-} - R_A\varepsilon_{434}^{HIn^-}}{R_A\varepsilon_{434}^{In^{2-}} - \varepsilon_{620}^{In^{2-}}} = \frac{R_A\varepsilon_{434}^{HIn^-} - \varepsilon_{620}^{HIn^-}}{\varepsilon_{620}^{In^{2-}} - R_A\varepsilon_{434}^{In^{2-}}}$$

(b) Mass balance for indicator: $[HIn^-] + [In^{2-}] = F_{In}$

Dividing both sides by $[HIn^-]$ gives

$$\frac{[HIn^-]}{[HIn^-]} + \frac{[In^{2-}]}{[HIn^-]} = \frac{F_{In}}{[HIn^-]} \Rightarrow 1 + R_{In} = \frac{F_{In}}{[HIn^-]} \Rightarrow [HIn^-] = \frac{F_{In}}{R_{In} + 1}$$

Acid dissociation constant of indicator:

$$K_{In} = \frac{[In^{2-}][H^+]}{[HIn^-]}$$

Substituting $F_{In}/(R_{In} + 1)$ for $[HIn^-]$ gives

$$K_{In} = \frac{[In^{2-}][H^+](R_{In} + 1)}{F_{In}} \Rightarrow [In^{2-}] = \frac{K_{In}F_{In}}{[H^+](R_{In} + 1)}$$

(c) Equation A in the problem defines R_{In} as $[In^{2-}]/[HIn^-]$. So,

$$K_{In} = \frac{[In^{2-}][H^+]}{[HIn^-]} = R_{In}[H^+] \Rightarrow [H^+] = K_{In}/R_{In}$$

(d) From the acid dissociation reaction of carbonic acid, we can write

$$K_1 = \frac{[HCO_3^-][H^+]}{[CO_2(aq)]} \Rightarrow [HCO_3^-] = \frac{K_1[CO_2(aq)]}{[H^+]}$$

From the acid dissociation reaction of bicarbonate, we can write

$$K_2 = \frac{[CO_3^{2-}][H^+]}{[HCO_3^-]} \Rightarrow [CO_3^{2-}] = \frac{K_2[HCO_3^-]}{[H^+]}$$

Substituting in the expression for $[HCO_3^-]$ gives

$$[CO_3^{2-}] = \frac{K_1K_2[CO_2(aq)]}{[H^+]^2}$$

(e) Charge balance:

$$[Na^+] + [H^+] = [OH^-] + [HIn^-] + 2[In^{2-}] + [HCO_3^-] + 2[CO_3^{2-}]$$

$$F_{Na} + [H^+] = $$
$$\frac{K_w}{[H^+]} + \frac{F_{In}}{R_{In} + 1} + 2\frac{K_{In}F_{In}}{[H^+](R_{In} + 1)} + \frac{K_1[CO_2(aq)]}{[H^+]} + 2\frac{K_1K_2[CO_2(aq)]}{[H^+]^2}$$

(f) From part (c) we know that $[H^+] = K_{In}/R_{In}$. We calculate R_{In} from part (a):

$$R_{In} = \frac{R_A\varepsilon_{434}^{HIn^-} - \varepsilon_{620}^{HIn^-}}{\varepsilon_{620}^{In^{2-}} - R_A\varepsilon_{434}^{In^{2-}}} = \frac{(2.84)(8.00 \times 10^3) - (0)}{(1.70 \times 10^4) - (2.84)(1.90 \times 10^3)} = 1.95_8$$

$$[H^+] = K_{In}/R_{In} = (2.0 \times 10^{-7})/1.95_8 = 1.0_2 \times 10^{-7}\ M$$

Substituting this value of $[H^+]$ into the mass balance in part (e) produces an equation in which the only unknown is $[CO_2(aq)]$:

$$F_{Na} + [H^+] = $$
$$\frac{K_w}{[H^+]} + \frac{F_{In}}{R_{In} + 1} + 2\frac{K_{In}F_{In}}{[H^+](R_{In} + 1)} + \frac{K_1[CO_2(aq)]}{[H^+]} + 2\frac{K_1K_2[CO_2(aq)]}{[H^+]^2}$$
$$92.0 \times 10^{-6} + 1.0_2 \times 10^{-7} = $$
$$\frac{(6.7 \times 10^{-15})}{(1.0_2 \times 10^{-7})} + \frac{(50.0 \times 10^{-6})}{1.95_8 + 1} + 2\frac{(2.0 \times 10^{-7})(50.0 \times 10^{-6})}{(1.0_2 \times 10^{-7})(1.95_8 + 1)}$$
$$+ \frac{(3.0 \times 10^{-7})[CO_2(aq)]}{(1.0_2 \times 10^{-7})} + 2\frac{(3.0 \times 10^{-7})(3.3 \times 10^{-11})[CO_2(aq)]}{(1.0_2 \times 10^{-7})^2}$$

$$9.21 \times 10^{-5} = 6.56 \times 10^{-8} + 1.69 \times 10^{-5} + 6.62 \times 10^{-5} +$$
$$+ 2.94 \, [CO_2(aq)] + 0.001 \, 9 \, [CO_2(aq)]$$
$$\Rightarrow [CO_2(aq)] = 3.0_4 \times 10^{-6} \, M$$

(g) The ions in solution are Na^+, HIn^-, In^{2-}, HCO_3^-, CO_3^{2-}, H^+, and OH^-. We know that $[Na^+] = 92.0 \, \mu M$ and $[H^+] = 0.10 \, \mu M$. If the total cation charge is 92.1 μM, the total anion charge must be 92.1 μM, and the ionic strength must be ~92 $\mu M \approx 10^{-4} \, M$. (The ionic strength is not exactly 92.1 μM because some anions have a charge of -2, which will increase the ionic strength from 92.1 μM.) An ionic strength of $10^{-4} \, M$ is low enough that the activity coefficients are close to 1.00.

We can calculate the exact ionic strength from the following expressions derived above:

$$[OH^-] = \frac{K_w}{[H^+]} = 0.07 \, \mu M$$

$$[HIn^-] = \frac{F_{In}}{R_{In} + 1} = 16.9 \, \mu M; \quad [In^{2-}] = \frac{K_{In}F_{In}}{[H^+](R_{In} + 1)} = 33.1 \, \mu M$$

$$[HCO_3^-] = \frac{K_1[CO_2(aq)]}{[H^+]} = 2.94 \, [CO_2(aq)] = 8.9 \, \mu M$$

$$[CO_3^{2-}] = \frac{K_1 K_2 [CO_2(aq)]}{[H^+]^2} = 0.001 \, 9 \, [CO_2(aq)] = 0.003 \, \mu M$$

$$\text{Ionic strength} = \frac{1}{2} \sum_i c_i z_i^2 =$$

$$\frac{1}{2} \left\{ [Na^+] \cdot 1^2 + [H^+] \cdot 1^2 + [OH^-] \cdot 1^2 + [HIn^-] \cdot 1^2 + [In^{2-}] \cdot 2^2 + [HCO_3^-] \cdot 1^2 + [CO_3^{2-}] \cdot 2^2 \right\}$$

$$= 125 \, \mu M$$

19-11. The Scatchard plot is a graph of $[PX]/[X]$ versus $[PX]$. Data in the following spreadsheet are plotted in the figure in the textbook. The slope is $-4.0 \times 10^9 \, M^{-1}$, giving a binding constant $K = 4.0 \times 10^9 \, M^{-1}$. The fraction of saturation in column E is $S = [PX]/P_0$, where $P_0 = 10.0 \, nM$. S ranges from 0.29 to 0.84.

	A	B	C	D	E	F
1	Data for Scatchard plot			$P_o =$	10	nM
2				Fraction of saturation		
3	[PX] (nM)	[X] (nM)	[PX]/[X]		$= [PX]/P_o$	
4	2.87	0.120	24.0		0.29	
5	3.80	0.192	19.8		0.38	
6	4.66	0.296	15.7		0.47	
7	5.54	0.450	12.3		0.55	
8	6.29	0.731	8.61		0.63	
9	6.77	1.22	5.54		0.68	
10	7.52	1.50	5.02		0.75	
11	8.45	3.61	2.34		0.84	

19-12. (a)

	A	B	C	D	E	F	G
1	Estradiol - Albumin Scatchard Plot						
2	abscissa	ordinate					
3	P (µM)	Xo/X		Highlight cells B15:C17			
4	6.3	1.26		Type			
5	10.0	1.62		"=LINEST(B4:B12,A4:A12,TRUE,TRUE)"			
6	20.0	2.16		Press CTRL+SHIFT+ENTER (on PC)			
7	30.0	2.51		Press COMMAND+RETURN (on Mac)			
8	40.0	3.34					
9	50.0	3.33		Student's t (95% confidence,			
10	60.0	4.19		7 degrees of freedom) =			
11	70.0	4.13		2.364624	=TINV(0.05,7)		
12	80.0	4.36		95% confidence interval =			
13				0.008248	$= t*u_m = D11*B16$		
14		LINEST output					
15	m	0.042885	1.243476	b			
16	u_m	0.003488	0.166224	u_b			
17	R^2	0.955742	0.259413	s_y			

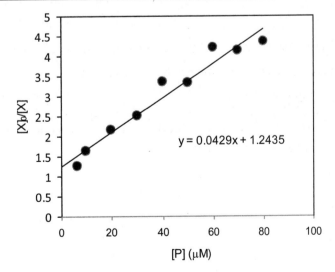

$y = 0.0429x + 1.2435$

The slope is 0.042 88 with a standard deviation of 0.003 49 in cells B15 and B16. However, the units on the abscissa are μM, so the slope is really $(0.042\ 88 \pm 0.003\ 49)/10^{-6}$ M^{-1} = $(4.288 \pm 0.349) \times 10^4$ M^{-1}. To find the 95% confidence interval, we need Student's t for $9 - 2 = 7$ degrees of freedom, which is $t = 2.365$ in cell D11. The 95% confidence interval is $(0.349)(2.365) = 0.825$. The final result is $K = (4.3 \pm 0.8) \times 10^4$ M^{-1}.

(b) Estradiol is X. The quotient $[X]_0/[X]$ is 1.26 at the first point and 4.36 at the last point. The fraction of free estradiol is $[X]/[X]_0 = 1/1.26 = 0.79$ at the first point and $1/4.36 = 0.23$ at the last point. The fraction of bound estradiol is $1 - 0.79 = 0.21$ at the first point and $1 - 0.23 = 0.77$ at the last point.

19-13. (a) We will make the substitutions $[complex] = A/\varepsilon$ and $[I_2] = [I_2]_{tot} - [complex]$ in the equilibrium expression:

$$K = \frac{[complex]}{[I_2][mesitylene]} = \frac{A/\varepsilon}{([I_2]_{tot} - [complex])\,[mesitylene]}$$

$$K[I_2]_{tot} - K[complex] = \frac{A}{\varepsilon[mesitylene]}$$

Making the substitution $[complex] = A/\varepsilon$ once more on the left-hand side gives

$$K[I_2]_{tot} - \frac{KA}{\varepsilon} = \frac{A}{\varepsilon[mesitylene]}$$

Multiplying both sides by ε and dividing by $[I_2]_{tot}$ gives the desired result:

$$\varepsilon K - \frac{KA}{[I_2]_{tot}} = \frac{A}{[I_2]_{tot}\,[mesitylene]}$$

(b) The graph of $A/([mesitylene][I_2]_{tot})$ versus $A/[I_2]_{tot}$ is an excellent straight line with a slope of -0.464 and an intercept of 4.984×10^3. Since slope $= -K$, the equilibrium constant is 0.464. The molar absorptivity is $\varepsilon =$ intercept/$K \Rightarrow \varepsilon = 1.074 \times 10^4$ M^{-1} cm^{-1}.

19-14. After running Solver, the average value of K in cell E10 is 0.464 and $\varepsilon = 1.073 \times 10^4$ M^{-1} cm^{-1}. The Scatchard plot in the previous problem gave $K = 0.464$ and $\varepsilon = 1.074 \times 10^4$ M^{-1} cm^{-1}.

	A	B	C	D	E
1	Equilibrium constant for reaction of I_2 with mesitylene				
2					
3	[Mesitylene] (M)	[I_2]tot (M)	A	[Complex] = A/ε	K_{eq}
4	1.6900	7.82E-05	0.369	3.437E-05	0.46443
5	0.9218	2.56E-04	0.822	7.657E-05	0.46349
6	0.6338	3.22E-04	0.787	7.331E-05	0.46439
7	0.4829	3.57E-04	0.703	6.549E-05	0.46473
8	0.3900	3.79E-04	0.624	5.813E-05	0.46480
9	0.3271	3.93E-04	0.556	5.179E-05	0.46353
10				Average =	0.46423
11	Guess for ε:			Standard Dev =	0.00058
12	1.073E+04			Stdev/Average =	0.00125
13					
14	D4 = C4/$A$12				
15	E4 = D4/(A4*(B4-D4)) = [complex]/([Mesitylene][Free I_2])				
16	E10 = AVERAGE(E4:E9)				
17	E11 = STDEV(E4:E9)				
18	E12 = E11/E10				
19	Use Solver to vary ε (cell A12) until cell E12 is minimized				

19-15. (a) Both solutions contain 1.00 mM reactant, so the mole fraction is the same as the solution volume fraction. For example, the solution made from 6.00 mL iron plus 24.00 mL thiocyanate has a mole fraction

$$X_{SCN^-} = \frac{\text{mol } SCN^-}{\text{mol } Fe^{3+} + \text{mol } SCN^-} = \frac{24.00 \text{ mL}}{6.00 \text{ mL} + 24.00 \text{ mL}} = 0.800$$

Maximum absorbance occurs at $X_{SCN^-} = 0.500$

$$\Rightarrow \text{stoichiometry} = 1:1 \ (n = 1)$$

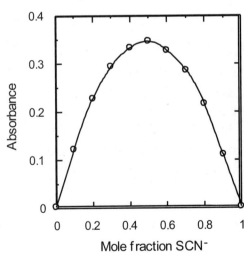

(b) The curved maximum indicates that the equilibrium constant is not very large.

(c) The different acid concentrations give both solutions the same ionic strength (= 16.0 mM).

19-16. The Job plot peak is at a xylenol orange mole fraction of 0.40, suggesting the stoichiometry (xylenol orange)$_2$Zr$_3$ which has a mole fraction of

$$\frac{\text{xylenol orange}}{\text{xylenol orange} + \text{Zr(IV)}} = \frac{2}{2+3} = 0.4$$

The plot could have been improved by obtaining data points at mole fractions of 0.35 and 0.45 to verify the location of the maximum.

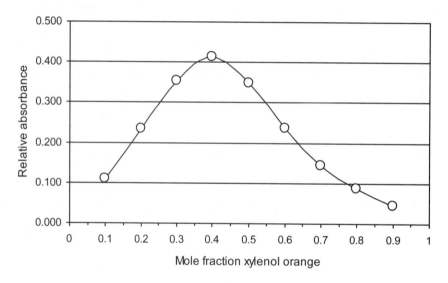

19-17. (a) Here are the results:

[A]$_{total}$	[B]$_{total}$	[AB$_2$] $K = 10^6$	$K = 10^7$	$K = 10^8$	Mole fraction A
1e-5	9e-5	8.01e-8	7.27e-7	4.02e-6	0.1
2e-5	8e-5	1.26e-7	1.14e-6	6.26e-6	0.2
2.5e-5	7.5e-5	1.39e-7	1.25e-6	6.83e-6	0.25
3e-5	7e-5	1.45e-7	1.30e-6	7.12e-6	0.3
3.33e-5	6.67e-5	1.46e-7	1.31e-6	7.17e-6	0.333
4e-5	6e-5	1.42e-7	1.28e-6	6.99e-6	0.4
5e-5	5e-5	1.23e-7	1.12e-6	6.20e-6	0.5
6e-5	4e-5	9.49e-8	8.66e-7	4.97e-6	0.6
7e-5	3e-5	6.24e-8	5.78e-7	3.51e-6	0.7
8e-5	2e-5	3.18e-8	3.00e-7	2.00e-6	0.8
9e-5	1e-5	8.97e-9	8.68e-8	6.70e-7	0.9

(b) The maximum occurs at a mole fraction of A = 1/3, since the stoichiometry is 1:2. The greater the equilibrium constant, the greater the extent of reaction and the steeper the curve. When the equilibrium constant is too small, the curve is so shallow that it does not at all resemble two intersecting lines.

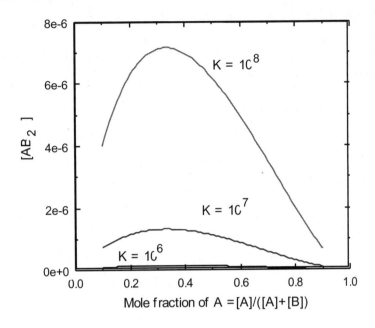

19-18. The mole fraction of thymine varies from 0.10 to 0.90 in increments of 0.10 as we go down the table. The Job plot reaches a broad peak at a mole fraction of 0.50, which is consistent with 1:1 complex formation. The Job plot gives us no information on the structure of the product except for its stoichiometry.

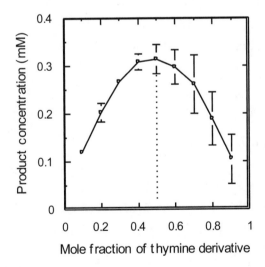

19-19. In *flow injection analysis*, unknown is injected into a continuously flowing stream that could contain a reagent which reacts with analyte. Alternatively, one or more reagents can be added to the flow downstream of sample injection. The sample and reagents travel through a coil in which mixing and reaction occur. Absorbance or fluorescence or some other property of the flowing stream is measured in the detector. In general, reactants and products do not reach equilibrium in flow injection analysis. The precision of the result depends on executing the same process in a very repeatable manner.

In *sequential injection*, unknown and one or more reagents are sequentially injected under computer control into a holding coil. There is not a continuous flow. At an appropriate time, flow is reversed and the mixture is directed through a detector. As in flow injection analysis, equilibrium is generally not reached and precision depends on repeatability of the sequence of steps.

The principal difference between flow injection and sequential injection is that there is continuous flow in the former and there is a sequence of steps without continuous flow in the latter. Sequential injection uses less reagent and generates less waste than flow injection. Sequential injection is called "lab-on-a-valve" because sample, reagents, and product all flow through a central multi-port valve.

19-20. Each molecule of analyte bound to antibody 1 also binds one molecule of antibody 2 that is linked to one enzyme molecule. Each enzyme molecule catalyzes many cycles of reaction in which a colored or fluorescent product is created. Therefore, many product molecules are created for every analyte molecule.

19-21. In time-resolved emission measurements, the short-lived background fluorescence decays to near zero prior to recording emission from the lanthanide ion. By reducing background signal, the signal-to-noise ratio is increased. Also, the wavelength of the lanthanide emission is longer than the wavelength of much of the background emission.

19-22. Many biological chromophores absorb visible light and some of them are fluorescent. Stimulation with visible light creates significant background visible fluorescence from the biological matrix in addition to the desired fluorescence from analyte. Few molecules absorb radiation in the near-infrared range 800 to 1 000 nm, so less background fluorescence occurs. With low background fluorescence, the signal/noise ratio for analyte fluorescence is high.

19-23.

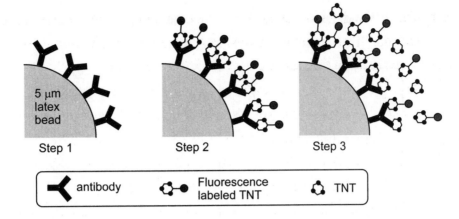

Step 1   Step 2   Step 3

| antibody | Fluorescence labeled TNT | TNT |

In Step 1, antibodies for TNT are attached to latex beads. In Step 2, the antibody is saturated with a fluorescent derivative of TNT. Excess fluorescent derivative is removed. In Step 3, the beads are incubated with TNT, which displaces some fluorescent derivative from binding sites on the antibodies. The suspension of beads is then injected into the flow cytometer. As each bead passes in front of the detector, it is excited by a laser and its fluorescence is measured. The graph in the textbook shows median bead fluorescence versus TNT concentration in a series of standards. The more TNT in the standard, the less fluorescence remains associated with the beads.

19-24. The graph of K_{sv} versus pH has a plateau at low pH near $K_{sv} \approx 100$ and a plateau at high pH near $K_{sv} \approx 1350$. The quencher, 2,6-dimethylphenol, is a weak acid whose pK_a is expected to be near 10. A logical interpretation is that the basic form, A^-, is a strong quencher with $K_{sv} \approx 1350$, and the acidic form, HA, is a weak quencher with $K_{sv} \approx 100$. We estimate pK_a as the midpoint in the curve at $K_{sv} \approx (1350 - 100)/2 = 625$. At this point, $pH \approx 10.8$, which is our estimate for pK_a. The literature value is 10.63.

The smooth curve in the graph is a least-squares fit to the equation

$$K_{sv} = K_{sv}^{HA} \underbrace{\left(\frac{[H^+]}{[H^+] + K_a} \right)}_{\substack{\text{Fraction in form HA}}} + K_{sv}^{A^-} \underbrace{\left(\frac{K_a}{[H^+] + K_a} \right)}_{\substack{\text{Fraction in form } A^-}}$$

$$\underbrace{\phantom{K_{sv} = K_{sv}^{HA} \left(\frac{[H^+]}{[H^+] + K_a} \right)}}_{\text{Quenching by HA}} \quad \underbrace{\phantom{K_{sv}^{A^-} \left(\frac{K_a}{[H^+] + K_a} \right)}}_{\text{Quenching by } A^-}$$

The least-squares fit gave $K_{sv}^{HA} = 69.0$ M^{-1}, $K_{sv}^{A^-} = 1\,370$ M^{-1}, and $pK_a = 10.78$.

19-25. (a) The following graph at the left shows that the Stern-Volmer equation is not obeyed. If it were obeyed, the graph would be linear.

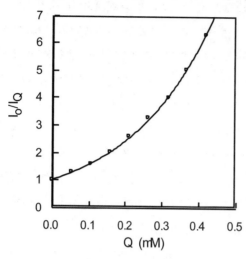 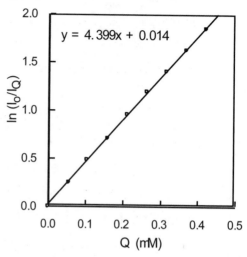

(b) The graph at the right above shows that Equation 5 is obeyed. Ideally, the intercept should be zero. The slope of the graph is $N_{av}/([S] - [CMC])$.

Given that [Q] was expressed in mM, we will express [S] and [CMC] in mM: $4.399 = N_{av} / ([20.8] - [8.1]) \rightarrow N_{av} = 55.9$

(c) $[M] = ([S] - [CMC]) / N_{av} = ([20.8] - [8.1]) / 55.9 = 0.227$ mM

$\overline{Q} = [Q]/[M] = 0.200$ mM $/ 0.227$ mM $= 0.881$ molecules per micelle

(d) $P_n = \dfrac{\overline{Q}^n}{n!} e^{-\overline{Q}}$. For $n = 0$, $P_0 = e^{-0.881} = 0.414$

$P_1 = \dfrac{(0.881)^1}{1!} e^{-0.881} = 0.365$ $\qquad$ $P_2 = \dfrac{(0.881)^2}{2!} e^{-0.881} = 0.161$

CHAPTER 20
SPECTROPHOTOMETERS

20-1. The light source provides ultraviolet, visible, or infrared radiation. The monochromator selects a narrow band of wavelengths to pass on to the sample. As the experiment progresses, the monochromator scans through a desired range of wavelengths. The beam chopper is a rotating mirror that alternately directs light to the sample or reference. The sample cuvet holds the sample of interest. The reference cuvet is an identical cell containing pure solvent or a reagent blank. The mirror after the reference cuvet and the semitransparent mirror after the sample cuvet pass both beams of light through to the detector, which could be a photomultiplier tube that generates an electric current proportional to the photon irradiance. The amplifier increases the detector signal for display.

20-2. An excited state of the lasing material is pumped to a high population by light, an electric discharge, or other means. Photons emitted when the excited state decays to a less populated lower state stimulate emission from other excited molecules. The stimulated emission has the same energy and phase as the incident photon. In the laser cavity, most light is retained by reflective end mirrors. Some light is allowed to escape from one end. Properties of laser light: monochromatic, bright, collimated, polarized, coherent.

20-3. Deuterium. Silicon carbide globar.

20-4. Resolution increases in proportion to the number of grooves that are illuminated and to the diffraction order. The number of grooves can be increased with a more finely ruled grating (closer grooves) and with a longer grating. The diffraction order is optimized by appropriate choice of the blaze angle of the grating.

Dispersion is proportional to the diffraction order and inversely proportional to the spacing between lines in the grating. The closer the lines, the greater the dispersion.

The optimum wavelength selected by a grating is the one for which the diffraction condition $n\lambda = d(\sin\theta + \sin\phi)$ is satisfied by specular reflection when the angle of incidence is equal to the angle of reflection ($\alpha = \beta$).

20-5. A filter removes higher order diffraction (different wavelengths) at the same angle as the desired diffraction.

20-6. Advantage - increased ability to resolve closely spaced spectral peaks. Disadvantage - more noise because less light reaches detector.

20-7. (a) A photomultiplier tube has a photosensitive cathode that emits an electron when struck by a photon. Electrons from the cathode are accelerated by a positive electric potential toward the first dynode. When an accelerated electron strikes the dynode, more electrons are emitted from the dynode. This multiplication process continues through several successive dynodes until $\sim 10^6$ electrons are finally collected at the anode for each photon striking the photocathode. The signal is the current measured at the anode.

(b) Each photodiode in a linear array has p-type silicon on a substrate of n-type silicon. Reverse bias draws electrons and holes away from the junction, which is a depletion region with few electrons and holes. The junction acts as a capacitor, with charge on either side of the depletion region. At the beginning of a measurement cycle, each diode is fully charged. Free electrons and holes created when radiation is absorbed in the semiconductor migrate to regions of opposite charge, partially discharging the capacitor. Charge left in each capacitor is measured at the end of a collection cycle by measuring the current needed to recharge each capacitor.

(c) A charge coupled device is made from p-doped Si on an n-doped substrate. The structure is capped with an insulating layer of SiO_2, on top of which is a two-dimensional pattern of conducting Si electrodes. When light is absorbed in the p-doped region, an electron is introduced into the conduction band and a hole is left in the valence band. The electron is attracted to the region beneath the positive electrode, where it is stored. The hole migrates to the n-doped substrate, where it combines with an electron. Each electrode can store $\sim 10^5$ electrons. After the desired observation time, electrons stored in each pixel of the top row of the array are moved into a serial register at the top and then moved, one pixel at a time, to the top right position, where the charge is read out. Then the next row is moved up and read out, and the sequence is repeated until the entire array has been read.

(d) The image intensified charge coupled device amplifies the signal by more than 10^6 prior to detection. Incoming radiation strikes a photocathode that converts photons to electrons. Electrons are multiplied by $>10^6$ by passage through a microchannel plate. Electrons exiting the plate strike a fluorescent screen that converts electrons back to photons that are detected by a charge coupled device.

20-8. DTGS has a permanent electric polarization. That is, one face of the crystal has a positive charge and the opposite face has a negative charge. When the temperature of the crystal changes by absorption of infrared light, the polarization (the voltage difference between the two faces) changes. The change in voltage is the detector signal.

20-9. (a) $n\lambda = d(\sin\theta + \sin\phi)$

$1 \cdot 600 \times 10^{-9}$ m $= d(\sin 40° + \sin(-30°)) \Rightarrow d = 4.20 \times 10^{-6}$ m

Lines/cm $= 1/(4.20 \times 10^{-4}$ cm$) = 2.38 \times 10^3$ lines/cm

(b) $\lambda = 1/(1\,000$ cm$^{-1}) = 10^{-3}$ cm $\Rightarrow d = 7.00 \times 10^{-3}$ cm $\Rightarrow 143$ lines/cm

20-10. 10^3 grooves/cm means $d = 10^{-5}$ m $= 10$ μm

Dispersion $= \dfrac{n}{d\cos\phi} = \dfrac{1}{(10\ \mu m)\cos 10°} = 0.102\ \dfrac{\text{radians}}{\mu m}$

$0.102\ \dfrac{\text{radians}}{\mu m} \times \dfrac{180°}{\pi\ \text{radians}} = 5.8°/\mu m$

20-11. (a) Resolution $= \dfrac{\lambda}{\Delta\lambda} = \dfrac{512.245}{0.03} = 1.7 \times 10^4$

(b) $\Delta\lambda = \dfrac{\lambda}{10^4} = \dfrac{512.23}{10^4} = 0.05$ nm

(c) Resolution $= nN = (4)(8.00$ cm $\times 1\,850$ cm$^{-1}) = 5.9 \times 10^4$

(d) 250 lines/mm $= 4$ μm/line $= d$

$\dfrac{\Delta\phi}{\Delta\lambda} = \dfrac{n}{d\cos\phi} = \dfrac{1}{(4\ \mu m)\cos 3°} = 0.250\ \dfrac{\text{radians}}{\mu m} = 14.3°/\mu m$

For $\Delta\lambda = 0.03$ nm, $\Delta\phi = (14.3°/\mu m)(3 \times 10^{-5}\ \mu m) = 4.3 \times 10^{-4}$ degrees.

For 30th order diffraction, the dispersion will be 30 times greater, or 0.013°.

20-12. (a) True transmittance $= 10^{-1.500} = 0.031\,6$. With 0.50% stray light, the apparent transmittance is

Apparent transmittance $= \dfrac{P+S}{P_0+S} = \dfrac{0.031\,6 + 0.005\,0}{1 + 0.005\,0} = 0.036\,4$

The apparent absorbance is $-\log 0.036\,4 = 1.439$.

(b) Apparent absorbance $= 1.999$

Apparent transmittance $= 10^{-1.999} = 0.010\,023\,05$

$$\text{Apparent transmittance} = \frac{P+S}{P_0+S} = \frac{0.010+S}{1+S} = 0.010\,023\,05$$

$$\Rightarrow S = 2.328 \times 10^{-5} \text{ or } 0.002\,328\%$$

(c) For true absorbance = 2,

$$\text{Apparent transmittance} = \frac{P+S}{P_0+S} = \frac{0.010+0.000\,000\,5}{1+0.000\,000\,5} = 0.010\,000\,49$$

Apparent absorbance is $-\log T = -\log(0.010\,000\,49) = 1.999\,978$

Absorbance error = $2 - 1.999\,978 = 0.000\,022$

For true absorbance = 3,

$$\text{Apparent transmittance} = \frac{P+S}{P_0+S} = \frac{0.001+0.000\,000\,5}{1+0.000\,000\,5} = 0.001\,000\,495$$

Apparent absorbance is $-\log T = -\log(0.001\,000\,495) = 2.999\,785$

Absorbance error = $3 - 2.999\,785 = 0.000\,215$

20-13. $\quad b = \dfrac{30}{2\cdot1}\left(\dfrac{1}{1\,906-698\text{ cm}^{-1}}\right) = 0.124\,2 \text{ mm}$

(Air between the plates has refractive index of 1.)

20-14. $\quad M = \sigma T^4 = [5.669\,8 \times 10^{-8} \text{ W}/(m^2K^4)]T^4$

$T(\text{K})$	$M(\text{W/m}^2)$
77	1.99
298	447

20-15. (a) $\quad M_\lambda = \dfrac{2\pi hc^2}{\lambda^5}\left(\dfrac{1}{e^{hc/\lambda kT}-1}\right)$

at $T = 1\,000$ K:

λ (μm)	M_λ (W/m^3)
2.00	8.79×10^9
10.00	1.164×10^9

(b) $M_\lambda\Delta\lambda = (8.79 \times 10^9 \text{ W/m}^3)(0.02 \times 10^{-6} \text{ m}) = 1.8 \times 10^2 \text{ W/m}^2$ at 2.00 μm

(c) $M_\lambda\Delta\lambda = (1.164 \times 10^9 \text{ W/m}^3)(0.02 \times 10^{-6} \text{ m}) = 2.3 \times 10^1 \text{ W/m}^2$ at 10.00 μm

(d) at $T = 100$ K:

λ (μm)	M_λ (W/m^3)
2.00	6.69×10^{-19}
10.00	2.111×10^3

$$\frac{M_{2.00\,\mu m}}{M_{10.00\,\mu m}} = \frac{8.79 \times 10^9 \text{ W/m}^3}{1.164 \times 10^9 \text{ W/m}^3} = 7.55 \text{ at } 1\,000 \text{ K}$$

$$\frac{M_{2.00\ \mu m}}{M_{10.00\ \mu m}} = \frac{6.69 \times 10^{-19}\ \text{W/m}^3}{2.111 \times 10^3\ \text{W/m}^3} = 3.17 \times 10^{-22}\ \text{at 100 K}$$

At 100 K, there is virtually no emission at 2.00 μm compared to 10.00 μm, whereas at 1 000 K, there is a great deal of emission at both wavelengths.

20-16. (a) $A = \dfrac{L}{c \ln 10} \left(\dfrac{1}{\tau} - \dfrac{1}{\tau_o} \right)$

$$= \frac{0.210\ \text{m}}{(3.00 \times 10^8\ \text{m/s}) \ln 10} \left(\frac{1}{16.06 \times 10^{-6}\ \text{s}} - \frac{1}{18.52 \times 10^{-6}\ \text{s}} \right) = 2.51 \times 10^{-6}$$

(b) (i) The ordinate is ringdown lifetime with sample in the cavity. The greater the absorbance, the shorter the ringdown time. The spectrum is somewhat analogous to a transmission spectrum in which transmittance is plotted on the ordinate.

(ii) $\tilde{\nu} =$ wavenumber $\equiv 1/\text{wavelength} = 1/\lambda \implies \lambda = 1/\tilde{\nu}$.

$\lambda = 1/(6\ 046.954\ 6\ \text{cm}^{-1}) = 1.653\ 7250\ 0 \times 10^{-4}\ \text{cm}$

$(1.653\ 7250\ 0 \times 10^{-4}\ \text{cm})(10^7\ \text{nm/cm}) = 1\ 653.725\ 00\ \text{nm}$

(iii) infrared (see Figure 18-2 for names of regions)

20-17. $n_1 \sin \theta_1 = n_2 \sin \theta_2$, where $n_1 = 1.50$ and $n_2 = 1.33$

(a) If $\theta_1 = 30°$, $\theta_2 = 34°$

(b) If $\theta_1 = 0°$, $\theta_2 = 0°$ (no refraction)

20-18. Light inside the fiber strikes the wall at an angle greater than the critical angle for total reflection. Therefore, all light is reflected back into the core and continues to be reflected from wall-to-wall as it moves along the fiber. If the bending angle is not too great, the angle of incidence will still exceed the critical angle and light will not leave the core.

20-19. When traveling from medium 1 into medium 2, the critical angle for total internal reflection is $\sin \theta_{critical} = n_2/n_1$, where n_i is the refractive index of medium i. For the solvent/silica interface, $n_1 = 1.50$ and $n_2 = 1.46$, so $\sin \theta_{critical} = 1.46/1.5 = 0.973\ 3$. $\theta_{critical} = \sin^{-1} (0.973\ 3) = 1.339$ radians from the Excel function ASIN(0.9733). Degrees $= 180 \times \dfrac{\text{radians}}{\pi} = 180 \times \dfrac{1.339}{\pi} = 76.7°$.

For the silica/air interface, $n_1 = 1.45$ and $n_2 = 1.00$, so $\sin \theta_{critical} = 1.00/1.46 = 0.684\ 9$. $\theta_{critical} = \sin^{-1} (0.684\ 9) = 0.754\ 5$ radians. Degrees $= 180 \times \dfrac{0.754\ 5}{\pi} = 43.2°$.

The angle in the photograph is ~55°, which exceeds the critical angle at the silica/air interface but does not exceed the critical angle at the solvent/silica interface. Total internal reflection in the photo must be from the silica/air interface.

20-20. Light is transmitted through the diamond crystal waveguide by total internal reflection at the upper and lower surfaces. The upper surface is in contact with a fluid channel containing sorbent beads that retain caffeine from a soft drink flowing through the channel. Caffeine in the beads absorbs some of the evanescent wave from the totally internally reflected radiation, decreasing the radiant power transmitted through the waveguide. The integrated area of the absorption spectrum of transmitted power is proportional to the concentration of caffeine in the soft drink.

20-21. Sensitivity increases as the number of reflections inside the waveguide increases, because there is some attenuation at each reflection. For a constant angle of incidence, the number of reflections increases as the thickness of the waveguide decreases.

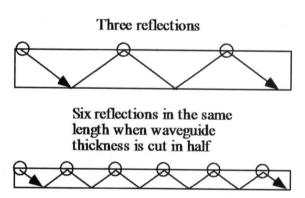

20-22. (a) The value of θ_i, called the critical angle (θ_c), is such that $(n_1/n_2)\sin\theta_c = 1$. For $n_1 = 1.52$ and $n_2 = 1.50$, $\theta_c = 80.7°$. That is, θ must be $\geq 80.7°$ for total internal reflection.

(b) $\dfrac{\text{power out}}{\text{power in}} = 10^{-\ell(\text{dB/m})/10} = 10^{-(20.0\text{ m})(0.010\,0\text{ dB/m})/10} = 0.955$

20-23. (a) $n_{\text{core}}\sin\theta_i = n_{\text{cladding}}\sin\theta_r$

For total reflection, $\sin\theta_r \geq 1 \Rightarrow \sin\theta_i \geq \dfrac{n_{\text{cladding}}}{n_{\text{core}}}$

For $n_{\text{cladding}} = 1.400$ and $n_{\text{core}} = 1.600$, $\sin\theta_i \geq \dfrac{1.400}{1.600} \Rightarrow \theta_i \geq 61.04°$

(b)   For $n_{\text{cladding}} = 1.400$ and $n_{\text{core}} = 1.800$, $\theta_i \geq 51.06°$

20-24.   Angle of incidence = angle of reflection = 45°. Angle of refraction $\equiv \theta$.
$n_{\text{prism}} \sin 45° = n_{\text{air}} \sin \theta$. If total reflection occurs, there is no refracted light.
This happens if $\sin \theta > 1$, or $\dfrac{n_{\text{prism}} \sin 45°}{n_{\text{air}}} > 1$. Using $n_{\text{air}} = 1$ gives
$n_{\text{prism}} > \sqrt{2}$. As long as $n_{\text{prism}} > \sqrt{2}$, no light will be transmitted through the
prism and all light will be reflected.

20-25.   (a)   The Teflon tube acts as an optical fiber because the internal solution has a
higher refractive index (1.33) than the walls (1.29). The tube is a 4.5-m-long
sample cell that can be conveniently coiled to fit in a reasonable volume and
guide the incident radiation all the way through the tube. The long
pathlength allows us to obtain a measurable absorbance for a very low
concentration of analyte.

(b) $n_{\text{core}} \sin \theta_i = n_{\text{cladding}} \sin \theta_r$

For total reflection, $\sin \theta_r \geq 1 \Rightarrow \sin \theta_i \geq \dfrac{n_{\text{cladding}}}{n_{\text{core}}}$

For $n_{\text{cladding}} = 1.29$ and $n_{\text{core}} = 1.33$, $\sin \theta_i \geq \dfrac{1.29}{1.33} \Rightarrow \theta_i \geq 76°$

(c) $A = \varepsilon bc = (4.5 \times 10^4 \text{ M}^{-1} \text{ cm}^{-1})(450 \text{ cm})(1.0 \times 10^{-9} \text{ M}) = 0.020$

20-26.   (a)   The following diagram shows the path of a light wave through the
waveguide. The length of one bounce, ℓ, satisfies the equation $(0.60 \text{ μm})/\ell$
$= \tan 20°$, giving $\ell = 1.648 \text{ μm}$. The hypotenuse of the triangle, h, satisfies
the equation $(0.60 \text{ μm})/h = \sin 20°$, giving $h = 1.754 \text{ μm}$. The number of
intervals of length ℓ in 3.0 cm is $(3.0 \text{ cm})/(1.648 \text{ μm}) = 1.820 \times 10^4$.
Therefore, the pathlength covered by the light is $h \times (1.820 \times 10^4) = 3.19$
cm.

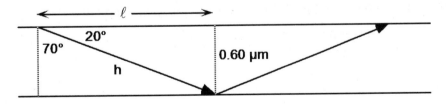

$\dfrac{\text{power out}}{\text{power in}} = 10^{-\ell(\text{dB/m})/10} = 10^{-(3.19 \text{ cm})(0.050 \text{ dB/cm})/10} = 0.964$

(b) Wavelength $= \lambda_o/n$, where λ_o is the wavelength in vacuum and n is the refractive index. Wavelength $= (514 \text{ nm})/1.5 = 343 \text{ nm}$.

The frequency is unchanged from that in vacuum:

$\nu = c/\lambda = (2.997\,9 \times 10^8 \text{ m/s})/(514 \times 10^{-9} \text{ m}) = 5.83 \times 10^{14} \text{ Hz}$

20-27. (a)

λ (μm)	n	λ (μm)	n
0.2	1.550 5	2	1.438 1
0.4	1.470 1	3	1.419 2
0.6	1.458 0	4	1.389 0
0.8	1.453 3	5	1.340 5
1	1.450 4	6	1.258 0

(b) $dn/d\lambda$ is greater for blue light (~ 400 nm) than for red light (~ 600 nm)

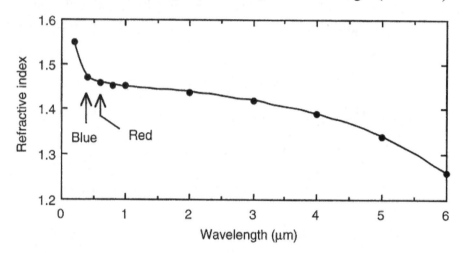

20-28. (a) $\Delta = \pm 2$ cm

(b) Resolution refers to the ability to distinguish closely spaced peaks.

(c) Resolution $\approx 1/\Delta = 0.5$ cm^{-1}

(d) $\delta = 1/(2\Delta\nu) = 1/(2 \times 2\,000 \text{ cm}^{-1}) = 2.5$ μm

20-29. The background transform gives the incident irradiance P_0. The sample transform gives the transmitted irradiance P. Transmittance is P/P_0, not $P - P_0$.

20-30. White noise is independent of frequency. One source is random motion of charge carriers in electronic circuits. $1/f$ noise decreases with increasing frequency. Drift and flicker of the lamp of a spectrophotometer are sources of $1/f$ noise. Line noise results from disturbances at discrete frequencies. The 60 Hz wall frequency is the most common electromagnetic disturbance seen in electronic instruments.

20-31. In beam chopping, the beam is alternately directed through the sample and reference cells. Beam chopping moves the analytical signal from zero frequency to the frequency of the chopper, which can be selected so that $1/f$ noise and line noise are minimal.

20-32. The difference voltage is ideally zero if the sample and reference are the same because the same lamp intensity goes through each compartment. If the sample absorbs some radiation, the difference voltage should respond to sample absorption with very little noise from source flicker.

20-33. To increase the ratio from 8 to 20 (a factor of $20/8 = 2.5$) requires $2.5^2 = 6.25 \approx 7$ scans.

20-34. (a) $(100 \pm 1) + (100 \pm 1) = 200 \pm \sqrt{2}$, since $e = \sqrt{e_1^2 + e_2^2} = \sqrt{1^2 + 1^2} = \sqrt{2}$

(b) $(100 \pm 1) + (100 \pm 1) + (100 \pm 1) + (100 \pm 1) = 400 \pm 2$, since $e = \sqrt{1^2 + 1^2 + 1^2 + 1^2} = 2$. The signal-to-noise ratio is $400:2 = 200:1$.

(c) The initial measurement has signal/noise $= 100/1$.

Averaging n measurements gives

$$\text{average signal} = \frac{n \cdot 100}{n} = 100$$

$$\text{average noise} = \frac{\sqrt{n}}{n} = 1/\sqrt{n}$$

$$\frac{\text{average signal}}{\text{average noise}} = \frac{100}{1/\sqrt{n}} = 100\sqrt{n},$$

which is $\sqrt{n}$ times greater than the original value of signal/noise.

20-35. Averaging 10 times as many signals should decrease the signal-to-noise ratio by $\sqrt{10} = 3.16$. Expected detection limit after 10 s of averaging is $37/\sqrt{10} = 11.7$. The observed limit is 12.5.

Averaging 40 times as many signals should decease the signal-to-noise ratio by $\sqrt{40} = 6.32$. Expected detection limit after 40 s of averaging is $37/\sqrt{10} = 5.8_5$. The observed limit is 5.6. The observed improvement in detection limit is following the theoretical prediction at least up to 40 s of signal averaging.

20-36. The theoretical signal-to-noise (S/N) ratio should increase in proportion to the square root of the number of cycles that are averaged.

Number of cycles $= n$	$\sqrt{n}$	Predicted relative S/N ratio
1	1	$\equiv 1$
100	10.00	10.00
300	17.32	17.32
1 000	31.62	31.62

If the observed S/N = 60.0 for the average of 1 000 cycles, then the predicted S/N for the other experiments are shown in the following table:

Number of cycles	Predicted S/N ratio	Observed S/N ratio
1 000	60.0 (observed)	60.0
300	$60.0 \left(\frac{17.32}{31.62}\right) = 32.9$	35.9
100	$60.0 \left(\frac{10.00}{31.62}\right) = 19.0$	20.9
1	$60.0 \left(\frac{1}{31.62}\right) = 1.90$	1.95

20-37.   (a)   For data in the table below

$$y_{12}\text{(5-pt cubic smoothing)} = \frac{-3y_{10} + 12y_{11} + 17y_{12} + 12y_{13} - 3y_{14}}{35}$$

$$= \frac{-3(0.869) + 12(1.956) + 17(6.956) + 12(5.000) - 3(6.086)}{35} = 5.167$$

$y_{12}\text{(9-pt cubic smoothing)} =$

$$\frac{-21y_8 + 14y_9 + 39y_{10} + 54y_{11} + 59y_{12} + 54y_{13} + 39y_{14} + 14y_{15} - 21y_{16}}{231} =$$

$$\frac{-21(0.000)+14(1.086)+39(0.869)+54(1.956)+59(6.956)+54(5.000)+39(6.086)+14(14.347)-21(14.130)}{231}$$

$$= 4.228$$

(b)   Partial data for 5-point cubic smoothing are shown in the spreadsheet. Final results are in the graph.

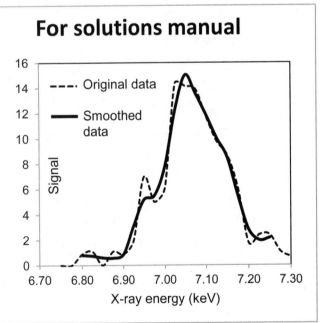

	A	B	C	D
1	Savitzky-Golay polynomial smoothing			
2				
3	x	y	y(smoothed)	
4	6.750	0.000		
5	6.775	0.000		
6	6.800	0.869	0.794	
7	6.825	1.086	0.732	
8	6.850	0.000	0.596	
9	6.875	1.086	0.565	
10	6.900	0.869	0.869	
11	6.925	1.956	3.111	
12	6.950	6.956	5.167	
13	6.975	5.000	5.503	
14	7.000	6.086	7.782	
15	7.025	14.347		
16	7.050	14.130		
17				
18	C6 =(-3*B4+12*B5+17*B6+12*B7-3*B8)/35			

CHAPTER 21
ATOMIC SPECTROSCOPY

21-1. Temperature is more critical in emission spectroscopy, because the small population of the excited state varies substantially as the temperature is changed. The population of the ground state does not vary much.

21-2. Furnaces give increased sensitivity and require smaller sample volumes, but give poorer reproducibility with manual sample introduction. Automated sample introduction gives good precision.

21-3. Drying (~20–100°C) removes water from the sample. Ashing (~100–500°C) is intended to remove as much matrix as possible without evaporating analyte. Atomization (~500–2 000°C) vaporizes analyte (and most of the rest of the sample) for the atomic absorption measurement. For some samples, ashing temperatures might be much higher than 500°C to remove more of the matrix, but it must be demonstrated that the ashing does not remove analyte.

21-4. The plasma operates at higher temperature than a flame and the environment is Ar, not combustion gases. The plasma decreases chemical interference (such as oxide formation) and allows emission instead of absorption to be used. Lamps are not required and simultaneous multi-element analysis is possible. Self-absorption is reduced in the plasma because the temperature is more uniform. Disadvantages of the plasma are increased cost of equipment and operation.

21-5. Doppler broadening occurs because an atom moving toward the radiation source sees a higher frequency than one moving away from the source. Increasing temperature gives increased speeds (more broadening) and increased mass gives decreased speeds (less broadening).

21-6. (a) A beam chopper alternately blocks or exposes the lamp to the flame and detector. When the lamp is blocked, signal is due to background. When the lamp is exposed, signal is due to analyte plus background. The difference is the desired analytical signal.

(b) The flame or furnace is alternately exposed to a D_2 lamp and the hollow-cathode lamp. Absorbance from the D_2 lamp is due to background. Absorbance from the hollow-cathode lamp is due to analyte plus background. The difference is the desired signal.

(c) When a magnetic field parallel to the viewing direction is applied to the furnace, the analytical signal is split into two components that are separated from the analytical wavelength, and one component at the analytical wavelength. The component at the analytical wavelength is not observed because of its polarization. The other two components have the wrong wavelength to be observed. Analyte is essentially "invisible" to the detector when the magnetic field is applied, and only background is seen. Corrected signal is that observed without a field minus that observed with the field.

21-7. Spectral interference refers to the overlap of analyte signal with signals due to other elements or molecules in the sample or with signals due to the flame or furnace. Chemical interference occurs when a component of the sample decreases the extent of atomization of analyte through some chemical reaction. Isobaric interference is the overlap of different species with nearly the same mass-to-charge ratio in a mass spectrum. Ionization interference refers to a loss of analyte atoms through ionization.

21-8. La^{3+} acts as releasing agent by binding tightly to PO_4^{3-} and freeing Pb^{2+}.

21-9. (a) A dynamic reaction cell contains a reactive gas such as NH_3, CH_4, N_2O, CO, or O_2 and its electric field is configured to select lower and upper masses of ions to pass through the cell. Plasma species that interfere with some elements can be reduced by as many as 9 orders of magnitude by reactions such as electron transfer ($^{40}Ar^{16}O^+ + NH_3 \rightarrow NH_3^+ + Ar + O$) and proton transfer ($^{40}ArH^+ + NH_3 \rightarrow NH_4^+ + Ar$). The reactive gas can also be used to shift the analyte signal from a position at which interference occurs (for example, $^{40}Ar^{16}O^+$ interferes with $^{56}Fe^+$) to one where there is no interference ($^{56}Fe^+ + N_2O \rightarrow ^{56}Fe^{16}O^+ + N_2$).

(b) When $^{87}Sr^+$ is converted to $^{87}Sr^{19}F^+$, which has a mass of 106, it no longer overlaps with $^{87}Rb^+$ in the mass spectrum.

21-10. The extent to which an element is ablated, transported to the plasma, and atomized depends on the matrix in which it is found. Different elements in a given matrix might not behave in the same manner. The most reliable calibration is for the analyte of interest to be measured in the same matrix as the unknown—if that is possible.

21-11. In the excitation spectrum, we are looking at emission over a band of wavelengths 1.6 nm wide, while exciting the sample with different narrow bands (0.03 nm) of laser light. The sample absorbs light only when the laser frequency coincides with the atomic frequency. Therefore, emission is observed only when the narrow laser line is in resonance with the atomic levels. In the emission spectrum, the sample is excited by a fixed laser frequency and then emits radiation. The monochromator bandwidth is not narrow enough to discriminate between emission at different wavelengths, so a broad envelope is observed.

21-12. For Pb:

$$\left(104 \pm 17 \; \frac{\text{pg Pb}}{\text{g snow}}\right)\left(11.5 \; \frac{\text{g snow}}{\text{cm}^2}\right) = 1\,196 \pm 196 \; \frac{\text{pg Pb}}{\text{cm}^2}$$

$$\left(1\,196 \pm 196 \; \frac{\text{pg Pb}}{\text{cm}^2}\right)\left(\frac{1 \; \text{ng}}{1\,000 \; \text{pg}}\right) = 1.2 \pm 0.2 \; \frac{\text{ng Pb}}{\text{cm}^2}$$

Similarly, we multiply each of the other concentrations by 11.5 g snow/cm^2 to find Tl: 0.005 ± 0.001; Cd: 0.04 ± 0.01; Zn: 2.0 ± 0.3; Al: $7 \, (\pm 2) \times 10^1 \, \text{ng/cm}^2$.

21-13. $\lambda = \dfrac{hc}{\Delta E} = \dfrac{(6.626 \times 10^{-34} \, \text{J·s})(2.998 \times 10^8 \, \text{m/s})}{3.371 \times 10^{-19} \, \text{J}} = 5.893 \times 10^{-7} \, \text{m} = 589.3 \; \text{nm}$

21-14. We derive the value for 6 000 K as follows:

$\Delta E = h\nu = \dfrac{hc}{\lambda} = \dfrac{(6.626\,1 \times 10^{-34} \, \text{J·s})(2.997\,9 \times 10^8 \, \text{m/s})}{500 \times 10^{-9} \, \text{m}} = 3.97 \times 10^{-19} \, \text{J}$

$\dfrac{N^*}{N_0} = \dfrac{g^*}{g_0} e^{-\Delta E/kT} = \dfrac{g^*}{g_0} e^{-(3.97 \times 10^{-19} \, \text{J})/(1.381 \times 10^{-23} \, \text{J/K})(6\,000\text{K})} = \dfrac{g^*}{g_0}(8.3 \times 10^{-3})$

If $g^*/g_0 = 3$, then $N^*/N_0 = 3 \, (8.3 \times 10^{-3}) = 0.025$.

21-15. Doppler linewidth: $\Delta\lambda = \lambda \, (7 \times 10^{-7}) \sqrt{T/M}$

For $\lambda = 589$ nm, $M = 23$ (sodium) at $T = 2\,000$ K,
$\Delta\lambda = (589 \, \text{nm})(7 \times 10^{-7}) \sqrt{(2\,000)/23} = 0.003_8 \; \text{nm}$

For $\lambda = 254$ nm, $M = 201$ (mercury) at $T = 2\,000$ K,
$\Delta\lambda = (254 \, \text{nm})(7 \times 10^{-7}) \sqrt{(2\,000)/201} = 0.000\,5_6 \; \text{nm}$

21-16. (a) $\Delta E = h\nu = \dfrac{hc}{\lambda} = \dfrac{(6.626\,1 \times 10^{-34}\,\text{J} \cdot \text{s})(2.997\,9 \times 10^{8}\,\text{m/s})}{422.7 \times 10^{-9}\,\text{m}}$

$= 4.699 \times 10^{-19}\,\text{J/molecule} = 283.0\,\text{kJ/mol}$

(b) $\dfrac{N^*}{N_0} = \dfrac{g^*}{g_0}\,e^{-\Delta E/kT} = 3e^{-(4.699 \times 10^{-19}\,\text{J})/(1.381 \times 10^{-23}\,\text{J/K})(2\,500\text{K})}$

$= 3.66_{57} \times 10^{-6}$

(c) At 2 515 K, $N^*/N_0 = 3.97_{58} \times 10^{-6} \Rightarrow 8.5\%$ increase from 2 500 to 2 515 K

(d) At 6 000 K, $N^*/N_0 = 1.03 \times 10^{-2}$

21-17.

Element:	Na	Cu	Br
Excited state energy (eV):	2.10	3.78	8.04
Wavelength (nm):	591	328	154
Degeneracy ratio (g^*/g_0):	3	3	2/3
N^*/N_0 at 2 600 K in flame:	2.6×10^{-4}	1.4×10^{-7}	1.8×10^{-16}
N^*/N_0 at 6 000 K in plasma:	5.2×10^{-2}	2.0×10^{-3}	1.2×10^{-7}

Calculations: wavelength $= hc/\Delta E$ $\qquad N^*/N_0 = (g^*/g_0)\,e^{-\Delta E/kT}$

Br is not readily observed in atomic absorption, because its lowest excited state requires far-ultraviolet radiation for excitation. Nitrogen and oxygen in the air absorb far-ultraviolet energy and would have to be excluded from the optical path. The excited state lies at such high energy that it is not sufficiently populated to provide adequate intensity for optical emission.

21-18. The dissociation energy of YC is greater than that of BaC, so the equilibrium BaC + Y $\rightleftharpoons$ Ba + YC is driven to the right, increasing the concentration of free Ba atoms in the gas phase.

21-19. Area of pit $= \pi(20 \times 10^{-4}\,\text{cm})^2 = 1.26 \times 10^{-5}\,\text{cm}^2$

Power $= 2.4 \times 10^{-3}\,\text{J}/10 \times 10^{-9}\,\text{s} = 2.4 \times 10^{5}\,\text{W}$

Power density $= 2.4 \times 10^{5}\,\text{W}/1.26 \times 10^{-5}\,\text{cm}^2 = 1.9 \times 10^{10}\,\text{W/cm}^2 = 20\,\text{GW/cm}^2$

Ablated mass $=$ volume $\times$ density $=$ depth $\times$ area $\times$ density $=$

$(1 \times 10^{-4}\,\text{cm})(1.26 \times 10^{-5}\,\text{cm}^2)(4\,\text{g/cm}^3) = 5\,\text{ng}$

21-20. Analyte and standard are lost in equal proportions, so their ratio remains constant.

21-21.

	A	B	C	D
1	Standard Addition Constant Volume Least-Squares			
2	x	y		
3	Volume added (mL)	Absorbance		
4	0.00	0.151		
5	1.00	0.185		
6	3.00	0.247		
7	5.00	0.300		
8	8.00	0.388		
9	10.00	0.445		
10	15.00	0.572		
11	20.00	0.723		
12	B14:C16 = =LINEST(B4:B11,A4:A11,TRUE,TRUE)			
13		LINEST output:		
14	m	0.0282	0.1579	b
15	u_m	0.0003	0.0031	u_b
16	R^2	0.9993	0.0057	s_y
17	x-intercept = -b/m =	-5.599		
18	n =	8	= COUNT(A4:A11)	
19	Mean y =	0.376	= AVERAGE(B4:B11)	
20	$\Sigma(x_i - mean\ x)^2$ =	343.5	= DEVSQ(A4:A8)	
21	Std uncertainty of			
22	x-intercept (u_x) =	0.1630		
23	B22=(C16/ABS(B14))*SQRT((1/B18) + B19^2/(B14^2*B20))			

The x-intercept is $-5.60 \pm 0.16_3$ mL. Standard [Ca] = 20.0 µg Ca/mL. The intercept corresponds to Ca = $(5.60 \pm 0.16_3$ mL$)(20.0$ µg Ca/mL$) = 112.0 \pm 3.2_6$ µg. This is the mass of Ca in 5.00 mL of unknown. The total volume of unknown was 100.0 mL, so mass of Ca in total unknown = (100.0 mL/5.00 mL)$(112.0 \pm 3.2_6$ µg Ca$) = 2240 \pm 65$ µg Ca.

$$\text{wt\% Ca in cereal} = \frac{100 \times (2240 \pm 65_2\ \text{µg Ca})}{0.521\ 6\ \text{g cereal}} = 0.429 \pm 0.012\ \text{wt\%}.$$

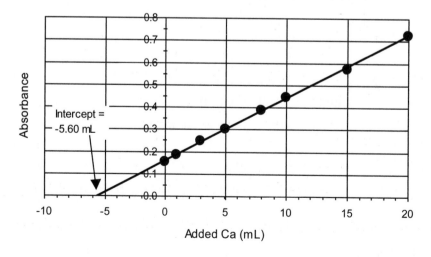

21-22. (a) [S] in unknown mixture $= (8.24\ \mu g/mL)\left(\dfrac{5.00}{50.0}\right) = 0.824\ \mu g/mL$

Standard mixture has equal concentrations of X and S:

$$\frac{A_X}{[X]} = F\left(\frac{A_S}{[S]}\right) \Rightarrow \frac{0.930}{[\cancel{X}]} = F\left(\frac{1.000}{[\cancel{S}]}\right) \Rightarrow F = 0.930$$

Unknown mixture:

$$\frac{A_X}{[X]} = F\left(\frac{A_S}{[S]}\right) \Rightarrow \frac{1.690}{[X]} = 0.930\left(\frac{1.000}{[0.824\ mg/mL]}\right) \Rightarrow [X] = 1.497\ \mu g/mL$$

But X was diluted by a factor of 10.00/50.0, so the original concentration in the unknown was $(1.49_7\ \mu g/mL)\left(\dfrac{50.0}{10.00}\right) = 7.49\ \mu g/mL$.

(b) Standard mixture has equal concentrations of X and S:

$$\frac{A_X}{[X]} = F\left(\frac{A_S}{[S]}\right) \Rightarrow \frac{0.930}{[3.42]} = F\left(\frac{1.000}{[1.00]}\right) \Rightarrow F = 0.271_9$$

Unknown mixture:

$$\frac{A_X}{[X]} = F\left(\frac{A_S}{[S]}\right) \Rightarrow \frac{1.690}{[X]} = 0.271_9\left(\frac{1.000}{[0.824\ \mu g/mL]}\right) \Rightarrow [X] = 5.12_2\ \mu g/mL$$

But X was diluted by a factor of 10.00/50.0, so the original concentration in the unknown was $(5.12_2\ \mu g/mL)\left(\dfrac{50.0}{10.00}\right) = 25.6\ \mu g/mL$.

21-23.

	A	B	C	D	E	F	G	H	I
1	Least-Squares Spreadsheet								
2		x =	y =						
3	Highlight cells B10:C12	µg/mL	signal						
4	Type "= LINEST(C4:C8,	0	0						
5	B4:B8,TRUE,TRUE)	5	124						
6	For PC, press	10	243						
7	CTRL+SHIFT+ENTER	20	486						
8	For Mac, press	30	712						
9	COMMAND+RETURN	LINEST output:							
10	m	23.7672	4.0259	b					
11	u_m	0.2256	3.8090	u_b					
12	R^2	0.9997	5.4338	s_y					
13									
14	n =	5	B16 = COUNT(B4:B8)						
15	Mean y =	313.00	B17 = AVERAGE(C4:C8)						
16	$\Sigma(x_i - \text{mean } x)^2 =$	580.00	B18 = DEVSQ(B4:B8)						
17									
18	Measured y =	417	Input						
19	k = Number of replicate measurements of y =	1	Input						
20	Derived x =	17.3758	B20 =(B18-C10)/B10						
21	Standard uncertainty $u_x =$	0.2539	B21=(C12/ABS(B10))*SQRT((1/B19)+(1/B14)+((B18-B15)^2)/(B10^2*B16))						

Chart (columns E–I): y = signal vs x = µg/mL, data points at (0,0), (5,124), (10,243), (20,486), (30,712), line $y = 23.77x + 4.03$

Cells B20 and B21 give us [unknown] $= 17.4 \pm 0.3\ \mu g/mL$ for an emission intensity of 417.

21-24. (a) CsCl provides Cs atoms which ionize to $Cs^+ + e^-$ in the plasma. Electrons in the plasma inhibit ionization of Sn. Therefore, emission from atomic Sn is not lost to emission from Sn^+.

(b)

	A	B	C	D	E	F
1	Tin in canned food - Anal. Bioanal. Chem. 2002, 374, 235					
2	Calibration data for 189.927 nm					
3						
4	Conc (µg/L)	Signal intensity			Output from LINEST	
5	0	4			slope	intercept
6	10	8.5		Parameter	0.7815257	0.862272
7	20	19.6		Std Dev	0.0184397	1.5510156
8	30	23.6		R^2	0.9966709	3.2018176
9	40	31.1				
10	60	41.7				
11	100	78.8				
12	200	159.1				
13						
14	Select cells E6:F8					
15	Enter the formula = LINEST(B5:B12,A5:A12,TRUE,TRUE)					
16	CONTROL+SHIFT+ENTER on PC or COMMAND+RETURN on Mac					

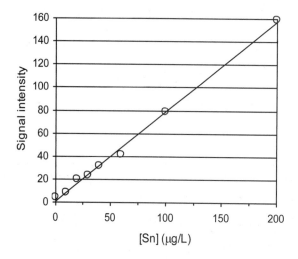

(c) For the 189.927 nm Sn emission line, spike recoveries are all near 100 µg/L, which is near 100%. None of the elements in the table appears to interfere significantly at 189.927 nm. For the 235.485 nm emission line, interference from an emission line of Fe is so serious that the Sn signal cannot be measured. Several other elements interfere enough to reduce the accuracy of the Sn measurement. These elements include Cu, Mn, Zn, Cr, and, perhaps, Mg. The 189.927 nm line is clearly the better of the two wavelengths for minimizing interference.

(d) Limit of detection = minimum detectable concentration = $3s/m$
where s is the standard deviation of the replicate samples and m is the slope of the calibration curve. Putting in the values $s = 2.4$ units and $m = 0.782$ units per (µg/L) gives

$$\text{limit of detection} = \frac{3s}{m} = \frac{3(2.4 \text{ units})}{0.782 \text{ units}/(\text{µg/L})} = 9.2 \text{ µg/L}$$

$$\text{limit of quantitation} = \frac{10s}{m} = \frac{10(2.4 \text{ units})}{0.782 \text{ units}/(\text{µg/L})} = 30.7 \text{ µg/L}$$

It would be reasonable to quote a limit of detection as 9 µg/L and a limit of quantitation as 31 µg/L.

(e) A 2-g food sample ends up in a volume of 50 mL. The limit of quantitation is 30.7 µg Sn/L for the solution. A 50-mL volume with Sn at the limit of quantitation contains $(0.050 \text{ L})(30.7 \text{ µg Sn/L}) = 1.54 \text{ µg Sn}$. The quantity of Sn per unit mass of food is

$$\frac{(1.54 \text{ µg Sn})(1 \text{ mg}/1\ 000 \text{ µg})}{(2.0 \text{ g food})(1 \text{ kg}/1\ 000 \text{ g})} = 0.77 \frac{\text{mg Sn}}{\text{kg food}} = 0.77 \text{ ppm}$$

21-25. Standard addition graph: plot signal versus Ti or S concentration.

Ti (ppm)	Signal		S (ppm)	Signal
0.00	0.86		0.0	0.0174
3.00	1.10		37.0	0.0221
6.00	1.34		74.0	0.0268
12.00	1.82		148.0	0.0362

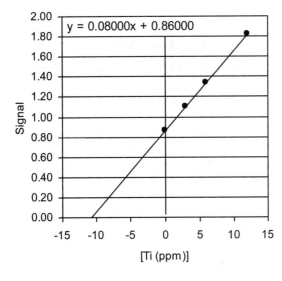

$y = 0.08000x + 0.86000$

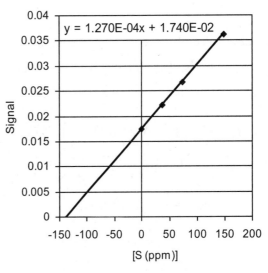

$y = 1.270\text{E}-04x + 1.740\text{E}-02$

Ti standard addition graph: negative intercept = $(0.860)/(0.0800 \text{ ppm}^{-1})$

= 10.75 ppm = 10.75 mg/L

S standard addition graph: negative intercept = $(0.017\ 4)/(0.000\ 127\ \text{ppm}^{-1})$

= 137.0 ppm = 137.0 mg/L

Ti atomic mass = 47.867   S atomic mass = 32.065

$[Ti] = (10.75 \text{ mg/L})/(47.867 \text{ g/mol}) = 2.246 \times 10^{-4}$ M

$[S] = (137.0 \text{ mg/L})/(32.065 \text{ g/mol}) = 4.273 \times 10^{-3}$ M

$[Transferrin] = [S]/39 = 1.096 \times 10^{-4}$ M

Ti/transferrin = $(2.246 \times 10^{-4} \text{ M})/(1.096 \times 10^{-4} \text{ M}) = 2.05$

21-26. Absorption of X-rays of sufficient energy ionizes electrons from atoms. When an electron is removed from an inner shell of an atom, an electron from an outer shell falls into the vacancy. The excess energy of the electron making the transition is emitted as an X-ray. Electronic energy levels are different for every element, so the signature (energies of emitted X-rays) is different for each element.

21-27. K_β peaks for Ti, Se, and Zr should have energies of 4.93, 12.50, and 17.67 keV, respectively. A weak peak is observed in the spectrum near 4.9 keV, which could be Ti K_β. Se K_β at 12.50 keV comes under the strong Pb L_β peak at 12.61 keV. There is a shoulder at the left side of the base of Pb L_β from Se K_β. Zr K_β at 17.67 keV is observed as a weak peak superimposed on the broad bremsstrahlung radiation.

21-28. We observe Pb L_α = 10.55 and Pb L_β = 12.61 keV in the fluorescence spectrum. Pb K_α (74.97) and K_β (84.94) keV are beyond the working energy range of the handheld analyzer.

21-29. $E = (392 \text{ eV})(1.602 \times 10^{-19} \text{ J/eV}) = 6.28 \times 10^{-17}$ J (conversion from Table 1-4) $(6.28 \times 10^{-17} \text{ J/atom})(6.022 \times 10^{23} \text{ atoms/mol}) = 3.78 \times 10^7 \text{ J/mol} = 3.78 \times 10^4$ kJ/mol. The K_α energy is $(37\ 800 \text{ kJ/mol})/(945 \text{ kJ/mol}) = 40$ times greater than the N≡N bond energy.

21-30.

Energy (keV)	Assignment	
6.40	Fe K_α	
7.05	Fe K_β	
7.50	Ni K_α	K_β peak would be at 8.26, but too small to see
8.07	Cu K_α	K_β peak would be at 8.90, but too small to see
8.62	Zn K_α	There is a possible K_β peak at 9.57
10.57	Pb L_α	
12.60	Pb L_β	
~14.1	Sr K_α?	K_β peak would be at 15.84
~14.8	?	
~15.78	Zr K_α?	K_β peak would be at 17.67
17.50	Mo K_α	Probably X-ray tube anode
~18.58	?	
19.59	Mo K_β	Probably X-ray tube anode

..

Energy (keV)	Assignment	
3.70	Ca K_α	
4.01	Ca K_β	
6.40	Fe K_α	
7.06	Fe K_β	
8.73	?	
9.20	?	
9.99	Hg L_α	
10.55	Pb L_α	
11.84	Hg L_β	
12.63	Pb L_β	
13.84	Hg L_γ	L_γ is observed because there is so much Hg in the sample. L_γ is not in the table in the text.
14.20	Sr K_α?	K_β would be at 15.84, where there is weak bump
14.76	Pb L_γ	L_γ is observed because there is so much Pb in the sample. L_γ is not in the table in the text.
15.17	?	
25.25	Sn K_α	
28.48	Sn K_β	

CHAPTER 22
MASS SPECTROMETRY

22-1. Gaseous molecules are ionized by collisions with 70-eV electrons in the ion source. The ions are accelerated out of the source by a voltage, V. All ions have nearly the same kinetic energy ($\frac{1}{2}mv^2$, where m is mass and v is velocity), so the heavier ions have lower velocity. Ions then enter a magnetic field (B) and are deflected so they travel through the arc of a circle whose radius is $(\sqrt{2V(m/z)/e})/B$, where z is the number of charges on the ion and e is the elementary charge. By varying the magnetic field, ions of different m/z are deflected through the slit leading to the detector. At the detector, ion impacts liberate electrons from a cathode. The electrons are amplified by a series of dynodes (as in a photomultiplier tube). The mass spectrum is a graph of detector signal versus m/z.

22-2. For the electron ionization spectrum, pentobarbital is bombarded by electrons with an energy of 70 electron volts. The molecular ion ($m/z = 226$) produced by the impact has enough energy to break into fragments and little $M^{+\bullet}$ is observed. Large peaks correspond to the most stable cation fragments. For chemical ionization, pentobarbital reacts with CH_5^+, which is a potent proton donor, but does not have excess kinetic energy. The dominant peak is usually MH^+ ($m/z = 227$). In the case of pentobarbital, some fragmentation is observed even in the chemical ionization spectrum.

22-3. 1 dalton (Da) $\equiv$ 1/12 of the mass of ^{12}C $= \left(\dfrac{(1/12) \times 12 \text{ g/mol (exactly)}}{6.022\,14 \times 10^{23} \text{ mol}^{-1}} \right)$

$\qquad = 1.660\,54 \times 10^{-24}$ g

$[5.03\ (\pm 0.14) \times 10^{10}\text{ Da}][1.660\,54 \times 10^{-24}\text{ g/Da}]$

$\qquad = 8.35\ (\pm 0.23) \times 10^{-14}$ g $= 83.5\ (\pm 2.3)$ fg

22-4. The atomic mass in the periodic table is a weighted average of the masses of all the isotopes of that element. We can estimate the relative abundance of the two major isotopes of Ni from the heights of their mass spectral peaks. The heights of the peaks that I measured from an earlier version of this illustration are 42.6 mm for ^{58}Ni and 17.1 mm for ^{60}Ni. The weighted average is atomic mass

$$= (^{58}\text{Ni mass})(\% \text{ abundance of } ^{58}\text{Ni}) + (^{60}\text{Ni mass})(\% \text{ abundance of } ^{60}\text{Ni})$$

$$= (57.935\ 3)\left(\frac{42.6}{42.6 + 17.1}\right) + (59.933\ 2)\left(\frac{17.1}{42.6 + 17.1}\right) = 58.51$$

The atomic mass in the periodic table is 58.69. This main reason for disagreement is that we neglected the existence of ^{61}Ni (1.13% natural abundance), ^{62}Ni (3.59%), and ^{64}Ni (0.90%).

22-5. Resolving power $= \dfrac{m}{m_{1/2}} = \dfrac{2\ 846.3}{0.19} = 1.5 \times 10^4$.

We should be able to barely distinguish two peaks differing by 1 Da at a mass of 1.5×10^4 Da. Therefore, we should be able to distinguish two peaks at 10 000 and 10 001 Da.

22-6. The overlap at the base of the peaks is approximately 10% in the mass spectrum. The resolving power is approximately $m/\Delta m \approx 31/0.010 \approx 3\ 100$.

22-7. Resolving power by 10% valley formula: $m/\Delta m = 906.49/0.000\ 45 = 2.0 \times 10^6$
Resolving power by half-width formula: $m/m_{1/2} = 906.49/0.000\ 27 = 3.4 \times 10^6$
The mass of an electron, 0.000 55 Da, is greater than the mass difference between the two compounds. The mass difference of the compounds is 82% of the mass of one electron.

22-8. $C_5H_7O^{+\bullet}$ $5 \times 12.000\ 00$
$+7 \times\ \ 1.007\ 825$
$+1 \times 15.994\ 91$
$- e^- \text{ mass }\ \underline{-1 \times\ \ 0.000\ 55}$
$83.049\ 14$

$C_6H_{11}^{+\bullet}$ $6 \times 12.000\ 00$
$+11 \times\ \ 1.007\ 825$
$- e^- \text{ mass }\ \underline{-1 \times\ \ 0.000\ 55}$
$83.085\ 52$

$C_6H_{11}^+$ is a closer match than $C_5H_7O^+$ to the observed mass of 83.086 5 Da.

22-9. $^{31}P^+ = {^{31}P} - e^- = 30.973\,76 - 0.000\,55 = 30.973\,21$ (observed: 30.973_5)

To measure m/z, I enlarged the figure and sketched a Gaussian curve over each signal by eye. I then measured the position of the center of the peak with a millimeter scale ruler.

$^{15}N^{16}O^+ = {^{15}N} + {^{16}O} - e^- = 15.000\,11 + 15.994\,91 - 0.000\,55$
$\quad\quad = 30.994\,47$ (observed: 30.994_6)

$^{14}N^{16}OH^+ = {^{14}N} + {^{16}O} + {^1H} - e^- =$
$\quad\quad 14.003\,07 + 15.994\,91 + 1.007\,82 - 0.000\,55 = 31.005\,25$
$\hfill$(observed: 31.005_6)

22-10.   (a) m/z 140: $^{12}C_7{}^1H_{10}{}^{14}N^{16}O_2{}^+$

$\quad\quad m/z$ 141: $^{13}C^{12}C_6{}^1H_{10}{}^{14}N^{16}O_2{}^+$ and $^{12}C_7{}^2H^1H_9{}^{14}N^{16}O_2{}^+$ and
$\quad\quad\quad ^{12}C_7{}^1H_{10}{}^{15}N^{16}O_2{}^+$ and $^{12}C_7{}^1H_{10}{}^{14}N^{17}O^{16}O^+$

Intensity at m/z 141 $= \underbrace{7 \times 1.08\%}_{^{13}C} + \underbrace{10 \times 0.012\%}_{^2H} + \underbrace{1 \times 0.369\%}_{^{15}N} + \underbrace{2 \times 0.038\%}_{^{17}O} = 8.1\%$

(b) m/z 142: $^{12}C_7{}^1H_{10}{}^{14}N^{18}O^{16}O^+$ and $^{13}C_2{}^{12}C_5{}^1H_{10}{}^{14}N^{16}O_2{}^+$ and
$\quad\quad\quad ^{13}C^{12}C_6{}^1H_{10}{}^{15}N^{16}O_2{}^+$

22-11.   Intensity at X+2 $= \underbrace{12 \times 11 \times 0.005\,8\%}_{^{13}C} + \underbrace{8 \times 0.205\%}_{^{18}O} = 2.4\%$

The observed intensity is nearly 100% because the predominant species is $[^{12}C_{12}{}^1H_{18}{}^{16}O_8{}^{35}Cl_2{}^{37}Cl]^-$.

22-12.   Mass differences in the bottom row of the spreadsheet are 0.2, 0.6, and 1.2 ppm

	mass	m/z 395 $C_{12}H_{18}O_8{}^{35}Cl_3$	m/z 397 $C_{12}H_{18}O_8{}^{35}Cl_2{}^{37}Cl$	m/z 399 $C_{12}H_{18}O_8{}^{35}Cl^{37}Cl_2$	m/z 401 $C_{12}H_{18}O_8{}^{37}Cl_3$
^{12}C	12	144	144	144	144
^1H	1.007825	18.14085	18.14085	18.14085	18.14085
^{16}O	15.99491	127.95928	127.95928	127.95928	127.95928
^{35}Cl	34.96885	104.90655	69.9377	34.96885	
^{37}Cl	36.9659		36.9659	73.9318	110.8977
e^-	0.00055	0.00055	0.00055	0.00055	0.00055
Calculated exact mass =		395.00723	397.00428	399.00133	400.99838
Observed mass =		395.0073	397.0045	399.0018	not measured
Difference (ppm) =		0.2	0.6	1.2	
Difference (ppm) = 10^6*(observed mass -calculated mass)/calculated mass					

22-13. (a) The base peak has a mass loss of $390 - 149 = 241$ from the molecular ion

Nominal mass of fragments that might be lost: OC_8H_{17} 129

C_8H_{17} 113

These two fragments have a total mass of 242. Loss of OC_8H_{17} and C_8H_{17}

with retention of one H atom gives a possible structure for the base peak:

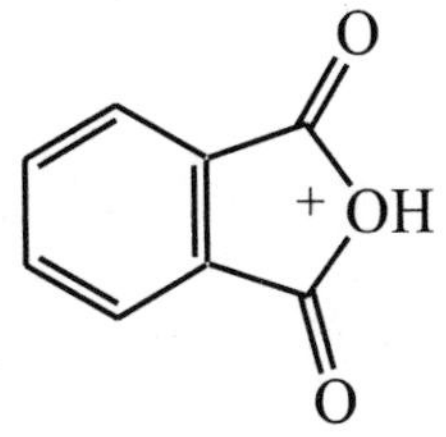

m/z 141

The *m/z* 279 peak corresponds to a mass loss of $390 - 279 = 111$ from $M^{+\bullet}$

Loss of C_8H_{17} with retention of two hydrogen atoms gives a possible

structure

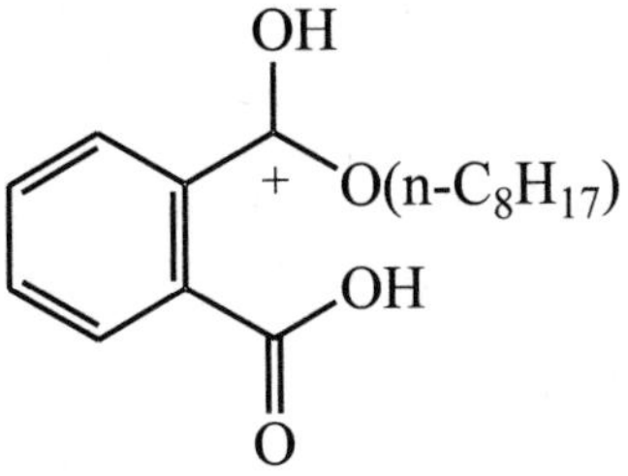

m/z 279

(b) *m/z* 277 has a mass loss of $390 - 277 = 113 =$ loss of C_8H_{17} from $M^{-\bullet}$

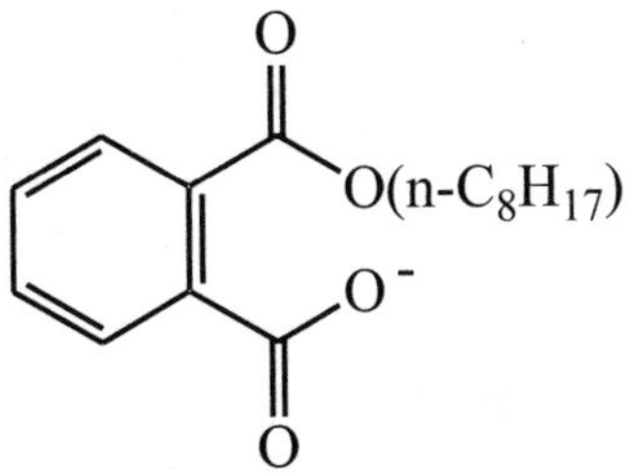

m/z 277

m/z 221 has a mass loss of $390 - 221 = 169 = 390 - 113 - 56$

$= 390 - C_8H_{17} - C_4H_8$ from $M^{-\bullet}$. A possible structure is

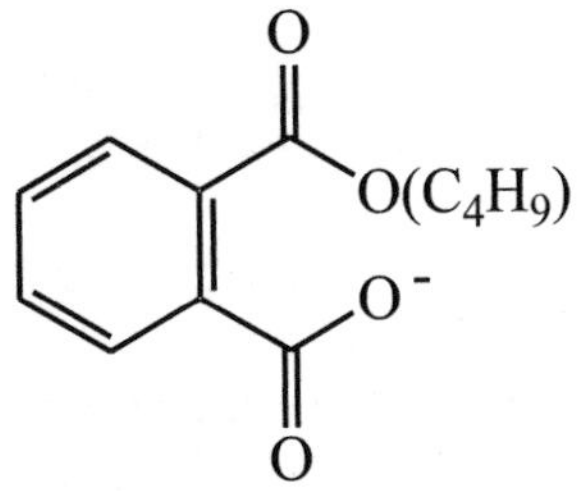

m/z 221

22-14. ^{79}Br abundance $\equiv a = 0.506\,9$ $\qquad$ ^{81}Br abundance $\equiv b = 0.493\,1$

Abundance of $C_2H_2{}^{79}Br_2 = a^2 = 0.256\,9_5$

Abundance of $C_2H_2{}^{79}Br^{81}Br = 2ab = 0.499\,9_0$

Abundance of $C_2H_2{}^{81}Br_2 = b^2 = 0.243\,1_5$

Relative abundances: $M^+ : M+1 : M+2 = 1 : 1.946 : 0.946\,3$

Figure 22-7 shows the stick diagram.

22-15. ^{10}B abundance $\equiv a = 0.199$ $\qquad$ ^{11}B abundance $\equiv b = 0.801$

Abundance of $^{10}B_2H_6 = a^2 = 0.039\,6_0$

Abundance of $^{10}B^{11}BH_6 = 2ab = 0.318_8$

Abundance of $^{11}B_2H_6 = b^2 = 0.641_6$

Relative abundances: $M^+ : M+1 : M+2 = 1 : 8.05 : 16.20$

22-16. ^{79}Br abundance $\equiv a = 0.506\,9$ $\qquad$ ^{81}Br abundance $+ b = 0.493\,1$

Abundance of $CH^{79}Br_3 = a^3 = 0.130\,2_5$

Abundance of $CH^{79}Br_2{}^{81}Br = 3a^2b = 0.380\,1_0$

Abundance of $CH^{79}Br^{81}Br_2 = 3ab^2 = 0.369\,7_5$

Abundance of $CH^{81}Br_3 = b^3 = 0.119\,9_0$

Relative abundances: $M^+ : M+1 : M+2 : M+3 = 0.342\,7 : 1 : 0.972\,8 : 0.315\,4$

22-17. (a) Nominal mass of $C_{72}H_{146} = 72 \times 12 + 146 \times 1 = 1\,010$ Da

(b) Monoisotopic mass of $^{12}C_{72}{}^1H_{146} = 72 \times 12 + 146 \times 1.007\,825$

$= 1\,011.14$ Da

(c) $(a + b)^n = (0.989\,3 + 0.010\,7)^{72}$

$= (0.989\,3)^{72} + \frac{72}{1!}(0.989\,3)^{71}(0.010\,7)^1 + \dots$

$= 0.460\,9 + 0.358\,9 + \dots$

Relative intensity $M/(M+1) = 0.460\,9/0.358\,9 = 1 : 0.778\,7$

22-18. (a)

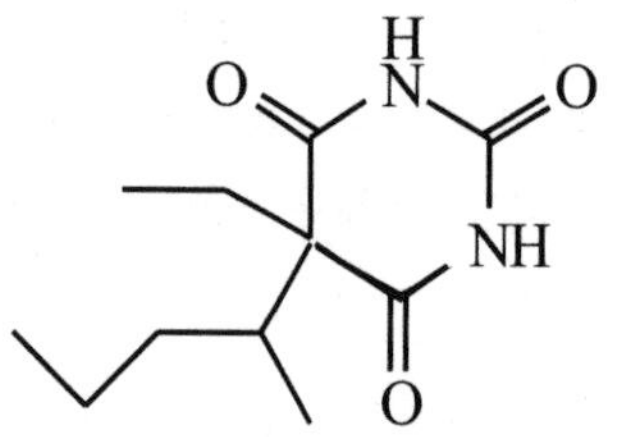

phenobarbital, $C_{11}H_{18}N_2O_3$

$R + DB = c - h/2 + n/2 + 1$

$R + DB = 11 - 18/2 + 2/2 + 1 = 4$

The molecule has one ring + three double bonds.

(b)

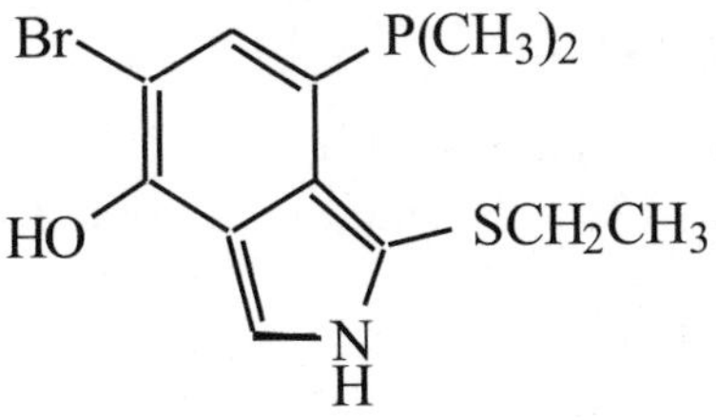

$C_{12}H_{15}BrNPOS$

$R + DB = c - h/2 + n/2 + 1$

$R + DB = 12 - \dfrac{15 + 1}{2} + \dfrac{1 + 1}{2} + 1 = 6$

The molecule has two rings + four double bonds. Note that h includes H + Br, and n includes N + P. S, like O, does not contribute to the count.

(c)

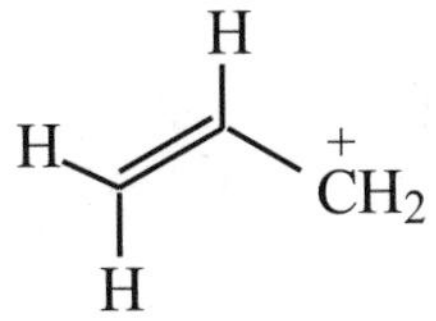

A fragment in a mass spectrum

$C_3H_5^+$

$R + DB = c - h/2 + n/2 + 1$

$R + DB = 3 - 5/2 + 1 = 1\frac{1}{2}$ Huh?

We come out with a fraction instead of an integer because the species is an ion in which one C makes three bonds instead of four.

22-19. (a)

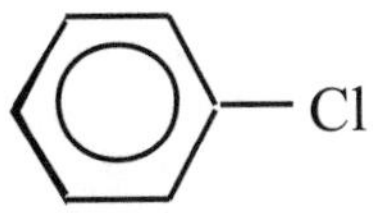

C_6H_5Cl: $M^{+\bullet} = 112$

The pair of peaks at $m/z = 112$ and 114 strongly suggest that the molecule contains 1 Cl.

rings + double bonds $= c - h/2 + n/2 + 1 = 6 - 6/2 + 1 = 4$

$\uparrow$

h includes H + Cl

Expected intensity of M+1 is $1.08(6) + 0.012(5) = 6.54\%$

carbon   hydrogen

Observed intensity of M+1 $= 69/999 = 6.9\%$

Expected intensity of M+2 $= 0.005\ 8(6)(5) + 32.0(1) = 32.2\%$

carbon   chlorine

Observed intensity of M+2 $= 329/999 = 32.9\%$

The M+3 peak is the isotopic partner of the M+2 peak. M+3 contains ^{37}Cl plus either 1 ^{13}C or 1 ^2H. Therefore, the expected intensity of M+3 (relative to M+2) is $1.08(6) + 0.012(5) = 6.54\%$ of predicted intensity of M+2 $= (0.065\ 4)(32.2) = 2.11\%$ of $M^{+\bullet}$.

Observed intensity of M+3 is $21/999 = 2.1\%$.

(b) Cl—⟨O⟩—Cl $C_6H_4Cl_2$: $M^{+\cdot} = 146$

The peaks at $m/z = 146$, 148, and 150 look like the isotope pattern from 2 Cl in Figure 22-7.

rings + double bonds $= c - h/2 + n/2 + 1 = 6 - 6/2 + 1 = 4$

Expected intensity of M+1 is $\underset{\text{carbon}}{1.08(6)} + \underset{\text{hydrogen}}{0.012(4)} = 6.53\%$

Observed intensity of M+1 $= 56/999 = 5.6\%$

Expected intensity of M+2 $= \underset{\text{carbon}}{0.005\ 8(6)(5)} + \underset{\text{chlorine}}{32.0(2)} = 64.2\%$

Observed intensity of M+2 $= 624/999 = 62.5\%$

The M+3 peak is the isotopic partner of the M+2 peak. M+3 contains 1 ^{35}Cl + 1 ^{37}Cl plus either 1 ^{13}C or 1 ^2H. Therefore, the expected intensity of M+3 (relative to M+2) is $1.08(6) + 0.012(4) = 6.53\%$ of predicted intensity of M+2 $= (0.065\ 3)(64.2) = 4.19\%$ of $M^{+\cdot}$.

Observed intensity of M+3 is $33/999 = 3.3\%$.

Expected intensity of M+4 from $C_6H_4{}^{37}Cl_2$ is $5.11(2)(1) = 10.22\%$ of $M^{+\cdot}$. The small contribution from $^{12}C_4{}^{13}C_2H_4{}^{35}Cl{}^{37}Cl$ is based on the predicted intensity of M+2. It is $0.005\ 8(6)(5) = 0.174\%$ of $64.2\% = 0.11\%$.

Total expected intensity of M+4 is $10.22\% + 0.11\% = 10.33\%$ of $M^{+\cdot}$

Observed intensity $= 99/999 = 9.9\%$.

Expected intensity of M+5 from $^{12}C_5{}^{13}CH_4{}^{37}Cl_2$ and $^{12}C_6H_3{}^2H{}^{37}Cl_2$ is based on the predicted intensity of M+4. M+5 should have $1.08(6) + 0.012(4) = 6.53\%$ of M+4 $= 6.53\%$ of $10.33\% = 0.67\%$.

Observed intensity $= 5/999 = 0.5\%$.

(c) ⟨O⟩—NH$_2$ C_6H_7N: $M^{+\cdot} = 93$

The peak at $m/z = 93$ was chosen as the molecular ion, because it is the tallest peak in the cluster and it has plausible isotope peaks at M+1 and M+2. The significant peak at M–1 could be from loss of 1 H. The tiny stuff at M–2 and M–3 could be noise or, possibly, loss of more than 1 H.

With an odd mass, the nitrogen rule tells us that there are an odd number of N atoms in the molecule.

rings + double bonds $= c - h/2 + n/2 + 1 = 6 - 7/2 + 1/2 + 1 = 4$

Expected intensity of M+1 is $1.08(6) + 0.012(7) + 0.369(1) = 6.93\%$

$\quad\quad\quad\quad\quad\quad\quad\quad\quad\quad$ carbon $\quad\quad$ hydrogen $\quad\quad$ nitrogen

Observed intensity of M+1 = $71/999 = 7.1\%$

Expected intensity of M+2 = $0.005\ 8(6)(5) = 0.17\%$

Observed intensity of M+2 = $2/999 = 0.2\%$

(d) $(CH_3)_2Hg \quad\quad C_2H_6Hg$: $M^{+\bullet} = 228$

There are six strong peaks in an unfamiliar pattern. Given that only elements from Table 22-1 are admissible, we notice that Hg has six significant isotopes. By convention, we take the lightest isotope, ^{198}Hg, for the molecular ion at $m/z = 228$. This leaves just 30 Da for the rest of the molecule, which could be composed of two methyl groups.

In computing rings + double bonds, we include Hg as a Group 6 atom (like O or S) because it makes 2 bonds.

rings + double bonds = $c - h/2 + n/2 + 1 = 2 - 6/2 + 1 = 0$.

The peak at $m/z = 228$ is $M^{+\bullet} = (CH_3)_2{}^{198}Hg$.

Small peaks at $m/z = 227$ and 226 could arise from loss of one or two H atoms. If $(CH_3)_2{}^{198}Hg$ loses H atoms, then all the species at higher mass, such as $(CH_3)_2{}^{199}Hg$, will also lose H atoms. That is, each isotopic molecule is going to contribute some intensity to peaks of lower mass. It makes no sense for us to get too carried away with the analysis of the isotopic pattern, because each peak derives intensity from peaks at lower and higher mass.

The peak at $m/z = 229$ is M+1, composed mainly of $(CH_3)_2{}^{199}Hg$, with some $(^{12}CH_3)(^{13}CH_3)^{198}Hg + {}^{12}C_2H_5{}^2H^{198}Hg$. Just considering Hg, the predicted intensity, based on M^+, is $\frac{16.87}{9.97} \times 100 = 169.2\%$ of $M^{+\bullet}$. The observed intensity is $215/130 = 165\%$ of $M^{+\bullet}$. In this calculation, the fraction $\frac{16.87}{9.97}$ is the ratio of the abundances of ^{199}Hg to ^{198}Hg. The peak at $m/z = 230$ is M+2, composed mainly of $(CH_3)_2{}^{200}Hg$. The predicted ^{200}Hg isotopic intensity, based on M^+, is $\frac{23.10}{9.97} \times 100 = 231.7\%$ of $M^{+\bullet}$.

Observed intensity of M+2 = $291/130 = 224\%$ of $M^{+\bullet}$.

Just considering Hg isotopes, we expect the peaks at M, M+1, M+2, M+3, M+4, and M+6 to have the ratios $9.97 : 16.87 : 23.10 : 13.18 : 29.86 : 6.87$ $= 1 : 1.69 : 2.32 : 1.32 : 2.99 : 0.69$.

Observed intensity ratio = $1 : 1.65 : 2.24 : 1.29 : 2.81 : 0.64$.

(e) CH_2Br_2: $M^{+\bullet} = 172$

The three peaks at $m/z = 172$, 174 and 176, with approximate ratios 1 : 2 : 1 looks like the pattern from 2 Br atoms in Figure 22-7.

rings + double bonds $= c - h/2 + n/2 + 1 = 1 - 4/2 + 1 = 0$

h includes H + Br

Expected intensity of M+1 is $1.08(1) + 0.012(2) = 1.10\%$

carbon   hydrogen

Observed intensity of M+1 = 12/531 = 2.3%. It is possible that this peak at $m/z = 173$ also has contributions from $CH^{79}Br^{81}Br$. We have no way to compute the intensity at $m/z = 173$ if some of this peak comes from $CH^{79}Br^{81}Br$. Given this ambiguity, we will just compare the theoretical pattern for 2 Br atoms to the observed pattern:

Theoretical intensity of M+2 = $97.3(2) = 194.6\%$

Observed intensity of M+2 = 999/531 = 188%

Theoretical intensity of M+4 = $47.3(2)(1) = 94.6\%$

Observed intensity of M+4 = 497/531 = 93.6%

(f) 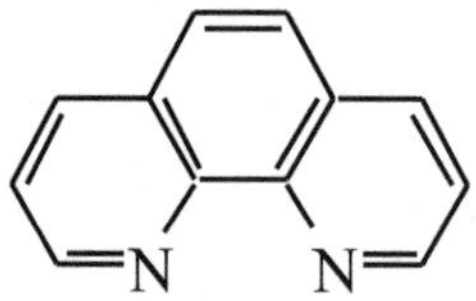   1,10-Phenanthroline, $C_{12}H_8N_2$: $M^{+\bullet} = 180$

The strongest peak in the high-mass cluster is at $m/z = 180$, which could be the molecular ion. It has plausible isotopic peaks at 181 and 182. The significant peak at $m/z = 179$ could be from loss of 1 H.

The intensity ratio M+1/$M^{+\bullet}$ = 138/999 = 13.8%. We estimate that the number of C atoms is 13.8/1.08 = 12.8.

If the molecule contains 13 C atoms, the formula might be $C_{13}H_8O$, which would have $13 - 8/2 + 1 = 10$ rings plus double bonds. The expected intensity of M+1 would be $1.08(13) + 0.012(8) + 0.038(1) = 14.2\%$. The expected intensity of M+2 would be $0.005\,8(13)(12) + 0.205(1) = 1.1\%$. Observed intensity of M+2 = 9/999 = 0.9%. The formula $C_{13}H_8O$ fits the data and a conceivable structure is

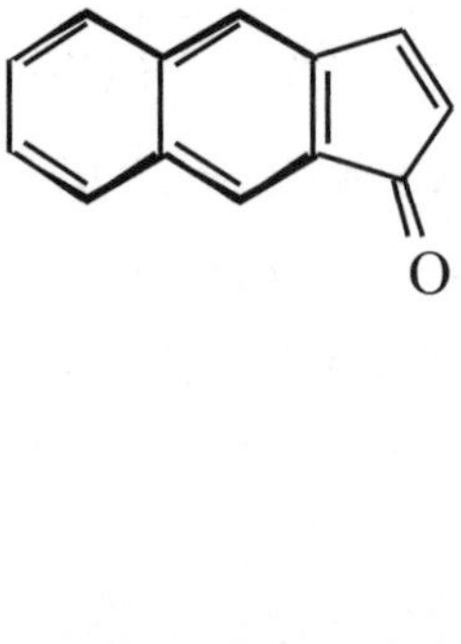

If the molecule contains 12 C atoms, the formula might be $C_{12}H_4O_2$, which would have $12 - 4/2 + 1 = 11$ rings plus double bonds. A molecule with this many rings + double bonds would be pretty implausible.

If the molecule contains nitrogen, it must contain an even number of N atoms because the molecule has an even mass. A possible formula is $C_{12}H_8N_2$, which would have $12 - 8/2 + 2/2 + 1 = 10$ rings plus double bonds. This turns out to be the correct formula, and the structure is shown at the beginning of this answer. The predicted intensity of M+1 is $1.08(12) + 0.012(8) + 0.369(2) = 13.8\%$, which is exactly equal to the observed intensity. The expected intensity of M+2 is $0.005\ 8(12)(11) = 0.8\%$. Observed intensity $= 0.9\%$.

(g)   Ferrocene, $C_{10}H_{10}Fe$: $M^{+\cdot} = 186$

The strongest peak at high mass is at $m/z = 186$, which could be the molecular ion. It has plausible isotopic peaks at 187 and 188. Significant peaks at $m/z = 184$ and 185 could be from loss of H. Calling $M^{+\cdot} = 186$, we find the following ratios of peak intensities:

M–2	M–1	$M^{+\cdot}$	M+1	M+2
8.3	1.6	100	13.2	1.0

From the intensity ratio $M+1/M^{+\cdot} = 13.2\%$, we could estimate that the number of C atoms 1s $13.8/1.08 = 12.8$. From this we could propose formulas like $C_{13}H_{14}O$ or $C_{12}H_{10}O_2$.

Alternatively, noting the significant intensity of M–2, we could propose that the molecule has Fe in it, which, in fact, it does. For the formula $C_{10}H_{10}Fe$, we predict that M–2 will have an intensity of $\frac{5.845}{91.754} \times 100 = 6.37\%$ of $M^{+\cdot}$, which is not terribly far from the observed value of 8.3%. The intensity at M+1 will have a contribution from ^{57}Fe and from ^{13}C and ^{2}H. The ^{57}Fe contribution is $2.119/91.754 = 2.31\%$ of $M^{+\cdot}$. The other contributions are $1.08(10) + 0.012(10) = 10.92\%$. The total intensity predicted at M+1 is 13.23% and the observed intensity is 13.2%. The predicted intensity at M+2 is $\frac{0.282}{91.754} \times 100$ (from Fe) $+ 0.005\ 8(10)(9)$ (from C) $= 0.83\%$, and the observed intensity is 1.0%.

22-20. The compound is dibromochloromethane:

212	$CH^{81}Br_2^{37}Cl$	94	$CH^{81}Br$
210	$CH^{81}Br_2^{35}Cl + CH^{79}Br^{81}Br^{37}Cl$	93	$C^{81}Br$
208	$CH^{79}Br^{81}Br^{35}Cl + CH^{79}Br_2^{37}Cl$	92	$CH^{79}Br$
206	$CH^{79}Br_2^{35}Cl$	91	$C^{79}Br$
175	$CH^{81}Br_2$	81	^{81}Br
173	$CH^{79}Br^{81}Br$	79	^{79}Br
171	$CH^{79}Br_2$	50	$CH^{37}Cl$
162	$^{81}Br_2$	49	$C^{37}Cl$
160	$^{79}Br^{81}Br$	48	$CH^{35}Cl$
158	$^{79}Br_2$	47	$C^{35}Cl$
131	$CH^{81}Br^{37}Cl$	37	^{37}Cl
129	$CH^{81}Br^{35}Cl + CH^{79}Br^{37}Cl$	35	^{35}Cl
127	$CH^{79}Br^{35}Cl$		

22-21. (a) For the formula $C_9H_4N_2Cl_6$,

rings + double bonds = $c - h/2 + n/2 + 1 = 9 - (4+6)/2 + 2/2 + 1 = 6$,

which agrees with the structure that has 2 rings + 4 double bonds.

(b) Nominal mass = integer mass of the species with the most abundant isotope of each of the constituent atoms. For $C_9H_4N_2Cl_6$, nominal mass = (9×12) $+ (4 \times 1) + (2 \times 14) + (6 \times 35) = 350$.

(c) The sequence m/z 350, 315, 280, 245, and 210 corresponds to successive losses of 35 Da. A logical assignment is $C_9H_4N_2{}^{35}Cl_6^+$, $C_9H_4N_2{}^{35}Cl_5^+$, $C_9H_4N_2{}^{35}Cl_4^+$, $C_9H_4N_2{}^{35}Cl_3^+$, $C_9H_4N_2{}^{35}Cl_2^+$.

22-22. The CO_2 that we exhale is derived from oxidation of the food we eat. The chart shows that the group of plants called C_3 plants has less ^{13}C than the groups called C_4 and CAM plants. If the diet in the United States contains more C_4 and CAM plants and the diet in Europe contains more C_3 plants, then the difference in ^{13}C content of exhaled CO_2 might be explained.

22-23. (a) Mass of proton + electron = 1.007 276 467 + 0.000 548 580

$\qquad$ =1.007 825 047 Da. To the number of significant digits in Table 1, the masses of the proton and electron are equal to the mass of ^1H.

(b) mass of proton + neutron + electron

$\qquad$ = 1.007 276 467 + 1.008 664 916 + 0.000 548 580 = 2.016 489 963 Da

mass of ^2H in table = 2.014 10 Da.

The ^2H atom is 0.002 39 Da lighter than the sum of its elementary particles.

(c) Mass difference = (0.002 39 Da) (1.660 5 × 10^{-27} kg/Da)

$$= 3.97 \times 10^{-30} \text{ kg}$$

$$E = mc^2 = (3.97 \times 10^{-30} \text{ kg})(2.997\ 9 \times 10^8 \text{ m/s})^2 = 3.57 \times 10^{-13} \text{ J}$$

mc^2 is the binding energy of a single nucleus. For a mole, the energy is

$(3.57 \times 10^{-13} \text{ J})(6.022 \times 10^{23} \text{ mol}^{-1}) = 2.15 \times 10^{11}$ J/mol = 2.15×10^8 kJ/mol.

(d) Binding energy for atom = (13.6 eV)(1.602 18 × 10^{-19} J/eV) = 2.18×10^{-18} J

To convert to a mole: $(2.18 \times 10^{-18} \text{ J})(6.022 \times 10^{23} \text{ mol}^{-1}) = 1.31 \times 10^6$

J/mol = 1.31×10^3 kJ/mol. The ratio of the nuclear binding energy to the electron binding energy is $(2.15 \times 10^8$ kJ/mol $)/(1.31 \times 10^3$ kJ/mol$)$

$$= 1.64 \times 10^5.$$

(e) $\dfrac{\text{nuclear binding energy}}{\text{bond energy}} \approx (2.15 \times 10^8$ kJ/mol $)/(400$ kJ/mol$) = 5 \times 10^5$

22-24. ^{28}Si abundance $\equiv a = 0.922\ 30$ ^{29}Si $\equiv b = 0.046\ 83$ ^{30}Si $\equiv c = 0.030\ 87$

$(a+b+c)^3 = a^3 + 3a^2b + 3a^2c + 3ab^2 + 6abc + 3ac^2 + b^3 + 3b^2c + 3bc^2 + c^3$

Slicon abundance	Polynomial expansion	Relative abundance				Total intensity
a =	a^3 =	= term value/a^3	Composition	mass		at mass
0.92230	0.784543	1.000000	28Si 28Si 28 Si	84		1
b =	3a^2b =					
0.04683	0.119506	0.152326	28Si 28Si 29 Si	85		0.152326
c =	3a^2c =					
0.03087	0.078778	0.100412	28Si 28Si 30 Si	86		0.108146
	3ab^2 =					
	0.006068	0.007734	28Si 29Si 29 Si	86		
	6abc =					
	0.008000	0.010197	28Si 29Si 30 Si	87		0.010328
	3ac^2 =					
	0.002637	0.003361	28Si 30Si 30 Si	88		0.003620
	b^3 =					
	0.000103	0.000131	29Si 29Si 29 Si	87		
	3b^2c =					
	0.000203	0.000259	29Si 29Si 30 Si	88		
	3bc^2=					
	0.000134	0.000171	29Si 30Si 30 Si	89		0.000171
	c^3 =					
	2.94178E-05	0.000037	30Si 30Si 30 Si	90		0.000037
Check: sum of terms in column B =						
	1					

mass:	84	85	86	87	88	89	90
intensity:	1	0.1523	0.1081	0.01033	0.00362	0.000171	0.000037

22-25. In a double-focusing mass spectrometer, ions ejected from the source pass through an electric sector that selects ions with a narrow band of kinetic energies to continue into the magnetic sector. The electric sector acts as an energy filter and the magnetic sector disperses the ions to form the mass spectrum.

22-26. The reflectron improves resolving power by ensuring that all ions of the same mass reach the detector grid at the same time. Ions from the ion source have some spread of kinetic energy. Faster ions penetrate deeper into the reflectron and therefore spend more time there before being turned around. The reflectron allows slower ions to catch up to faster ions of the same mass.

22-27. At m/z 100, 2 ppm = $(100)(2 \times 10^{-6})$ = 0.000 2 Da.
At m/z 20 000, 2 ppm = $(20\ 000)(2 \times 10^{-6})$ = 0.04 Da.

22-28. From Box 22-2, we know that an ion of $m/z = 500$ accelerated through a potential difference of V volts attains a velocity of $\sqrt{2zeV/m}$. We need to express the mass in kg. The footnote of Table 22-1 gives the conversion factor.

$$500\ \text{Da} \times 1.661 \times 10^{-27}\ \text{kg/Da} = 8.30 \times 10^{-25}\ \text{kg}$$

$$\text{velocity} = \sqrt{\frac{2zeV}{m}} = \sqrt{\frac{2(1)(1.602 \times 10^{-19}\ \text{C})(5.00 \times 10^3\ \text{V})}{8.30 \times 10^{-25}\ \text{kg}}}$$

$$= 4.39 \times 10^4\ \text{m/s}$$

To figure out the units, remember that work (joules) = $E \cdot q$ = volts·coulombs. So the product $C \times V = J = m^2 kg/s^2$. Putting these units into the square root gives velocity in m/s.

The time needed to travel 2.00 m is (2.00 m)/(4.39 × 10⁴ m/s) = 45.6 μs. If we repeated a cycle each time this heaviest ion reaches the detector, we could collect 1/(45.6 μs) = 2.20×10^4 spectra per second.

If we double the mass in the square root to get up to 1 000 Da, the velocity decreases by $1/\sqrt{2}$ and the frequency goes down by $1/\sqrt{2}$ to 1.56×10^4 spectra per second.

22-29. We use the equation from the previous problem. The masses of the two ions are

$$m_{100} = 100\ \text{Da} \times 1.661 \times 10^{-27}\ \text{kg/Da} = 1.661 \times 10^{-25}\ \text{kg}$$

$$m_{1\ 000\ 000} = 10^6\ \text{Da} \times 1.661 \times 10^{-27}\ \text{kg/Da} = 1.661 \times 10^{-21}\ \text{kg}$$

$$\text{velocity} = \sqrt{\frac{2zeV}{m}} = \sqrt{\frac{2(1)(1.602 \times 10^{-19}\ \text{C})(20.0 \times 10^3\ \text{V})}{m}}$$

$$= 1.964 \times 10^5 \text{ m/s for 100 Da and } 1.964 \times 10^3 \text{ m/s for 1 000 000 Da}$$

$$\text{transit time} = 2.00 \text{ m/velocity} = 10.2 \text{ µs for 100 Da}$$

$$= 1.02 \text{ ms for 1 000 000 Da}$$

22-30. (a) $\lambda = \dfrac{kT}{(\sqrt{2}\sigma P)} = \dfrac{(1.38 \times 10^{-23} \text{ J/K})(300 \text{ K})}{(\sqrt{2}(\pi(10^{-9} \text{ m})^2)(10^{-5} \text{ Pa}))} = 93 \text{ m}$

(The answer is in meters if you substitute $m^2 \cdot kg \cdot s^{-2}$ for J and $kg \cdot m^{-1} \cdot s^{-2}$ for Pa from Table 1-2.)

(b) $\lambda = \dfrac{kT}{(\sqrt{2}\sigma P)} = \dfrac{(1.38 \times 10^{-23} \text{ J/K})(300 \text{ K})}{(\sqrt{2}(\pi(10^{-9} \text{ m})^2)(10^{-8} \text{ Pa}))} = 93 \text{ km}$

22-31. Ions seen in electrospray usually existed in solution prior to electrospray. Atmospheric pressure chemical ionization creates ions in the corona discharge around the high voltage needle.

22-32. The mean free path for an ion in a mass spectrometer must be greater than the distance it must travel in the spectrometer. Therefore, adequate vacuum must be maintained. A mass spectrometer must be able to pump away gas or solvent from chromatography to maintain adequate vacuum. Gas chromatography produces a relatively small flow of gas compared with the flow of solvent from most liquid chromatography columns. Vacuum pumps can keep up with flow rates in gas chromatography, but not with flow rates from most liquid chromatography columns. Vacuum pumps can handle flow rates of ~100 to 500 nL/min from capillary liquid chromatography columns.

22-33. The photon energy is ~10 eV, which is just enough for ionization of many molecules. There is little extra energy available for fragmentation. A 70 eV electron carries far more energy than required for ionization.

22-34. In collisionally activated dissociation, ions are accelerated through an electric field and directed into a region with a significant pressure of N_2 or Ar. Collisions transfer enough energy to break molecules into fragments. Collisionally activated dissociation can be conducted at the entrance to the mass spectrometer or in a collision cell in the middle part of tandem mass spectrometry.

22-35.   A reconstructed total ion chromatogram shows the current from all ions above a selected mass displayed as a function of time. The chromatogram is "reconstructed" by summing the intensities for all observed values of m/z. The total ion chromatogram shows everything coming off the column. An extracted ion chromatogram displays detector current for just one or a few values of m/z as a function of time. The intensity displayed is extracted from the full mass spectrum recorded at each time interval. A selected ion chromatogram also displays detector current for just one or a small number of m/z values. However, for a selected ion chromatogram, the detector is not measuring the signal for all values of m/z in each time interval. The detector is set at just the desired values of m/z and collects that information for the whole time. The extracted ion chromatogram and the selected ion chromatogram are selective for an analyte of interest (plus anything else that gives a signal at the same m/z). The selected ion chromatogram has improved signal-to-noise ratio because all of the time is spent detecting signal at the selected mass.

22-36.   In selected reaction monitoring, an ion of one m/z value is selected by the first mass separator. This ion is directed to a collision cell in which it undergoes collisionally activated dissociation to produce fragment ions. One of those fragment ions is then selected by a second mass separator and passed through to the detector. The detector is responding to just one product ion from the selected precursor ion. This technique is called MS/MS because it involves two consecutive mass separation steps. The signal/noise ratio is improved because noise (extraneous signals) is very low. There are few sources of the precursor ion other than the desired analyte, and it is unlikely that other precursor ions of the selected m/z can decompose to give the same product ion being monitored.

22-37.   (a)   Ibuprofen can readily dissociate to form a carboxylate anion, so I would choose the negative ion mode. It would be harder to form a cation.

The carboxylate anion should exist in neutral solution, since pK_a is probably around 4. In sufficiently acidic solution, the carboxylate will be protonated. I would use a neutral chromatography solvent to ensure a good supply of analyte anions.

(b) The formula of the molecular ion, M^-, is $C_{13}H_{17}O_2^-$. The intensity expected

at M+1 is $\underset{\text{carbon}}{1.08(13)} + \underset{\text{hydrogen}}{0.012(17)} + \underset{\text{oxygen}}{0.038(2)} = 14.32$.

22-38. The analysis follows the same steps as Table 22-3. The work is set out in the following table. Peaks A and B give $n_A = 12$ and peaks H and I give $n_H = 19$. The combination of peaks G and H give $n_G \approx 21$, which makes no sense and will be ignored. Assigning peaks A, B, C... as $n = 12, 13, 14...$ gives the sensible, constant molecular masses in the last column of the table. The mean value, disregarding peak G, is 15 126.

Analysis of electrospray mass spectrum of α-chain of hemoglobin

Peak	Observed m/z $\equiv m_n$	$m_{n+1} - 1.008$	$m_n - m_{n+1}$	Charge $= n =$ $\dfrac{m_{n+1} - 1.008}{m_n - m_{n+1}}$	Molecular mass $= n \times (m_n - 1.008)$
A	1 261.5	1 163.6	96.9	$12.0_1 \approx 12$	15 126
B	1 164.6	—	—	[13]	15 127
C	—	—	—	[14]	—
D	—	—	—	[15]	—
E	—	—	—	[16]	—
F	—	—	—	[17]	—
G	834.3	796.1	37.2	21.4 [18]	14 999
H	797.1	756.2	39.9	$18.9_5 \approx 19$	15 126
I	757.2			[20]	15 124
					mean = 15 100
					mean without peak G = 15 126

22-39. The separation between adjacent peaks is 0.27, 0.28, 0.25, 0.24, 0.24, 0.24, 0.27, 0.23, 0.24, 0.25, 0.26, and 0.24 m/z units, giving a mean value of 0.25_1. If species differing by 1 Da are separated by 0.25_1 m/z unit, the species must carry 4 charges ($z = 4$). The mass of the tallest peak must be 4(1 962.12) = 7 848.48 Da.

22-40. (a)

m/z	z	molecular ion mass (Da)	
38 152.7	2	$2 \times 38\ 152.7 = 76\ 305.4$	
25 433.3	3	$3 \times 25\ 433.3 = 76\ 299.9$	mean =
19 075.2	4	$4 \times 19\ 075.2 = 76\ 300.8$	76 302.0 Da
15 260.4	5	$5 \times 15\ 260.4 = 76\ 302.0$	

(b) $R = n\text{-}C_{12}H_{25}$ molecular ion mass $= 3 \times 27\ 243 = 81\ 729$ Da

$R = CH_2CH_2C_6H_5$ molecular ion mass $= 3 \times 25\ 440 = 76\ 320$

$R = n\text{-}C_6H_{13}$ molecular ion mass $= 3 \times 24\ 876 = 74\ 628$

Molecular mass of thiol ligands		Difference in SR mass from $SC_{12}H_{25}$
$-SC_{12}H_{25}$	201.395	0
$-SCH_2CH_2C_6H_5$	137.225	64.170
$-SC_6H_{13}$	117.235	84.159

The difference in nanoparticle mass between $Au_x(SC_{12}H_{25})_y$ and $Au_x(SCH_2CH_2C_6H_5)_y$ is $81\ 729 - 76\ 320 = 5\ 409$ Da, corresponding to the difference in mass for y thiol ligands.

The difference in mass for one ligand is 64.169 Da.

The number of ligands is therefore $y = 5\ 409$ Da/64.17- Da = 84.3.

The difference in nanoparticle mass between $Au_x(SC_{12}H_{25})_y$ and $Au_x(SC_6H_{13})_y$ is $81\ 729 - 74\ 628 = 7\ 101$ Da, corresponding to the difference in mass for y thiol ligands.

The difference in mass for one ligand is 84.159 Da.

The number of ligands is therefore $y = 7\ 101$ Da/84.159 Da = 84.4.

There appear to be $y = 84$ thiol ligands in each nanoparticle. The mass of thiol ligand is found in the first column below. Subtracting the mass of thiol from the mass of the nanoparticle gives the mass of Au. Dividing the mass of Au by the atomic mass of Au gives $x =$ number of Au atoms in particle

Mass of thiol = $84 \times$ FM of thiol	Au mass = nanoparticle mass – thiol mass	$x =$ Au atoms = Au mass/atomic mass
$84 \times 201.395 = 16\ 917$	$81\ 729 - 16\ 917 = 64\ 812$	329.1
$84 \times 137.225 = 11\ 527$	$76\ 320 - 11\ 527 = 64\ 793$	329.0
$84 \times 117.235 = 9\ 848$	$74\ 628 - 9\ 848 = 64\ 780$	328.9

Conclusion: the formula of the nanoparticle is $Au_{329}(SR)_{84}$

22-41. Expected intensities for 37:3, whose formula is $[MNH_4]^+ = C_{37}H_{72}ON$

$X + 1 = 0.012n_H + 1.08n_C + 0.369n_N + 0.038n_O$

$= 0.012(72) + 1.08(37) + 0.369(1) + 0.038(1) = 41.2\%$ (observed = 35.8%)

$X + 2 = 0.005\ 8n_C(n_C - 1) + 0.205n_O$

$= 0.005\ 8(37)(36) + 0.205(1) = 7.9\%$ (observed = 7.0%)

Expected intensities for 37:3, whose formula is $[MH]^+ = C_{37}H_{69}O$

$X + 1 = 0.012n_H + 1.08n_C + 0.038n_O$

$= 0.012(69) + 1.08(37) + 0.038(1) = 40.8\%$ (observed = 23.0%)

$X + 2 = 0.005\ 8n_C(n_C - 1) + 0.205n_O$

$= 0.005\ 8(37)(36) + 0.205(1) = 7.9\%$ (observed = 8.0%)

Expected intensities for 37:2, whose formula is $[MNH_4]^+ = C_{37}H_{74}ON$

$X + 1 = 0.012n_H + 1.08n_C + 0.369n_N + 0.038n_O$

$= 0.012(74) + 1.08(37) + 0.369(1) + 0.038(1) = 41.3\%$ (observed = 40.8%)

$X + 2 = 0.005\ 8n_C(n_C - 1) + 0.205n_O$

$= 0.005\ 8(37)(36) + 0.205(1) = 7.9\%$ (observed = 3.7%)

Expected intensities for 37:2, whose formula is $[MH]^+ = C_{37}H_{71}O$

$X + 1 = 0.012n_H + 1.08n_C + 0.369n_N + 0.038n_O$

$= 0.012(71) + 1.08(37) + 0.038(1) = 40.8\%$ (observed = 33.4%)

$X + 2 = 0.005\ 8n_C(n_C - 1) + 0.205n_O$

$= 0.005\ 8(37)(36) + 0.205(1) = 7.9\%$ (observed = 8.4%)

22-42. Selected reaction monitoring chooses the molecular ion ClO_3^- ($m/z = 83$) with the mass separator Q1. In collision cell Q2, this species could possibly undergo the following decomposition:

$$^{35}ClO_3^- \xrightarrow[\text{collisions}]{\text{high energy}} {}^{35}ClO_2^- + {}^{35}ClO^- + {}^{35}Cl^-$$

$$m/z = 83 \qquad\qquad m/z = 67 \qquad m/z = 51 \qquad m/z = 35$$

Quadrupole Q3 selects only $m/z = 67$. The measurement is specific for ClO_3^- because there are probably few compounds in water producing ions at $m/z = 83$, and *very few* of them are likely to decompose into $m/z = 67$. None of the species ClO_2^-, BrO_3^-, or IO_3^- can produce $m/z = 83$ to be selected by Q1.

22-43. (a) Consider the term $A_x C_x m_x$, which applies to the unknown:

$A_x C_x m_x$

$$= \left(\frac{\mu\text{mol isotope A}}{\mu\text{mol isotope A} + \mu\text{mol isotope B}}\right)\left(\frac{\mu\text{mol V}}{\text{g unknown}}\right)(\text{g unknown})$$

$$= \left(\frac{\mu\text{mol isotope A}}{\mu\text{mol isotope A} + \mu\text{mol isotope B}}\right)(\mu\text{mol V})$$

$$= \left(\frac{\mu\text{mol isotope A}}{\mu\text{mol isotope A} + \mu\text{mol isotope B}}\right)(\mu\text{mol isotope A} + \mu\text{mol isotope B})$$

$$= \mu\text{mol isotope A in the unknown.}$$

Similarly, $B_x C_x m_x$ = μmol isotope B in the unknown, $A_s C_s ms_x$ = μmol isotope A in the spike, and $B_s C_s m_s$ = μmol isotope B in the unknown.

When we mix the unknown and the spike, the isotope ratio is

$$R = \frac{\mu\text{mol A}}{\mu\text{mol B}} = \frac{\mu\text{mol A in unknown} + \mu\text{mol A in spike}}{\mu\text{mol B in unknown} + \mu\text{mol B in spike}}$$

$$= \frac{A_x C_x m_x + A_s C_s m_s}{B_x C_x m_x + B_s C_s m_s}.$$

(b) Cross-multiplying Equation A gives

$$R(B_x C_x m_x + B_s C_s m_s) = A_x C_x m_x + A_s C_s m_s$$

$$R B_x C_x m_x + R B_s C_s m_s = A_x C_x m_x + A_s C_s m_s$$

$$R B_x C_x m_x - A_x C_x m_x = A_s C_s m_s - R B_s C_s m_s$$

$$C_x = \frac{A_s C_s m_s - R B_s C_s m_s}{R B_x m_x - A_x m_x} = \left(\frac{C_s m_s}{m_x}\right)\left(\frac{A_s - R B_s}{R B_x - A_x}\right)$$

(c) $A = {}^{51}V$ and $B = {}^{50}V$

Atom fractions in unknown: $A_x = 0.9975$ and $B_x = 0.0025$

Atom fractions in spike: $A_s = 0.6391$ and $B_s = 0.3609$

$$C_x = \left(\frac{C_s m_s}{m_x}\right)\left(\frac{A_s - R B_s}{R B_x - A_x}\right)$$

$$= \left(\frac{(2.2435\ \mu\text{mol V/g})(0.41946\ \text{g})}{0.40167\ \text{g}}\right)\left(\frac{0.6391 - (10.545)(0.3609)}{(10.545)(0.0025) - 0.9975}\right)$$

$$= 7.6394\ \mu\text{mol V/g}$$

(d) $$C_x = \left(\frac{(2.2435\ \mu\text{mol V/g})(0.41946\ \text{g})}{0.40167\ \text{g}}\right)\left(\frac{0.6391 - (10.545)(0.3609)}{(10.545)(0.0025 - 0.9975)}\right)$$

$$= \left(\frac{(2.2435\ \mu\text{mol V/g})(0.41946\ \text{g})}{0.40167\ \text{g}}\right)\left(\frac{0.6391 - 3.8057}{0.02636 - 0.9975}\right)$$

$$= (2.3429)\left(\frac{-3.166}{-0.9711}\right)$$

$$= 7.639\ \mu\text{mol V/g}$$

CHAPTER 23
INTRODUCTION TO ANALYTICAL SEPARATIONS

23-1. Three extractions with 100 mL are more effective than one extraction with 300 mL.

23-2. (a) No, the partition coefficient is unaffected. Shaking increases the rate at which partitioning reaches equilibrium, but does not affect the position of that equilibrium.

(b) Extraction requires mass transfer (i.e., physical movement) of the solute between the two phases. Shaking the separatory funnel forms small droplets of one phase in the other phase (dispersion). Formation of these small droplets shortens the physical distance that the solute must move to transfer from one phase to the other. Thus it enhances the mass transfer.

23-3. Adjust the pH to 3 so the acid is in its neutral form (CH_3CO_2H), rather than its anionic form ($CH_3CO_2^-$).

23-4. (a) The EDTA complex is anionic (AlY^-), whereas the 8-hydroxyquinoline complex is neutral (AlL_3).

(b) The EDTA complex is anionic (AlY^-), so we need a hydrophobic cation such as $(C_8H_{17})_3NH^+$ to try to bring hydrophobic AlY^- into the organic solvent.

23-5. The complexation reaction $mHL + M^{m+} \rightleftharpoons ML_m + mH^+$ is driven to the right at high pH by consumption of H^+. This consumption increases the fraction of metal in the form ML_m, which is extracted into organic solvent.

23-6. The form that is extracted into organic solvent is ML_n. The formation of ML_n is favored by increasing the formation constant (β). ML_n is also favored by increasing K_a, which increases the fraction of ligand in the form L^-. Increasing K_L decreases the fraction of ligand in the aqueous phase, thereby decreasing the formation of ML_n. Increasing $[H^+]$ decreases the concentration of L^- available for complexation.

23-7. When pH > pK_{BH^+}, the predominant form is B, which is extracted into the organic phase. When pH > pK_a for HA, the predominant form is A^-, which is extracted into the aqueous phase.

23-8. (a) $S_{H_2O} \rightleftharpoons S_{CHCl_3}$ $\qquad K = [S]_{CHCl_3}/[S]_{H_2O} = 4.0$

$\qquad [S]_{CHCl_3} = K[S]_{H_2O} = (4.0)(0.020\text{ M}) = 0.080\text{ M}$

(b) $\dfrac{\text{mol S in CHCl}_3}{\text{mol S in H}_2O} = \dfrac{(0.080\text{ M})(10.0\text{ mL})}{(0.020\text{ M})(80.0\text{ mL})} = 0.50$

23-9. Fraction remaining $= \left(\dfrac{V_1}{V_1 + KV_2}\right)^n = \left(\dfrac{80.0}{80.0 + (4.0)(10.0)}\right)^6 = 0.088$

23-10. (a) $D = \dfrac{\text{total conc of all forms in organic}}{\text{total conc of all forms in aqueous}}$

For a weak base, the neutral form B would be extracted into benzene, but the ionic form BH^+ would remain in water. Both B and BH^+ would be present in the aqueous phase.

$D = \dfrac{[B]_{C_6H_6}}{[B]_{H_2O} + [BH^+]_{H_2O}}$

(b) D (distribution coefficient) is the quotient of the total concentrations in the phases. Thus, it is a function of secondary equilibria such as acid/base reactions.

K (partition coefficient) is the quotient of the concentrations of the neutral species (B) in the phases. It is an intrinsic measure of the hydrophobicity of a compound.

(c) B has $K_b = 1.0 \times 10^{-5}$. Therefore, BH^+ has a $K_a = 1.0 \times 10^{-9}$ or $pK_a = 9.0$

$D = \dfrac{K \cdot K_a}{K_a + [H^+]} = \dfrac{(50.0)(1.0 \times 10^{-9})}{(1.0 \times 10^{-9}) + (1.0 \times 10^{-8})} = 4.5$

(d) D will be greater at pH 10 because a greater fraction of B is neutral.

23-11. From Equation 23-12, $D \approx \dfrac{[ML_n]_{org}}{[M^{n+}]_{aq}} = K_{extraction} \dfrac{[HL]^n_{org}}{[H^+]^n_{aq}}$

Comparing this result to Equation 23-13 gives $K_{extraction} = \dfrac{K_M \beta K_a^n}{K_L^n}$

Constant	Effect on $K_{extraction}$	Reason
K_M	increase	ML_n is more soluble in organic phase.
β	increase	Ligand binds metal more tightly and ML_n is the organic-soluble form.
K_a	increase	Ligand dissociates to L^- more easily, increasing ML_n formation.
K_L	decrease	HL is more soluble in organic phase, where it is not available to react with $M^{n+}(aq)$.

23-12. (a) $D = K[H^+]/([H^+] + K_a) = 3 \cdot 10^{-4.00}/(10^{-4.00} + 1.52 \times 10^{-5}) = 2.60$ at pH 4.00. Fraction remaining in water $= q = V_1/(V_1 + DV_2) = 100/[100 + 2.60(25)] = 0.606$. Therefore, the molarity in water is $0.606\,(0.10\text{ M}) = 0.060\,6$ M. The total moles of solute in the system is $(0.100\text{ L})(0.10\text{ M}) = 0.010$ mol. The fraction of solute in benzene is 0.394, so the molarity in benzene is $0.394\,(0.010\text{ mol})/0.025\text{ L} = 0.16$ M.

(b) At pH 10.0: $D = 1.97 \times 10^{-5}$, $q = 0.999\,995\,1$, molarity in water $= 0.10$ M, and molarity in benzene $= 2 \times 10^{-6}$ M.

23-13. $D = C/[H^+]^n$, where $C = K_M \beta K_a^n [HL]^n_{org}/K_L^n$
$D_1 = 0.01 = C/[H^+]_1^2$ and $D_2 = 100 = C/[H^+]_2^2$
$D_2/D_1 = 10^4 = [H^+]_1^2/[H^+]_2^2 \Rightarrow [H^+]_1/[H^+]_2 = 10^2 \Rightarrow \Delta pH = 2$ pH units

23-14. (a) Since there is so much more dithizone than Cu, it is safe to say that $[HL]_{org} = 0.1$ mM.

$$D = \dfrac{K_M \beta K_a^n}{K_L^n} \dfrac{[HL]^n_{org}}{[H^+]^n_{aq}} = \dfrac{(7 \times 10^4)(5 \times 10^{22})(3 \times 10^{-5})^2}{(1.1 \times 10^4)^2} \dfrac{(1 \times 10^{-4})^2}{[H^+]^2}$$

$= 2.6 \times 10^4$ at pH 1 and 2.6×10^{10} at pH 4

(b) $q = V_1/(V_1 + DV_2) = 100/[100 + 2.6 \times 10^4\,(10)] = 3.8 \times 10^{-4}$

23-15. (a) $D = \dfrac{[ML_2]_{org}}{[ML_2]_{aq}} = \dfrac{C_{org}V_{org}}{C_{aq}V_{aq}} \Rightarrow C_{org} = D\,C_{aq}\,\dfrac{V_{aq}}{V_{org}}$

$$\%\ \text{extracted} = \frac{100\,C_{org}}{C_{aq} + C_{org}} = \frac{100\,D\,C_{aq}\dfrac{V_{aq}}{V_{org}}}{C_{aq} + D\,C_{aq}\dfrac{V_{aq}}{V_{org}}} = \frac{100\,D\,\dfrac{V_{aq}}{V_{org}}}{1 + D\,\dfrac{V_{aq}}{V_{org}}}$$

(b) Spreadsheet for pH dependence of dithizone extraction

	A	B	C	D	E
1	K(M) =	pH	H	D = Dist.coeff	% extracted
2	70000	1	1.00E-01	2.60E-02	0.05
3	Beta =	2	1.00E-02	2.60E+00	4.95
4	5E+18	2.2	6.31E-03	6.54E+00	11.57
5	Ka =	2.4	3.98E-03	1.64E+01	24.73
6	0.00003	2.6	2.51E-03	4.13E+01	45.21
7	K(L) =	2.8	1.58E-03	1.04E+02	67.46
8	11000	3	1.00E-03	2.60E+02	83.89
9	[HL]org =	3.2	6.31E-04	6.54E+02	92.90
10	0.00001	3.4	3.98E-04	1.64E+03	97.05
11	V(org) =	3.6	2.51E-04	4.13E+03	98.80
12	2	3.8	1.58E-04	1.04E+04	99.52
13	V(aq) =	4	1.00E-04	2.60E+04	99.81
14	100	5	1.00E-05	2.60E+06	100.00
15					
16	C2 = 10^-B2				
17	D2 = ($A$2*$A$4*$A$6^2*$A$10^2)/($A$8^2*C2^2)				
18	E2 = (D2*$A$12/$A$14)/(1+(D2*$A$12/$A$14))*100				

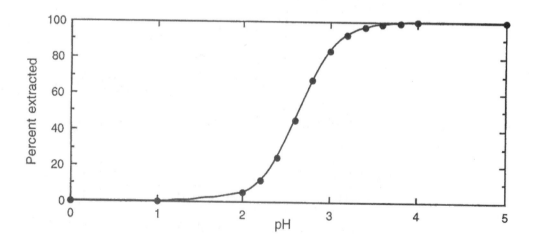

23-16.

	A	B	C	D	E	F
1	Liquid-liquid extraction efficiency					
2						
3	V_2 =	50	mL (volume of extraction solvent)			
4	V_1 =	50	mL (volume to be extracted)			
5	K =	2	(partition coefficient = $[S]_2/[S]_1$)			
6	Divide V_2 into n equal portions for n extractions					
7	Theoretical maximum fraction extracted = $1-q_{limit}$ = $1-\exp([V_2/V_1]K)$					
8		$1-q_{limit}$ =	0.864665			
9						
10		V_2/n				
11		individual	q =	1-q =	% of limiting	
12		extraction	fraction	fraction	fraction	
13	n	volume	remaining	extracted	extracted	
14	1	50.0	0.333	0.667	77.1	
15	2	25.0	0.250	0.750	86.7	
16	3	16.7	0.216	0.784	90.7	
17	4	12.5	0.198	0.802	92.8	
18	5	10.0	0.186	0.814	94.1	
19	6	8.3	0.178	0.822	95.1	
20	7	7.1	0.172	0.828	95.7	
21	8	6.3	0.168	0.832	96.2	
22	9	5.6	0.164	0.836	96.6	
23	10	5.0	0.162	0.838	97.0	
24	C14 = (\$B\$4/(\$B\$4+B14*\$B\$5))^A14					
25	q = $[V_1/(V_1 + (V_2/n)K)]^n$					

The theoretical limit for fraction extracted is in cell C8. 95% of the theoretical fraction extracted is $(0.95)(0.864\ 6) = 0.821\ 4$. This fraction is exceeded with $n = 6$ equal extractions.

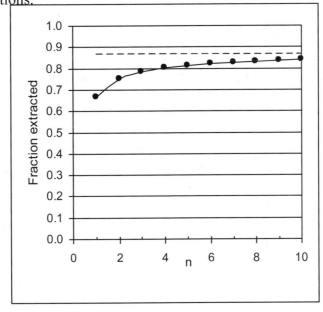

23-17.   1-C, 2-D, 3-A, 4-E, 5-B

23-18.   The larger the partition coefficient, the greater the fraction of solute in the stationary phase, and the smaller the fraction that is moving through the column.

23-19.   (a) $k = \dfrac{\text{time solute spends in stationary phase}}{\text{time solute spends in mobile phase}} = \dfrac{t_r - t_m}{t_m} = \dfrac{t_s}{t_m}$

(b) Fraction of time in mobile phase $= \dfrac{t_m}{t_m + t_s} = \dfrac{t_m}{t_m + kt_m} = \dfrac{1}{1 + k}$

(c) $R = \dfrac{t_m}{t_r} = \dfrac{t_m}{t_m + t_s} = \dfrac{1}{1 + k}$. Parts (b) and (c) together tell us that

$\dfrac{\text{time for solvent to pass through column}}{\text{time for solute to pass through column}} = \dfrac{\text{time spent by solute in mobile phase}}{\text{total time on column}}$

23-20.   (a) Volume per cm of length $= \pi r^2 \times \text{length} = \pi \left(\dfrac{0.461 \text{ cm}}{2}\right)^2 (1 \text{ cm}) = 0.167 \text{ mL}$

mobile phase volume $= (0.390)(0.167 \text{ mL}) = 0.065\,1 \text{ mL per cm of column}$

linear velocity $= u_x = \dfrac{1.13 \text{ ml/min}}{0.065\,1 \text{ mL/cm}} = 17.4 \text{ cm/min}$

(b) $t_m = (10.3 \text{ cm}) / (17.4 \text{ cm/min}) = 0.592 \text{ min}$

(c) $k = \dfrac{t_r - t_m}{t_m} \Rightarrow t_r = kt_m + t_m = 10(0.592) + 0.592 = 6.51 \text{ min}$

23-21.   (a) Linear velocity $= (30.1 \text{ m}) / (2.16 \text{ min}) = 13.9 \text{ m/min}$.

Inner diameter of open tube $= 530 \text{ μm} - 2(3.1 \text{ μm}) = 523.8 \text{ μm}$

$\Rightarrow$ radius $= 261.9 \text{ μm}$.

Volume $= \pi r^2 \times \text{length} = \pi (261.9 \times 10^{-4} \text{ cm})^2 (30.1 \times 10^2 \text{ cm}) = 6.49 \text{ mL}$

Volume flow rate $= (6.49 \text{ mL}) / (2.16 \text{ min}) = 3.00 \text{ mL/min}$

(b) $k = \dfrac{t_r - t_m}{t_m} = \dfrac{17.32 - 2.16}{2.16} = 7.02$

$k = t_s/t_m$ (where $t_s = $ time in stationary phase)

Fraction of time in stationary phase $= \dfrac{t_s}{t_s + t_m} = \dfrac{kt_m}{kt_m + t_m} =$

$\dfrac{k}{k + 1} = \dfrac{7.02}{7.02 + 1} = 0.875$

(c) Volume of coating $\approx 2\pi r \times$ thickness $\times$ length

$$= 2\pi \, [(261.9 + 1.55) \times 10^{-4} \text{ cm}](3.1 \times 10^{-4} \text{ cm})(30.1 \times 10^2 \text{ cm}) = 0.154 \text{ mL}$$

$$k = K \frac{V_s}{V_m} \Rightarrow 7.02 = K \frac{0.154 \text{ mL}}{6.49 \text{ mL}} \Rightarrow K = \frac{c_s}{c_m} = 295$$

23-22. (a) $\dfrac{\text{Large load}}{\text{Small load}} = \left(\dfrac{\text{Large column radius}}{\text{Small column radius}}\right)^2$

$\dfrac{100 \text{ mg}}{4.0 \text{ mg}} = \left(\dfrac{\text{Large column diameter}}{0.85 \text{ cm diameter}}\right)^2 \Rightarrow$ large column diameter = 4.25 cm

Use a 40-cm-long column with a diameter near 4.25 cm.

(b) The linear velocity should be the same. Since the cross-sectional area of the column is increased by a factor of 25, the volume flow rate should be increased by a factor of 25 $\Rightarrow u_v = 5.5$ mL/min.

(c) Volume of small column $= \pi r^2 \times$ length $= \pi(0.85/2 \text{ cm})^2(40 \text{ cm}) = 22.7$ mL

Mobile phase volume = 35% of column volume = 7.94 mL

$$\text{Linear velocity} = \frac{40 \text{ cm}}{(7.94 \text{ mL})/(0.22 \text{ mL/min})} = 1.11 \text{ cm/min for both}$$

columns

23-23. (a) $k = \dfrac{9.0 - 3.0}{3.0} = 2.0$

(b) Fraction of time solute is in mobile phase $= \dfrac{t_m}{t_r} = \dfrac{3.0}{9.0} = 0.33$

(c) $K = k \dfrac{V_m}{V_s} = (2.0) \dfrac{V_m}{0.10 \, V_m} = 20$

23-24. Solvent volume per cm of column length $= (0.15)\pi\left(\dfrac{0.30 \text{ cm}}{2}\right)^2 = 0.010\,6$ mL/cm.

A volume flow rate of 0.20 mL/min corresponds to a linear velocity of $\left(\dfrac{0.20 \text{ mL/min}}{0.010\,6 \text{ mL/cm}}\right) = 19$ cm/min.

23-25. $k = K \dfrac{V_s}{V_m} = 3\left(\dfrac{1}{5}\right) = \dfrac{3}{5}$ For $K = 30$, $k = 30\left(\dfrac{1}{5}\right) = 6$.

23-26. $k = \dfrac{V'_r}{V_m} = \dfrac{V_r - V_m}{V_m} = \dfrac{76.2 - 16.6}{16.6} = 3.59$

$K = k\dfrac{V_m}{V_s} = (3.59)\dfrac{16.6}{12.7} = 4.69$

23-27. $K = k \dfrac{V_m}{V_s}$

$k = \dfrac{t_r - t_m}{t_m} = \dfrac{433 - 63}{63} = 5.87$

$\dfrac{V_m}{V_s} = \dfrac{\cancel{\pi} r^2 \times \cancel{\text{length}}}{2 \cancel{\pi} r \times \text{thickness} \times \cancel{\text{length}}} = \dfrac{(103)^2}{2(103.25) \times 0.5} = 102.8$

(In the numerator, r refers to the radius of the open tube $= {}^1\!/_2 \,(207 - 1.0)$ μm $= 103$ μm. In the denominator, r is the radius at the center of the stationary phase, which is $103 + \frac{1}{2}\,(0.5) = 103.25$ μm.)

Therefore, the partition coefficient is $K = k \dfrac{V_m}{V_s} = 5.87\,(102.8) = 603$.

Fraction of time in stationary phase $= \dfrac{t_s}{t_s + t_m} = \dfrac{k t_m}{k t_m + t_m} = \dfrac{k}{k + 1}$

$= \dfrac{5.87}{5.87 + 1} = 0.854$

23.28. (a) Column 2.

(b) Column 3.

(c) Column 3.

(d) Column 4 since it provides the highest resolution.

23-29. Resolution $= \dfrac{\sqrt{N}}{4}\,\dfrac{(\alpha - 1)}{\alpha}\,\dfrac{k_2}{(1 + k_2)}$. Plots relating the individual factors N, α, and k_2 versus resolution are shown below.

Resolution = $(\sqrt{N})/4 \times ((\alpha - 1)/\alpha) \times k_2/(1+k_2)$

N	resolution $\propto \sqrt{N}$		α	resolution $\propto (\alpha - 1)/\alpha$		k_2	resolution $\propto k_2/(1 + k_2)$
0	0.0		1.00	0.00		0	0.000
1000	31.6		1.05	0.05		1.5	0.600
2000	44.7		1.10	0.09		3.0	0.750
3000	54.8		1.15	0.13		4.5	0.818
4000	63.2		1.20	0.17		6.0	0.857
5000	70.7		1.25	0.20		7.5	0.882
6000	77.5		1.30	0.23		9.0	0.900
7000	83.7		1.35	0.26		10.5	0.913
8000	89.4		1.40	0.29		12.0	0.923
9000	94.9		1.45	0.31		13.5	0.931
10000	100.0		1.50	0.33		15.0	0.938

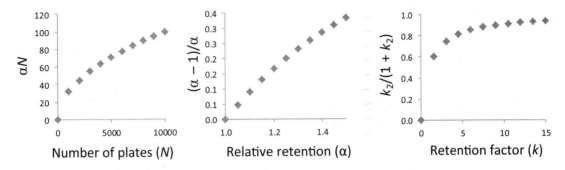

Key observations:

- Resolution increases with the number of plates, but only gradually due to the square root relationship.

- Resolution increases near proportionally to the amount that the relative retention is greater than 1.

- Resolution increases dramatically with a little retention (e.g., up to $k_2=2$), and then plateaus at large ($k_2>10$) retention factors. Large retention factors mean long separation times. Therefore chromatography method development tries to get all compounds in the mixture to have $0.5 \leq k \leq 10$.

23-30. (a) Based on the Stokes-Einstein equation, diffusion coefficients are inversely dependent on the radius of the diffusing molecule. CH_3OH (FM = 32) is smaller than sucrose (FM = 342), and so would be expected to diffuse faster.

(b) Diffusion coefficients in gases are about 10^4 times those in liquids. The diffusion coefficient of H_2O in water is 2.3×10^{-9} m^2/s. The diffusion coefficient of H_2O in air is approximately 10^{-5} m^2/s.

23-31. (a) After 10 cycles, the compounds have passed through a length $10L$ containing $10N$ theoretical plates. We are told that ⊠ = 1.03 and $k_2 = 1.62$.

$$\text{resolution} = \frac{\sqrt{N}}{4} \frac{(\alpha - 1)}{\alpha} \frac{k_2}{(1 + k_2)}$$

$$1.60 = \frac{\sqrt{10N}}{4} \frac{(1.03 - 1)}{1.03} \frac{1.62}{(1 + 1.62)} \Rightarrow N = 1.2_6 \times 10^4$$

(b) Plate height $= H = L/N = 25 \text{ cm}/1.2_6 \times 10^4 = 2.0 \times 10^{-3}$ cm $= 20$ μm

(c) Resolution is proportional to $\sqrt{N}$ or $\sqrt{\text{number of cycles}}$

$$\frac{\text{resolution in 2 cycles}}{\text{resolution in 10 cycles}} = \sqrt{\frac{2}{10}} = 0.447$$

resolution in 2 cycles = 0.447(resolution in 10 cycles) = 0.447(1.6) = 0.72

(observed resolution = 0.71)

23-32.   (a) Column 1 (sharper peaks)

(b) Column 2 (large plate height (H) means fewer plates (N) means broader peaks)

(c) Column 1 (less overlap between peaks because they are sharper)

(d) Neither (relative retention, $\alpha = (t'_{rB}/t'_{rA})$) is equal for the two columns

(e) Compound B (longer retention time)

(f) Compound B (longer retention time means greater affinity for stationary phase)

(g) $k_A = \dfrac{8.0 - 1.3}{1.3} = 5.2$

(h) $k_A = \dfrac{10.0 - 1.3}{1.3} = 6.7$

(i) $\alpha = \dfrac{10.0 - 1.3}{8.0 - 1.3} = 1.3$

23-33.   (a) Increasing flow rate decreases the time for longitudinal diffusion to occur, thereby narrowing the bands, decreasing H, increasing N, and increasing resolution.

(b) Decreasing flow rate will narrow the bands and increase resolution.

(c) Multiple flow paths broadening depends on the geometry of the column and the quality of the packing, but does not depend on flow rate. So H, N, and resolution will remain unchanged.

23-34.   The linear velocity at which solution goes past the stationary phase determines how completely the equilibrium between the two phases is established. This determines the size of the mass transfer term (Cu_x) in the van Deemter equation. The extent of longitudinal diffusion depends on the time spent on the column, which is inversely proportional to linear velocity.

23-35.   Smaller plate height gives less band spreading: 0.1 mm

23-36.   Diffusion coefficients of gases are 10^4 times greater than those of liquids. Therefore, longitudinal diffusion occurs much faster in gas chromatography than in liquid chromatography.

23-37. The optimum flow rate is that at which longitudinal diffusion (B/u_x) and mass transfer broadening (Cu_x) are equal. The smaller the particle size, the more rapid is equilibration between mobile and stationary phases, and so the smaller is Cu_x. Hence the optimum flow rate will be at a higher flow rate so that B/u_x is decreased to match the smaller Cu_x.

23-38. Minimum plate height is at 33 mL/min.

23-39. 1-C_m, 2-A, 3-EC, 4-B, 5-C_s

23-40. Silanization caps hydroxyl groups to which strong hydrogen bonding can occur.

23-41. Isotherms and band shapes are given in Figure 23-21. When a column is overloaded, the excessive amount of solute injected alters the nature of the stationary phase.

In gas chromatography overloading, fronting is observed as the solute becomes more soluble in the stationary phase as solute concentration increases. Consider injection of high concentrations of an alcohol such as hexanol onto a nonpolar stationary phase like polydimethylsiloxane. Partitioning of hexanol into polydimethylsiloxane makes the stationary phase more polar, and thus more attractive for partitioning of more hexanol. This change in polarity of the solute zone in the stationary phase results in greater retention of the high concentration of hexanol, and gives a non-Gaussian fronting shape.

In liquid chromatography overloading, tailing is observed as the excess injected solute saturates a portion of the retention sites on the column. As more excess solute is injected, more of the retention sites are saturated and the peak maximum becomes less retained, and gives a non-Gaussian tailing shape.

23-42. (a) Injection of a larger volume of sample will require a plug width (Δt). This increases the variance caused by injection ($=(\Delta t)^2/12$), which will increase the peak width and thus lower the resolution.

(b) With 5.0 mg, the column may be overloaded. That is, the quantity of solute may be too great for the capacity of the stationary phase. Depending on the nature of the interaction, this could lead to the upper nonlinear isotherm in Figure 23-21 which causes fronting or to the lower nonlinear isotherm which causes tailing. Either type of overload broadens bands and decreases resolution.

23-43. Equation 23-28 says that the standard deviation of the band is proportional to $\sqrt{t}$. Here is what we know of the rate of diffusion:

time	standard deviation
t_1	$\sigma_1 = 1$
$t_2 = t_1 + 20$	$\sigma_2 = 2$
$t_3 = t_1 + 40$	$\sigma_3 = ?$

From the bandwidths at times t_1 and t_2, we can write

$$\frac{\sigma_2}{\sigma_1} = \sqrt{\frac{t_2}{t_1}} \Rightarrow \frac{2}{1} = \sqrt{\frac{t_1 + 20}{t_1}} \Rightarrow t_1 = \frac{20}{3} \text{ min}$$

For time t_3: $\quad \dfrac{\sigma_3}{\sigma_1} = \sqrt{\dfrac{t_3}{t_1}} \Rightarrow \dfrac{\sigma_3}{1} = \sqrt{\dfrac{\frac{20}{3} + 40}{\frac{20}{3}}} \Rightarrow \sigma_3 = 2.65 \text{ mm}$

23-44. (a) $N = \dfrac{5.55\, t_r^2}{w_{1/2}^2} = \dfrac{5.55\, (9.0 \text{ min})^2}{(2.0 \text{ min})^2} = 1.1_2 \times 10^2 \text{ plates}$

(b) $(10 \text{ cm})/(1.1_2 \times 10^2 \text{ plates}) = 0.89 \text{ mm}$

23-45. (a) Measurement with a ruler gives $B/A = 2.1$

(b) $N = \dfrac{41.7\, (t_r/w_{0.1})^2}{(B/A) + 1.25} = \dfrac{41.7\, (900 \text{ s}/44 \text{ s})^2}{2.1 + (1.25)} = 5.2 \times 10^3 \text{ plates}$

(c) To use the equation $N = (t_r/\sigma)^2$, we need to find the standard deviation of the peak. The width at 1/10 height is $22 + 22 = 44$ s, which we are told is equal to 4.297σ. Therefore, $\sigma = (44 \text{ s})/4.297 = 10.24$ s. $N = (t_r/\sigma)^2 = (900 \text{ s}/10.24 \text{ s})^2 = 7.72 \times 10^3$ plates.

The equation for an asymmetric peak from (b) gives

$$N = \frac{41.7\, (t_r/w_{0.1})^2}{(B/A) + 1.25} = \frac{41.7\, (900 \text{ s}/44 \text{ s})^2}{(22 \text{ s}/22 \text{ s}) + (1.25)} = 7.75 \times 10^3 \text{ plates}$$

23-46. (a) Resolution $= \dfrac{\Delta t_r}{w} = \dfrac{5 \text{ min}}{6 \text{ min}} = 0.83$. This is most like the diagram for a resolution of 0.75.

(b) At resolution $=1.5$ the valley between the peaks does not quite return to the baseline signal. The baseline separation described has resolution > 1.5.

23-47. Since $w = 4V_r/\sqrt{N}$, w is proportional to V_r (if N is constant).

$w_2/w_1 = V_2/V_1 = 127/49 \Rightarrow w_2 = (127/49)(4.0) = 10.4 \text{ mL}$.

23-48. $w_{1/2} = (39.6 \text{ s} / 60 \text{ s/min}) \times 0.66 \text{ mL/min} = 0.436 \text{ mL}$

$$\sigma_{obs}^2 = \left(\frac{w_{1/2}}{2.35}\right)^2 = \left(\frac{0.436 \text{ mL}}{2.35}\right)^2 = 0.034\ 4 \text{ mL}^2$$

$$\sigma_{injection}^2 = \frac{\Delta V_{injection}^2}{12} = \frac{(0.40 \text{ mL})^2}{12} = 0.013\ 3 \text{ mL}^2$$

$$\sigma_{detector}^2 = (\Delta V)_{detector}^2/12 = (0.25 \text{ mL})^2/12 = 0.005\ 2 \text{ mL}^2$$

From Table 23-1, sucrose has $D = 0.52 \times 10^{-9} \text{ m}^2/\text{s} = 3.12 \times 10^{-4} \text{ cm}^2/\text{min}$

$$\sigma_{tubing}^2 = \frac{\pi d_t^4 l_t u_v}{384 D} = \frac{\pi (0.050 \text{ cm})^4 (20 \text{ cm})(0.66 \text{ cm}^3/\text{min})}{384(3.12 \times 10^{-4} \text{ cm}^2/\text{min})} = 0.002\ 2 \text{ cm}^6$$

$$= 0.002\ 2 \text{ mL}^2$$

$$\sigma_{obs}^2 = \sigma_{column}^2 + \sigma_{injection}^2 + \sigma_{detector}^2 + \sigma_{tubing}^2$$

$$0.034\ 4 \text{ mL}^2 = \sigma_{column}^2 + 0.013\ 3 \text{ mL}^2 + 0.005\ 2 \text{ mL}^2 + 0.002\ 2 \text{ mL}^2$$

$$\Rightarrow \sigma_{column} = 0.117 \text{ mL}$$

$$w_{1/2} = 2.35\ \sigma_{column} = 0.275 \text{ mL}$$

$$w_{1/2} = (0.275 \text{ mL})(60 \text{ s/min})/0.66 \text{ mL/min} = 25.0 \text{ s}$$

23-49. $\alpha = \dfrac{t_{r2}'}{t_{r1}'} = \dfrac{k_2}{k_1} = \dfrac{K_2}{K_1} = \dfrac{18}{15} = 1.2_0$

$$k_2 = K_2 \frac{V_s}{V_m} = 18\left(\frac{1}{3.0}\right) = 6.0$$

$$\text{Resolution} = \frac{\sqrt{N}}{4} \frac{(\alpha - 1)}{\alpha} \frac{k_2}{(1 + k_2)}$$

$$1.5 = \frac{\sqrt{N}}{4} \frac{(1.2_0 - 1)}{1.2_0} \frac{6.0}{(1 + 6.0)} = \frac{\sqrt{N}}{4}(0.14_3) \Rightarrow 1.8 \times 10^3 \text{ plates}$$

23-50. (a) $\text{Resolution} = 2.0 = \dfrac{\sqrt{N}}{4} \dfrac{(1.05 - 1)}{1.05} \dfrac{5.0}{(1 + 5.0)} \Rightarrow N = 4.1 \times 10^4 \text{ plates}$

(b) $\text{Resolution} = 2.0 = \dfrac{\sqrt{N}}{4} \dfrac{(1.10 - 1)}{1.10} \dfrac{5.0}{(1 + 5.0)} \Rightarrow N = 1.1 \times 10^4 \text{ plates}$

(c) $\text{Resolution} = 2.0 = \dfrac{\sqrt{N}}{4} \dfrac{(1.05 - 1)}{1.05} \dfrac{10.0}{(1 + 10.0)} \Rightarrow N = 3.4 \times 10^4 \text{ plates}$

(d) N can be increased by increasing the column length ($N \propto L$), by decreasing the capillary radius (r) in an open tubular column, or by adjusting the flow rate so that H is minimized. α can be increased by changing solvent in liquid chromatography or by changing the stationary phase in both liquid and gas chromatography. k_2 can be increased by increasing the volume of the stationary phase in an open tubular gas chromatography column or increasing the surface area in a packed liquid chromatography column.

23-51. (a) C_6HF_5: $t'_r = 12.98 - 1.06 = 11.92$ min. $k = 11.92/1.06 = 11.25$

C_6H_6: $t'_r = 13.20 - 1.06 = 12.14$ min. $k = 12.14/1.06 = 11.45$

(b) $\alpha = 11.24/11.25 = 1.018$

(c) $w_{1/2}$ $(C_6HF_5) = 0.124$ min; $w_{1/2}$ $(C_6H_6) = 0.121$ min

C_6HF_5: $N_1 = \dfrac{5.55\, t_r^2}{w_{1/2}^2} = \dfrac{5.55\, (12.98)^2}{0.124^2} = 6.08 \times 10^4$ plates

Plate height $= \dfrac{30.0\text{ m}}{6.08 \times 10^4 \text{ plates}} = 0.493$ mm

C_6H_6: $N_2 = \dfrac{5.55\, (13.20)^2}{0.121^2} = 6.60 \times 10^4$ plates

Plate height $= \dfrac{30.0\text{ m}}{6.60 \times 10^4 \text{ plates}} = 0.455$ mm

(d) w $(C_6HF_5) = 0.220$ min; w $(C_6H_6) = 0.239$ min

C_6HF_5: $N_1 = \dfrac{16\, t_r^2}{w^2} = \dfrac{16\, (12.98)^2}{0.220^2} = 5.57 \times 10^4$ plates

C_6H_6: $N_2 = \dfrac{16\, (13.20)^2}{0.239^2} = 4.88 \times 10^4$ plates

(e) Resolution $= \dfrac{\Delta t_r}{w_{av}} = \dfrac{13.20 - 12.98}{0.229} = 0.96$

(f) $N_{av} = \sqrt{(5.57 \times 10^4)(4.88 \times 10^4)} = 5.21 \times 10^4$ plates

Resolution $= \dfrac{\sqrt{N}}{4} \dfrac{(\alpha - 1)}{\alpha} \dfrac{k_2}{(1 + k_2)} = \dfrac{\sqrt{5.21 \times 10^4}}{4} \dfrac{(1.018 - 1)}{1.018} \dfrac{11.45}{(1 + 11.45)} =$

0.93

23-52. Concentration ($c = \text{mol/m}^3$) is computed as a function of distance (x) and time (t) from the center of the band with the equation

$$c = \dfrac{m}{\sqrt{4\pi Dt}}\, e^{-x^2/(4Dt)}$$

where D is the diffusion coefficient (m^2/s) and initial concentration per unit *area* (m) = 10 nmol/(1.96×10^{-3} m^2) = 5.09×10^{-6} mol/m^2. Diffusion will be symmetric about the origin. Only diffusion in the positive direction is computed below for $t = 60$ s. Other conditions in the graphs are obtained by changing t and the diffusion coefficient D.

	A	B	C
1	Diffusion problem		
2		x (m)	c(mol/m^3)
3	moles =	0	4.637E-03
4	1.00E-08	0.0001	4.518E-03
5	diameter (m) =	0.0002	4.178E-03
6	0.05	0.0003	3.668E-03
7	x-sectional area (m^2)	0.0004	3.057E-03
8	0.001963495	0.0005	2.418E-03
9	m (mol/m^2)=	0.0006	1.816E-03
10	5.093E-06	0.0007	1.294E-03
11	D (m^2/s) =	0.0008	8.758E-04
12	1.600E-09	0.0009	5.625E-04
13	t (s) =	0.001	3.430E-04
14	60	0.0012	1.090E-04
15		0.0014	2.815E-05
16		0.0016	5.901E-06
17		0.0018	1.004E-06
18	A10 = A4/A8	0.002	1.388E-07
19			
20	C3 = ($A$10/(SQRT(4*PI()*$A$12*$A$14)))		
21		*EXP(-(B3^2)/(4*$A$12*$A$14))	

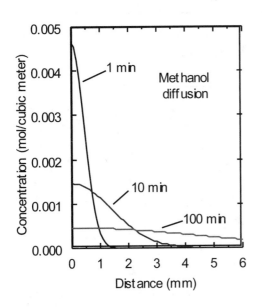

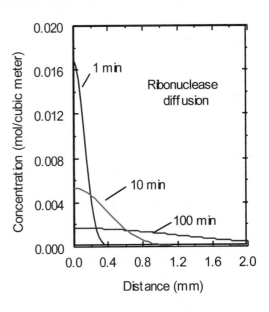

23-53. Plate height $= H_{\text{longitudinal diffusion}} + H_{\text{mass transfer}} = \dfrac{B}{u_x} + (C_s + C_m)u_x$

$$= \frac{2D_m}{u_x} + \left(\frac{2kd^2}{3(k+1)^2 D_s} + \frac{1 + 6k + 11k^2 r^2}{24(k+1)^2 D_m} \right) u_x$$

Parameters for 0.25 μm thick stationary phase:

$D_m = 1.0 \times 10^{-5}$ m^2/s $\qquad\qquad D_s = 1.0 \times 10^{-9}$ m^2/s

$d = 2.5 \times 10^{-7}$ m $\qquad\qquad\quad r = 12.5 \times 10^{-4}$ m

$k = 10$

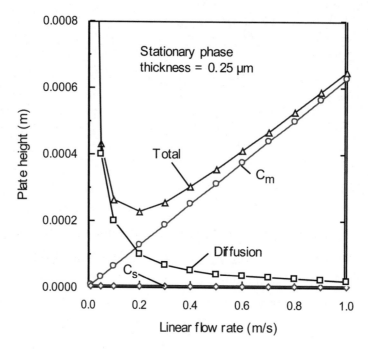

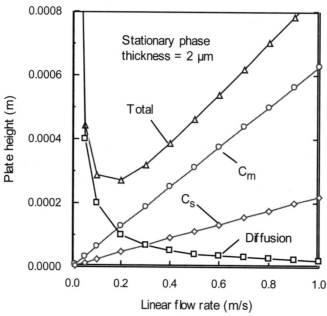

	A	B	C	D	E	F
1	Plate height calculation for 0.25-μm-thick stationary phase					
2				H(mass transfer)		
3	D_m =	u_x (m/s)	H(diffusion)	C_s term	C_m term	H (total)
4	0.00001	0.01	2.00E-03	3.44E-08	6.25E-06	2.01E-03
5	D_s =	0.05	4.00E-04	1.72E-07	3.12E-05	4.31E-04
6	1E-09	0.1	2.00E-04	3.44E-07	6.25E-05	2.63E-04
7	k =	0.2	1.00E-04	6.89E-07	1.25E-04	2.26E-04
8	10	0.3	6.67E-05	1.03E-06	1.87E-04	2.55E-04
9	d (m) =	0.4	5.00E-05	1.38E-06	2.50E-04	3.01E-04
10	2.50E-07	0.5	4.00E-05	1.72E-06	3.12E-04	3.54E-04
11	r (m) =	0.6	3.33E-05	2.07E-06	3.75E-04	4.10E-04
12	1.25E-04	0.7	2.86E-05	2.41E-06	4.37E-04	4.68E-04
13		0.8	2.50E-05	2.75E-06	5.00E-04	5.27E-04
14		0.9	2.22E-05	3.10E-06	5.62E-04	5.88E-04
15		1	2.00E-05	3.44E-06	6.25E-04	6.48E-04
16	C4 = 2*$A$4/B4					
17	D4 = 2*$A$8*$A$10^2*B4/(3*($A$8+1)^2*$A$6)					
18	E4 = (1+6*$A$8+11*$A$8^2)*$A$12^2*B4/(24*($A$8+1)^2*$A$4)					
19	F4 = C4+D4+E4					

For stationary phase thickness = 0.25 μm, plate height contribution from mass transfer in the stationary phase is negligible, as shown in the first graph. If the stationary phase is 2.0 μm thick, plate height from mass transfer in the stationary phase is not negligible, but it is still less than plate height from mass transfer in the mobile phase. C_s and total plate height in the second graph are greater than in the first graph. C_m and longitudinal diffusion terms are unaffected.

23-54. Inspection of Equation 4-3 shows that the general form of a Gaussian curve is $y = Ae^{-(x-x_0)^2/2\sigma^2}$, where A is a constant proportional to the area under the curve, x_0 is the abscissa of the center of the peak, and σ is the standard deviation. We can arbitrarily let $\sigma = 1$, which means that the width at the base ($w = 4\sigma$) is 4. A peak with an area of 1 centered at the origin is $y = 1*e^{-(x)^2/2}$. A curve of area 4 is $y = 4*e^{-(x-x_0)^2/2}$. The resolution is $\Delta x/w$. For a resolution of 0.5, $\Delta x = 0.5*w = 2$. That is, the second peak is centered at $x = 2$ if the resolution is 0.5. Its equation is $y = 4*e^{-(x-2)^2/2}$. Similarly, for a resolution of 1, $\Delta x = 1*w = 4$ and the second peak is centered at $x = 4$. For a resolution of 2, the second peak is centered at $x = 8$. The equations of the curves plotted below are:

Resolution = 0.5: $y = 1*e^{-(x)^2/2} + 4*e^{-(x-2)^2/2}$

Resolution = 1: $y = 1*e^{-(x)^2/2} + 4*e^{-(x-4)^2/2}$

Resolution = 2: $y = 1*e^{-(x)^2/2} + 4*e^{-(x-8)^2/2}$

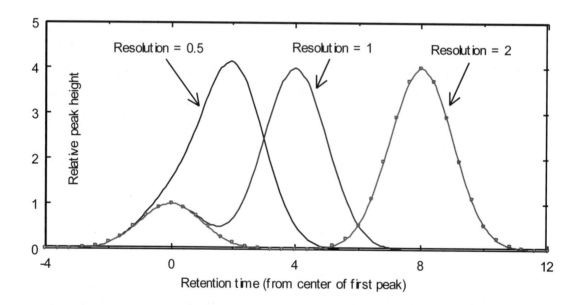

CHAPTER 24
GAS CHROMATOGRAPHY

24-1. (a) Packed columns offer high sample capacity, while open tubular columns give better separation efficiency (smaller plate height), shorter analysis time, and increased sensitivity to small quantities of analyte.

(b) Wall-coated: liquid stationary phase bonded to the wall of column
Porous-layer: solid stationary phase on wall of column

(c) Bonding or cross-linking the stationary phase reduces the tendency for the stationary phase to bleed from the column during use.

24-2. Resolution is proportional to $\sqrt{N}$. $N = L/H$. Open tubular columns eliminate the multiple path term (A) from the van Deemter equation, decreasing plate height H and increasing the number of plates, N. Also, the lower resistance to gas flow allows longer columns L to be used with the same elution time.

24-3. (a) Advantages: A narrower column increases the rate of mass transfer between the mobile and stationary phases by decreasing the time needed for solute in the mobile phase to diffuse to the stationary phase. The increased rate of mass transfer lowers plate height H, which increases the number of plates N and increases resolution. With lower plate height, a shorter column could be used to obtain the same resolution in a shorter time.
Disadvantages: lower sample capacity; requires high sensitivity detector (not mass spectrometry).

(b) Advantages: larger N and greater resolution.
Disadvantage: longer analysis time.

(c) Advantages: greater retention; greater sample capacity.
Disadvantages: lower N (because the rate of mass transfer in the stationary phase decreases); long retention times for high-boiling point compounds; increased bleeding.

24-4. (a) Poly(dimethylsiloxane) is nonpolar. Based on "like dissolves like," it is most suitable for nonpolar solutes such as nonpolar organic solvents or petroleum products.

(b) Poly(ethylene glycol) is highly polar and so suited for separation of amines and fatty acid methyl esters.

(c) Porous layer open tubular columns are used for separation of compounds such as O_2, N_2, or CO that are too volatile to be adequately retained by a wall coated open tubular column.

24-5. (a) Advantages: Compounds with a wide range of retention characteristics (such as a mixture of low and high boiling compounds of similar polarity) can all have sufficient retention for resolution and yet all be eluted in a reasonable time. Peaks are sharper, which improves sensitivity. Low initial temperature allows use of splitless or on-column injection.

Disadvantages: There may be a sloping baseline if there is stationary phase bleed (decomposition or evaporation of stationary phase at high temperature). Time is needed to cool the column oven between separations.

(b) Higher pressure gives higher flow rate. If pressure is increased during a separation, retention times of late-eluting peaks are reduced. The effect is the same as increasing temperature, but high temperatures are not required. Pressure programming reduces the likelihood of decomposing thermally sensitive compounds.

24-6. (a) Ideal carrier gas should be: inert (not react with samples); low viscosity; compatible with detector.

(b) Diffusion of solute in H_2 and He is more rapid than in N_2. Therefore, equilibration of solute between mobile phase and stationary phase is faster.

24-7. (a) Split injection is the ordinary mode for open tubular columns. It is best for high concentrations of analyte, gas analysis, high resolution, and dirty samples (with an adsorbent packing in the injection liner). Splitless injection is useful for trace analysis (dilute solutions) and for compounds with moderate thermal stability. On-column injection is best for quantitative analysis and for thermally sensitive solutes that might decompose during a high-temperature injection.

(b) In solvent trapping, the initial column temperature is low enough to condense solvent at the beginning of the column. Solute is very soluble in the solvent and is trapped in a narrow band at the start of the column. In cold trapping, the initial column temperature is 150°C lower than the boiling points of solutes, which condense in a narrow band at the start of the column. In both cases, elution occurs as the column temperature is raised.

24-8.   (a) All analytes

      (b) Carbon atoms bearing hydrogen atoms

      (c) Molecules with halogens, CN, NO_2, conjugated C=O

      (d) P and S and other elements selected by wavelength

      (e) P and N (and to a much lesser extent hydrocarbons)

      (f) Aromatic and unsaturated compounds

      (g) S

      (h) Most elements (selected individually by wavelength)

      (i) All analytes

      (j) All analytes

24-9.   The thermal conductivity detector measures changes in the thermal conductivity of the gas stream exiting the column. Any substance other than the carrier gas will change the conductivity of the gas stream. Therefore, the detector responds to all analytes. The flame ionization detector burns eluate in an H_2/O_2 flame to create CH radicals from carbon atoms (except carbonyl and carboxyl carbons), which then go on to be ionized to a small extent in the flame: $CH + O \rightarrow CHO^+ + e^-$. Most other kinds of molecules do not create ions in the flame and are not detected.

24-10.   A *reconstructed total ion chromatogram* is created by summing all ion intensities (above a selected value of *m/z*) in each mass spectrum at each time interval during a chromatography experiment. The technique responds to essentially everything eluted from the column and has no selectivity at all.

In *selected ion monitoring*, intensities at just one or a few values of *m/z* are plotted versus elution time. Only species with ions at those *m/z* values are detected, so the selectivity is much greater than that of the reconstructed total ion chromatogram. The signal-to-noise ratio is increased because ions are collected at each *m/z* for a longer time than would be allowed if the entire spectrum were being scanned.

Selected reaction monitoring is most selective. One ion from the first mass separator is passed through a collision cell, where it breaks into several product ions that are separated by a second mass separator. The intensities of one or a few of these product ions are plotted as a function of elution time. The selectivity is high because few species from the column produce the first selected ion and even fewer break into the same fragments in the collision cell. This technique is so selective that it can transform a poor chromatographic separation into a highly specific determination of one component with virtually no interference.

24-11.   Derivatization uses a chemical reaction to convert analyte into a form that is more convenient to separate or easier to detect. For example, carboxylic acids that might not have enough volatility for gas chromatography can be treated with chlorotrimethylsilane to make the volatile trimethylsilyl ester:

$$\underset{\text{Carboxylic acid}}{\text{RCOH}} + \text{ClSi(CH}_3)_3 \xrightarrow{-\text{HCl}} \underset{\text{Volatile derivative}}{\text{RCOSi(CH}_3)_3}$$

24-12.   (a)   In solid-phase microextraction, analyte is extracted from a liquid or gas into a thin coating on a silica fiber extended from a syringe. After extraction, the fiber is withdrawn into the syringe. To inject sample into a chromatograph, the metal needle is inserted through the septum and the fiber is extended into the injection port. Analyte slowly evaporates from the fiber in the high-temperature port. Cold trapping is required to condense analyte at the start of the column during slow evaporation from the fiber. If cold trapping were not used, the peaks would be extremely broad because of the slow evaporation from the fiber. During solid-phase microextraction, analyte equilibrates between the unknown and the coating on the fiber. Only a fraction of analyte is extracted into the fiber.

(b)   In stir-bar sorptive extraction, a thick coating on the outside of a glass-coated stirring bar is used in place of a thin coating on a fiber. After extraction, the bar is placed in a thermal desorption tube where analyte is vaporized and cold trapped for chromatography. The volume of the coating is ~100 times greater in stir-bar sorptive extraction, so the sensitivity is ~100 times higher.

24-13. The goal of purge and trap is to collect *all* of the analyte from the unknown and to inject *all* of the analyte into the chromatography column. Splitless injection is required so analyte is not lost during injection. Any unknown loss of analyte would lead to an error in quantitative analysis.

24-14. The order of decisions is: (1) goal of the analysis, (2) sample preparation method, (3) detector, (4) column, and (5) injection method.

24-15. (a) $t'_r = 8.4 - 3.7 = 4.7$ min; $k = 4.7$ min/3.7 min $= 1.3$

(b) Column radius $= 0.16$ mm
$\beta = r/2d_f = (0.16\ \text{mm})/(2 \times 1.0\ \mu\text{m} \times 10^{-3}\ \text{mm}/\mu\text{m}) = 80$

(c) $k = K/\beta \Rightarrow K = k\beta = (1.3)(80) = 104$

(d) The stationary phase and temperature are the same, so the partition coefficient would be 104, as in part (c).
$\beta = r/2d_f = (0.16\ \text{mm})/(2 \times 0.5\ \mu\text{m} \times 10^{-3}\ \text{mm}/\mu\text{m}) = 160$
$k = K/\beta = 104/160 = 0.65$
$k = t'_r/t_m \Rightarrow t'_r = kt_m = (0.65)(3.7\ \text{min}) = 2.4\ \text{min}$
$t_r = t'_r + t_m = 2.4\ \text{min} + 3.7\ \text{min} = 6.1\ \text{min}$

24-16. (a) $I = 100 \cdot \left[8 + (9 - 8)\ \dfrac{\log(12.0) - \log(11.0)}{\log(14.0) - \log(11.0)} \right] = 836$

(b) No change. Based on $k = K/\beta$ the retention of all peaks would be similarly affected by the change in phase ratio.[x]

(c) No change. The adjusted retention times of both the reference compounds and the unknown would be similarly affected by the change in length.

24-17. Kovats retention indexes are defined for alkanes as: hexane 600; heptane 700; octane 800. The retention indexes for the other compounds allows prediction of their elution on the columns.
Column (a): hexane < butanol = benzene < 2-pentanone < heptane < octane
Column (b): hexane < heptane < butanol < benzene < 2-pentanone < octane
Column (c): hexane < heptane < octane < benzene < 2-pentanone < butanol

24-18. We reason by analogy with the order of elution of the compounds in the table in the text. For poly(dimethylsiloxane) column (a), the order of elution is

C_6 alkane $<$ C_4 alcohol $<$ C_5 ketone $<$ C_7 alkane

Incrementing the chain length by one CH_2 unit, we predict the elution order

C_7 alkane $<$ C_5 alcohol $<$ C_6 ketone $<$ C_8 alkane

Column (a): 3, 1, 2, 4, 5, 6

For p(diphenyl)$_{0.35}$(dimethyl)$_{0.65}$polysiloxane column (b), the order of elution is

C_7 alkane $<$ C_4 alcohol $<$ C_5 ketone $<$ C_8 alkane

Incrementing the chain length by one CH_2 unit, we predict the elution order

C_8 alkane $<$ C_5 alcohol $<$ C_6 ketone $<$ C_9 alkane

Column (b): 3, 4, 1, 2, 5, 6

For poly(ethylene glycol) column (b), the order of elution is

C_9 alkane $<$ C_5 ketone $<$ C_{10} alkane $<$ C_{11} alkane $<$ C_4 alcohol

Incrementing the chain length by one CH_2 unit, we predict the elution order

C_{10} alkane $<$ C_6 ketone $<$ C_{11} alkane $<$ C_{12} alkane $<$ C_5 alcohol

Column (c): 3, 4, 5, 6, 2, 1

24-19. (a) $\Delta H_{vap}^o \approx (88 \text{ J mol}^{-1} \text{ K}^{-1}) \cdot T_{bp} = (88 \text{ J mol}^{-1} \text{ K}^{-1})(126° + 273°)$

$= 3.5 \times 10^4 \text{ J mol}^{-1}$

(b) $\ln\left(\dfrac{P_1}{P_2}\right) = -\left(\dfrac{\Delta H_{vap}}{R}\right)\left(\dfrac{1}{T_1} - \dfrac{1}{T_2}\right)$

T_{bp} is the temperature at which the vapor pressure of a solute equals atmospheric pressure (101 325 Pa).

$\ln\left(\dfrac{P_1}{101\ 325 \text{ Pa}}\right) = -\left(\dfrac{3.5 \times 10^4 \text{ J mol}^{-1}}{8.314 \text{ J mol}^{-1} \text{ K}^{-1}}\right)\left(\dfrac{1}{70° + 273°} - \dfrac{1}{126° + 273°}\right) = -1.73$

$\ln P_1 = -1.73 + \ln(101\ 325) = 9.80 \implies P_1 = e^{9.80} = 1.8 \times 10^4 \text{ Pa} = 0.18 \text{ bar}$

(c) $H_{vap}^o \approx (88 \text{ J mol}^{-1} \text{ K}^{-1})(69° + 273°) = 3.0 \times 10^4 \text{ J mol}^{-1}$

$\ln(P_1) = -\left(\dfrac{3.0 \times 10^4 \text{ J mol}^{-1}}{8.314 \text{ J mol}^{-1} \text{ K}^{-1}}\right)\left(\dfrac{1}{70° + 273°} - \dfrac{1}{69° + 273°}\right) + \ln(101\ 325) = 11.56$

$P_1 = e^{11.56} = 1.05 \times 10^5 \text{ Pa} = 1.05 \text{ bar (fully vaporized)}$

(d) The lower the vapor pressure, the greater the retention. Fully vaporized compounds (e.g., hexane in part c) are weakly retained on wall coated open tubular columns.

(e) The "gas" in "gas chromatography" refers to the mobile phase (N_2, He or H_2). Solutes are only partially in the vapor phase and spend only a fraction of the time in the mobile phase.

24-20.

$$\left.\begin{array}{l} \log(15.0) = \dfrac{a}{373} + b \\[2mm] \log(20.0) = \dfrac{a}{363} + b \end{array}\right\} \Rightarrow a = 1.69_2 \times 10^3 \text{ K} \qquad b = -3.36$$

To solve for a, subtract one equation from the other to eliminate b. Once you have a, substitute it back into either equation and solve for b.

At 353 K: $\log t'_r = \dfrac{1.69_2 \times 10^3 \text{ K}}{353 \text{ K}} - 3.36 \Rightarrow t'_r = 27.1 \text{ min}$

24-21. At 70°C, butanol has a retention time of 9.1 min. Increasing temperature decreases retention. The boiling point of butanol at 1 bar is 117°C. At some temperature above 117°C, butanol will all be in the gas phase in the chromatography column and therefore will have little retention. Its retention time will approach t_m. Since $\log k$ is related to $1/T$, k decreases in an exponential manner from 9.1 min at 70°C toward t_m at some temperature above 117°C.

24-22. (a) $N = 16(t_r/w)^2 = 16(17.0 \text{ min}/0.34 \text{ min})^2 = 4.0 \times 10^4$

$H = L/N = 30 \text{ m} / 4.0 \times 10^4 = 0.000\,75 \text{ m} = 0.75 \text{ mm}$

If your N was lower, you may have incorrectly used the integration tick marks in the figure. To calculate N, the baseline width must be determined by drawing tangents to the sides of the peak.

(b) Resolution $= \Delta t_r/w_{av}$ with width measured by extending tangents to the baseline. Based on measurements with my copy of the figure:

Resolution $= 0.68 \text{ cm} / 0.40 \text{ cm} = 1.7$

24-23. (a) $k = (t_r - t_m)/t_m = (49.6 \text{ min} - 6.7 \text{ min})/6.7 \text{ min} = 6.4$

(b) $N = 5.54(t_r/w_{1/2})^2 = 5.54(49.6 \text{ min}/0.26 \text{ min})^2 = 2.0 \times 10^5$

$H = L/N = 100 \text{ m}/(2.0 \times 10^5) = 0.50 \text{ mm}$

(c) Resolution $= \Delta t_r/w_{av}$ with width measured by extending tangents to the baseline. Based on measurements with my copy of the figure:

Resolution $= 0.24 \text{ cm} / 0.22_5 \text{ cm} = 1.1$

This resolution makes sense as the signal does not return to the baseline between the peaks. Thus the resolution must be less than 1.5.

24-24. (a) A thin stationary phase permits rapid equilibration of analyte between the mobile and stationary phases, which reduces the C term in the van Deemter equation. A thin stationary phase in a narrow-bore column gives small plate height and high resolution. In a wide-bore column, the large diameter of the column slows down the rate of mass transfer between the mobile and stationary phases (because it takes time for analyte to diffuse across the diameter of the column), which defeats the purpose of the thin stationary phase.

(b) Narrow-bore column: plate height = $1/(5\,000$ m$^{-1}) = 2.0 \times 10^{-4}$ m = 200 μm. The area of a length (ℓ) of the inside wall of the column is $\pi d\ell$, where d is the column diameter. The volume of stationary phase in this length is $\pi d\ell t$, where t is the thickness of the stationary phase. For $d = 250$ μm, $\ell = 200$ μm, and $t = 0.10$ μm, the volume is $1.5_7 \times 10^4$ μm^3. A density of 1.0 g/mL is 1.0 g/cm^3 = 1.0 g/$(10^4$ μm$)^3$ = 1.0 g/10^{12} μm^3 = 1 pg/μm^3. The mass of stationary phase in one theoretical plate is $(1.5_7 \times 10^4$ μm$^3)(1$ pg/μm$^3) = 1.5_7 \times 10^4$ pg. 1.0 % of this mass is = 0.16 ng.

Wide-bore column: For $d = 530$ μm, $\ell = 667$ μm, and $t = 5.0$ μm, the volume is $5.5_5 \times 10^6$ μm^3. Mass of stationary phase is $(5.5_5 \times 10^6$ μm$^3)(1$ pg/μm$^3) = 5.5_5 \times 10^6$ pg. 1.0% of this mass is = 56 ng.

24-25. (a) The column on a chip is part of a system intended to be an autonomous environmental monitor. Therefore, it needs to be compact and to require little power and consumables. Air is selected as carrier gas because it can be taken from the atmosphere. Any other carrier gas would require a supply tank which would be heavy, bulky, and would run out of gas. Oxygen from air could degrade the column at elevated temperature. Therefore, the temperature must be kept below the point at which oxidation would occur. Air has impurities which must be removed by a filtration system. The filter is most likely a consumable which eventually needs replacement.

(b) The optimum velocity gives the lowest plate height. It is the minimum in each curve. Optimum velocity = 9.3 cm/s for air and 17.6 cm/s for H$_2$. Plate height at optimum velocity = 0.036 cm for air and 0.051 for H$_2$. (Values come from the original publication. You will probably measure somewhat different values from the figure.)

(c) Plates = column length/plate height = 3.0 m/0.036 cm = 8 300 for air and 5 900 for H_2

(d) Time = column length/optimum velocity = 3.0 m/9.3 cm/s = 32 s for air and 17 s for H_2

(e) The two terms describe broadening due to the finite time for solute to diffuse through the stationary phase and the mobile phase. If the stationary phase is sufficiently thin, the time for diffusion through the stationary phase (the C_s term) becomes negligible.

(f) Acceptable flow rates for H_2 are higher than for air because solutes diffuse through H_2 faster than they diffuse through air. With H_2 carrier, solutes can diffuse from the center of the column to the wall more rapidly than they can with air carrier.

24-26. (a) Resolution $= \dfrac{\sqrt{N}}{4} \dfrac{(\alpha - 1)}{\alpha} \left(\dfrac{k_2}{1 + k_2} \right)$. This equation provides guidance.

Increasing the column length or using a narrower column would increase the number of plates N and thereby increase the resolution.

Increasing the film thickness (decreasing the phase ratio) would increase the retention factor k, and so increase resolution.

Changing polarity of the stationary phase would alter the separation factor α and the retention factor k, and so would change the resolution. But it is less predictable whether that change would increase or decrease in the resolution.

Adjusting the flow rate to the optimum in the van Deemter curve will give the maximum N, and thus better resolution.

The retention factor k is inversely dependent on the column temperature. Decreasing temperature would increase the last term in the resolution equation, and thus increase resolution. Lowering the temperature would have a large effect on resolution if the original k_2 was less than 3.

(b) Increasing column length, using a narrower column, or using a thicker film would all require purchase of a new column. Changing the flow rate or temperature only involve adjusting instrument settings, and so these would be the approaches to explore first.

24-27. (a) $S = [\text{pentanol}] = \dfrac{234 \text{ mg} / 88.15 \text{ g/mol}}{10.0 \text{ mL}} = 0.265_5 \text{ M}$

$X = [2,3\text{-dimethyl-2-butanol}] = \dfrac{237 \text{ mg} / 102.17 \text{ g/mol}}{10.0 \text{ mL}} = 0.232_0 \text{ M}$

$\dfrac{A_X}{[X]} = F\left(\dfrac{A_S}{[S]}\right) \Rightarrow \dfrac{1.00}{[0.232_0 \text{ M}]} = F\left(\dfrac{0.913}{[0.265_5 \text{ M}]}\right) \Rightarrow F = 1.25_3$

(b) I estimate the areas by measuring the height and $w_{1/2}$ in millimeters. Your answer will be different from mine if the figure size in your book is different from that in my manuscript. However, relative peak areas should be the same.

pentanol: height = 40.1 mm; $w_{1/2}$ = 3.7 mm;

area = 1.064 × peak height × $w_{1/2}$ = 15_8 mm^2

2,3-dimethyl-2-butanol:

height = 77.0 mm; $w_{1/2}$ = 2.0 mm; area = 16_4 mm^2

(c) $\dfrac{164}{2,3\text{-dimethyl-2-butanol}} = 1.25_3 \left(\dfrac{158}{[93.7 \text{ mM}]}\right)$

$\Rightarrow [2,3\text{-dimethyl-2-butanol}] = 77._6 \text{ mM}$

24-28. $\dfrac{A_X}{[X]} = F\left(\dfrac{A_S}{[S]}\right) \Rightarrow \dfrac{395}{[63 \text{ nM}]} = F\left(\dfrac{787}{[200 \text{ nM}]}\right) \Rightarrow F = 1.59$

The concentration of internal standard mixed with unknown is
$\dfrac{0.100 \text{ mL}}{10.00 \text{ mL}} (1.6 \times 10^{-5} \text{ M}) = 0.16 \text{ μM}$

$\dfrac{633}{[\text{iodoacetone}]} = 1.59 \left(\dfrac{520}{[0.16 \text{ μM}]}\right) \Rightarrow [\text{iodoacetone}] = 0.12_2 \text{ μM}$

[iodoacetone] in original unknown $= \dfrac{10.00}{3.00} (0.12_2 \text{ μM}) = 0.41 \text{ μM}$

24-29. $I = 100 \left[(7 + (10-7) \dfrac{\log(20.0) - \log(12.6)}{\log(22.9) - \log(12.6)}\right] = 932$

24-30. (a) Multiple small extractions are more effective than a single large extraction. The fraction remaining in phase 1 after four extractions would be q^4, where q is the fraction remaining after one extraction.

(b) On-column injection is better for thermally unstable molecules. Also, use of on-column injection allows higher sample loading on the column which would aid detection of the odors.

(c) Many of the odorants are polar molecules such as alcohols, esters, aldehydes, etc. Based on like dissolves like, a polar column such as poly(ethylene glycol) would provide best retention and resolution of these polar solutes.

(d) $\beta = r/2d_f = [(0.53 \text{ mm}/2) \times 1000 \text{ µm/mm}]/(2 \times 1.0 \text{ µm}) = 132$

(e) A wide bore column enables greater sample loading onto the column. This higher loading will yield more odorant eluting off the column, and thus increase the chance of smelling it.

24-31. (a) NaCl lowers the solubility of moderately nonpolar compounds, such as ethers, in water. Adding NaCl increases the fraction of the organic compounds that will be transferred to the extraction fiber.

(b) Selected ion monitoring is measuring ion abundance for m/z 73. Only three compounds in the extract have appreciable intensity at m/z 73.

(c) The base peak for both MTBE and TAME is at m/z 73. This mass corresponds to M – 15 (loss of CH_3) for MTBE and M – 29 (loss of C_2H_5) for TAME. Loss of the ethyl group bound to carbon in TAME suggests that the methyl group lost from MTBE is also bound to carbon, not to oxygen. If methyl bound to oxygen were easily lost from MTBE and TAME, we would expect to see the ethyl group bound to oxygen lost from ETBE. There is no significant peak at M – 29 (m/z 73) in ETBE. The following structures are suggested:

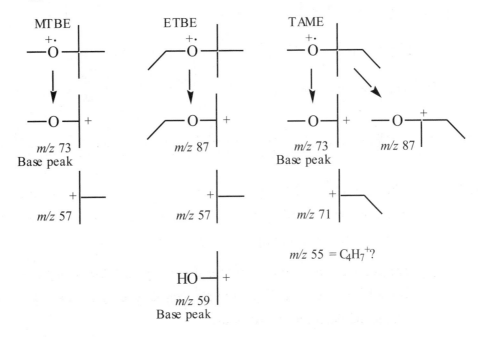

24-32. (a) The vial was heated to increase the vapor pressure of the analyte and the internal standard, so there would be enough in the gas phase (the headspace) to extract a significant quantity with the microextraction fiber.

(b) At 60ºC the analyte and internal standard are cold trapped at the beginning of the column. Since desorption from the fiber takes many minutes, we do not want chromatography to begin until desorption is complete.

(c) H—$\overset{+}{\underset{\underset{CH_3}{|}}{N}}$ $C_5H_{10}N^+$, m/z 84

For 5-aminoquinoline, m/z 144 is the molecular ion, $C_9H_8NC_5H_{10}N_2^+$

(d)

	A	B	C	D	E
1	Least-Squares Spreadsheet				
2					
3		x	y		
4		12	0.056		
5		12	0.059		
6		51	0.402		
7		51	0.391		
8		102	0.684		
9	Highlight cells B16:C18	102	0.669		
10	Type "= LINEST(C4:C13,	157	1.011		
11	B4:B13,TRUE,TRUE)	157	1.063		
12	For PC, press	205	1.278		
13	CTRL+SHIFT+ENTER	205	1.355		
14	For Mac, press				
15	COMMAND+RETURN	LINEST output:			
16	m	0.006401	0.0222	b	
17	u_m	0.000185	0.0234	u_b	
18	R^2	0.9933	0.0409	s_y	
19					
20	n =	10	B20 = COUNT(B4:B13)		
21	Mean y =	0.6968	B21 = AVERAGE(C4:C13)		
22	$\Sigma(x_i - \text{mean } x)^2$ =	48554.4	B22 = DEVSQ(B4:B13)		
23					
24	Measured y =	1.25	Input		
25	k = Number of replicate measurements of y =	2	Input		
26	Derived x =	191.83	B26 = (B24-C16)/B16		
27	u_x =	5.54			
28	B27 = (C18/ABS(B16))*SQRT((1/B25)+(1/B20)+((B24-B21)^2)/(B16^2*B22))				

Least-squares parameters are computed in the block B16:C18. In cell B24, we insert the mean y value (1.25) for 2 replicate unknowns. The number of replicates is entered in cell B25. The derived value of x is computed in cell B26 and the uncertainty is computed with Equation 4-27 in cell B27.

Answers for the unknowns:

nonsmoker: 78 ± 5 µg/L

nonsmoker with smoking parents: 192 ± 6 µg/L

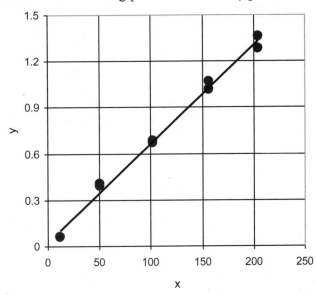

24-33. Nitrite: $[^{14}NO_2^-] = [^{15}NO_2^-](R - R_{blank}) = [80.0 \text{ µM}](0.062 - 0.040) = 1.8$ µM

Nitrate: $[^{14}NO_3^-] = [^{15}NO_3^-](R - R_{blank}) = [800.0 \text{ µM}](0.538 - 0.058) = 384$ µM

24-34. (a) The A term describing multiple flow paths is 0 for an open tubular column. Multiple paths arise in a packed column when liquid takes different paths through the column.

(b) $B = 2D_m$, where D_m is the diffusion coefficient of solute in the mobile phase.

(c) $C = C_s + C_m$

$$C_s = \frac{2k}{3(k+1)^2} \frac{d^2}{D_s} \qquad C_m = \frac{1 + 6k + 11k^2}{24(k+1)^2} \frac{r^2}{D_m}$$

where k = retention factor
d = thickness of stationary phase
r = column radius
D_s = diffusion coefficient of solute in the stationary phase
D_m = diffusion coefficient of solute in the mobile phase

(d) $H = B/u_x + Cu_x$ (u_x = linear velocity)

Plate height is a minimum at the optimum velocity:

$$\frac{dH}{du_x} = -\frac{B}{u_x^2} + C = 0 \implies u_x(\text{optimum}) = \sqrt{\frac{B}{C}}$$

The minimum plate height is found by plugging this value of u_x (optimum) back into the van Deemter equation:

$$H_{min} = B/u_x + Cu_x = B\sqrt{\frac{C}{B}} + C\sqrt{\frac{B}{C}} = 2\sqrt{BC} = 2\sqrt{B(C_s + C_m)}$$

$$H_{min} = 2\sqrt{(2D_m)\left(\frac{2k}{3(k+1)^2}\frac{d^2}{D_s} + \frac{1+6k+11k^2}{24(k+1)^2}\frac{r^2}{D_m}\right)}$$

$$H_{min} = 2\sqrt{\frac{4k}{3(k+1)^2}\frac{d^2 D_m}{D_s} + \frac{(1+6k+11k^2)\,2r^2}{24(k+1)^2}}$$

24-35. (a) As $k \to 0$, $H_{min}/r = \sqrt{1/3} = 0.58$

As $k \to \infty$, $H_{min}/r = \sqrt{\frac{1+6k+11k^2}{3(1+k)^2}} \to \sqrt{\frac{11k^2}{3k^2}} = \sqrt{\frac{11}{3}} = 1.9$

(b) As $k \to 0$, $H_{min} = 0.58\,r = 0.058$ mm

As $k \to \infty$, $H_{min} = 1.9\,r = 0.19$ mm

(c) For $k = 5.0$, $H_{min} = r\sqrt{\frac{1+6\cdot5.0+11\cdot25}{3(36)}} = 1.68\,r = 0.168$ mm

Number of plates $= \dfrac{50 \times 10^3 \text{ mm}}{0.168 \text{ mm/plate}} = 3.0 \times 10^5$

(d) $\beta = V_m/V_s$. For a length of column, ℓ, the volume of mobile phase is $\pi r^2 \ell$ and the volume of stationary phase is $2\pi r d_f \ell$. Substituting these volumes into the equation for β gives $\beta = (\pi r^2 \ell)/(2\pi r d_f \ell) = r/2d_f$.

(e) $k = K/\beta = 1000/\left(\dfrac{100 \text{ μm}}{2(0.20 \text{ μm})}\right) = 4.0$

24-36. The van Deemter equation has the form

$$H = B/u_x + Cu_x = B/u_x + (C_s + C_m)u_x$$

$$B = 2D_m \qquad C_s = \frac{2k}{3(k+1)^2}\frac{d^2}{D_s} \qquad C_m = \frac{1+6k+11k^2}{24(k+1)^2}\frac{r^2}{D_m}$$

where k = retention factor = 8.0

d = thickness of stationary phase = 3.0×10^{-6} m

r = column radius = 2.65×10^{-4} m

D_s = diffusion coefficient of solute in the stationary phase

D_m = diffusion coefficient of solute in the mobile phase

Experimentally, we find $H = (6.0 \times 10^{-5} \text{ m}^2/\text{s})/u_x + (2.09 \times 10^{-3} \text{ s})u_x$.

Therefore, $B = 2D_m = (6.0 \times 10^{-5} \text{ m}^2/\text{s})$, or $D_m = 3.0 \times 10^{-5} \text{ m}^2/\text{s}$.

From the second term of the experimental van Deemter equation, we know that

$$C_s + C_m = 2.09 \times 10^{-3} \text{ s} = \frac{2k}{3(k+1)^2} \frac{d^2}{D_s} + \frac{1 + 6k + 11k^2}{24(k+1)^2} \frac{r^2}{D_m}$$

Inserting the known values of all parameters allows us to solve for D_s:

$$2.09 \times 10^{-3} \text{ s} =$$
$$\frac{2(8.0)}{3((8.0)+1)^2} \frac{(3.0 \times 10^{-6} \text{ m})^2}{D_s} + \frac{1 + 6(8.0) + 11(8.0)^2}{24((8.0)+1)^2} \frac{(2.65 \times 10^{-4} \text{ m})^2}{(3.0 \times 10^{-5} \text{ m}^2/\text{s})}$$
$$\Rightarrow D_s = 5.0 \times 10^{-10} \text{ m}^2/\text{s}$$

The diffusion coefficient in the mobile phase is $(3.0 \times 10^{-5} \text{ m}^2/\text{s})/(5.0 \times 10^{-10} \text{ m}^2/\text{s}) = 6.0 \times 10^4$ times greater than the diffusion coefficient in the stationary phase. This makes sense, because it is easier for solute to diffuse through He gas than through a viscous liquid phase.

24-37. (a)

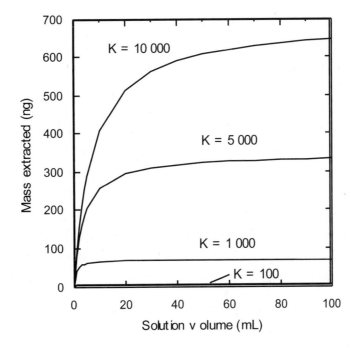

	A	B	C
1	Mass of analyte extracted by solid-phase microextraction		
2			
3	K = partition coefficient =	V_s (mL)	m (ng)
4	1.00E+02	0	0.000
5	V_f = volume of film (mL) =	1	6.455
6	6.90E-04	2	6.670
7	c_o = initial concentration	3	6.745
8	of analyte in solution (µg/mL) =	4	6.783
9	0.1	5	6.806
10		10	6.853
11		50	6.890
12		100	6.895
13		1000	6.900
14	C4 = 1000*$A$4*$A$6*$A$9*B4/($A$4*$A$6+B4)		

(b) $m = \dfrac{KV_f c_0 V_s}{KV_f + V_s}$ If $V_s \gg KV_f$, $m = KV_f c_0$

For $V_f = 6.9 \times 10^{-4}$ mL and $c_0 = 0.1$ µg/mL, $m \to (6.9 \times 10^{-5})(K)$ µg

For $K = 100$, $m \to 6.9$ ng, which agrees with the graph.

For $K = 10\,000$, $m \to 690$ ng, which is where the graph is heading, but it will require ~1 liter of solution to attain the limiting concentration in the fiber.

(c) The spreadsheet tells us that when $K = 100$, 6.85 ng have been extracted into the fiber and when $K = 10\,000$, 408 ng have been extracted into the fiber. The total analyte in 10.0 mL is (0.10 µg/mL)(10.0 mL) = 1.0 µg. The fraction extracted for $K = 100$ is 6.86 ng/1.0 µg = 0.006 9 (or 0.69%). The fraction extracted for $K = 10\,000$ is 0.41 (or 41%).

24-38. (a) H. Li, Y. Liu, Q. Zhang, and H. Zhan, "Determination of the Oxalate Content in Food by Headspace Gas Chromatography," *Anal. Methods* **2014**, *6*, 3720.

(b) In the "Apparatus and operation" section, the paper states a thermal conductivity detector was used. Oxalate is converted into CO_2. A flame ionization detector does not respond to CO_2. CO_2 does not contain the elements to which the nitrogen-phosphorus, flame photometric or chemiluminescence detectors selectively respond.

(c) "Method calibration, precision and validation" section states the relative standard deviation of repeated measurements was 0.84% and the limit of quantification (LOQ) was 1.95 µmol. Figure 6 is a calibration curve linear to 30 µmol. Therefore the linear range is from 1.95 µmol (LOQ) to 30 µmol.

(d) Five food samples were analyzed by both the gas chromatography method and an enzymatic method.

(e) The introduction of paper lists the following alternate methods: ion chromatography, high performance liquid chromatography, enzymatic methods, capillary electrophoresis, and chemiluminescence, and gives leading references for each. An enzymatic method is also used in the paper to validate the gas chromatography results.

(f) In the "Apparatus and operation" section the paper states a 30-m-long by 0.53-mm-diameter "model GS-Q (J&W Scientific)," but does not detail the nature of that column. CO_2 is a permanent gas and so would not be retained on a wall coated open tubular column. Therefore it must be a porous layer open tubular column. The manufacturer's literature states that the GS-Q is a porous divinylbenzene polymer.

24-39. (a) M. Peng, H. Wen, J. Le, and Y. Yang, "Trace-Level Analysis of Mesityl Oxide in Enalapril Maleate by Gas Chromatography with Electron Ionization Mass Spectrometry,"*Anal. Methods* **2012**, *4*, 4063.

(b) In the last paragraph of the "Introduction" section, the paper states that the method used is gas chromatography-electron impact-mass spectrometry with selective ion monitoring is used.

(c) In the "Experimental" section under "Instrumentation and GC/MS conditions" it states that an HP-1ms fused silica capillary column of 30 m length x 0.25 mm diameter and 0.25 μm film thickness was used. The first paragraph of "Results and Discussion" elaborates that the column is non-polar 100% dimethylsiloxane.

(d) In "Experimental" under "Method Validation," selected ion monitoring mass spectrometry is used. The sum of the area of selected ions *m/z* 55, 83 and 98 for mesityl oxide are divided by the selected ions *m/z* 58, 85 and 100 of the internal standard methyl isobutyl ketone.

(e) From the "Abstract" or "Results and Discussion," the precision is better than 2.9% (also Table 2); limit of detection is 0.05 ng mesityl oxide injected or 0.5 μg/g in enalapril maleate; limit of quantification is 0.18 ng mesityl oxide injected or 1.8 μg/g in enalapril maleate; linear range is 0.18 to 7.17 μg/mL in standards corresponding to 1.8 to 72 μg/g in the drug.

(f) In the "Results and Discussion" under "Stability test" the reference standard solutions were found to be stable for 24 hours at room temperature.

CHAPTER 25
HIGH-PERFORMANCE LIQUID CHROMATOGRAPHY

25-1. (a) In reversed-phase chromatography, the solutes are nonpolar and more soluble in a nonpolar mobile phase. In normal-phase chromatography, the solutes are polar and more soluble in a polar mobile phase.

(b) A gradient of increasing pressure gives increasing solvent density, which gives increasing eluent strength in supercritical fluid chromatography.

25-2. Solvent is competing with solute for adsorption sites. The strength of the solvent-adsorbent interaction is independent of solute.

25-3. In hydrophilic interaction chromatography, solute equilibrates between the mobile phase and an aqueous layer on the surface of the polar stationary phase. The more water in the eluent, the better can eluent compete with the stationary aqueous layer to dissolve polar solute and elute it from the column.

25-4. (a) Small particles give increased resistance to flow. High pressure is required to obtain a usable flow rate.

(b) Efficiency increases because solute equilibrates between phases more rapidly if the distance that it has to diffuse is smaller. This effect decreases the C term in the van Deemter equation. Also, flow paths between small particles are more uniform, decreasing the multiple path (A) term.

(c) A bonded stationary phase is covalently attached to the support.

25-5. (a) $L(\text{cm}) \approx \dfrac{N d_p(\mu m)}{3\,000}$

If $N = 1.0 \times 10^4$ and $d_p = 10.0\ \mu m$, $L = 33$ cm

$d_p = 5.0\ \mu m \Rightarrow L = 17$ cm; $\qquad d_p = 3.0\ \mu m \Rightarrow L = 10$ cm

$d_p = 1.5\ \mu m \Rightarrow L = 5$ cm

(b) Retention time is proportional to column length. If $t_r = 20.0$ minutes on the 33 cm long column, linear velocity was 33 cm/20.0 min = 1.6_5 cm/min.

$d_p = 5.0\ \mu m$ and $L = 17$ cm $\Rightarrow t_r = (17\ \text{cm}/33\ \text{cm})(20\ \text{min}) = 10.3$ min

$d_p = 3.0\ \mu m$ and $L = 10$ cm $\Rightarrow t_r = (10\ \text{cm}/33\ \text{cm})(20\ \text{min}) = 6.1$ min

$d_p = 1.5\ \mu m$ and $L = 5$ cm $\Rightarrow t_r = (5\ \text{cm}/33\ \text{cm})(20\ \text{min}) = 3.0$ min

(c) $P = $ (constant) $\times L/d_p^2$ If $P = 4.4$ MPa for $L = 33$ cm and $d_p = 10$ μm, then

the constant is 13.33 MPa μm^2 cm^{-1}

$d_p = 5.0$ μm, $L = 17$ cm: $P = (13.33) \times L/d_p^2 = (13.33)(17)/(5.0)^2 = 9.1$ MPa

$d_p = 3.0$ μm, $L = 10$ cm: $P = (13.33)(10)/(3.0)^2 = 14.8$ MPa

$d_p = 1.5$ μm, $L = 5$ cm: $P = (13.33)(5)/(1.5)^2 = 29.6$ MPa

(d) $N = \dfrac{16\, t_r^2}{w_b^2} \Rightarrow w_b(\text{minutes}) = \dfrac{4 t_r}{\sqrt{N}}$

$\Rightarrow w_b(\text{μL}) = w_b(\text{min}) \times 2.0$ mL/min $\times 1000$ μL/mL $= w_b(\text{min}) \times 2\,000$

10 μm: $w_b\,(\text{min}) = \dfrac{4(20.0\ \text{min})}{\sqrt{10^4}} = 0.80_0$ min; $w_b(\text{μL}) = 1\,600$ μL

5 μm: $w_b\,(\text{min}) = \dfrac{4(10.3\ \text{min})}{\sqrt{10^4}} = 0.41_2$ min; $w_b(\text{μL}) = 820$ μL

3 μm: $w_b\,(\text{min}) = \dfrac{4(6.1\ \text{min})}{\sqrt{10^4}} = 0.24_4$ min; $w_b(\text{μL}) = 490$ μL

1.5 μm: $w_b\,(\text{min}) = \dfrac{4(3.0\ \text{min})}{\sqrt{10^4}} = 0.12_0$ min; $w_b(\text{μL}) = 240$ μL

(e) HPLC instruments can operate in the 7-40 MPa pressure range, but are most reliable below 20 MPa. UHPLC systems can operate at pressures as high as 100 MPa. The 3–10 μm particle columns are within the pressure limits of an HPLC system, but the 1.5 μm particle column should be run on a UHPLC system. Also the variance contribution from extra-column components should be small compared to that of the column.

25-6. (a) The highly porous nature of HPLC particles greatly increases the surface area of the particles, which in turn increases the mass of sample that may be loaded onto the column without causing overload.

(b) The vast majority of the surface area of an HPLC particle is within the pores of the particle. Pore opening must be large enough to allow the solute to diffuse into the pore. Larger pores are needed to allow access of macromolecules such as proteins or polypeptides to the surface within the pores. However, the larger the pore, the smaller the surface area (and thus the capacity) of the column. Therefore narrower pores with high surface area are used for smaller molecules.

25-7. Plates $(N) = (15\text{ cm})/(5.0 \times 10^{-4}\text{ cm/plate}) = 3.0 \times 10^4$

$$N = \frac{5.55\,t_r^2}{w_{1/2}^2} \Rightarrow w_{1/2} = t_r\sqrt{\frac{5.55}{N}} = (10.0\text{ min})\sqrt{\frac{5.55}{3.0 \times 10^4}} = 0.13_6\text{ min}$$

If plate height = 25 μm, plates = 6 000 and $w_{1/2} = 0.30_4$ min

25-8. (a) $N = \dfrac{5.55\,t_r^2}{w_{1/2}^2} = \dfrac{5.55(0.63\text{ min} \times 60\text{ s/min})^2}{(2.3\text{ s})^2} = 1\,500$

   $H = L/N = (50\text{ mm})/(1\,500) = 0.033\text{ mm} = 33$ μm

   Number of particles side-by-side equaling one plate = 33 μm/1.7 μm = 19

   (b) Particles equal to one theoretical plate = 4 μm/1.7 μm = 2.4

   The column in (a) is being run for maximum speed.

25-9. Silica dissolves above pH 8 and the siloxane bond to the stationary phase hydrolyzes below pH 2. Bulky isobutyl groups hinder the approach of H_3O^+ to the Si–O–Si bond, so the rate of acid-catalyzed hydrolysis is decreased.

25-10. The high concentration of additive binds to the sites on the stationary phase that would otherwise hold on tightly to solutes and cause tailing.

25-11. (a) Your sketch should look like Figure 23-14, in which the asymmetry factor is $B/A = 1.8$, measured at one tenth of the peak height.

   (b) Tailing, particularly of ionic compounds, may be due to overload of the column. Try reducing the concentration of sample injected. If all peaks are affected, tailing could be caused by a clogged frit which you might be able to clean by reversing the flow direction. If only amines are tailing, try switching to a Type B silica column or add 30 mM triethylamine to the mobile phase. Tailing of acidic compounds might be eliminated by adding 30 mM ammonium acetate. For unknown mixtures, 30 mM triethylammonium acetate is useful. If tailing persists, 10 mM dimethyloctylamine or dimethyloctylammonium acetate might be effective.

25-12. (a)

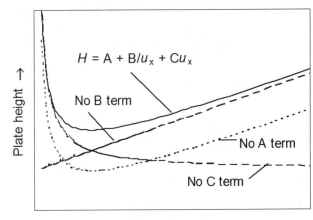

$$H = A + B/u_x + Cu_x$$

No B term

No A term

No C term

Plate height →

Linear flow rate →

(b) For 1.8-μm particle size, the experimental van Deemter curve looks almost like the curve with no C term in (a) (that is, finite equilibration time ≈ 0). When particle size is small enough, equilibration between the mobile and stationary phases is very rapid and this process contributes little to peak broadening. The experimental curve for 1.8-μm particles levels off at a smaller plate height than the curves for 5- and 3.5-μm particles. This behavior suggests that the A term (multiple flow paths) is smaller for the smaller particles.

(c) A superficially porous particle has a thin porous shell on a solid inner core. A 2.7-μm superficially porous particle provides resolution similar to that of a 1.8-μm totally porous particle mainly because of decreased eddy diffusion broadening arising from more uniform column packing of superficially porous particles. Resistance to flow depends on the width of the channels between the particles which scales with the overall diameter of the particle. The overall diameter of the superficially porous particle is larger, so its resistance to fluid flow is not as high as that of a small particle.

25-13. (a) $N = \dfrac{5.55\,t_r^2}{w_{1/2}^2} = \dfrac{5.55\,(4.70\ \text{min})^2}{(0.28\ \text{min})^2} = 1\,560$ for l enantiomer

$N = \dfrac{5.55\,(5.37\ \text{min})^2}{(0.35\ \text{min})^2} = 1\,310$ for d enantiomer

(b) $w_{1/2\text{av}} = \frac{1}{2}(0.28\ \text{min} + 0.35\ \text{min}) = 0.31_5\ \text{min}$

Resolution $= \dfrac{0.589\Delta t_r}{w_{1/2\text{av}}} = \dfrac{0.589\,(5.37\ \text{min} - 4.70\ \text{min})}{0.315\ \text{min}} = 1.25$

(c) Retention factor for *d* enantiomer: $k = \dfrac{(5.37\ \text{min} - 1.62\ \text{min})}{1.62\ \text{min}} = 2.31$

Relative retention: $\alpha = \dfrac{t'_{r2}}{t'_{r1}} = \dfrac{5.37\ \text{min} - 1.62\ \text{min}}{4.70\ \text{min} - 1.62\ \text{min}} = 1.22$

Average N $= \frac{1}{2}(1\,560 + 1\,310) = 1\,435$

Resolution $= \dfrac{\sqrt{N}}{4}\dfrac{(\alpha - 1)}{\alpha}\left(\dfrac{k_2}{1 + k_2}\right) = \dfrac{\sqrt{1\,435}}{4}\dfrac{(1.22 - 1)}{1.22}\left(\dfrac{2.31}{1 + 2.31}\right) = 1.19$

25-14. (a) $P \propto \dfrac{1}{d_p^2}$ For two conditions (1 and 2),

$\dfrac{P_2}{P_1} \propto \dfrac{d_1^2}{d_2^2} = \left(\dfrac{3\ \mu\text{m}}{0.7\ \mu\text{m}}\right)^2 = 18.$ Pressure must be 18 times greater.

(b) $u_x \propto P$, so if pressure is increased by a factor of 10, then linear velocity should increase by a factor of 10.

(c) Mass transfer between the mobile and stationary phase is faster for small particles than for large particles. The optimum velocity for maximum efficiency (highest plate number) increases as the rate of mass transfer increases. In the example cited, the high flow rate is closer to the optimum flow rate than is the low flow rate.

25-15. (a) Normal-phase chromatography

(b) Bonded reversed-phase

(c) Bonded reversed-phase with buffered mobile phase

(d) Hydrophilic interaction liquid chromatography (HILIC)

(e) Ion-exchange or ion chromatography

(f) Molecular-exclusion chromatography

(g) Ion-exchange chromatography with wide pore stationary phase

(h) Molecular-exclusion chromatography

25-16. 10-μm-diameter spheres: volume $= \frac{4}{3}\pi r^3 = \frac{4}{3}\pi(5 \times 10^{-4}\ \text{cm})^3 = 5.24 \times 10^{-10}\ \text{cm}^3$

Mass of one sphere $= (5.24 \times 10^{-10}\ \text{mL})(2.2\ \text{g/mL}) = 1.15 \times 10^{-9}\ \text{g}$

Number of particles in 1 g $= 1\ \text{g} / (1.15 \times 10^{-9}\ \text{g/particle}) = 8.68 \times 10^8$

Surface area of one particle $= 4\pi r^2 = 4\pi(5 \times 10^{-6}\ \text{m})^2 = 3.14 \times 10^{-10}\ \text{m}^2$

Surface area of 8.68×10^8 particles $= 0.27\ \text{m}^2$

Since the observed surface area is 300 m^2, the particles must be highly porous.

25-17. (a) Since the nonpolar compounds should become more soluble in the mobile phase, the retention time will be shorter in 90% methanol.

(b) At pH 3, the predominant forms are neutral RCO_2H and cationic RNH_3^+. The amine will be eluted first, since RNH_3^+ is insoluble in the nonpolar stationary phase.

(c) Polar compounds become less soluble in the mobile phase as the amount of water is decreased, so retention time will be greater in 90% acetonitrile.

(d) 2-Propanol has a higher eluent strength (0.60) than methyl t-butyl ether (0.48). Changing to 60% 2-propanol will increase the eluent strength, so retention times will be shorter in 60% 2-propanol.

25-18. (a) Unretained component travels at the solvent velocity, u_x.

$$u_x = \frac{\text{column length}}{\text{transit time}} = \frac{4\,400 \text{ mm}}{(41.7 \text{ min})(60 \text{ s/min})} = 1.76 \text{ mm/s}$$

(b) $k = \dfrac{t_r - t_m}{t_m} = \dfrac{188.1 \text{ min} - 41.7 \text{ min}}{41.7 \text{ min}} = 3.51$

(c) $N = \dfrac{5.55\,t_r^2}{w_{1/2}^2} = \dfrac{5.55\,(188.1 \text{ min})^2}{(1.01 \text{ min})^2} = 192\,000$

$H = \dfrac{(440 \text{ cm})(10^4 \text{ μm/cm})}{192\,000} = 22.9 \text{ μm}$

(d) Resolution $= \dfrac{0.589 \Delta t_r}{w_{1/2av}} = \dfrac{0.589(1.01 \text{ min})}{1.01 \text{ min}} = 0.589$

(e) $\alpha = \dfrac{t_{r2}'}{t_{r1}'} = \dfrac{194.3 \text{ min} - 41.7 \text{ min}}{193.3 \text{ min} - 41.7 \text{ min}} = 1.006_6$

(f) Increasing column length does not change α or k

$$\text{resolution} = \frac{\sqrt{N}}{4} \frac{(\alpha - 1)}{\alpha} \left(\frac{k_2}{1 + k_2}\right) \Rightarrow \frac{R_2}{R_1} = \frac{\sqrt{N_2}}{\sqrt{N_1}}$$

$\dfrac{1.000}{0.589} = \dfrac{\sqrt{N_2}}{\sqrt{192\,000}} \Rightarrow N_2 = 5.5_3 \times 10^5$

A column length of 440 cm gave $N = 1.92 \times 10^5$ plates. To obtain $5.5_3 \times 10^5$ plates, the column must be longer by a factor of $\dfrac{5.5_3 \times 10^5 \text{ plates}}{1.92 \times 10^5 \text{ plates}} = 2.8_8$.

Required length $= (2.8_8)(4.40 \text{ m}) = 12.7 \text{ m}$

(g) Adjust the flow rate to the optimum velocity in the van Deemter plot to obtain the lowest H and thereby increase N.

Decrease the mobile phase strength to increase the retention factor k.

Change the solvent to change the relative retention.

(h) Resolution $\dfrac{\sqrt{N}}{4}\dfrac{(\alpha-1)}{\alpha}\left(\dfrac{k_2}{1+k_2}\right) = \dfrac{\sqrt{192\ 000}}{4}\dfrac{(1.008_8-1)}{1.008_8}\left(\dfrac{17.0}{1+17.0}\right) = 0.90$

25-19. (a) On (R,R)-stationary phase, (S)-gimatecan is eluted at 6.10 min. On (S,S)-stationary phase, (S)-gimatecan is retained more strongly and is eluted at 6.96 min. (R)-gimatecan *must have the exact opposite behavior*. It will be eluted at 6.96 min from (R,R)-stationary phase and at 6.10 min from (S,S)-stationary phase.

(b) With (S,S)-stationary phase, we observe a small peak at 6.10 min for (R)-gimatecan. This peak is well separated from the front of the big (S)-gimatecan peak centered at 6.96 min, so the two areas can be integrated and compared with each other. With (R,R)-stationary phase, we see the (S)-gimatecan peak at 6.10 min with no evidence of the minor (R)-gimatecan peak at 6.96 min. The minor peak is lost beneath the tail of (S)-gimatecan. If one enantiomer is in low concentration compared to the other, it is desirable to have the trace enantiomer eluted first. Chromatography on each enatiomer of the stationary phase enables us to unambiguously locate where each enantiomer of gimatecan is eluted, even though we do not have a standard sample of (R)-gimatecan.

(c) For the (S,S)-stationary phase, we have the following information:

(S)-gimatecan: $t_r = 6.96$ min $k = 1.50$

(R)-gimatecan: $t_r = 6.10$ min $k = 1.22$

Relative retention: $\alpha = \dfrac{k_2}{k_1} = \dfrac{1.50}{1.22} = 1.23$

(d) Resolution $= \dfrac{\sqrt{N}}{4}\dfrac{(\alpha-1)}{\alpha}\left(\dfrac{k_2}{1+k_2}\right) = \dfrac{\sqrt{6\ 800}}{4}\dfrac{(1.23-1)}{1.23}\left(\dfrac{1.50}{1+1.50}\right) = 2.3$

which is more than adequate for "baseline" separation. Tailing of the peaks creates a little overlap, but it should not be very serious for an equal mixture of the enantiomers.

25-20. Peak areas will be proportional to molar absorptivity, since the number of moles of A and B are equal.

$$\frac{\text{Area of A}}{\text{Area of B}} = \frac{2.26 \times 10^4}{1.68 \times 10^4} = \frac{1.064 \times h_A w_{1/2}}{1.064 \times h_B w_{1/2}} = \frac{(128)(10.1)}{h_B (7.6)}$$

$$\Rightarrow h_B = 126 \text{ mm}$$

25-21. (a) $V_m \approx L d_c^2/2 = (5.0 \text{ cm})(0.46 \text{ cm})^2/2 = 0.53 \text{ cm}^3 = 0.53 \text{ mL}$

$t_m = V_m/F = (0.53 \text{ mL})/(1.4 \text{ mL/min}) = 0.38 \text{ min for column A}$

$= (0.53 \text{ mL})/(2.0 \text{ mL/min}) = 0.26_5 \text{ min for column B}$

(b) Morphine 3-β-D-glucuronide is more polar than morphine because of the added hydroxyl groups and the carboxylic acid. The more polar compound is less retained by the nonpolar reversed-phase column.

(c) $k = \dfrac{t_r - t_m}{t_m} = \dfrac{1.5 - 0.65}{0.65} = 1.3$ for morphine 3-β-D-glucuronide

$k = \dfrac{t_r - t_m}{t_m} = \dfrac{2.8 - 0.65}{0.65} = 3.3$ for morphine

(d) Bare silica is a polar, hydrophilic surface. Morphine should not be retained as strongly as the more polar morphine 3-β-D-glucuronide. The gradient goes to increasing H_2O for increasing polarity (that is, increasing solvent strength) to remove the more strongly adsorbed, more polar compound.

(e) $V_m = F t_m = (2.0 \text{ mL/min})(0.50 \text{ min}) = 1.0 \text{ mL}$

$k^* = \dfrac{t_G F}{\Delta\Phi V_m S} = \dfrac{(5.0 \text{ min})(2.0 \text{ mL/min})}{(0.4)(1.0 \text{ mL})(4)} \quad (= 6.2_5) = 6.2$

25-22. (a) Electrical power = current × voltage. Current is the rate of flow of charge through a circuit. It is analogous to the rate of flow of liquid through a column. Voltage is the potential difference driving charge through the wire. It is analogous to the pressure difference driving liquid through a column.

(b) $1 \text{ mL} = 1 \text{ cm}^3 = (10^{-2} \text{ m})^3 = 10^{-6} \text{ m}^3$

$1 \text{ mL/min} = 10^{-6} \text{ m}^3/60 \text{ s} = 1.67 \times 10^{-8} \text{ m}^3/\text{s}.$

$3\,500 \text{ bar} = 3\,500 \times 10^5 \text{ Pa} = 3.5 \times 10^8 \text{ Pa}$

power = volume flow rate × pressure drop

$= (1.67 \times 10^{-8} \text{ m}^3/\text{s})(3.5 \times 10^8 \text{ Pa}) = 5.8 \text{ W}$

25-23. (a) Below 210 nm near universal, above 210 nm selective for molecules with an absorbing chromophore

(b) almost any molecule, but not very sensitive

(c) near universal for non-volatile compounds

(d) near universal for non-volatile compounds

(e) compounds that can be oxidized or reduced

(f) molecules that fluoresce

(g) molecules that contain N

(h) ionic or ionizable compounds

25-24. (a) Many chromophores absorb ultraviolet radiation below 210 nm, so the detector is nearly universal below 210 nm. Only some molecules absorb at longer wavelengths. At 254 nm, the ultraviolet detector is selective for those molecules that absorb.

(b) The separation uses gradient elution with increasing methanol in CO_2. CO_2 is transparent above 190 nm. The ultraviolet cutoff for methanol is 205 nm. At 210 nm methanol absorbs some light, and absorbance increases with increasing methanol concentration. At 254 nm in the lower trace, methanol does not absorb, and so the baseline absorbance is unaffected by the mobile phase composition.

(c) The dip in the baseline occurs at 1.23 minutes.

$$t_m \approx \frac{L d_c^2}{2F} \approx \frac{25 \text{ cm}(0.46 \text{ cm})^2}{2 \times 2 \text{ mL/min}} \approx 1.32 \text{ min}$$

25-25. (a)

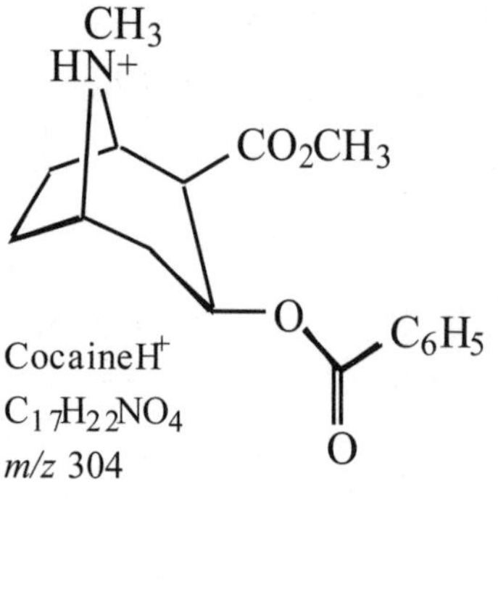

CocaineH^+
$C_{17}H_{22}NO_4$
m/z 304

(b) The $C_6H_5CO_2$ group has a mass of 121 Da. Subtracting 121 from 304 gives 183 Da. The peak at m/z 182 probably represents cocaine minus $C_6H_5CO_2H$. The structure might be the one below or some rearranged form of it.

(c) The ion at m/z 304 was selected by mass filter Q1. Its isotopic partner containing ^{13}C at m/z 305 was blocked by Q1. Because the species at m/z 304 is isotopically pure, there is no ^{13}C-containing partner for the collisionally activated dissociation product at m/z 182.

(d) For selected reaction monitoring, the mass filter Q1 selects just m/z 304, which eliminates components of plasma that do not give a signal at m/z 304. Then this ion is passed to the collision cell, in which it breaks into a major fragment at m/z 182 which passes through Q3. Few other components in the plasma that give a signal at m/z 304 also break into a fragment at m/z 182. The 2-step selection process essentially eliminates everything else in the sample and produces just one clean peak in the chromatogram.

(e) The phenyl group must be labeled with deuterium because the labeled product gives the same fragment at m/z 182 as unlabeled cocaine.

(f) First, we need to construct a calibration curve to get the response factor for cocaine compared to $^2H_5$-cocaine. We expect this response factor to be close to 1.00. We would prepare a series of solutions with known concentration ratios [cocaine]/[$^2H_5$-cocaine] and measure the area of each chromatographic peak in the chromatography/atmospheric chemical ionization/selected reaction monitoring experiment. A graph would be constructed, in which (peak area of cocaine)/(peak area of $^2H_5$-cocaine) is plotted versus [cocaine]/[$^2H_5$-cocaine]. The slope of this line is the response factor.

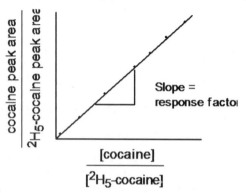

For quantitative analysis, a known amount of the internal standard $^2H_5$-cocaine is injected into the plasma. From the calibration curve, the relative peak areas tell us the relative concentrations of cocaine and the internal standard. From the known quantity of internal standard injected into the plasma, we can calculate the quantity of cocaine.

25-26. (a) Atmospheric pressure chemical ionization gives a prominent peak at m/z 234, which must be MH^+. The peak at m/z 84 is probably the fragment $C_5H_{10}N^+$, which might have the structure shown below.

In selected reaction monitoring, m/z 234 is selected by mass filter Q1 and m/z 84 is selected by mass filter Q3 in a triple quadrupole spectrometer.

(b) Deuterated internal standard has the formula $C_{14}H_{16}{}^2H_3O_2N$, with a nominal mass of 236. The protonated molecule is m/z 237. Cleavage of the C-C bond gives the same $C_5H_{10}N^+$ fragment as unlabeled Ritalin. The transition to monitor is m/z 237 → 84.

m/z 237 → m/z 84

25-27. (a) To find k, measure the retention time for the peak of interest (t_r) and the elution time for an unretained solute (t_m). Then use the formula $k = (t_r - t_m)/t_m$. The resolution between neighboring peaks is the difference in their retention time divided by their average width at the baseline.

(b) (i) Unretained solutes such as uracil or sodium nitrate could be run and observed with an ultraviolet detector. (ii) t_m is usually the time when the first baseline disturbance is observed. (iii) Alternatively, the formula $t_m \approx Ld_c^2/(2F)$ can be used, where L is the length of the column (cm), d_c is the column diameter (cm), and F is the flow rate (mL/min).

(c) (i) Unretained solutes such as toluene could be run and observed with an ultraviolet detector. (ii) and (iii) from part (b) will also work in HILIC.

(d) $t_m \approx Ld_c^2/(2F) = (15)(0.46)^2/(2 \cdot 1.5) = 1.0_6$ min

t_m does not depend on particle size. The estimate is 1.0_6 min for both 5.0- and 3.5-μm particles.

24-28. *Extra-column volume* is the volume of the system (not including the chromatography column) from the point of injection to the point of detection. *Dwell volume* is the volume of the system from the point of mixing solvents to the beginning of the column. Excessive extra-column volume causes peak broadening, particularly of early eluting peaks. In gradient elution, dwell volume determines the time from the initiation of a gradient until the gradient reaches the column. The greater the dwell volume, the more the delay between initiating a gradient and the actual increase of solvent strength on the column.

25-29. A rugged procedure should not be seriously affected by gradual deterioration of the column, *small* variations in solvent composition, pH, and temperature, or use of a different batch of the same stationary phase. A procedure should be rugged so that inevitable, small variations in conditions do not substantially affect the outcome of the separation.

25-30. $0.5 \leq k \leq 20$; resolution ≥ 2; operating pressure ≤ 15 MPa; $0.9 \leq$ asymmetry factor ≤ 1.5

25-31. Run a wide gradient (such as 5%B to 100%B) in gradient time t_G selected to produce $k^* \approx 5$ in Equation 25-12. Measure the difference in retention time (Δt) between the first and last peaks. Use a gradient if $\Delta t/t_G > 0.40$ and use isocratic elution if $\Delta t/t_G < 0.25$. For $0.25 < \Delta t/t_G < 0.40$, either may be appropriate.

25-32. The first steps are to (1) determine the goal of the analysis, (2) select a method of sample preparation, and (3) choose a detector that allows you to observe the desired analytes in the mixture. The next step could be a wide gradient elution to determine whether or not an isocratic or gradient separation is more appropriate. If isocratic separation is chosen, vary %B until criterion for good retention ($0.5 \leq k \leq 20$) is met. If adequate resolution is not attained, try minor adjustments in %B, a different organic solvent of equivalent mobile phase strength, or a different column. Finally, select column length or particle size, either to increase plate number if resolution < 2 or to shorter separation time if resolution >> 2.

25-33. (a) 53% tetrahydrofuran in water

(b) Mix 530 mL tetrahydrofuran and 470 mL H_2O. The total volume of the mixture will not equal 1 L. Do not add any additional tetrahydrofuran or water, as this would alter the composition of the mixture.

(c) Ultraviolet detection might be affected by the high wavelength cutoff (212 nm) of tetrahydrofuran. Tetrahydrofuran is also incompatible with polyether ether ketone plastic components of many HPLC systems.

25-34. Chromatography is conducted with four conditions: (A) high %B, low T, (B) high %B, high T, (C) low %B, high T, and (D) low %B, low T. Based on the appearance of the chromatograms, combinations between the points A, B, C, and D can be explored for further improvement in the separation.

25-35. Peak 5 has a retention time (t_r) of 11.0 min for 50% B. The retention factor is $k = (t_r - t_m)/t_m = (11.0 - 2.7)/2.7 = 3.1$. When B is reduced to 40%, the rule of three predicts $k = 3(3.1) = 9.3$. Rearranging the definition of retention factor, we find $t_r = t_m k + t_m = t_m(k + 1)$. We predict for 40% B $t_r = t_m(k + 1) = (2.7)(9.3 + 1) = 27.8$ min. Observed retention time at 40% B is 20.2 min.

25-36. (a)

%B	Retention time (min)		
	peak 6	peak 7	peak 8
90	4.4	4.4	4.9
80	4.5	4.5	5.1
70	5.6	5.6	7.3
60	8.2	8.2	12.2
50	13.1	13.6	24.5
40	24.8	27.5	65.1
35	37.6	44.2	125.2

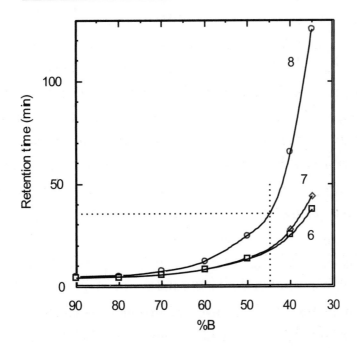

At 45% B, we could estimate that Peak 8 will be eluted halfway between the times for 40% B and 50% B, which is about 45 min. The fit to the curve above suggests that 36 min is a more realistic estimate.

(b) The table shows the measurement of retention factor k for Peaks 6-8.

				$t_m =$	2.7 min				
	retention time t_r (min)			retention factor $k = (t_r - t_m)/t_m$			log k		
Φ	Peak 6	Peak 7	Peak 8	Peak 6	Peak 7	Peak 8	Peak 6	Peak 7	Peak 8
0.9	4.4	4.4	4.9	0.630	0.630	0.815	-0.201	-0.201	-0.089
0.8	4.5	4.5	5.1	0.667	0.667	0.889	-0.176	-0.176	-0.051
0.7	5.6	5.6	7.3	1.074	1.074	1.704	0.031	0.031	0.231
0.6	8.2	8.2	12.2	2.037	2.037	3.519	0.309	0.309	0.546
0.5	13.1	13.6	24.5	3.852	4.037	8.074	0.586	0.606	0.907
0.4	24.8	27.5	65.1	8.185	9.185	23.111	0.913	0.963	1.364
0.35	37.6	44.2	125.2	12.926	15.370	45.370	1.111	1.187	1.657

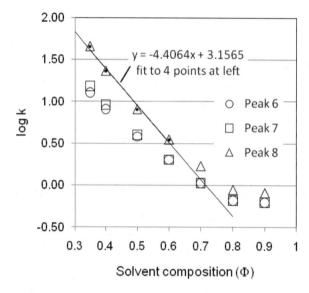

The first obvious point is that log k versus Φ does not follow a straight line over a wide range of solvent composition. The straight line going through the four points for Peak 8 from $\Phi = 0.35$ to 0.6 is log $k = -4.4064\Phi + 3.1565$. At $\Phi = 0.45$, we compute log $k = 1.1736$ and $k = 14.92$. We compute $t_r = t_m$ $(k+1) = 43.0$ min. If we had only taken the first three points ($\Phi = 0.35$ to 0.5), we would find log $k = -4.9364\Phi + 3.3660$. At $\Phi = 0.45$, we compute log $k = 1.11447$, $k = 13.95$, and $t_r = 40.4$ min.

(c) The gradient goes from 40-70% acetonitrile over 30 minutes. Therefore the mobile phase changes 10% every 10 minutes.

Time (min)	%B	Retention factor peak 6	peak 8
0	40	8.185	23.111
10	50	3.852	8.074
20	60	2.037	3.519
30	70	1.074	1.704

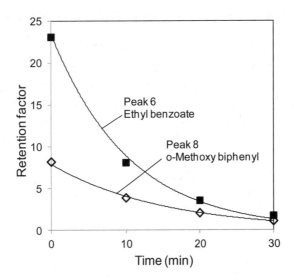

The retention factor of both compounds is very high at the start of the gradient. As the gradient proceeds, retention factors decrease exponentially. At the end of the gradient the compounds are almost unretained.

(d) Compounds spend much of the gradient retained near the inlet of the column. But by the time a compound is eluted, it is almost unretained. At the time of elution, the peak width is similar to that of a weakly retained compound.

25-37. In the nomograph in Figure 25-28, a vertical line at 48% methanol intersects the acetonitrile line at 38%.

25-38. (a) Lower solvent strength usually increases the difference in retention between different compounds. Use a lower percentage of acetonitrile.

(b) In normal-phase chromatography, solvent strength increases as the solvent becomes more polar, which means increasing the methyl t-butyl ether concentration. We need a higher concentration of hexane to lower the solvent strength, increase the retention times, and probably improve resolution.

25-39. (a) A is a weak acid with $pK_a = 7.7$ and $k_{HA} = 6.5$. B is a weak acid with $pK_a = 3.7$ and $k_{HA} = 5.9$. C is a weak base with $pK_a = 4.7$ and $k_B = 3.2$.

(b) Chromatography of C would be least rugged at pH $\approx$ 3 to 6.5 because retention changes with pH in the vicinity of pK_a.

(c) Mixtures of an acid and its conjugate base have buffering capacity within $pK_a \pm 1$. To study pH 2-11 buffers of different pK_a are required.

25-40. (a) pH 2.0 provides retention within $0.5 \leq k \leq 20$. Minor changes in pH would not significantly affect retention.

(b) $pH \approx 4$. Lower pH would result in insufficient retention of 4-methylaniline. Higher pH would not provide rugged methods, as retention of benzoic acid and 3-methyl benzoic acid would vary significantly if the pH changed slightly. However, the retention factors for all three compounds would be influenced by small changes in pH.

(c) pH 7.5. pH greater than 8 would provide stable retention of 4-methylaniline and codeine, but little retention of 4-nitrophenol. Also, silica bonded phase columns are typically limited to the pH 2–8 range. pH below 7 would yield inadequate retention of codeine.

25-41. Acetophenone is neutral at all pH values. Its retention is nearly unaffected by pH. For salicylic acid, we expect the neutral molecule, HA, to have some affinity for the C_8 nonpolar stationary phase and the ion, A^-, to have little affinity for C_8. Salicylic acid is predominantly HA below pH 2.97 and A^- above pH 2.97. At pH 3, there is nearly a 1:1 mixture of HA and A^-, which is moderately retained on the nonpolar column. At pH 5 and 7, more than 99% of the molecules are A^-, so retention is weak (small retention factor).

Ionic forms of nicotine ought to have low affinity for the nonpolar stationary phase and the neutral molecule would have some affinity. Abbreviating nicotine as B, the form B is dominant above $pH = pK_2 = 7.85$. BH^+ is dominant between pH 3.15 and 7.85. BH_2^{2+} is dominant below pH 3.15. B does not become appreciable until $pH \approx 7$, so the retention factor is low below pH 7 and increases at pH 7.

25-42. (a) $\Delta t/t_G = (50 - 22)/60 = 0.47$. Because $\Delta t/t_G > 0.40$, gradient elution is suggested.

(b) At $t = 22$ min, the solvent composition entering the column can be calculated by linear interpolation: $5 + \frac{22}{60}(100 - 5) = 39.8\%$. At 50 minutes, the composition is $5 + \frac{50}{60}(100 - 5) = 84.2\%$. A reasonable gradient for the second experiment is from 40 to 85% acetonitrile in 60 min.

25-43. (a) (1) Change the solvent strength by varying the fraction of each solvent. (2) Change the temperature. (3) Change the pH (in small steps). (4) Use a different solvent. (5) Use a different kind of stationary phase.

(b) Use a slower flow rate, a different temperature, a longer column, or a smaller particle size.

25-44. (a) Start with conditions to give $k^* = 5$ and assume that $S = 4$ for molecules in the mixture. $V_m \approx L\,d_c^2/2 = (15\text{ cm})(0.46\text{ cm})^2/2 = 1.5_9$ mL. Particle size does not come into the calculation.

$$t_G = \frac{k^* \Delta\Phi\, V_m\, S}{F} = \frac{(5)(0.9)(1.59\text{ mL})(4)}{(1.0\text{ mL/min})} = 29 \text{ min}$$

(b) $k^* = \dfrac{t_G\, F}{\Delta\Phi\, V_m\, S} = \dfrac{(11.5\text{ min})(1.0\text{ mL/min})}{(0.14)(1.59\text{ mL})(4)} = 12.9$

The large column has the same length as the small column, but the diameter is increased from 0.46 to 1.0 cm. The volume increases by a factor of $(1.0/0.46)^2 = 4.7$. Therefore, we increase the flow rate and the sample loading by a factor of 4.7. Flow rate = 4.7 mL/min and sample load = 4.7 mg. The gradient time is unchanged at 11.5 min. For the large column, $V_m \approx L\,d_c^2/2 = (15\text{ cm})(1.0\text{ cm})^2/2 = 7.5$ mL and

$$k^* = \frac{t_G\, F}{\Delta\Phi\, V_m\, S} = \frac{(11.5\text{ min})(4.7\text{ mL/min})}{(0.14)(7.5\text{ mL})(4)} = 12.9$$

25-45. Spreadsheet and figure are the same as in the textbook.

25-46. (a) To find the items start in the top left of the screen and go counter-clockwise. On the image in the question, the values are (your values may differ):

Solvent A: water   Solvent B: acetonitrile

Solvent B fraction: 40%   Temperature: 40°C

Injection volume: 5 μL   Flow rate: 2.0 mL/min

Plate height (HETP): 6.068×10^{-4} cm   Number of plates: 16480

Backpressure: 147.98 bar   First compound in list: acetophenone

Its retention factor and time: 1.0671 and 1.0993

Time required for separation: 4.8 min (right side of x-axis of chromatogram)

(b) Below left is a graph of my results for acetophenone. Your exact data may differ, but the general shape of the curve will be the same. Retention is high for low amounts of acetonitrile, and low for high amounts of acetonitrile. At 100% acetonitrile, all compounds are eluted in the time t_m taken for mobile phase to pass through the column.

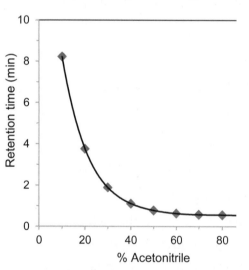
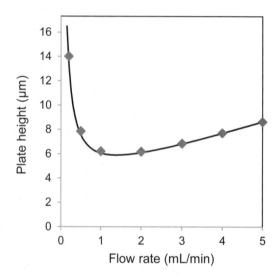

(c) The graph of my results for acetophenone with 20% acetonitrile at 40°C is shown above right. Your exact data may differ, but the general shape of the curve will be the similar.

The C-term broadening at the right half of the curve increases gradually, which is similar to the 3.5 μm particle behavior in Figure 25-3. The optimum plate height is about 0.0006 cm = 6 μm. Optimum plate heights are ideally 2-3 times the particle size. Both of these characteristics suggest my column is packed with 3 μm particles.

Move the slide bar on the left side of the screen down to see the "Column Properties" which includes the Particle size. Explore the effect of particle size on the van Deemter curve.

(d) My results might differ from yours. I found 40% acetonitrile gave $0.73 \leq k \leq 7.8$, with poor resolution for many peaks. 52% methanol gave comparable retention of $0.86 \leq k \leq 7.5$. This composition agrees well with the 50% methanol predicted by Figure 25-28 to be equivalent to 40% acetonitrile. In my case, the resolution was much improved with methanol.

(e) Changing the gradient time t_G alters k* which is the average retention factor of each solute at the midpoint of the column. The order of elution and resolution showed a complex dependence on gradient time.

25-47. (a) A. Brancaccio, P. Maresca, S. Albrizio, M. Fattore, M. Cozzolino, and S. Seccia, "Development and Validation of a Diode Array High Performance Liquid Chromatography Method to Determine Seized Street Cocaine Sample Purity," *Anal. Methods* **2013**, *5*, 2584.

(b) From the second column of the Introduction on the first page, gas chromatography with flame ionization, nitrogen phosphorus detection, and mass spectrometric detection and HPLC with electrochemical, UV or mass spectrometric detection have been used.

(c) From the Experimental on the second page, at the top of the second column, street cocaine purity can range from 95 to 5%.

(d) From "HPLC-DAD conditions" on the second page, the column was an Agilent Zorbax Eclipse XDB-RP-C18 of dimensions 150 mm × 4.6 mm inner diameter with 5 μm particles. A guard column was also used.

(e) In the Results and Discussion, the method validation is detailed in a section of that name. International Conference on Harmonization (ICH) guidelines were followed and the method was validated based on the specificity, linearity, limit of detection (LOD), limit of quantification (LOQ), precision, accuracy, and robustness.

(f) Robustness is defined on page 5 of the article, within the Results and Discussion section, in the section titled "Robustness." Robustness is a measure of an analytical method's ability to remain unaffected by small but deliberate variations in the method parameters.

25-48.   (a)   F. Eertmans, V. Bogaert, and B. Puype, "Development and Validation of a High-Performance Liquid Chromatography (HPLC) Method for the Determination of Human Serum Albumin (HSA) in Medical Devices,"*Anal. Methods* **2011**, *3*, 1296.

(b)   The Abstract mentions colorimetric, electrophoretic, and immunological assays. The second column of the Introduction on the first page goes into more detail, also mentioning a fluorometric method.

(c)   From the Materials and Methods section on the second page, under the subtitle Equipment, the column is a C_4 (butyl) bonded phase of dimensions 250 mm × 4.6 mm inner diameter with 5 μm particles with a pore diameter of 300 Å. A guard column was also used.

(d)   Table 1 shows that the gradient went from 20% of mobile phase B to 60% of B over 10 minutes. There was another 10 minutes devoted to washing the column with 100% mobile phase B and re-equilibrating the column with 20% B.

(e)   From the Materials and Methods section on the second page, under the subtitle Method Validation, peak identity, linearity, range, precision, accuracy, robustness, specificity, and sensitivity and were all assessed.

(f)   In the Discussion section at the top left of the sixth page it is noted that human serum albumin is a relatively large protein (66-67 kDa). The pores must be big enough to allow entry of such a large molecule. Most of the surface area of the stationary phase is inside the pores.

CHAPTER 26

CHROMATOGRAPHIC METHODS AND CAPILLARY ELECTROPHORESIS

26-1. Increased cross-linking gives decreased swelling, increased exchange capacity and selectivity, but longer equilibration time.

26-2. One way is to rinse a column containing a weighed amount of resin extensively with NaOH to regenerate the columns so that all ion-exchange sites are loaded with OH^-. After thoroughly washing with water to remove excess NaOH, elute the column with a large quantity of aqueous NaCl to displace OH^-. Then titrate the eluate with standard HCl to measure the moles of displaced OH^-.

26-3. Particles pass through 200 mesh (75 μm) sieve and are retained by 400 mesh (38 μm) sieve. 200/400 mesh particles are smaller than 100/200 mesh particles.

26-4. (a) Ion-exchange retention is largely governed by the charge of the ion. Pyruvate is bracketed by F^- and Cl^-, so probably –1; 2-oxovalerate is bracketed by Cl^- and NO_2^-, and so also –1; maleate is between CO_3^{2-} and SO_4^{2-} and so is probably –2.

(b) In addition to the charge of an ion, ion exchange also depends on the polarizability of the ion. Iodide is a very polarizable anion. Its electron cloud is greatly deformed by the positive charges on the anion-exchange resin. The resultant induced dipole strongly binds iodide to the resin.

26-5. (a) As pH is lowered the anionic protein becomes protonated, so the magnitude of the negative charge decreases. The protein becomes less strongly retained by the anion-exchange gel.

(b) As the ionic strength of eluent is increased, the protein will be displaced from the gel by the increasing concentration of anions in the eluent.

26-6. The pK_a values are: NH_4^+ (9.24), $CH_3NH_3^+$ (10.64), $(CH_3)_2NH_2^+$ (10.77), and $(CH_3)_3NH^+$ (9.80). If the four ammonium ions are adsorbed on a cation exchange resin at, say, pH 7, they might be separated by elution with a gradient of increasing pH. The anticipated order of elution is $NH_3 < (CH_3)_3N < CH_3NH_2 < (CH_3)_2NH$. We should not be surprised if the elution order were different, since steric and hydrogen bonding effects could be significant determinants of the selectivity coefficients. It is also possible that elution with a constant pH (of, say, 8) might separate all four species from each other.

26-7. Deionized water has had cations and anions removed. Deionized water has been passed through ion exchangers to convert cations to H^+ and anions to OH^-, making H_2O. Nonionic impurities (such as organic compounds) are not removed by this process, but can be removed by activated carbon.

26-8. (a) $V_{seawater} = 10$ mL/min $\times$ 17 h $\times$ 60 min/h $= 10$ L

The Fe^{3+} from 10 L of seawater was eluted from the column using 10 mL of acid. The concentration increased by a factor of

$$\frac{10 \text{ L} \times 1\,000 \text{ mL/L}}{10 \text{ mL}} = 1\,000$$

(b) $[Fe^{3+}]_{seawater} = \dfrac{57 \text{ nM} \times 10 \text{ mL}}{10 \text{ L} \times 1\,000 \text{ mL/L}} = 0.057$ nM $= 57$ pM

(c) 0.2 ppm Fe $= \dfrac{(0.2 \text{ mg/L})/(1\,000 \text{ mg/g})}{55.845 \text{ g/mol}} = 3.6 \times 10^{-6}$ M

$[Fe^{3+}]_{1.5M\ HNO_3} \leq \dfrac{1.5 \text{ M}}{15.7 \text{ M}} (3.6 \times 10^{-6} \text{ M}) = 3.4 \times 10^{-7}$ M $= 340$ nM

apparent $[Fe^{3+}]_{seawater} \leq$ (concentration factor) $\times [Fe^{3+}]_{1.5M\ HNO_3}$

$\leq 1\,000 \times 340$ nM $= 340\,000$ nM (6 million times greater than $[Fe^{3+}]_{seawater}$)

High-purity acids and reagents are essential when performing trace analysis.

26-9. The sum of anion charge in the spreadsheet is $-0.001\,59$ M, and the sum of cation charge is $0.002\,02$ M. Either some of the ion concentrations are inaccurate, or there are other ions in the pondwater that were not detected. For example, there could be large organic anions derived from living matter (such as humic acid from plants) that are not detected in this experiment.

	A	B	C	D	E	F
1	Ion	Formula mass	Concentration		Ion	Charge
2		(g/mol)	(µg/mL)	(mol/L)	charge	(mol/L)
3	Fluoride	18.998	0.26	1.37E-05	-1	-1.37E-05
4	Chloride	35.453	43.6	1.23E-03	-1	-1.23E-03
5	Nitrate	62.005	5.5	8.87E-05	-1	-8.87E-05
6	Sulfate	96.064	12.6	1.31E-04	-2	-2.62E-04
7						
8				Sum of anion charge =		-0.00159
9						
10	Sodium	22.990	2.8	1.22E-04	1	1.22E-04
11	Ammonium	18.038	0.2	1.11E-05	1	1.11E-05
12	Potassium	39.098	3.5	8.95E-05	1	8.95E-05
13	Magnesium	24.305	7.3	3.00E-04	2	6.01E-04
14	Calcium	40.078	24.0	5.99E-04	2	1.20E-03
15						
16				Sum of cation charge =		0.00202

26-10. At pH 2 (0.01 M HCl), TCA is more dissociated than DCA, which is more dissociated than MCA. The greater the average charge of the compound, the more it is excluded from the negatively charged ion-exchange resin and the more rapidly it is eluted.

26-11. (a) The hydrophilic stationary phase is a zwitterion with fixed positive and negative charges. Anions are retained by positive charges and cations are retained by negative charges. In hydrophilic interaction chromatography, the stationary phase is polar and there is thought to be a thin layer of aqueous phase on the surface of the stationary phase. Solvent must be made more polar to compete with the stationary phase to elute polar solutes. Eluent strength is increased when the acetonitrile content is decreased.

(b) Eluate from the column is nebulized and the mist is dried to produce an aerosol of the nonvolatile analytes. A high voltage electrode transfers a charge to the aerosol. The total charge measured by a collector is proportional to the mass of analyte. All nonvolatile analytes yield nearly the same response per unit mass.

26-12. The separator column separates ions by ion exchange, while the suppressor exchanges the counterion with either H^+ or OH^- to neutralize the eluent and reduce its conductivity. After separating cations in the cation-exchange column, the suppressor exchanges Cl^- for OH^- to convert H^+Cl^- eluent into H_2O.

26-13. (a) K^+ in reservoir $= (0.75)(1.5 \text{ L})\left(2.0 \, \frac{\text{mol K}_2\text{PO}_4}{\text{L}}\right)\left(2.0 \, \frac{\text{mol K}^+}{\text{mol K}_2\text{PO}_4}\right) = 4.5$ mol

Flow rate $= \left(20 \times 10^{-3} \, \frac{\text{mol KOH}}{\text{L}}\right)\left(0.001 \, 0 \, \frac{\text{L}}{\text{min}}\right) = 2.0 \times 10^{-5} \, \frac{\text{mol KOH}}{\text{min}}$

Time available $= \dfrac{4.5 \text{ mol K}}{2.0 \times 10^{-5} \, \frac{\text{mol KOH}}{\text{min}}} = 2.25 \times 10^5$ min

$\dfrac{2.25 \times 10^5 \text{ min}}{60 \text{ min/h}} = 3.8 \times 10^2$ h

(b) A flow of 5.0 mM KOH at 1.0 mL/min provides

$(5.0 \times 10^{-3} \text{mol KOH/L})(0.001 \, 0 \text{ L/min}) = 5.0 \times 10^{-6}$ mol KOH/min

$\dfrac{5.0 \times 10^{-6} \text{mol KOH/min}}{60 \text{ s/min}} = 8.33 \times 10^{-8}$ mol KOH/s

One electron provides one OH^- at the cathode, so the current must provide 8.33×10^{-8} mol e^-/s. We multiply by the Faraday constant to convert moles of electrons into coulombs:

$$(8.33 \times 10^{-8} \text{ mol e}^-/\text{s})(9.648\ 5 \times 10^4 \text{ C/mol e}^-)$$

$$= 8.0 \times 10^{-3} \text{ C/s} = 8.0 \times 10^{-3} \text{ A} = 8.0 \text{ mA}.$$

To produce 0.10 M KOH at 1.0 mL/min requires 20 times as much current, because the concentration of KOH is 20 times higher than 5.0 mM. The current at the end of the gradient will be $(20)(8.0 \text{ mA}) = 160 \text{ mA} = 0.16 \text{ A}$.

26-14.

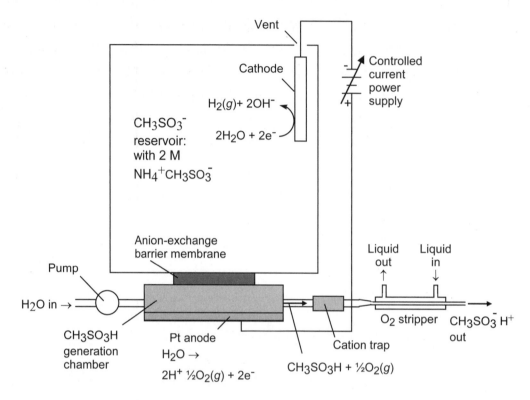

26-15. (a) Suppression converts analyte ions X^- into their conjugate acids HX. If HX is a strong acid such as HCl or HBr, it fully dissociates to yield a high conductivity signal. If HX is a very weak acid, such as HCN ($pK_a = 9.21$) or boric acid ($pK_a = 9.237$), very little HX deprotonates and so the conductivity signal is low.

(b) In the suppressor, H^+ replaces Na^+, to yield the protonated form of the eluent. For carbonate and bicarbonate, the suppression product is H_2CO_3. Carbonic acid is a weak acid ($pK_a = 10.329$), which dissociates to sufficient ions to increase the background conductivity.

26-16. This is an example of *indirect detection*. Eluent contains naphthalenetrisulfonate, which absorbs at 280 nm. Charge balance dictates that when one of the analyte anions is emerging from the column, there must be less naphthalenetrisulfonate anion emerging. Since analytes do not absorb as strongly at 280 nm, the absorbance is negative with respect to the steady baseline.

26-17. (a) Sodium octyl sulfate partitions into the nonpolar stationary phase making the particles cation exchangers. The surface charge forms an ion-pair with NE or DHBA. Other ions in the eluent compete with NE or DHBA, and slowly elute them from the column by ion exchange.

(b) Construct a graph of (peak height ratio) vs. (added concentration of NE). The *x*-intercept gives [NE] = 29 ng/mL.

Added NE	signal
0	0.298
12	0.414
24	0.554
36	0.664
48	0.792

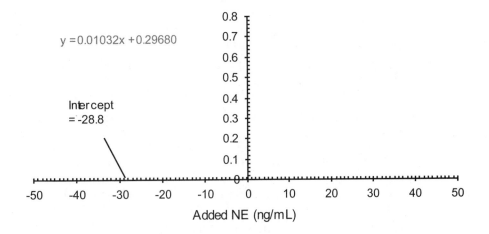

$y = 0.01032x + 0.29680$

Intercept
= -28.8

26-18. (a) Affinity chromatography. The strong and specific interaction between antibodies and antigens yields strong retention on an affinity column which allows the antibody to be cleanly separated from matrix components.

(b) Molecular exclusion. The protein is too large to fit into the pores of the gel and flows quickly through the column. Small molecules such as salts diffuse into the pores of the gel, and so are eluted later at volume V_m of the column.

(c) Molecular exclusion. Different polymer chain lengths are eluted at different volumes. Longer chains emerge earlier and shorter chains emerge later. There are so many possible chain lengths that individual chain lengths are not resolved. Only a single broad peak is observed. The width of this unresolved peak reflects the mass distribution of the polymer.

(d) Hydrophobic interaction. The difference in surface hydrophobicity of the two proteins suggests they can be separated by hydrophobic interaction chromatography. The difference in the protein masses is too small to yield a separation by molecular exclusion chromatography.

26-19. (a) There is a range in which retention volume is logarithmically related to molecular mass. The unknown is compared with standards of known molecular mass.

(b) Molecular mass 10^5 is near the middle range of the 25 nm pore size column.

26-20. (a) $V_t = \pi(0.80 \text{ cm})^2 (20.0 \text{ cm}) = 40.2 \text{ mL}$

(b) $K_{av} = \dfrac{V_r - V_0}{V_m - V_0} = \dfrac{27.4 - 18.1}{35.8 - 18.1} = 0.53$

26-21. Ferritin maximum is in tube 22 ($= 22 \times 0.65$ mL) $= 14.3$ mL $= V_0$
Ferric citrate maximum is in tube 84 ($= 84 \times 0.65$ mL) $= 54.6$ mL $= V_m$
Does this value of V_m make sense? The total column volume is $V_t = \pi r^2 \times$ length $= \pi(0.75 \text{ cm})^2(37 \text{ cm}) = 65.4$ mL, so $V_m = 54.6$ mL is plausible.

Transferrin maximum = tube 32 $= 20.8$ mL $\Rightarrow K_{av} = \dfrac{20.8 - 14.3}{54.6 - 14.3} = 0.16$

26-22. (a) $V_0 = 4.7$ mL. The vertical line begins at $\approx 10^6 = 1\,000\,000$ Da.

(b) A vertical line at 9.7 mL intersects the 12.5-nm calibration line at (molecular mass) $\approx 10^4 = 10\,000$ Da.

(c) The vertical drop on the 45-nm curve begins at $\approx 10^4 = 10\,000$ Da.

26-23. (a) The total column volume is $\pi r^2 \times$ length $= \pi(0.39)^2 (30) = 14.3$ mL. Totally excluded molecules do not enter the pores and are eluted in the solvent volume (the interstitial volume) outside the particles. Interstitial volume = 40% of 14.3 mL = 5.7 mL.

(b) The smallest molecules that completely penetrate pores will be eluted in a volume that is the sum of the volumes between particles and within pores = 80% of 14.3 mL = 11.5 mL.

(c) These solutes must be adsorbed on the polystyrene resin. Otherwise, they would all be eluted between 5.7 and 11.5 mL.

26-24. A graph of log (molecular mass, MM) vs. V_r should be constructed.

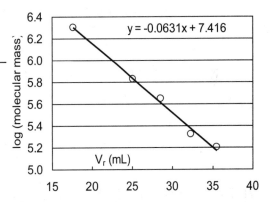

	log(MM)	V_r(mL)
aldolase	5.199	35.6
catalase	5.322	32.3
ferritin	5.643	28.6
thyroglobulin	5.825	25.1
Blue Dextran	6.301	17.7
unknown	?	30.3

The equation of the graph of K_{av} vs. log (MM) is $y = -0.063\,1\,x + 7.416$. Inserting $x = 30.3$ gives $y = \log(MM) = 5.50 \Rightarrow$ molecular mass $= 320\,000$

26-25. Hydrophobic regions of the protein are less soluble in water as the salt concentration in the water increases. This decrease in solubility of nonpolar substances in water with increasing salt concentration is known as "salting out." By decreasing the salt concentration, the protein becomes more soluble in the aqueous phase and can be eluted from the column. Eluent strength increases as the salt concentration decreases.

26-26. (a) Electrophoretic mobility is governed by $\mu_{ep} = q/6\pi\eta r$, where r is the radius of the solute. Addition of an alkyl group increases the size of phenol, which increases r, and decreases the electrophoretic mobility. An ethyl group is larger than a methyl group, so the magnitude of the mobility decreases in the order phenol > 4-methylphenol > 4-ethylphenol.

(b) Predicted electrophoretic mobility is $\mu_{ep\ predicted} = (\alpha_{A-})(\mu_{ep\ A-})$, where $\alpha_{A-} = K_a/([H^+] + K_a)$

	A	B	C	D	E
1		pK_a	μ_{A-}	α_{A-}	$\mu_{ep,\ predicted}$
2			$m^2/(V\cdot s)$	pH 10.0	pH 10.0
3	Phenol	9.98	-2.99E-08	0.512	-1.53E-08
4	4-Methylphenol	10.27	-2.59E-08	0.349	-9.05E-09
5	4-Ethylphenol	10.22	-2.39E-08	0.376	-8.99E-09
6	2,4,5-Trichlorophenol	6.83	-2.85E-08	0.999	-2.85E-08
7	$\alpha_{A-} = 10^{-pKa}/(10^{-pH} + 10^{-pKa})$			D3 = 10^-B3/(10^-10.0 + 10^-B3)	
8	$\mu_{ep,\ predicted} = (\alpha_{A-})(\mu_{A-})$				E3 = D3*C3

The pH is well above the pK_a of trichlorophenol and so it is fully ionized and its predicted electrophoretic mobility equals its μ_{A-}. The pH is near the pK_a of phenol, and so its average charge is ~ -0.5, and its predicted electrophoretic mobility is ~50% of its μ_{A-}. The pK_a of the alkyl phenols are higher, and so they are less ionized at pH 10. So their predicted mobilities are lower.

(c) Order of elution is cations, then neutrals, and then anions with highest mobility anions last. From part (b), ethylphenol would be eluted first, followed closely by methylphenol, then phenol, and finally the highest mobility trichlorophenol.

26-27. Electroosmosis is the bulk flow of fluid in a capillary caused by migration of the dominant ion in the diffuse part of the double layer toward the anode or cathode.

26-28. At pH 10, the wall of the bare capillary is negatively charged with $-Si-O^-$ groups and there is strong electroosmotic flow toward the cathode. At pH 2.5, the wall is nearly neutral with $-Si-OH$ groups and there is almost no electroosmotic flow. The few $-Si-O^-$ groups left give slight flow toward the cathode. The aminopropyl capillary also has positive flow at pH 10, but the rate is only about half as great as that of the bare capillary. The negative charge might be reduced because there are fewer $-Si-O^-$ groups (because some of them have been converted to $-Si-CH_2CH_2CH_2NH_2$) or because some of the aminopropyl groups are protonated ($-Si-CH_2CH_2CH_2NH_2^+$) at pH 10. At pH 2.5, all the aminopropyl groups are protonated. The net charge on the wall is *positive* and the flow is *reversed*.

26-29. (a) The unknown appears before the electroosmotic flow (methanol), and so must be a cation.

(b) $\mu_{app} = \dfrac{u_{net}}{E} = \dfrac{0.400 \text{ m}/86.0 \text{ s}}{5.00 \times 10^4 \text{ V/m}} = 9.30 \times 10^{-8} \dfrac{\text{m}^2}{\text{V} \cdot \text{s}}$

$\mu_{ep} = \mu_{app} - \mu_{eo}$

$\mu_{eo} = 4.26 \times 10^{-8} \dfrac{\text{m}^2}{\text{V} \cdot \text{s}}$ was calculated in the Example Mobilities Calculation.

$\mu_{ep} = \mu_{app} - \mu_{eo} = (9.30 - 4.26) \times 10^{-8} \dfrac{\text{m}^2}{\text{V} \cdot \text{s}} = +5.04 \times 10^{-8} \dfrac{\text{m}^2}{\text{V} \cdot \text{s}}$

26-30. Arginine is the only amino acid listed with a positively charged side chain. All of the derivatized amino acids have a negative charge because the fluorescent group and the terminal carboxyl group are both negative. Arginine is least negative, so its electrophoretic mobility toward the anode is slowest and its net migration toward the cathode (from electroosmosis) is fastest.

26-31. Under ideal conditions, longitudinal diffusion is the principle source of zone broadening. Even under ideal conditions, the finite length of the injected sample and, possibly, the finite length of the detector contribute to zone broadening. In real electrophoresis, adsorption on the capillary wall and irregular flow paths due to imperfections in the capillary could contribute to zone broadening. For an experimental study of zone broadening, see D. Xiao, T. V. Le, and M. J. Wirth, "Surface Modification of the Channels of Poly(dimethylsiloxane) Microfluidic Chips with Polyacrylamide for Fast Electrophoretic Separations of Proteins," *Anal. Chem.* **2004**, *76*, 2055.

26-32. (a) At pH 2.8, electroosmotic flow will be very small. Anionic analyte will migrate from negative to positive polarity with little effect from the small electroosmotic flow. Reverse polarity places the detector at the positive end of the capillary.

(b) The conductivity of the buffer needs to be higher than the conductivity of the sample so that the sample will stack. At lower buffer concentration, analyte bands will be broader and resolution of heparin from its impurities would be diminished.

(c) High buffer concentration gives high conductivity, high current, and high heat generation. The narrow column reduces the current and the heat generation and makes it easier to cool the entire volume inside the capillary.

(d) Li^+ has lower mobility than Na^+, so the conductivity of lithium phosphate solution will be lower than the conductivity of sodium phosphate solution at the same pH. The lower the conductivity, the higher the electric field required to generate the same current.

High field strength reduces the migration time to shorten the analysis. Also, according to Equation 26-14, the number of plates increases in proportion to applied voltage.

26-33. Electroosmotic flow can be reduced by (a) lowering the pH, so the charge on the capillary wall is reduced; (b) adding ions such as $^+H_3NCH_2CH_2CH_2CH_2NH_3^+$ that adhere to the capillary wall and effectively neutralize its charge; and (c) covalently attaching silanes with neutral, hydrophilic substituents to the Si—O$^-$ groups on the walls. A cationic surfactant can form a bilayer, which effectively reverses the charge on the wall.

26-34. In the absence of micelles, neutral molecules are all swept through the capillary at the electroosmotic velocity. Negatively charged micelles swim upstream with some electrophoretic velocity, so they take longer than neutral molecules to reach the detector. A neutral molecule spends some time free in solution and some time dissolved in the micelles. Therefore, the net velocity of the neutral molecule is reduced from the electroosmotic velocity. Because different neutral molecules have different partition coefficients between the solution and the micelles, each type of neutral molecule has its own net migration speed. We say that micellar electrokinetic chromatography is a form of chromatography because the micelles behave as a "stationary" phase in the capillary because their concentration is uniform throughout the capillary. Analyte partitions between the mobile phase and the micelles as the analyte travels through the capillary. Micelles are called a pseudostationary phase because they are not stationary and their concentration is constant in the capillary because they are part of the run buffer.

26-35. (a) Volume of sample $=$ cross-sectional area $\times$ length
$$= \pi r^2(\text{length}) = \pi(25 \times 10^{-6} \text{ m})^2(0.006\,0 \text{ m}) = 1.18 \times 10^{-11} \text{ m}^3$$
$$\Delta P = \frac{128\eta L_t(\text{Volume})}{t\pi d^4} = \frac{128(0.001\,0 \text{ kg/(m·s)})(0.600 \text{ m})(1.18\times10^{-11} \text{ m}^3)}{(4.0 \text{ s})\pi(50\times10^{-6} \text{ m})^4}$$
$$= 1.15 \times 10^4 \text{ Pa} \ (= 1.15 \times 10^4 \text{ kg/(m·s}^2))$$

(b) $\Delta P = h\rho g \Rightarrow h = \dfrac{\Delta P}{\rho g} = \dfrac{1.15 \times 10^4 \text{ kg/(m·s}^2)}{(1\,000 \text{ kg/m}^3)(9.8 \text{ m/s}^2)} = 1.17 \text{ m}$

Since the column is only 0.6 m long, we cannot raise the inlet to 1.17 m. Instead, we could use pressure at the inlet (1.15×10^4 Pa = 0.114 atm) or an equivalent vacuum at the outlet.

26-36. (a) Volume = $\pi r^2(\text{length}) = \pi(12.5 \times 10^{-6} \text{ m})^2(0.006\,0 \text{ m}) = 2.95 \times 10^{-12} \text{ m}^3 = 2.95$ nL. Moles = $(10.0 \times 10^{-6} \text{ M})(2.95 \times 10^{-9} \text{ L}) = 29.5$ fmol.

(b) Moles injected = $\mu_{app}\left(E\,\dfrac{\kappa_b}{\kappa_s}\right)t\pi r^2 C = \mu_{app}\left(\dfrac{V}{L_t}\dfrac{\kappa_b}{\kappa_s}\right)t\pi r^2 C$

In order for the units to work out, we need to express the concentration, C, in mol/m^3: $(10.0 \times 10^{-6} \text{ mol/L})(1\,000 \text{ L/m}^3) = 1.00 \times 10^{-2}$ mol/m^3

$V = \dfrac{(\text{moles})L_t(\kappa_s/\kappa_b)}{\mu_{app}\,t\pi r^2 C}$

$= \dfrac{(29.5 \times 10^{-15} \text{ mol})(0.600 \text{ m})(1/10)}{(3.0 \times 10^{-8} \text{ m}^2/(\text{V·s}))(4.0 \text{ s})\pi(12.5 \times 10^{-6} \text{ m})^2(1.00 \times 10^{-2} \text{ mol/m}^3)}$

$= 3.00 \times 10^3 \text{ V}$

26-37. Electrophoretic peak: $N = \dfrac{16\,t_r^2}{w^2} = \dfrac{16\,(6.08 \text{ min})^2}{(0.080 \text{ min})^2} = 9.2 \times 10^4$ plates

Chromatographic peak: $N \approx \dfrac{41.7(t_r/w_{0.1})^2}{(B/A + 1.25)}$

$= \dfrac{41.7(6.03 \text{ min}/0.37 \text{ min})^2}{(1.45 + 1.25)} = 4.1 \times 10^3$ plates

(According to my measurements, both plate counts are about 1/3 lower than the values labeled in the figure from the original source.)

26-38. (a) Fumarate is a longer molecule than maleate, so we guess that fumarate has a greater friction coefficient than maleate. Electrophoretic mobility is (charge)/(friction coefficient). Both ions have the same charge, so we predict that maleate will have the greater electrophoretic mobility.

(b) Since maleate moves upstream faster than fumarate, fumarate is eluted first.

(c) Since the anions move faster than the electroosmotic flow, the faster anion (maleate) is eluted first.

26-39. (a) pH 2: $u_{neutral} = \mu_{eo}E = \left(1.3 \times 10^{-8}\ \dfrac{m^2}{V \cdot s}\right)\left(\dfrac{27 \times 10^3\ V}{0.62\ m}\right) = 5.6_6 \times 10^{-4}\ m/s$

Migration time $= (0.52\ m)/(5.6_6 \times 10^{-4}\ m/s) = 9.2 \times 10^2\ s$

pH 12: $u_{neutral} = \mu_{eo}E = \left(8.1 \times 10^{-8}\ \dfrac{m^2}{V \cdot s}\right)\left(\dfrac{27 \times 10^3\ V}{0.62\ m}\right) = 3.5_3 \times 10^{-3}\ m/s$

Migration time $= (0.52\ m)/(3.5_3 \times 10^{-3}\ m/s) = 1.4_7 \times 10^2\ s$

(b) pH 2: $\mu_{app} = \mu_{ep} + \mu_{eo} = (-1.6 + 1.3) \times 10^{-8}\ \dfrac{m^2}{V \cdot s} = -0.3 \times 10^{-8}\ \dfrac{m^2}{V \cdot s}$

The anion will not migrate toward the detector at pH 2.

pH 12: $\mu_{app} = \mu_{ep} + \mu_{eo} = (-1.6 + 8.1) \times 10^{-8}\ \dfrac{m^2}{V \cdot s} = 6.5 \times 10^{-8}\ \dfrac{m^2}{V \cdot s}$

$u_{anion} = \mu_{app}E = \left(6.5 \times 10^{-8}\ \dfrac{m^2}{V \cdot s}\right)\left(\dfrac{27 \times 10^3\ V}{0.62\ m}\right) = 2.8_3 \times 10^{-3}\ m/s$

Migration time $= (0.52\ m)/(2.8_3 \times 10^{-3}\ m/s) = 1.8_4 \times 10^2\ s$

26-40. (a) The net speed of an ion moving through the capillary by electroosmosis plus electrophoresis is proportional to electric field ($u_{net} = \mu_{app}E$), which, in turn, is proportional to voltage. Increasing voltage by 120 kV/28 kV = 4.3 should increase the speed by 4.3 and decrease the migration time by 4.3. Peak 1 has a migration time of 211.3 min at 28 kV and 54.36 min at 120 kV. The ratio is 211.3 min/54.36 min = 3.9.

(b) Plate count is proportional to voltage ($N = \dfrac{\mu_{app}V}{2D}\dfrac{L_d}{L_t}$). Increasing voltage by a factor of 4.3 should increase the plate count by 4.3.

(c) Bandwidth is proportional to the $1/\sqrt{N}$ ($N = L_d^2/\sigma^2 \Rightarrow \sigma = L_d/\sqrt{N}$). Increasing voltage by 4.3 should increase N by 4.3 and decrease bandwidth by $1/\sqrt{4.3} = 0.48$. Bandwidth at 120 kV should be 48% as great as bandwidth at 28 kV.

(d) Increasing voltage makes the ions move faster, which gives the peaks less time to undergo longitudinal diffusion broadening. Therefore, the bandwidth is reduced and resolution is increased.

26-41. At low voltage (low electric field), the number of plates increases in proportion to voltage. Above ~25 000 V/m, the capillary is probably overheating, which produces band broadening and decreases the number of plates.

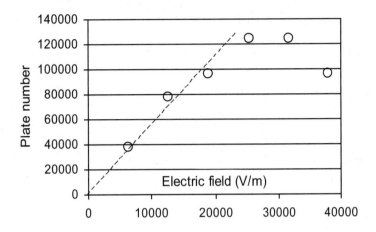

26-42. $N = \dfrac{5.55\, t_r^2}{w_{1/2}^2} = \dfrac{5.55\,(39.9\ \text{min})^2}{(0.81\ \text{min})^2} = 1.3_5 \times 10^4$ plates

Plate height $= 0.400\ \text{m}/(1.3_5 \times 10^4\ \text{plates}) = 30\ \mu\text{m}$

26-43. $t = \dfrac{L_d}{u_{net}} = \dfrac{L_d}{\mu_{app}E}$ (t = migration time, L = length to detector, u = speed, E = field)

$\Rightarrow \mu_{app} = \dfrac{L_d}{tE} = \dfrac{L_d/E}{17.12}$ for Cl^- and $\mu_{app} = \dfrac{L_d/E}{17.78}$ for I^-

Therefore, we can write that the difference in mobilities is

$\Delta\mu_{app}(\text{I-Cl}) = \dfrac{L_d/E}{17.12} - \dfrac{L_d/E}{17.78}$ (L_d/E is an unknown constant)

But we know that $\Delta\mu_{app}(\text{I-Cl}) = [\mu_{eo} + \mu_{ep}(I^-)] - [\mu_{eo} + \mu_{ep}(Cl^-)] =$
$\mu_{ep}(I^-) - \mu_{ep}(Cl^-) = 0.05 \times 10^{-8}\ \text{m}^2/(\text{s}\cdot\text{V})$ in Table 15-1.

For the difference between Cl^- and Br^- we can say

$\Delta\mu_{app}(\text{Br-Cl}) = \dfrac{L_d/E}{17.12} - \dfrac{L_d/E}{x}$

and we know that $\Delta\mu_{app}(\text{Br-Cl}) = 0.22 \times 10^{-8}\ \text{m}^2/(\text{s}\cdot\text{V})$ in Table 15-1.

Therefore, we can set up a proportion:

$\dfrac{\Delta\mu_{app}(\text{Br-Cl})}{\Delta\mu_{app}(\text{I-Cl})} = \dfrac{0.22}{0.05} = \dfrac{\dfrac{L_d/E}{17.12} - \dfrac{L_d/E}{x}}{\dfrac{L_d/E}{17.12} - \dfrac{L_d/E}{17.78}} \Rightarrow x = 20.5\ \text{min}$

The observed migration time is 19.6 min. Considering the small number of significant digits in the $\Delta\mu$ values, this is a reasonable discrepancy.

26-44.

	A	B	C	D	E	F
1	Molecular mass by SDS/capillary gel electrophoresis					
2					Relative	
3		Molecular		Migration	migration	
4	Protein	mass (MM)	log(MM)	time (min)	time (t_{rel})	$1/t_{rel}$
5	Marker dye	low		13.17		
6	a-Lactalbumin	14200	4.152	16.46	1.250	0.8001
7	Carbonic anhydrase	29000	4.462	18.66	1.417	0.7058
8	Ovalbumin	45000	4.653	20.16	1.531	0.6533
9	Bovine serum albumin	66000	4.820	22.36	1.698	0.5890
10	Phosphorylase B	97000	4.987	23.56	1.789	0.5590
11	b-Galactosidase	116000	5.064	24.97	1.896	0.5274
12	Myosin	205000	5.312	28.25	2.145	0.4662
13	Ferritin light chain	?		17.07	1.296	0.7715
14	Ferritin heavy chain	?		17.97	1.364	0.7329

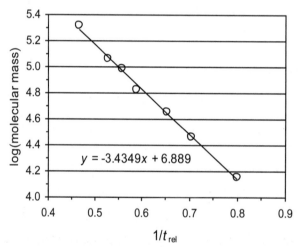

$$\log(MM) = (-3.434\ 9)/t_{rel} + 6.889 = 4.239 \text{ for } t_{rel} = 1.296 \text{ (ferritin light chain)}$$
$$= 4.372 \text{ for } t_{rel} = 1.364 \text{ (ferritin heavy chain)}$$

$$\text{Molecular mass} = 10^{\log(MM)} = 17\ 300 \text{ (ferritin light chain)}$$
$$= 23\ 500 \text{ (ferritin heavy chain)}$$

Molecular masses observed from amino acid sequences are 19 766 and 21 099 Da

26-45. $\text{Resolution} = \dfrac{\sqrt{N}}{4} \dfrac{\Delta\mu_{app}}{\overline{\mu}_{app}} \Rightarrow N = \left(4\,(\text{Resolution})\dfrac{\overline{\mu}_{app}}{\Delta\mu_{app}}\right)^2$

SO_4^{2-}: $\mu_{ep} = -8.27 \times 10^{-8} \text{ m}^2/(\text{s·V})$ in Table 15-1

$\mu_{app} = \mu_{eo} + \mu_{ep} = 16.1 \times 10^{-8} - 8.27 \times 10^{-8} = 7.8_3 \times 10^{-8} \text{ m}^2/(\text{s·V})$

Br^-: $\mu_{ep} = -8.13 \times 10^{-8} \text{ m}^2/(\text{s·V})$ in Table 15-1

$\mu_{app} = \mu_{eo} + \mu_{ep} = 16.1 \times 10^{-8} - 8.13 \times 10^{-8} = 7.9_7 \times 10^{-8} \text{ m}^2/(\text{s·V})$

$\overline{\mu}_{app} = \dfrac{1}{2}(7.8_3 + 7.9_7 \times 10^{-8}) = 7.9_0 \times 10^{-8} \text{ m}^2/(\text{s·V})$

$\Delta\mu_{app} = (8.27 - 8.13) \times 10^{-8} = 0.14 \times 10^{-8} \text{ m}^2/(\text{s·V})$

$N = \left(4\,(\text{Resolution})\dfrac{\overline{\mu}_{app}}{\Delta\mu_{app}}\right)^2 = \left(4\,(2.0)\dfrac{7.9_0}{0.14}\right)^2 = 2.0 \times 10^5 \text{ plates}$

26-46. In the absence of micelles, the expected order of elution is cations before neutrals before anions: thiamine < (niacinamide + riboflavin) < niacin. Since thiamine is eluted last, it must be most soluble in the micelles.

26-47. Carbon atoms labeled with black circles in cyclobarbital and thiopental are chiral, with four different substitutents. These compounds are not superimposable on their mirror images. The carbon atom indicated by the diamond in phenobarbital is not chiral because two of its substituents are identical. Cyclodextrin has a chiral pocket, in which these compounds can bind. The equilibrium constant for association of each of the enantiomers of cyclobarbital and thiopental with cyclodextrin will not be the same. Each enantiomer spends a different fraction of time associated with cyclodextrin as it migrates through the capillary. Therefore, cyclobarbital and thiopental will each separate into two peaks. Phenobarbital will only give one peak because it does not have enantiomers.

Cyclobarbital   Thiopental   Phenobarbital

26-48. (a) Plate height rises sharply at low velocity because bands broaden by diffusion when they spend more time in the capillary. This is the effect of the B term in the van Deemter equation, and it always operates in capillary electrophoresis. Plate height rises gradually at high velocity because solutes require a finite time to equilibrate with the micelles on the column. This is the effect of the C term in the van Deemter equation, and it is absent in capillary electrophoresis but present to a small extent in micellar electrokinetic capillary chromatography.

(b) There should be no irregular flow paths because the micelles are nanosized structures in solution. The large A term most likely arises from extra-column effects, such as the finite size of the injection plug and the finite width of the detector zone.

26-49. For the acid H_2A, the average charge is $\alpha_{HA^-} + 2\alpha_{A^{2-}}$, where α is the fraction in each form. From our study of acids and bases, we know that

$$\alpha_{HA^-} = \frac{K_1[H^+]}{[H^+]^2 + K_1[H^+] + K_1K_2} \text{ and } \alpha_{A^{2-}} = \frac{K_1K_2}{[H^+]^2 + K_1[H^+] + K_1K_2}$$

where K_1 and K_2 are acid dissociation constants of H_2A. The following spreadsheet finds the average charge of malonic acid (H_2M) and phthalic acid (H_2P) and finds that the maximum difference between them occurs at pH 5.55.

Charge Difference Between Malonic and Phthalic Acids

	A	B	C	D	E	F	G	H	I	J
1	Malonic:			Alpha	Alpha	Alpha	Alpha	Average charges		Charge
2	K1 =	pH	[H+]	HM-	M2-	HP-	P2-	Malonate	Phthalate	Difference
3	1.42E-03	5.52	3.0E-06	0.600	0.399	0.436	0.563	-1.398	-1.562	-0.16392
4	K2 =	5.53	3.0E-06	0.594	0.405	0.430	0.569	-1.403	-1.567	-0.16405
5	2.01E-06	5.54	2.9E-06	0.589	0.410	0.425	0.574	-1.409	-1.573	-0.16413
6	Phthalic:	5.55	2.8E-06	0.583	0.416	0.419	0.580	-1.415	-1.579	-0.16418
7	K1 =	5.56	2.8E-06	0.577	0.421	0.413	0.585	-1.420	-1.584	-0.16417
8	1.12E-03	5.57	2.7E-06	0.572	0.427	0.408	0.591	-1.426	-1.590	-0.16413
9	K2 =	5.58	2.6E-06	0.566	0.433	0.402	0.597	-1.432	-1.596	-0.16404
10	3.90E-06	5.59	2.6E-06	0.561	0.438	0.397	0.602	-1.437	-1.601	-0.16391
11										
12	D3 = $A$3*C3/(C3^2+$A$3*C3+$A$3*$A$5)						C3 = 10^-B3			
13	E3 = $A$3*$A$5/(C3^2+$A$3*C3+$A$3*$A$5)						H3 = -D3-2*E3			
14	F3 = $A$8*C3/(C3^2+$A$8*C3+$A$8*$A$10)						I3 = -F3-2*G3			
15	G3 = $A$8*$A$10/(C3^2+$A$8*C3+$A$8*$A$10)						J3 =I3-H3			

26-50. (a) In ion mobility spectrometry, gaseous ions are generated by irradiating analyte plus reagent gas (such as acetone in air) with high energy electrons (β emission) from radioactive ^{63}Ni. Periodically, ions are admitted into a drift tube by a short voltage pulse applied to an electronic gate (a grid). In the drift tube, ions experience a constant electric field that causes either cations or anions to migrate from the gate to a detector at the other end of the tube. The time to reach the detector is the drift time. Ions drift at a constant speed governed by the driving force of the electric field and the retarding force of friction (drag) by the atmosphere of gas (usually dry air) in the drift tube. Also, gas in the drift tube flows from the detector to the source, further decreasing the migration speed of an ion.

The electric field in ion mobility spectrometry causes ions to migrate from the source to the detector, just as the electric field in electrophoresis causes ions to migrate. Drift time in ion mobility spectrometry is the same quantity as migration time in electrophoresis. The mobility of an ion in liquid or in gas is governed by the charge-to-size ratio. The greater the charge and the smaller the size, the greater the mobility. In liquid or gas, the retarding force is caused by collisions with solvent or gas molecules.

(b)

	A	B	C	D	E	F
1	Ion Mobility Spectrometry					
2						
3	k =		Volts	t_d (s)	$w_{1/2}$ (s)	N
4	1.38065E-23	J/K	100	5.0000	2.68E-01	1.94E+03
5	e =		1000	0.5000	8.47E-03	1.94E+04
6	1.60218E-19	C	2000	0.2500	2.99E-03	3.87E+04
7	T =		3000	0.1667	1.63E-03	5.80E+04
8	300	K	4000	0.1250	1.06E-03	7.73E+04
9	μ =		5000	0.1000	7.59E-04	9.64E+04
10	0.00008	m^2/(sV)	6000	0.0833	5.78E-04	1.15E+05
11	z =		7000	0.0714	4.60E-04	1.34E+05
12	1		8000	0.0625	3.77E-04	1.52E+05
13	L =		9000	0.0556	3.18E-04	1.70E+05
14	0.2	m	10000	0.0500	2.72E-04	1.87E+05
15	t_g =		12000	0.0417	2.10E-04	2.19E+05
16	5.00E-05	s	14000	0.0357	1.69E-04	2.47E+05
17	16kT(ln2)/ez =		16000	0.0313	1.41E-04	2.71E+05
18	2.86707E-01		18000	0.0278	1.22E-04	2.90E+05
19			20000	0.0250	1.07E-04	3.03E+05
20	D4 = $A$14^2/($A$10*C4)					
21	E4 = SQRT($A$16^2+($A$18/C4)*D4^2)					
22	F4 = 5.55*(D4/E4)^2					

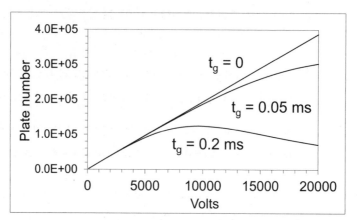

Increasing V increases N by decreasing the drift time, and therefore decreasing the time for diffusion to broaden the peak. Increasing the time that the ion gate is open increases the initial width of the peak, and therefore decreases the plate number. The peak can never be narrower than the pulse that is admitted by the gate. At high voltage, the effect of t_g on plate number overwhelms the effect of t_d. The disadvantage of using a short gate opening time is that fewer ions are admitted to the drift cell and the signal will be weaker.

(c) Decreasing T increases N because diffusional broadening decreases with decreasing temperature.

(d) $N = 5.55 (t_d/w_{1/2})^2 = 5.55 (0.024\,925\text{ s}/0.000\,154\text{ s})^2 = 1.45 \times 10^5$ plates

Theoretical $w_{1/2}^2 = t_g^2 + \left(\dfrac{16kT \ln2}{Vez}\right)t_d^2$

$\quad = (5.0 \times 10^{-5}\text{ s})^2 + \left(\dfrac{16(1.38 \times 10^{-23}\text{ J/K})(300\text{ K}) \ln2}{(12\,500\text{ V})(1.602 \times 10^{-19}\text{ C})(1)}\right)(0.0249\,25\text{ s})^2$

$\quad = 1.674 \times 10^{-8}\text{ s}^2$

Theoretical $N = 5.55 (t_d^2/w_{1/2}^2) = 5.55 (0.024\,925\text{ s})^2/(1.674 \times 10^{-8}\text{ s}^2)$

$\quad = 2.06 \times 10^5$ plates

(e) Resolution $= \dfrac{\sqrt{N}}{4}\dfrac{\Delta\mu_{app}}{\bar{\mu}_{app}}$ and $t = \dfrac{L_d}{u_{net}} = \dfrac{L_d}{\mu_{app}E} \Rightarrow \mu_{app} = \dfrac{L_d}{tE}$

$\mu_{app,leucine} = \dfrac{L_d}{tE} = \dfrac{10.0\text{ cm}}{0.0225\text{ s} \times 200\text{ V/cm}} = 2.22\text{ cm}^2/(\text{s·V})$

$\mu_{app,isoleucine} = \dfrac{10.0\text{ cm}}{0.0220\text{ s} \times 200\text{ V/cm}} = 2.27\text{ cm}^2/(\text{s·V})$

$\bar{\mu}_{app} = \frac{1}{2}(2.22 + 2.27) = 2.24\text{ cm}^2/(\text{s·V})$

$\Delta\mu_{app} = (2.27 - 2.22) = 0.05\text{ cm}^2/(\text{s·V})$

Resolution $= \dfrac{\sqrt{N}}{4}\dfrac{\Delta\mu_{app}}{\bar{\mu}_{app}} = \dfrac{\sqrt{80\,000}}{4}\dfrac{0.05}{2.24} = 1.6$

CHAPTER 27
GRAVIMETRIC AND COMBUSTION ANALYSIS

27-1. (a) In **ad**sorption, a substance becomes bound to the surface of another substance. In **ab**sorption, a substance is taken up inside another substance.

(b) An inclusion is an impurity that occupies lattice sites in a crystal. An occlusion is an impurity trapped inside a pocket in a growing crystal.

27-2. An ideal gravimetric precipitate should be insoluble, easily filterable, pure, and possess a known, constant composition.

27-3. High relative supersaturation often leads to formation of colloidal product with a large amount of impurities.

27-4. Relative supersaturation can be decreased by increasing temperature (for most solutions), mixing well during addition of precipitant, and using dilute reagents. Homogeneous precipitation is also an excellent way to control relative supersaturation.

27-5. Washing with electrolyte preserves the electric double layer and prevents peptization.

27-6. HNO_3 evaporates during drying. $NaNO_3$ is nonvolatile and will lead to a high mass for the precipitate.

27-7. During the first precipitation, the concentration of unwanted species in the solution is high, giving a relatively high concentration of impurities in the precipitate. In the reprecipitation, the level of solution impurities is reduced, giving a purer precipitate.

27-8. In thermogravimetric analysis, the mass of a sample is measured as the sample is heated. The mass lost during decomposition provides some information about the composition of the sample.

27-9. A quartz crystal microbalance consists of a specially cut, thin, disk-shape slice of quartz with gold electrodes on each of the two faces. Application of an oscillating electric field causes the crystal to oscillate at a characteristic frequency. Binding of small masses to the gold electrodes lowers the oscillation frequency. From the change in frequency, we can deduce how much mass was bound.

27-10. $\dfrac{0.214\,6 \text{ g AgBr}}{187.772 \text{ g AgBr/mol}} = 1.142\,9 \times 10^{-3} \text{ mol AgBr}$

$[\text{NaBr}] = \dfrac{1.142\,9 \times 10^{-3} \text{ mol}}{50.00 \times 10^{-3} \text{ L}} = 0.022\,86 \text{ M}$

27-11. $\dfrac{0.104 \text{ g CeO}_2}{172.114 \text{ g CeO}_2/\text{mol}} = 6.043 \times 10^{-4} \text{ mol CeO}_2 = 6.043 \times 10^{-4} \text{ mol Ce}$

$= 0.084\,66 \text{ g Ce}$

weight % Ce $= \dfrac{0.084\,66 \text{ g}}{4.37 \text{ g}} \times 100 = 1.94 \text{ wt\%}$

27-12. Formula mass of AgCl $= 107.8 + 35.4 = 143.2$

mol AgCl $= \dfrac{0.088\,90 \text{ g AgCl}}{143.2 \text{ g AgCl/mol AgCl}} = 6.208_1 \times 10^{-4} \text{ mol}$

mol radium $= \dfrac{6.208_1 \times 10^{-4} \text{ mol Cl}^-}{2 \text{ mol Cl}^-/\text{mol Ra}} = 3.104_0 \times 10^{-4} \text{ mol}$

$3.104_0 \times 10^{-4} \text{ mol RaCl}_2 = \dfrac{0.091\,92 \text{ g RaCl}}{x \text{ g RaCl}_2/\text{mol RaCl}_2}$

$x = \dfrac{0.091\,92 \text{ g RaCl}}{3.104_0 \times 10^{-4} \text{ mol RaCl}_2} = 296.1_3 \text{ g RaCl}_2/\text{mol RaCl}_2$

formula mass of RaCl$_2$ = atomic mass of Ra + 2(35.4 g/mol) $= 296.1_3$ g/mol

$\Rightarrow$ atomic mass of Ra $= 225.3$ g/mol

27-13. One mole of product (206.240 g) comes from one mole of piperazine (86.136 g). Grams of piperazine in sample $=$
(0.712 9 g of piperazine / g of sample) $\times$ (0.050 02 g of sample) $= 0.035\,66$.
Mass of product $= \left(\dfrac{206.240}{86.136}\right)(0.035\,66) = 0.085\,38$ g.

27-14. 2.500 g bis(dimethylglyoximate) nickel (II) $= 8.653\,2 \times 10^{-3} \text{ mol Ni} =$
0.507 85 g Ni $= 50.79\%$ Ni.

27-15. Formula masses: $CaC_{14}H_{10}O_6 \cdot H_2O$ (332.32), $CaCO_3$ (100.09), CaO (56.08). At 550°, $CaC_{14}H_{10}O_6 \cdot H_2O$ is converted to $CaCO_3$ (calcium carbonate).

332.32 g of starting material will produce 100.09 g of CaO.

Mass at 550° = (100.09/332.32)(0.635 6 g) = 0.191 4 g. At 1 000°C, the product is CaO (calcium oxide) and the mass is (56.08/332.32)(0.635 6 g) = 0.107 3 g.

27-16. 2.378 mg CO_2 / (44.009 g/mol) = $5.403\ 4 \times 10^{-5}$ mol CO_2 = $5.403\ 4 \times 10^{-5}$ mol C = $6.489\ 9 \times 10^{-4}$ g C

ppm C = 10^6 ($6.489\ 9 \times 10^{-4}$ / 6.234) = 104.1 ppm

27-17. 2.07% of 0.998 4 g = 0.020 67 g of Ni = 3.521×10^{-4} mol of Ni.

This requires $(2)(3.521 \times 10^{-4})$ mol of DMG = 0.081 77 g.

A 50.0% excess is (1.5)(0.081 77 g) = 0.122 7 g. The mass of solution containing 0.122 7 g is 0.122 7 g DMG / (0.021 5 g DMG/g solution) = 5.705 g of solution. The volume of solution is 5.705 g/(0.790 g/mL) = 7.22 mL.

27-18. Moles of Fe in product (Fe_2O_3) = moles of Fe in sample.

Because 1 mole of (Fe_2O_3) contains 2 moles of Fe, we can write the equation

$$\frac{2\ (0.264\ g)}{159.69\ g/mol} = 3.306 \times 10^{-3} \text{ mol of Fe.}$$

This many moles of Fe equals 0.919 2 g of $FeSO_4 \cdot 7\ H_2O$. Because we analyzed just 2.998 g out of 22.131 g of tablets, the $FeSO_4 \cdot 7\ H_2O$ in the 22.131 g sample is $\frac{22.131\ g}{2.998\ g}$ (0.919 2 g) = 6.786 g. This is the $FeSO_4 \cdot 7\ H_2O$ content of 20 tablets.

The content in one tablet is (6.786 g)/20 = 0.339 g.

27-19. (a) Mass of product ($CaCO_3$) = 18.546 7 g – 18.231 1 g = 0.315 6 g

$$\text{mol } CaCO_3 = \left(\frac{0.315\ 6\ g}{100.087\ g/mol} \right) = 3.153 \times 10^{-3} \text{ mol}$$

The product contains 3.153 mmol Ca = $(3.153 \times 10^{-3} \text{ mol})(40.078\ g/mol)$ = 0.1264 g Ca.

$$\text{wt\% Ca} = \frac{0.1264\ g\ Ca}{0.632\ 4\ g\ mineral} \times 100 = 19.98\ \%$$

(b) The solutions are heated before mixing to increase the solubility of the product that will precipitate. If the solution is less supersaturated during the precipitation, crystals form more slowly and grow to be larger and purer than if they precipitate rapidly. The larger crystals are easier to filter.

(c) $(NH_4)_2C_2O_4$ provides oxalate ion to prevent CaC_2O_4 from redissolving. Also, the ammonium and oxalate ions provide an ionic atmosphere that prevents the precipitate from peptizing (breaking into colloidal particles).

(d) $AgNO_3$ solution is added to the filtrate to test for Cl^- in the filtrate. If Cl^- is present, $AgCl(s)$ will precipitate when Ag^+ is added. The source of Cl^- is the HCl used to dissolve the mineral. All the original solution needs to be washed away, so no extra material is present that would increase the mass of final product, which should be pure $CaCO_3(s)$.

27-20. (a) $70 \text{ kg} \left(\dfrac{6.3 \text{ g P}}{\text{kg}} \right) = 441 \text{ g P in } 8.00 \times 10^3 \text{ L}$. This corresponds to

$$\frac{441 \text{ g P}}{8.00 \times 10^3 \text{L}} = 0.055\,1 \text{ g/L or } 5.5_1 \text{ mg/100 mL}.$$

(b) Fraction of P in one formula mass is $\dfrac{2(30.974)}{3\,596.46} = 1.722\%$.

P in $0.338\,7$ g of $P_2O_5 \cdot 24\,MoO_3 = (0.017\,22)(0.338\,7) = 5.834$ mg

This is near the amount expected from a dissolved man.

27-21. Let $x =$ mass of NH_4Cl and $y =$ mass of K_2CO_3.

For the first part, 1/4 of the sample (25 mL) gave 0.617 g of precipitate containing both products:

$$\frac{1}{4}\left[\overbrace{\left(\frac{x}{53.491} \right)(337.27)}^{\text{mol } NH_4Cl} + \overbrace{\left(\frac{2y}{138.21} \right)(358.33)}^{\text{mol } K_2CO_3 \times 2} \right] = 0.617 \text{ g}$$

$$\underbrace{\hphantom{\left(\frac{x}{53.491} \right)(337.27)}}_{\text{g } \phi_4BNH_4} \qquad \underbrace{\hphantom{\left(\frac{2y}{138.21} \right)(358.33)}}_{\text{g } \phi_4BK} \quad (\phi = \text{phenyl} = C_6H_5)$$

We multiplied moles of K_2CO_3 by 2 because one mole of K_2CO_3 gives 2 moles of ϕ_4BK. In the second part, half of the sample (50 mL) gave 0.554 g of ϕ_4BK:

$$\frac{1}{2}\underbrace{\overbrace{\left(\frac{2y}{138.21} \right)}^{\text{mol } K_2CO_3 \times 2}(358.33)}_{\text{g } \phi_4BK} = 0.554 \text{ g} \quad \Rightarrow \quad y = 0.213_7 \text{ g} = 14.5 \text{ wt\% } K_2CO_3$$

Putting this value of y into the first equation gives $x = 0.215_7$ g $= 14.6$ wt% NH_4Cl

27-22.

$$Fe_2O_3 + Al_2O_3 \xrightarrow[H_2]{\text{Heat}} Fe + Al_2O_3$$

$$\underbrace{2.019 \text{ g}} \qquad\qquad \underbrace{1.774 \text{ g}}$$

The mass of oxygen lost is $2.019 - 1.774 = 0.245$ g, which equals $0.015\,31$ moles of oxygen atoms. For every 3 moles of oxygen there is 1 mole of Fe_2O_3, so moles of $Fe_2O_3 = \frac{1}{3}(0.015\,31) = 0.005\,105$ mol of Fe_2O_3. This much Fe_2O_3 equals 0.815 g, which is 40.4 wt% of the original sample.

27-23. Let $x = $ g of $FeSO_4 \cdot (NH_4)_2 SO_4 \cdot 6H_2O$ and $y = $ g of $FeCl_2 \cdot 6H_2O$.

We can say that $x + y = 0.548\,5$ g. The moles of Fe in the final product (Fe_2O_3) must equal the moles of Fe in the sample.

The moles of Fe in $Fe_2O_3 = 2$ (moles of Fe_2O_3) $= 2\left(\dfrac{0.167\,8}{159.69}\right) = 0.002\,101\,6$ mol.

Mol Fe in $FeSO_4 \cdot (NH_4)_2 SO_4 \cdot 6\,H_2O = x / 392.13$ and
mol Fe in $FeCl_2 \cdot 6H_2O = y / 234.84$.

$$0.002\,101\,6 = \frac{x}{392.13} + \frac{y}{234.84} \tag{1}$$

Substituting $x = 0.548\,5 - y$ into Eq. (1) gives $y = 0.411\,46$ g of $FeCl_2 \cdot 6H_2O$.

Mass of Cl $= 2\left(\dfrac{35.453}{234.84}\right)(0.411\,46) = 0.124\,23$ g $= 22.65$ wt%

27-24. (a) Let $x =$ mass of $AgNO_3$ and $(0.432\,1 - x) =$ mass of $Hg_2(NO_3)_2$ in unknown. Each mol of $AgNO_3$ gives 1/3 mol $Ag_3[Co(CN)_6]$ and each mol of $Hg_2(NO_3)_2$ gives 1/3 mol $(Hg_2)_3[Co(CN)_6]_2$. Mass of both products must equal 0.451 5 g:

$$\underbrace{\frac{1}{3}\left(\frac{x}{169.873}\right)(538.643)}_{\text{mass of } Ag_3Co(CN)_6} + \underbrace{\frac{1}{3}\left(\frac{0.4321-x}{525.19}\right)(1\,633.62)}_{\text{mass of } (Hg_2)_3[Co(CN)_6]_2} = 0.4515$$

Above brackets: $\overbrace{\text{mol } Ag_3Co(CN)_6}$ $\qquad$ $\overbrace{\text{mol } (Hg_2)_3[Co(CN)_6]_2}$

$$\Rightarrow x = 0.173\,1 \text{ g} = 40.05 \text{ wt%}$$

(b) 0.30% error in 0.451 5 g $= \pm 0.001\,35$ g. This changes the equation of (a) to:

$$\frac{1}{3}\left(\frac{x}{169.873}\right)(538.643) + \frac{1}{3}\left(\frac{0.432\,1-x}{525.19}\right)(1\,633.62) = 0.451\,5\,(\pm 0.001\,35)$$

$$1.056\,952\,x + 0.448\,020 - 1.036\,844\,x = 0.451\,5\,(\pm 0.001\,35)$$

$$0.020\,109\,x = 0.451\,5\,(\pm 0.001\,35) - 0.448\,020$$

$$0.020\,109\,x = 0.003\,480\,(\pm 0.001\,35)$$

$$x = \frac{0.003\,480(\pm 0.001\,35)}{0.020\,109} = \frac{0.003\,480(\pm 38.8\%)}{0.020\,109} = 0.17 \text{ g} \pm 39\%$$

27-25. (a) Balanced equation for overall (31.8%) mass loss:

$$\text{Y}_2(\text{OH})_5\text{Cl} \cdot x\text{H}_2\text{O} \xrightarrow{\text{31.8\% mass loss}} \text{Y}_2\text{O}_3 + \underbrace{x\text{H}_2\text{O} + 2\text{H}_2\text{O}}_{} + \text{HCl}$$

FM 298.30 + x(18.015)   FM 225.81   FM (2+x)(18.015))   FM 36.461

$$\underbrace{(2+x)(18.015) + 36.461}_{\text{mass lost}} = \underbrace{(0.318)[298.30 + x(18.015)]}_{\text{31.8\% of original mass}} \Rightarrow x = 1.82$$

(b) Logical molecular units that could be lost are H_2O and HCl. At ~8.1% mass loss, the product is $\text{Y}_2(\text{OH})_5\text{Cl}$. Loss of 2 more H_2O would give a total mass loss of

$$\frac{1.82\text{H}_2\text{O} + 2\text{H}_2\text{O}}{\text{Y}_2(\text{OH})_5\text{Cl} \cdot 1.82\text{H}_2\text{O}} = \frac{68.82}{331.09} = 20.8\%$$

Loss of HCl from $\text{Y}_2(\text{OH})_5\text{Cl}$ would give a total mass loss of

$$\frac{1.82\text{H}_2\text{O} + \text{HCl}}{\text{Y}_2(\text{OH})_5\text{Cl} \cdot 1.82\text{H}_2\text{O}} = \frac{69.25}{331.09} = 20.9\%$$

The composition at the ~19.2% plateau could be either $\text{Y}_2\text{O}_2(\text{OH})\text{Cl}$ (from loss of $2\text{H}_2\text{O}$) or $\text{Y}_2\text{O}(\text{OH})_4$ (from loss of HCl).

27-26. (a) $\alpha = \dfrac{\text{mass of KPO}_3}{\text{mass of K}(\text{D}_x\text{H}_{1-x})_2\text{PO}_4} = \dfrac{118.070\,3}{136.085\,3 + 2.012\,55x}$

Cross-multiply: $(136.085\,3)\alpha + (2.012\,55)\,\alpha x = 118.070\,3$

$(2.012\,55)\alpha x = 118.070\,3 - (136.085\,3)\alpha$

Divide by $(2.012\,55)\alpha$:

$$x = \frac{118.070\,3}{(2.012\,55)\alpha} - \frac{(136.085\,3)\alpha}{(2.012\,55)\alpha} \qquad\qquad x = \frac{58.667\,0}{\alpha} - 67.618\,3$$

For fully deuterated material, $x = 1$ and $\alpha = \dfrac{58.667\,0}{x + 67.618\,3} = 0.854\,976$

(b) $x = \dfrac{58.667\,0}{0.856\,7_7} - 67.618\,3 = 0.856\,3_2$

(c) For the function $x = f(\alpha)$, we can write

$$e_x = \sqrt{\left(\frac{\partial F}{\partial \alpha}\right)^2 e_\alpha^{\,2}}$$

For $x = \dfrac{58.667\,0}{\alpha} - 67.618\,3$, $\dfrac{\partial F}{\partial \alpha} = -\dfrac{58.667\,0}{\alpha^2}$ giving

$$e_x = \sqrt{\left(-\frac{58.667\,0}{\alpha^2}\right)^2 e_\alpha^{\,2}} = \frac{58.667\,0\, e_\alpha}{\alpha^2}$$

(d) For $e_\alpha = 0.000\,1$, $e_x = \dfrac{(58.667\,0)(0.000\,1)}{(0.856\,7_7)^2} = 0.008$

D:H stoichiometry $= x \pm e_x = 0.856 \pm 0.008$

If e_x were 0.001, then $e_x = \dfrac{(58.667\,0)(0.001)}{(0.856\,7_7)^2} = 0.08$ and

D:H stoichiometry $= 0.86 \pm 0.08$

27-27. (a) Formula mass of $YBa_2Cu_3O_{7-x} = 666.19 - (16.00)\,x$

mmol of $YBa_2Cu_3O_{7-x}$ in experiment $= \dfrac{34.397\text{ mg}}{[666.19 - (16.00)x]\text{mg/mmol}}$

mmol of oxygen atoms lost in experiment $= \dfrac{(34.397 - 31.661)\text{ mg}}{16.00\text{ mg/mmol}}$

$= 0.171\,00$ mmol

From the stoichiometry of the reaction, we can write

$$\frac{\text{mmol oxygen atoms lost}}{\text{mmol } YBa_2Cu_3O_{7-x}} = \frac{3.5 - x}{1}$$

$$\frac{0.171\,00}{34.397\,/\,[666.19 - (16.00)x]} = 3.5 - x \Rightarrow x = 0.204\,2$$

(without regard to significant figures)

(b) Now let the uncertainty in each mass be 0.002 mg and let all atomic and molecular masses have negligible uncertainty.

The mmol of oxygen atoms lost are:

$$\frac{[34.397(\pm0.002) - 31.661(\pm0.002)]\text{mg}}{16.00\text{ mg / mmol}} = \frac{2.736(\pm0.002\,8)}{16.00}$$

$$= 0.171\,00\,(\pm0.102\%)$$

The relative error in the mass of starting material is $\dfrac{0.002}{34.397} = 0.005\,8\%$

The master equation becomes

$$\frac{0.171\,00\,(\pm0.102\%)}{34.397\,(\pm0.005\,8\%)/[666.19 - (16.00)\,x]} = 3.5 - x$$

$$0.171\,00\,(\pm0.102\%)[666.19 - (16.00)x] = (3.5 - x)[34.397\,(\pm0.005\,8\%)]$$

$113.918\,(\pm0.116) - [2.736\,(\pm0.002\,79)]\,x$

$$= 120.389\,5\,(\pm0.006\,98) - [34.397\,(\pm0.002)]\,x$$

$[31.66\,(\pm0.003\,46)]\,x = 6.471\,5\,(\pm0.116)$

$$= 0.204\,4\,(\pm1.79\%) = 0.204 \pm 0.004$$

27-28. In *combustion*, a substance is heated in the presence of excess O_2 to convert carbon to CO_2 and hydrogen to H_2O. In *pyrolysis*, the substance is decomposed by heating in the absence of added O_2. All oxygen in the sample is converted to CO by passage through a suitable catalyst.

27-29. WO_3 catalyzes the complete combustion of C to CO_2 in the presence of excess O_2. Cu converts SO_3 to SO_2 and removes excess O_2.

27-30. The tin capsule melts and is oxidized to SnO_2 to liberate heat and crack the sample. Tin uses the available oxygen immediately, ensures that sample oxidation occurs in the gas phase, and acts as an oxidation catalyst.

27-31. By dropping the sample in before very much O_2 is present, pyrolysis of the sample to give gaseous products occurs prior to oxidation. This minimizes the formation of nitrogen oxides.

27-32.
$$C_6H_5CO_2H + \frac{15}{2} O_2 \quad \rightarrow \quad 7\,CO_2 + 3\,H_2O$$
$$\text{FM } 122.121 \qquad\qquad\qquad 44.009 \quad\; 18.015$$

One mole of $C_6H_5CO_2H$ gives 7 moles of CO_2 and 3 moles of H_2O.
4.635 mg of benzoic acid = 0.037 95 mmol, which gives 0.265 7 mmol CO_2
(= 11.69 mg CO_2) and 0.113 9 mmol H_2O (= 2.051 mg H_2O).

27-33. $C_8H_7NO_2SBrCl + 9\frac{1}{4} O_2 \rightarrow 8CO_2 + \frac{5}{2} H_2O + \frac{1}{2} N_2 + SO_2 + HBr + HCl$

27-34. 100 g of compound contains 46.21 g C, 9.02 g H, 13.74 g N, and 31.03 g O. The atomic ratios are C : H : N : O =
$$\frac{46.21\text{ g}}{12.010\,6\text{ g/mol}} : \frac{9.02\text{ g}}{1.007\,98\text{ g/mol}} : \frac{13.74\text{ g}}{14.006\,8\text{ g/mol}} : \frac{31.03\text{ g}}{15.999\,4\text{ g/mol}}$$
$$= 3.847 : 8.94_9 : 0.981\,0 : 1.939$$

Dividing by the smallest factor (0.981 0) gives the ratios C : H : N : O = 3.922 : 9.12 : 1 : 1.977. The empirical formula is probably $C_4H_9NO_2$.

27-35.
$$C_6H_{12} + C_2H_4O \rightarrow CO_2 + H_2O$$
FM 84.159 44.052 44.009

Let x = mg of C_6H_{12} and y = mg of C_2H_4O

$x + y = 7.290.$

We also know that moles of $CO_2 = 6$ (moles of C_6H_{12}) + 2 (moles of C_2H_4O),

by conservation of carbon atoms.

$$6\left(\frac{x}{84.159}\right) + 2\left(\frac{y}{44.052}\right) = \frac{21.999}{44.009}$$

Making the substitution $x = 7.290 - y$ allows us to solve for y.

$y = 0.767$ mg $= 10.5$ wt%.

27-36. The atomic ratio H : C is

$$\frac{\left(\dfrac{6.76 \pm 0.12 \text{ g}}{1.007\,98 \text{ g/mol}}\right)}{\left(\dfrac{71.17 \pm 0.41 \text{ g}}{12.010\,6 \text{ g/mol}}\right)} = \frac{6.706 \pm 0.119}{5.926 \pm 0.034\,1} = \frac{6.706 \pm 1.78\%}{5.926 \pm 0.576\%} = 1.132 \pm 0.021$$

If we define the stoichiometry coefficient for C to be 8, then the stoichiometry coefficient for H is $8(1.132 \pm 0.021) = 9.06 \pm 0.17$.

The atomic ratio N:C is

$$\frac{\left(\dfrac{10.34 \pm 0.08 \text{ g}}{14.006\,8 \text{ g/mol}}\right)}{\left(\dfrac{71.17 \pm 0.41 \text{ g}}{12.010\,6 \text{ g/mol}}\right)} = \frac{0.738\,2 \pm 0.005\,7}{5.926 \pm 0.034\,1} = \frac{0.738\,2 \pm 0.774\%}{5.925 \pm 0.576\%}$$

$= 0.124\,6 \pm 0.001\,2$

If we define the stoichiometry coefficient for C to be 8, then the stoichiometry coefficient for N is $8(0.124\,6 \pm 0.001\,2) = 0.996\,8 \pm 0.009\,6$.

The empirical formula is reasonably expressed as $C_8H_{9.06\pm0.17}N_{0.997\pm0.010}$.

27-37. The reaction between H_2SO_4 and NaOH can be written

$$H_2SO_4 + 2NaOH \rightarrow 2H_2O + Na_2SO_4$$

One mole of H_2SO_4 requires two moles of NaOH. In 3.01 mL of 0.015 76 M NaOH there are $(0.003\,01 \text{ L})(0.015\,76 \text{ mol/L}) = 4.74_4 \times 10^{-5}$ mol of NaOH. The moles of H_2SO_4 must have been $(\frac{1}{2})(4.74_4 \times 10^{-5}) = 2.37_2 \times 10^{-5}$ mol. Because one mole of H_2SO_4 contains one mole of S, there must have been $2.37_2 \times 10^{-5}$ mol of S ($= 0.760_6$ mg). The percentage of S in the sample is

$$\frac{0.760_6 \text{ mg S}}{6.123 \text{ mg sample}} \times 100 = 12.4 \text{ wt%}.$$

27-38.   (a)   Experiment 1: $\bar{x} = 10.16_0$ µmol Cl⁻ $s = 2.70_7$ µmol Cl⁻

95% confidence interval $= \bar{x} \pm \dfrac{ts}{\sqrt{n}}$

$= 10.16_0 \pm \dfrac{(2.262)(2.70_7)}{\sqrt{10}} = 10.16_0 \pm 1.93_6$ µmol Cl⁻

Experiment 2: $\bar{x} = 10.77_0$ µmol Cl⁻ $s = 3.20_5$ µmol Cl⁻

95% confidence interval $= \bar{x} \pm \dfrac{ts}{\sqrt{n}}$

$= 10.77_0 \pm \dfrac{(2.262)(3.20_5)}{\sqrt{10}} = 10.77_0 \pm 2.29_3$ µmol Cl⁻

(b) $s_{pooled} = \sqrt{\dfrac{s_1^2\,(n_1-1)+s_2^2\,(n_2-1)}{n_1+n_2-2}} = \sqrt{\dfrac{2.70_7^2\,(10-1)+3.20_5^2\,(10-1)}{10+10-2}}$

$= 2.96_6$

$t_{calculated} = \dfrac{|\bar{x}_1 - \bar{x}_2|}{s_{pooled}}\sqrt{\dfrac{n_1 n_2}{n_1+n_2}} = \dfrac{|10.16_0 - 10.77_0|}{2.96_6}\sqrt{\dfrac{(10)(10)}{10+10}}$

$= 0.46_0 < t_{tabulated}$ for 18 degrees of freedom for

95% confidence level (or even for 50% confidence level)

Therefore, the <u>difference is not significant</u>. The result means that addition of

excess Cl⁻ prior to precipitation does not lead to additional coprecipitation of

Cl⁻ under the conditions of these experiments. (In general, under other

conditions we would expect extra Cl⁻ to lead to extra coprecipitation.)

(c)   10.0 mg of $SO_4^{2-} = 0.104_{10}$ mmol $= 24.2_{96}$ mg $BaSO_4$

(d)   In Experiment 1, the precipitate includes an additional 10.16_0 µmol Cl⁻ $=$

5.08 µmol $BaCl_2 = 1.05_8$ mg $BaCl_2$. The increase in mass is

$(1.05_8)/(24.2_{96}) = 4.35\%$. This represents a large error in the analysis.

CHAPTER 28
SAMPLE PREPARATION

28-1. There is no point analyzing a sample if you do not know that it was selected in a sensible way and stored so that its composition did not change after it was taken.

28-2. "Analytical quality" refers to the accuracy and precision of the method applied to the sample that was analyzed. High quality means that the analysis is accurate and precise. "Data quality" means that the sample that was analyzed is representative and appropriate for the question being asked and that the analytical quality is adequate for the intended purpose. If an accurate and precise analysis is performed on an unrepresentative or contaminated or decomposed sample, the results are meaningless.

28-3. (a) $s_o^2 = s_a^2 + s_s^2 = 3^2 + 4^2 \Rightarrow s_o = 5\%$.

      (b) $s_s^2 = s_o^2 - s_a^2 = 4^2 - 3^2 \Rightarrow s_s = 2.6\%$.

28-4. $mR^2 = K_s$. $m(6^2) = 36 \text{ g} \Rightarrow m = 1.0 \text{ g}$

28-5. Pass the powder through a 120 mesh sieve and then through a 170 mesh sieve. Sample retained by 170 mesh sieve has a size between 90 and 125 μm. It would be called 120/170 mesh.

28-6. 11.0×10^2 g will contain 10^6 total particles, since 11.0 g contains 10^4 particles. $n_{KCl} = np = (10^6)(0.01) = 10^4$.
Relative standard deviation $= \sqrt{npq}/n_{KCl} = \sqrt{(10^6)(0.01)(0.99)}/10^4 = 0.99\%$.

28-7. (a) $\sqrt{(10^3)(0.5)(0.5)} = 15.8$.

      (b) We are looking for the value of z, whose area is 0.45 (since the area from $-z$ to $+z$ is 0.90). The value lies between $z = 1.6$ and 1.7, whose areas are 0.445 2 and 0.455 4, respectively. Linear interpolation:
$$\frac{z - 1.6}{1.7 - 1.6} = \frac{0.45 - 0.445\,2}{0.455\,4 - 0.445\,2} \Rightarrow z = 1.647.$$

      (c) Since $z = (x - \bar{x})/s$, $x = \bar{x} \pm zs = 500 \pm (1.647)(15.8) = 500 \pm 26$. The range 474–526 will be observed 90% of the time.

28-8.   Use Equation 28-7, with $s_s = 0.05$ and $e = 0.04$. The initial value of t for 95% confidence in Table 4-4 is 1.960. $n = t^2 s_s^2 / e^2 = 6.0$ For $n = 6$, there are 5 degrees of freedom, so $t = 2.571$, which gives $n = 10.3$. For 9 degrees of freedom, $t = 2.262$, which gives $n = 8.0$. Continuing, we find $t = 2.365$ $n = 8.74$. This gives $t = 2.306 \Rightarrow n = 8.30$. Use 8 samples. For 90% confidence, the initial t is 1.645 in Table 4-4 and the same series of calculations gives $n = 6$ samples.

28-9.   (a) $mR^2 = K_s$. For $R = 2$ and $K_s = 20$ g, we find $m = 5.0$ g.

   (b) Use Equation 28-7 with $s_s = 0.02$ and $e = 0.015$. The initial value of t for 90% confidence in Table 4-4 is 1.645. $n = t^2 s_s^2 / e^2 = 4.8$.

   For $n = 5$, there are 4 degrees of freedom, so $t = 2.132$, which gives $n = 8.1$.
   For 7 degrees of freedom, $t = 1.895$, which gives $n = 6.4$.
   Continuing, we find $t = 2.015 \Rightarrow n = 7.2$. This gives $t = 1.943 \Rightarrow n = 6.7$.
   Use 7 samples.

28-10.

	A	B	C	D
1	Evaluation of the relation $mR^2 = K_s$			
2				
3	m (pg)	R %	R^2	$K_s = mR^2$
4	57	0.057	0.00325	0.185
5	68	0.069	0.00476	0.324
6	110	0.049	0.00240	0.264
7	110	0.045	0.00203	0.223
8	506	0.035	0.00123	0.620
9	515	0.027	0.00073	0.375
10	916	0.018	0.00032	0.297
11	955	0.022	0.00048	0.462
12			average	0.344
13			std dev	0.141

Average value of $K_s = 0.34 \pm 0.14$ pg

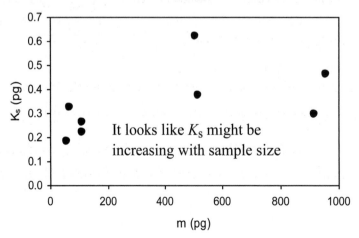

It looks like K_s might be increasing with sample size

The confidence interval is $\pm ts/\sqrt{n}$, where t is Student's t, s is the standard deviation, and n is the number of replicate measurements. If n is the same for all points, then t is the same for all points and the confidence interval is proportional to s. The equation in the text is expressed in terms of s. If the confidence interval is proportional to s, then the same equation should hold for the confidence interval.

28-11. (a) Volume = $(4/3)\pi r^3$, where $r = 0.075$ mm = 7.5×10^{-3} cm.

Volume = 1.767×10^{-6} mL.

Na_2CO_3 mass = $(1.767 \times 10^{-6}$ mL$)(2.532$ g/mL$)$ = 4.474×10^{-6} g.

K_2CO_3 mass = $(1.767 \times 10^{-6}$ mL$)(2.428$ g/mL$)$ = 4.291×10^{-6} g.

Number of particles of Na_2CO_3 = $(4.00$ g$)/(4.474 \times 10^{-6}$ g/particle$)$

$= 8.941 \times 10^5$.

Number of particles of K_2CO_3 = $(96.00$ g$)/(4.291 \times 10^{-6}$ g/particle$)$

$= 2.237 \times 10^7$.

The fraction of each type (which we will need for part c) is

$p_{Na_2CO_3} = (8.941 \times 10^5)/(8.941 \times 10^5 + 2.237 \times 10^7) = 0.038\ 4$

$q_{K_2CO_3} = (2.237 \times 10^7)/(8.941 \times 10^5 + 2.237 \times 10^7) = 0.962$.

(b) Total number of particles in 0.100 g is $n = 2.326 \times 10^4$.

(c) Expected number of Na_2CO_3 particles in 0.100 g is 1/1000 of number in 100 grams = 8.94×10^2.

Expected number of K_2CO_3 particles in 0.100 g is 1/1 000 of number in 100 grams = 2.24×10^4.

Sampling standard deviation = $\sqrt{npq}$ = $\sqrt{(2.326 \times 10^4)(0.038\ 4)(0.962)}$

$= 29.3$.

Relative sampling standard deviation for Na_2CO_3 = $\dfrac{29.3}{8.94 \times 10^2}$ = 3.28%.

Relative sampling standard deviation for K_2CO_3 = $\dfrac{29.3}{2.24 \times 10^4}$ = 0.131%.

28-12. Metals with reduction potentials below zero [for the reaction $M^{n+} + ne^- \rightarrow M(s)$] are expected to dissolve in acid. These are Zn, Fe, Co, and Al.

28-13. HNO_3 was used first to oxidize any material that could be easily oxidized. This helps prevent the possibility that an explosion will occur when $HClO_4$ is added.

28-14. Barbital has a higher affinity for the octadecyl phase than for water, so it is retained by the column. The drug dissolves readily in acetone/chloroform, which elutes it from the column.

28-15. Cocaine is an amine base. It will be a cation at low pH and neutral in ammonia. The cation at pH 2 is retained by the cation-exchange resin. The neutral molecule is easily eluted by methanol. Benzoylecgonine has an amine and carboxylate functionality. At pH 2, the amine will be protonated and the carboxylic acid should be neutral, so the molecule will be retained by the cation exchange column. At elevated pH, the amine will be neutral and the carboxylate will be negative. The anion is not retained by the cation-exchanger and is eluted with methanol.

28-16. The product gas stream is passed through an anion-exchange column, on which SO_2 is absorbed by the following reactions:

$$SO_2 + H_2O \rightarrow H_2SO_3$$

$$2Resin^+OH^- + H_2SO_3 \rightarrow (Resin^+)_2SO_3^{2-} + H_2O$$

The sulfite is eluted with Na_2CO_3/H_2O_2, which oxidizes it to sulfate that can be measured by ion chromatography.

28-17. Large particle size allows sample to drain through the solid-phase extraction column without applying high pressure. In chromatography, small particle size increases the efficiency of separation, but high pressure is necessary to force solvent through the column.

28-18. (a) Solid-phase extraction retains acrylamide while passing many other components in the aqueous extract of the french fries. We want to remove as many other components as possible to simplify the chromatographic analysis. The strong acid of the ion-exchange resin protonates acrylamide and retains it by ionic attraction:

$$R{-}SO_3H + CH_2{=}CHCONH_2 \rightarrow R{-}SO_3^- + CH_2{=}CHCONH_3^+$$

   (b) There are many ultraviolet-absorbing components in addition to acrylamide in the acrylamide fraction obtained from the ion-exchange column. Ultraviolet absorbance is not specific for acrylamide.

(c) For acrylamide, m/z 72 is selected by the mass filter Q1 of the mass spectrometer. This ion dissociates by collisions in Q2. The product m/z 55 is selected in Q3 for passage to the detector.

$$CH_2=CHCONH_3^+ \rightarrow CH_2=CHC\equiv O^+$$
$$\quad \text{m/z 72} \qquad\qquad\qquad \text{m/z 55}$$

$$CD_2=CDCONH_3^+ \rightarrow CD_2=CDC\equiv O^+$$
$$\quad \text{m/z 75} \qquad\qquad\qquad \text{m/z 58}$$

(d) Even though many compounds are applied to the chromatography column, acrylamide is the only one with m/z 72 that gives a significant reaction product at m/z 55.

(e) Acrylamide gives one peak by selected reaction monitoring of the transition m/z 72→55. The internal standard gives just one peak for $^2H_3$-acrylamide monitored by the transition m/z 75→58 with the same retention time as acrylamide. The transition m/z 72→55 does not respond to the internal standard, and the transition m/z 75→58 does not respond to unlabeled acrylamide. We know the concentration of internal standard added to the aqueous extract of the french fries, and we measure the integrated area of the m/z 75→58 peak for the internal standard. We also measure the integrated area of the m/z 72→55 peak for acrylamide. The concentration of acrylamide in the aqueous extract is found by the proportion

$$\frac{[\text{acrylamide}]}{[\text{internal standard}]} = \frac{[\text{area of } m/z \text{ 72→55 peak}]}{[\text{area of } m/z \text{ 75→58 peak}]}$$

(f) With ultraviolet absorption, the internal standard appears at the same elution time as acrylamide. The molar absorptivity of deuterated internal standard is probably very similar to that of acrylamide, so equal concentrations of internal standard and acrylamide contribute approximately the same integrated area in the chromatogram. With selected reaction monitoring by mass spectrometry, the detector sees either acrylamide or the internal standard, with no interference from the other, even though they are eluted at the same time.

(g) The internal standard is mixed with the aqueous extract from the french fries prior to solid-phase extraction. We expect little isotope effect on the binding of acrylamide to the solid-phase sorbent or the HPLC stationary phase. Therefore, the fraction of acrylamide and the fraction of internal standard that bind to and are recovered from the solid-phase extraction column are equal. Even though neither one is bound or eluted quantitatively, equal fractions of each are bound and eluted. The ratio of acrylamide and internal standard should remain constant throughout the entire procedure.

28-19. (a) *Extraction:* Mix homogenized or powdered, hydrated sample with acetonitrile, NaCl, MgSO$_4$, and buffer. Salts create two phases and drive organic materials out of the aqueous phase. Buffer protects pH-sensitive substances. After centrifugation, collect the organic phase. *Sample cleanup:* Add 1 mL of organic phase to anhydrous MgSO$_4$ (to absorb residual H$_2$O), "primary secondary amine" sorbent (to absorb anions such as fatty acids), and optional sorbents such as C$_{18}$-silica and graphitized carbon black (to absorb nonpolar and aromatic substances). After shaking and centrifugation, analyze supernatant liquid by chromatography.

(b) Internal standard is intended to suffer the same losses as analyte in the different steps of the procedure.

(c) Total ion chromatogram shows total signal from all ions eluted at any given short time interval.

(d) The extracted ion chromatogram shows *m/z* 312 signal taken from the full mass spectrum recorded at any short time interval during elution. In selected ion monitoring, only the signal at *m/z* 312 would be measured and other masses would not be measured. Since the detector spends only a small fraction of the time measuring *m/z* 312 in an extracted ion chromatogram, but full time measuring *m/z* 312 in a selected ion chromatogram, the signal:noise ratio is greater in the selected ion chromatogram.

(e) Selected reaction monitoring.

28-20. (a) Liquid-liquid extraction uses a large volume ($\gtrsim$100 mL) of organic solvent to extract an aqueous phase by continuous distillation. In dispersive liquid-liquid microextraction, a cloudy emulsion is created by a small volume (~10–100 μL) of immiscible organic phase plus ~0.5–1 mL of dispersant solvent in an aqueous sample. After centrifugation, the phases separate and the organic phase is collected.

   (b) Disperser solvent is miscible with both the aqueous and organic phases. It lowers the interfacial energy between the two phases, permitting a high-surface-area emulsion with a high rate of mass transfer to be formed. In the absence of disperser solvent, the two phases would just separate.

28-21. In solid-supported liquid-liquid extraction, the aqueous phase is suspended in a microporous medium through which organic solvent is passed to extract analytes. In solid-phase extraction, aqueous sample is passed through a small column containing a chromatographic stationary phase that retains analytes. Impurities and analytes are eluted by a series of washes with small volumes organic liquid with increasing solvent strength.

28-22. (a) Highest concentration of Ni $\approx$ 80 ng/mL. A 10 mL sample contains 800 ng Ni = 1.36×10^{-8} mol Ni. To this is added 50 μg Ga = 7.17×10^{-7} mol Ga. Atomic ratio Ga/Ni = $(7.17 \times 10^{-7})/(1.36 \times 10^{-8}) = 53$.

   (b) Apparently all the Ni is in solution because filtration does not decrease its total concentration. Since filtration removes most of the Fe, it must be present as a suspension of solid particles.

28-23. One-fourth of the sample (25 mL out of 100 mL) required
$(0.011\,44\ \text{M})\,(0.032\,49\ \text{L}) = 3.717 \times 10^{-4}$ mol EDTA $\Rightarrow (3.717 \times 10^{-4})\,(4) = 1.487 \times 10^{-3}$ mol Ba^{2+} in sample = 0.204 2 g Ba = 64.90 wt%.

28-24.   (a)   From the acid dissociation constants of Cr(III), we see that the dominant forms at pH 8 are $Cr(OH)_2^+$ and $Cr(OH)_3(aq)$. The dominant form of Cr(VI) is CrO_4^{2-}.

(b)   The anion exchanger retains the anion, CrO_4^{2-}, but permits the $Cr(OH)_2^+$ cation and neutral $Cr(OH)_3(aq)$ to pass through, thereby separating Cr(VI) from Cr(III).

(c)   A "weakly basic" anion exchanger contains a protonated amine $(-^+NHR_2)$ that might lose its positive change in basic solution. A "strongly basic" anion exchanger $(-^+NR_3)$ is a stable cation in basic solution.

(d) CrO_4^{2-} is eluted from the anion exchanger when the concentration of sulfate in the buffer is increased from 0.05 M in step 3 to 0.5 M in step 4.

28-25.   One possible cost-saving scheme is to monitor wells 8, 11, 12, and 13 individually, but to pool samples from the other sites. For example, a composite sample could be made with equal volumes from wells 1, 2, 3, and 4. Other composites could be constructed from (5, 6, 7), (9, 10), (14, 15, 16, 17), and (18, 19, 20, 21). If no warning level of analyte is found in a composite sample, we would assume that each well in that composite is free of the analyte. If analyte is found in a composite sample, then each contributor to the composite would be separately analyzed. The disadvantage of pooling samples from n wells is that the sensitivity of the analysis for analyte in any one well is reduced by $1/n$.